ELECTRICITY ISSUES

HEARINGS

BEFORE THE

SUBCOMMITTEE ON ENERGY AND POWER

OF THE

COMMITTEE ON ENERGY AND COMMERCE HOUSE OF REPRESENTATIVES

ONE HUNDRED THIRD CONGRESS

SECOND SESSION

JULY 13, 14, AND 21, 1994

Serial No. 103–146

Printed for the use of the Committee on Energy and Commerce

U.S. GOVERNMENT PRINTING OFFICE

86-340CC WASHINGTON : 1995

For sale by the U.S. Government Printing Office
Superintendent of Documents, Congressional Sales Office, Washington, DC 20402
ISBN 0-16-046677-6

COMMITTEE ON ENERGY AND COMMERCE

JOHN D. DINGELL, Michigan, *Chairman*

HENRY A. WAXMAN, California
PHILIP R. SHARP, Indiana
EDWARD J. MARKEY, Massachusetts
AL SWIFT, Washington
CARDISS COLLINS, Illinois
MIKE SYNAR, Oklahoma
W.J. "BILLY" TAUZIN, Louisiana
RON WYDEN, Oregon
RALPH M. HALL, Texas
BILL RICHARDSON, New Mexico
JIM SLATTERY, Kansas
JOHN BRYANT, Texas
RICK BOUCHER, Virginia
JIM COOPER, Tennessee
J. ROY ROWLAND, Georgia
THOMAS J. MANTON, New York
EDOLPHUS TOWNS, New York
GERRY E. STUDDS, Massachusetts
RICHARD H. LEHMAN, California
FRANK PALLONE, JR., New Jersey
CRAIG A. WASHINGTON, Texas
LYNN SCHENK, California
SHERROD BROWN, Ohio
MIKE KREIDLER, Washington
MARJORIE MARGOLIES-MEZVINSKY, Pennsylvania
BLANCHE M. LAMBERT, Arkansas

CARLOS J. MOORHEAD, California
THOMAS J. BLILEY, JR., Virginia
JACK FIELDS, Texas
MICHAEL G. OXLEY, Ohio
MICHAEL BILIRAKIS, Florida
DAN SCHAEFER, Colorado
JOE BARTON, Texas
ALEX McMILLAN, North Carolina
J. DENNIS HASTERT, Illinois
FRED UPTON, Michigan
CLIFF STEARNS, Florida
BILL PAXON, New York
PAUL E. GILLMOR, Ohio
SCOTT KLUG, Wisconsin
GARY A. FRANKS, Connecticut
JAMES C. GREENWOOD, Pennsylvania
MICHAEL D. CRAPO, Idaho

ALAN J. ROTH, *Staff Director and Chief Counsel*
DENNIS B. FITZGIBBONS, *Deputy Staff Director*
MARGARET A. DURBIN, *Minority Chief Counsel and Staff Director*

SUBCOMMITTEE ON ENERGY AND POWER

PHILIP R. SHARP, Indiana, *Chairman*

EDWARD J. MARKEY, Massachusetts
RICHARD H. LEHMAN, California
CRAIG A. WASHINGTON, Texas
MIKE KREIDLER, Washington
BLANCHE M. LAMBERT, Arkansas
AL SWIFT, Washington
MIKE SYNAR, Oklahoma
W.J. "BILLY" TAUZIN, Louisiana
RALPH M. HALL, Texas
RICK BOUCHER, Virginia
JIM COOPER, Tennessee
JOHN D. DINGELL, Michigan (Ex Officio)

MICHAEL BILIRAKIS, Florida
J. DENNIS HASTERT, Illinois
CLIFF STEARNS, Florida
PAUL E. GILLMOR, Ohio
SCOTT KLUG, Wisconsin
GARY A. FRANKS, Connecticut
MICHAEL D. CRAPO, Idaho
CARLOS J. MOORHEAD, California (Ex Officio)

SHELLEY N. FIDLER, *Staff Director*
SUE D. SHERIDAN, *Counsel*
ALVIE LEON LOWERY, *Special Assistant*
KAREN K. HUNSICKER, *Minority Counsel*

CONTENTS

ELECTRICITY ISSUES

WEDNESDAY, JULY 13, 1994

House of Representatives,
Committee on Energy and Commerce,
Subcommittee on Energy and Power,
Washington, DC.

The subcommittee met, pursuant to notice, at 10:20 a.m., in room 2359, Rayburn House Office Building, Hon. Philip R. Sharp (chairman) presiding.

Mr. Sharp. The subcommittee will please come to order.

Even before the passage of the Energy Policy Act of 1992, wholesale electricity markets had begun to change. Although the States' implementation of PURPA was often controversial, our experience under the Act also proved that nonutility generators can be reliable sources of electricity. Similarly, while the lack of readily available transmission service stymied some wholesale power purchasers, it was also true that an increasing number of wholesale purchases were taking place on a voluntary basis.

Today the pace of change in electricity markets and in thinking about the industry's future has outpaced most expectations. It is hard to remember how heated the debate was about whether PUHCA should be amended to permit EWG's to generate power and whether FERC should be given clear authority to prevent discrimination in transmission services.

The FERC deserves enormous credit for carrying out its responsibilities so far under EPAct very efficiently and for keeping its eye firmly on Congress's purpose of increasing competition so as to lower consumer prices. I am pleased that the President has renominated Chair Moler and that we will have the benefit of a stable and a very capable team of FERC Commissioners during this important period.

However, the debate about electricity policy is by no means an "inside the beltway" phenomenon. Indeed, as everyone in this room knows, some of the most exciting discussions, perhaps terrifying to some, are occurring at the State level as public utility commissions grapple with the future of regulation.

We will hear from several State commissioners at next week's hearing, and I am sure that a good deal of our other witnesses' testimony will address various State proposals today. Many of our witnesses will comment on very specific issues: How a particular retail wheeling proposal should be implemented, whether FERC is on the right track in its recent proposed rule on stranded investment, or where the line between State and Federal jurisdiction should be drawn.

But this series of hearings on electricity also offers an opportunity to speculate about the big picture and a market in evolution. I am particularly interested in the following broad questions:

To what degree have portions of the electric utility industry, entered a period of irreversible change, and how widespread will the change be?

How should regulators handle the transition to a more competitive market?

Can PUC's afford to be reactive, or should State commissions jump into the fray before change overtakes them?

How can environmental objectives be met in a restructured industry in a world in which services are increasingly unbundled? Is there an inherent conflict in the making?

How can the purposes of demand side management be served in a more competitive world?

What are the most important questions involving State and Federal jurisdiction? How are they likely to be resolved, and how will jurisdictional uncertainties affect the market's evolution?

I look forward to exploring these issues over the 3 days of our hearings and am pleased to have so many able witnesses here to join us in these discussions.

Several of our colleagues were here until the vote and I am sure will be returning, but since we have no further opening statements—I am sorry. Mr. Franks is here.

Do you have an opening statement?

Mr. FRANKS. No, Mr. Chairman.

Mr. SHARP. All right. We are delighted to have you here.

The Chair would now introduce our first panel. We are very pleased to have from the administration the Honorable Susan F. Tierney, the assistant secretary in the Office of Policy with the U.S. Department of Energy; Mr. David Moskowitz with the Regulatory Assistance Project; Dr. John Anderson, the executive director of the Electricity Consumers Resources Council, or ELCON; Mr. Scott Hempling, representing the Environmental Action Foundation; and Mr. Irwin Popowsky, Consumer Advocate for the State of Pennsylvania.

Ladies and gentlemen, we are very pleased to have you with us. I think all of you have been with us at one time or other and are familiar with our processes. We will be happy to make your written statements a part of our record and are going to ask each of our witnesses today—the customary rule is 5 minutes, but we want to allocate 10 minutes to everybody's oral summary and hope that you can help us efficiently move through this.

Ms. Tierney, we will be happy to hear from you now.

STATEMENTS OF SUSAN F. TIERNEY, ASSISTANT SECRETARY, OFFICE OF POLICY, DEPARTMENT OF ENERGY; DAVID MOSKOWITZ, DIRECTOR, REGULATORY ASSISTANCE PROJECT; JOHN A. ANDERSON, EXECUTIVE DIRECTOR, ELECTRICITY CONSUMERS RESOURCES COUNCIL; SCOTT HEMPLING, ATTORNEY, ON BEHALF OF ENVIRONMENTAL ACTION FOUNDATION; AND IRWIN A. POPOWSKY, CONSUMER ADVOCATE OF PENNSYLVANIA

Ms. TIERNEY. Thank you, Mr. Chairman. On behalf of Secretary Hazel O'Leary and the Department of Energy, I want to tell you how pleased we are that you have decided to hold this hearing.

As you know from the number of people in attendance today, this is an issue of enormous importance for the country and for States, and we share your interest in what is going on.

Thank you for giving us 10 minutes. As you know, all of us are hackers. I am here on a panel of big hackers in fact, and 10 minutes is going to be hard to keep to, but we appreciate that extra time.

As you know, for several years the electric industry in this country has been undergoing a transition from an extensively regulated industry to one that responds to an increasing extent on market forces. Some of those competitive forces have themselves been advanced by impressive leadership—you here in this committee as well as Congress as a whole and the States—and we appreciate the pro-competition forces that have been brought to bear by your enactment of the Public Utilities Regulatory Policies Act as well as the Energy Policy Act of 1992.

As you know very well, electricity rates in this country are high in many parts of the country. As you well know, neighboring utilities often have different rates, and those high rates and those different rates are creating a lot of pressure for calls for increased cost control as well as competition.

Many observers of the electric industry are attributing the high rates to a number of things, including the nature of retail rate regulation itself, including traditional cost of service regulation. There are other reasons, such as technological advances in generation markets that are enabling increased economies of scale and size, that are improving environmental impacts associated with technology and are decreasing the time associated with constructing new generating plants.

These new technologies enable new plants, whether operated or owned by an independent power producer or utility or, in fact, an industrial customer itself, to produce electricity at a price lower than the average cost of utilities' tariffed rates in many jurisdictions.

Utilities, customers, and regulators in various States are responding to these pressures for greater retail competition in a number of ways. You know them. Utilities are taking strong actions, restructuring, mergers, downsizing to control costs and lower rates. States are encouraging utilities to deliver least cost electric supplies, to increase end use efficiency so that customers' use of electricity goes down on average even when in some instances electric rates go up and therefore we experience lower bills. That kind of least cost supply that you have encouraged through the Energy

Policy Act is showing up in dramatically improved efficiencies over the past decade in terms of the country's productivity per unit of electric energy input.

States are also experimenting with alternatives to traditional cost of service regulation as a way to control utility rate increases. They are in many instances allowing utilities to negotiate special retail electric service contracts with large industrial customers.

Many customers are investing in generating plants within their own corporate boundaries so that they can minimize the amount of electricity that they buy from the electric utility itself. And they are getting approval in some instances to become municipal electric companies as well, so that they can enter into wholesale electric markets and obtain power from sources other than the local utility.

Now quite recently, as you well know, several States have begun to address the question of retail wheeling directly. On April 11 of this year, the Michigan commission approved a limited experimental retail wheeling program for certain large customers of Detroit Edison and Consumers Power.

On April 20 the California Public Utility Commission proposed by the year 2002 all of the customers of the State's investor-owned utilities would have the right to purchase power from the supplier of their choice. Other State commissions including Connecticut, Illinois, Montana, Pennsylvania, Texas, Minnesota, and New Mexico are considering or have considered retail wheeling questions.

Now as we at the Department of Energy observe these trends, we are interested in concerns about issues of national energy policy that may be affected both positively and negatively by changes occurring in the electric industry. These issues include U.S. energy productivity and economic competitiveness, the issue I mentioned a minute ago when I said we have had improved trends in terms of the amount of energy we use for each output of productivity. We are interested in ensuring that as we move to a more competitive market we do so in a way that ensures cost-effective energy efficiency investments are maintained in order to keep us as productive as possible.

We are concerned, as everyone is, about ensuring electric system reliability as we look to see more players in the marketplace. We know that we have seen good trends on that as wholesale generating markets are increasingly competitive, and we want to ensure that that continues to take place.

We are concerned about environmental impacts of energy production and use. Unless competition is suitably constructed we can imagine scenarios in which we would have increased emissions associated with ground level pollutants as well as global greenhouse gas emissions.

We are extremely interested in designing and seeing that the country continues to design its portfolio of resources in competitive ways, in ways that ensure renewable resources have a place in the portfolio.

We are interested in ensuring that technology research and development, advanced technology research and development, has a home in an increasingly competitive marketplace. We see in other industries that as competition gets more cutthroat, long-term research dollars and investments by companies go down. We think that has implications for keeping the U.S. technological edge.

Finally, we want to call to your attention the U.S. treaty commitments and obligations that we have approved in this country. I mention in particular the NAFTA treaty as well as our Climate Change Convention commitments that we see are affected both negatively and positively or at least are implicated by the way in which we design our markets.

As an indication of the Department's concerns about the trends in this industry and our strong interest in ensuring that the issues are subject to the appropriate level of debate, Secretary O'Leary recently announced her decision to sponsor, in conjunction with the National Association of Regulatory Utility Commissioners, a public forum in which we would invite members of the public, stakeholders, the hackers, if you will, as well as members of the general public to participate in discussions of these numerous and important questions associated with electric industry restructuring and regulatory reform.

Additionally, the Department of Energy has entered into an agreement with NARUC to jointly sponsor a body of research to examine a number of unanswered questions, many of which are exactly the questions you have asked here of the panel today.

Now I mentioned a minute ago the Department's interest in ensuring that the way in which we move forward in an increasingly competitive market does so in a way that supports our treaty commitments. I want to bring to your attention one that the Department is particularly interested in, and that is of course the framework Convention on Climate Change.

As you know, the President announced last October a package of 50 actions, principally voluntary actions, under which the Nation would meet its commitments to reduce greenhouse gas emissions to 1990 levels by the year 2000. Many of those emission reductions would come from the electric utility sector. Many of them come in the form of climate challenge actions that are voluntary commitments by electric companies to reduce greenhouse gas emissions.

I note that other players in the electric market have not yet made such voluntary commitments. I can envision the squeeze that will be placed on electric companies in a more competitive environment. We want to ensure that as we move forward we place a fair playing field amongst players in the market in terms of their responsibility for coming forward in meeting our country's cost-effective reductions of greenhouse gas emissions.

We strongly support the increased competition in the electric industry, and we are trying to involve ourselves in ensuring that the many stakeholders involved and the many decisionmakers and customers of the electric industry continue to design and formulate competitive designs of systems in ways that produce the kinds of benefits in the policy areas that I mentioned a minute ago. We are committed to working in a number of ways to do what we can to ensure that those cooperative efforts occur amongst stakeholders.

We think competition in wholesale markets is not only possible, it is extremely desirable, and it is well on its way, although wholesale electric competitive markets are not even themselves perfectly operating today.

We believe that careful consideration should be given by Federal and State regulators to creating an efficient structure for electric markets and to developing market-driven regulatory institutions and policies to address the imperfections in the energy markets that we see apparent both in the demand and the supply side still today.

We recognize the need for limited and flexible regulation to ensure that social goals are met where the market has not been forthcoming in producing those goals. Those are societal goals that I would include as environmental externalities of energy supply and use, energy efficiency, service availability to all sorts of consumers, and the exercise of market power by regulated firms who have unregulated subsidiaries or affiliates.

Competition in electricity, as in other markets, is not an end in itself. Increased competition should be pursued to achieve benefits for consumers as well as for the country as a whole. I mention again our particular interest in ensuring that the design of our competitive regulatory schemes are developed in a way to support economic efficiency, including energy efficiency, environmental protection, equity, system reliability, fuel diversity, especially with regard to renewable resource development and research and development on advanced technologies.

Now finally just let me mention the question of tensions between Federal and State jurisdictional issues as players look to a more competitive market at the retail level in addition to the wholesale level.

Many of the proposals that we have seen discussed in State jurisdictions will call into question and will raise for the courts' attentions, we believe, significant questions with regard to jurisdictional boundaries. The Department of Energy called to the attention of the California Public Utility Commission our concern about this regulatory and jurisdictional tension. We do not think that anyone today knows how those jurisdictional questions will be resolved. We are sure that no matter who steps into the retail wheeling fray and makes decisions in that, there will be actions taken to court and that we are assured of seeing many years over which these questions are litigated.

We think that there are things that States can do, short of crossing that line that takes on the retail wheeling question, to look aggressively and proactively at improving the procompetitive forces that they seek to harness to control costs.

There are things that States can do along with Federal energy retail—Federal Energy Regulatory Commission—excuse me; I didn't mean to have that Freudian slip—to ensure that wholesale market competition is further advanced. There is much that States can do to improve performance-based rate regulation that would send proper incentives to utilities to procure and operate a least cost energy supply.

There are things that States can do under existing jurisdiction to change rate design in ways that improve prize signals to customers. And there is much that States can do under their existing jurisdiction today to improve integrated resource planning and improve integrated resource least cost delivery of supplies in ways that get rid of the cumbersome practices of many State jurisdictions associated with what many call integrated resource planning. We think those reforms should be aggressively moved forward by States.

Because of the many years of court litigation that we foresee associated with retail wheeling, we would urge that a lot of States look carefully and work cooperatively as regions and with the Federal Government to decide where we can work together to come up with solutions that don't drive stakeholders to the courts to resolve those questions.

Finally, the breadth and complexity of the changes we see taking place in the electric industry as well as what we think is the strategic position of the electric industry in this country caution against moving as quickly as we see many States are interested in moving.

In so saying that, we don't advocate inaction. I mentioned the many ways in which we look forward to aggressive action by States and Federal agencies to ensure that we harness competitive forces as productively as possible. We recognize that State public utility commissions and legislators have responsibility to develop and implement policies on many of the issues that you too share jurisdiction over. We think efforts to revise State regulation of the industry to encourage competition raise these critical legal questions that will either come to the courts or you to decide before too long.

We urge that we coordinate on these matters, and it is for that reason we are extremely pleased that you have held this hearing, and I look forward to the testimony of my colleagues in the days ahead.

Thank you.

[The prepared statement of Ms. Tierney and responses to subcommittee questions follow:]

STATEMENT OF SUSAN F. TIERNEY, ASSISTANT SECRETARY, OFFICE OF POLICY, DEPARTMENT OF ENERGY

Chairman Sharp and Members of the Committee: On behalf of Secretary Hazel R. O'Leary and the Department of Energy, I am pleased to appear before you today to testify about the enormously important changes underway in the Nation's electric industry. The Department of Energy appreciates the committee's leadership in holding this hearing to explore the dimensions of the changes that are underway, along with their important implications for the Nation's consumers, industries and policymakers. You have posed salient questions, and we are pleased to offer our observations and comments at this time, although answers may not yet exist for all of the questions the committee raises.

For several years, the Nation's electric utility industry has been undergoing a transition from a highly structured and extensively regulated industry to one that responds to an increasing extent to market forces. Some of those competitive forces have been themselves advanced by impressive leadership by public policymakers in States and in the Federal Government, including significant Congressional leadership in enacting such laws as the Public Utilities Regulatory Policy Act of 1978 and the Energy Policy Act of 1992.

Observers of the electric industry attribute high electricity rates and high electricity bills to a number of things, including the nature of retail rate regulation. Under traditional cost-based rate regulation (at the State or Federal level); regulators permit an electric utility to charge rates sufficient to recover its prudently incurred expenses such as fuel costs or purchased power costs and to earn a return on its prudently incurred investments in generating, transmission and distribution facilities. Hence, electric rates reflect a utility's cost of providing service rather than the interaction of supply and demand in a competitive market. Some utilities have been more effective at controlling costs than others. Some have used innovative means of controlling costs that mimic competitive markets, such as competitive wholesale power procurements or least-cost purchases of energy efficiency improvements. These differences in utilities' decisions and practices (and their regulators' policies) account for some of the disparities that exist in retail electric rates across the country and sometimes even between neighboring utility service territories.

The increased interest in competition, including retail wheeling, also is being driven by new generation technologies that increase efficiency and reduce the size, cost, environmental impact, and time to construct new generating plants. Additionally, advanced communications technologies and improved access to transmission facilities enable new and existing plants to dispatch and move power with greater efficiency than in the past. These new technologies enable a new plant, whether owned by an independent power producer, an industrial self-generator, or a utility, to produce electricity at a price below the tariffed rate of some utilities, since those rates typically reflect yesterday's investments to meet today's (and often, tomorrow's) electric power needs.

Utilities, customers and regulators in various States are responding to these pressures for greater retail competition in different ways. Many utilities are taking strong actions, such as restructuring, mergers, and downsizing, to control costs and hence control rates. Some States are encouraging their utilities to deliver least-cost electricity and services to increase end-use efficiency, so that customers' use of electricity goes down, enabling lower average electricity bills even when the average electric rate goes up, which it sometimes does. Some States are experimenting with alternatives to traditional cost-based regulation, as a way to control utility rate increases. Some regulators are allowing their utilities to negotiate special retail electric service contracts with large industrial customers. Some customers are investing in generating plants within their own corporate boundaries so that they can minimize the amount of electricity they must buy from the utility. Some customers are getting the needed approvals from State and local decision makers to enable them to become municipal or other types of utilities, so that they can enter into competitive wholesale electric markets and obtain power from some other source beside the local utility.

Quite recently, several States have begun to address retail wheeling. In April, 1994, both Michigan and California issued proposals to allow retail wheeling in their States. The April llth interim order by the Michigan Public Service Commission approved a limited, experimental retail wheeling program for certain large customers of Detroit Edison and Consumers Power. On April 20th, the California Public Utility Commission proposed that by the year 2002, all of the customers of the State's investor-owned utilities would have the right to purchase power from a supplier other than the local utility. Other State commissions, including those in Connecticut, Illinois, Montana, Pennsylvania, Texas, Minnesota, and New Mexico, have considered or are considering retail wheeling issues.

The advent of competition also raises questions with important implications for national policy, which we discuss below.

The Department of Energy is concerned about matters of national importance that are likely to be affected—some positively, some negatively—by changes occurring in the electricity industry.

These include U.S. energy productivity and economic competitiveness; economic efficiency in electricity markets; electric system reliability; environmental impacts of energy production and use; security and diversity of energy resources; technology research and development; equity; energy efficiency; integrated resource planning; U.S. treaty commitments and obligations; and jurisdiction over electric regulation. One indication of our concern about these matters is the Secretary's recent decision to sponsor—jointly with the National Association of Regulatory Utility Commissioners—a National Electricity Forum which will meet about twice a year to discuss fundamental questions related to the future of the industry.

As you know, the Department of Energy has primary responsibility for energy policy within the executive branch of the U.S. Government. Clearly, DOE's policy role stands alongside the clear authority held by the Federal Energy Regulatory Commission to establish the rates, terms, and conditions under which wholesale electricity markets operate, including the terms and conditions of transmission service. Another matter of Federal interest involves the interstate and international energy and environmental obligations that are affected by regulation by the States.

For example, the President has announced the Nation's commitment to reduce its greenhouse gas emissions to 1990 levels by the year 2000, and issued a comprehensive Climate Change Action Plan to meet that goal. The plan includes nearly 50 separate actions to reduce emissions of $C02$ and other greenhouse gases. A significant portion of this reduction will come from the electric power sector and electricity consumers. To achieve this part of the goal, the Department has worked with the utility industry and its trade associations to develop the Climate Challenge Program, through which utilities commit themselves voluntarily to the development of innovative programs to reduce, limit, avoid, or offset emissions of C02 and other greenhouse gases. Scores of investor-owned utilities and hundreds of publicly-owned utilities have expressed interest in joining the program. Thus, it is necessary to assess the impact of changes in the industry on the ability of the Department and the industry to pursue such reductions. Achieving the Action Plan's objectives will be possible only with the cooperation and support of electric utilities and State regulatory authorities.

The Department recognizes that State public service commissions and legislatures have the responsibility to develop and implement policies regarding the regulation and operation of the electricity services industry within State borders. However, efforts to revise State regulation of the industry to encourage competition raise legal and jurisdictional issues that need to be resolved. Furthermore, given the strategic nature of electricity services and the interstate nature of electricity markets in the United States, it is important for Federal and State policymakers to coordinate on energy policy matters.

The Department of Energy wishes to indicate its support for efforts to increase competition in the electricity sector. The Department is convinced that increased competition in this industry, suitably formulated can produce benefits for all, and that the problems that must be addressed to achieve these benefits will yield to sustained cooperative efforts by affected government agencies and stakeholders.

The Department's policies relating to the electricity industry are rooted in principles established in the DOE Organization Act, the Public Utility Regulatory Policies Act, the Clean Air Act, the Energy Policy Act of 1992 (EPAct or Energy Policy Act), and most recently in the Department's Strategic Plan. The Department's primary energy policy goals are to:

Enhance energy productivity to strengthen the U.S. economy and improve living standards, through policies and programs that support more open and efficient energy markets that incorporate social costs, increase energy efficiency in all sectors of the economy, and develop and deploy new energy technologies and services both domestically and in international markets.

Ensure reliable, secure energy services by facilitating greater diversity of competitive energy resources and technologies in the marketplace, increasing reliance on risk sharing mechanisms, and promoting flexibility in the energy sector.

Reduce adverse environmental impacts associated with energy production, delivery and use, through efforts to reduce energy-related greenhouse gas emissions and other pollutants, support research, development, demonstration and commercialization of energy efficiency and alternative energy technologies, and support regulatory and incentive approaches to encourage energy efficiency and greenhouse gas emission reductions.

Promote economic, regional, and inter-generational equity by promoting universal access to affordable energy services, helping to eliminate disproportional adverse environmental impacts of energy systems on geographic regions and minority and low-income groups, promoting fairness in the distribution of economic benefits of energy programs among geographic regions and groups, and avoiding the transfer of costs and risks of energy systems to future generations.

The Department of Energy is working to incorporate these goals into national energy policies and programs. We hope that these principles and goals will resonate positively at the Federal and State level, and we seek to work with Federal, States and local agencies to achieve those objectives we hold in common.

Most electricity analysts agree that the generation sector of the industry is no longer a natural monopoly. Therefore, competition in wholesale markets is not only possible but desirable. The administration and the Department support efforts to reduce regulation in markets that can become reasonably competitive. Replacing regulation with competition in markets that no longer display monopolistic characteristics can—with attention by policymakers to the design of appropriate market mechanisms—provide for increased operational efficiency in the electricity supply system; reduce rates to consumers; expand consumer choice; increase diversity of fuel supply; and reward innovative ways to reduce costs, provide energy efficiency services, and protect the environment. At the same time, the administration and the Department also recognize the need for limited and flexible regulation to achieve social goals which may not be fully attained due to market imperfections, in areas such as environmental protection, energy efficiency, service availability, and exercise of market power by regulated firms with unregulated subsidiaries.

The breadth and complexity of the changes taking place in the electricity industry, as well as the strategic importance of the industry, caution against acting in haste. In so saying, the Department of Energy does not recommend inaction. We believe that many of the benefits of increased competition and regulatory reform can be achieved through existing Federal and State authorities.

The following elements of a restructured electricity market appear feasible under existing authorities of Federal and State regulatory agencies:

—Performance-Based Regulation. Basing rates on performance rather than cost can reduce regulatory burdens, provide incentives for economic and energy efficiency, promote environmental objectives, incorporate the principles of integrated resource planning, and provide flexibility for utilities to respond to customer needs and changing market conditions.

—Rate Design Improvements. Rate design principles based on marginal cost would send better price signals about the cost of providing (or value of receiving) electricity service. Better rate design would promote economic efficiency, discourage uneconomic bypass, and contribute to revenue stability for utilities.

—Integrated Resource Planning/Demand-Side Management. Rate reform programs should recognize the role that integrated resource planning (IRP) and demand-side management (DSM) play in addressing market imperfections and incorporate these processes into reform scenarios.

—Promote Wholesale Competition. Effective competition in wholesale electricity markets is a necessary precursor to greater competition in retail electricity markets. Federal and State regulators and agencies should promote development and implementation of competitive State-wide wholesale markets, regional transmission groups, and regulatory policies on transmission and other matters that are consistent with wholesale competition.

Competition in electricity or other markets is not an end in itself. Increased competition should be pursued to achieve benefits for consumers, such as greater economic efficiency and reduced prices that are not possible, or at least unlikely, through regulation. The Department of Energy supports efforts by Federal and State entities to create competitive electricity markets, but believes that such efforts should reflect concern for the following policy goals:

Economic Efficiency and Industrial Competitiveness. Proposals to increase competition in electricity markets should promote these goals, whether through regulatory methods such as performance-based regulation or marginal cost pricing, or through more far-reaching changes in the structure and regulation of the industry.

—Environmental Protection. Environmental protection must be adequately sustained under any regulatory reform proposal. Competitive markets may not always incorporate adequately the environmental impacts of electricity resource development and consumption decisions. Supplemental actions may be needed to ensure that environmental goals will be protected and effectively incorporated in complex proposals for regulatory reform.

—Equity. Proposals to increase competition in electricity markets must address recovery of stranded costs and other cost-shifting and cross-subsidy issues, as well as protect the financial health of utilities. Care should be taken to develop and implement consistent and comprehensive proposals that minimize the potential for distortions and inequities and fairly allocate the benefits and costs of expanded competition.

—System Reliability. Assurance of continued system reliability under changing circumstances must be a cornerstone of any regulatory reform approach. Markets generally can be relied upon to maintain necessary system reliability, given broad participation by market participants and support from regulators. Regulators should ensure that the necessary mechanisms are in place, or are developed, to maintain system reliability standards.

—Fuel Diversity and Renewables. Fuel diversity provides important energy security and reliability benefits, and renewables help reduce environmental impacts. Therefore, we stress that regulatory reform proposals should not jeopardize diversity of supply. Promotion of energy efficiency and commercialization of clean energy and renewable sources should remain priorities of any reform program.

—Research and Development and Use of New Technologies. Explicit consideration should be given to means of minimizing adverse impacts of reform proposals on utility and non-utility company investment in research, development and commercialization for new technologies, such as fuel cells, renewable energy generation, and energy efficient equipment.

The Department of Energy filed comments on the California Public Utilities Commission's proposal to restructure California's electricity industry and reform regulation of the industry in that State. The Department's comments addressed many of the issues of concern to this subcommittee. In general the Department's comments to the California commission stress that given the strategic nature of electricity services and the interstate nature of electricity markets in the United States, it is important for Federal and State policymakers to coordinate on energy policy matters.

The Department also recognizes that the California commission and the State legislature have the responsibility to develop and implement policies regarding the regulation and operation of the electricity services industry within California. However, elements of the California commission's proposal raise legal and jurisdictional issues that need to be resolved. The Department expressed its willingness to work with the California commission to achieve the goals of its proposal in a manner consistent with national interests and to help resolve legal and jurisdictional issues that may arise.

Furthermore, the Department of Energy expressed its strong support for the California commission's efforts to increase competition in the electricity sector. The Department stated its support for the California commission's proposal, provided it adequately addresses and resolves a number of technical, policy, and jurisdictional issues. Before the California commission's restructuring proposal is implemented, DOE suggests that effective institutions be developed to protect environmental, energy efficiency, and resource diversity objectives within a market-based paradigm.

The Department also urged the California commission to consider implementing methods and processes that reduce uncertainty and risk and increase the likelihood that the California commission will be able to achieve it's goals. Although there are many uncertainties in the California commission's proposal as presented to date, perhaps the one which will most affect the successful achievement of the California commission's goals is the uncertainty that exists with regard to the division of regulatory jurisdiction between Federal and State governments.

DOE urged that if the California commission seeks to implement change to reach its goals, it might consider a strategy less certain to be clouded by legal challenges over the next few years.

The Department believes that either these legal uncertainties should be resolved before the California commission goes forward, or the California commission should consider a strategy that would allow it to rely on its existing authorities to achieve its goals in the near term while jurisdictional and other uncertainties are being resolved.

Proposals to promote competition in the electricity industry and to reform regulation of the industry often raise Federal/State jurisdictional issues. This is particularly true when proposals, such as that made by the California commission, involve direct access to generation services by retail customers. In commenting on the California commission's proposal, the Department stressed that allowing all consumers of investor-owned utilities within California to have access to competitive power markets raises a number of issues where the boundaries of State regulatory jurisdiction are unclear and untested.

DOE argued that since no one knows precisely the boundaries of State and Federal jurisdiction in the post-Energy Policy Act era, and different constituencies have an interest in different outcomes on the jurisdictional questions, the courts will ultimately be asked to settle the matter. Further, we expect that it will take several years for these issues to be resolved.

The Department urged the California commission to attempt to resolve the critical jurisdictional questions before going forward. Alternatively, the Department recommended that the commission consider a transition strategy that included adopting those elements of the proposal for which there is relatively high jurisdictional certainty and then working to resolve the remaining elements before adopting them.

The California restructuring proposal is predicated on the assumption that the California commission can order utilities to provide retail wheeling services. However, whether State commissions actually have this authority is uncertain. Section 201 of the Federal Power Act grants the Federal Energy Regulatory Commission jurisdiction to regulate "transmission of electric energy in interstate commerce."[1] 16 U.S.C. 824. This grant of jurisdiction does not appear to be limited to transmission related to wholesale sales—the statutory language appears to extend to both wholesale and retail wheeling. According to one view, Congress occupied the field with regard to regulation of transmission with the Federal Power Act, thus precluding States from ordering wholesale or retail wheeling.[2]

---

[1] The courts have interpreted the word "interstate" to include transactions within a single State, if the transaction involves interconnected lines in which the potential exists for commingling with electricity from any out-of-State source. FPC v. Florida & Light Co., 404 U.S. 453 (1972).

[2] Because the FERC is specifically barred from ordering transmission of electric energy directly to an ultimate consumer (i.e, retail wheeling) 16 U.S.C. 824k(h), neither the States nor the FERC would have authority to order retail wheeling under this construction. However, retail

Continued

Proponents of this view argue that the courts have held that FERC's ratemaking jurisdiction is plenary and extends to all transmission and wholesale sales of power in interstate commerce which are not expressly exempted by the FPA itself. Federal Power Commission v. Southern California Edison Co., 376 U.S. 205 (1964). Thus, if transmission of electric energy over the electrical system of a electric utility falls within the definition of transmission of electric energy in interstate commerce," it is subject to FERC's exclusive authority to establish rates, terms, and conditions. Any effort by a State to order retail wheeling would interfere with FERC's plenary authority to set rates, terms, and conditions for transmission service, and must fail since it would "conflict with or interfere with Federal authority over the same activity." Mississippi Power and Light Co. v. Moore, 487 U.S. 352, 377 (1988).

On the other hand, States have exercised regulatory authority over transmission for retail sales as a part of bundled service for many years, suggesting the existence of extensive State authority in this area. The wholesale transmission provisions of the Federal Power Act, as amended by the Energy Policy Act of 1992, contain a savings clause that preserves "any authority" States had with regard to transmission of electric energy directly to an ultimate consumer (i.e., retail wheeling). 16 U.S.C. 824k(h)[3] Because FERC lacked direct authority to order retail wheeling prior to EPAct, the argument that States were preempted from taking the same action was undercut. On these bases, some have argued that States have authority to order retail wheeling.

DOE does not take a position on how the law should be construed, nor do we predict the outcome of litigation that may occur concerning State authority to order retail wheeling. Instead, we simply observe that there is significant legal uncertainty and that it may take several years before there is a conclusive judicial interpretation.

Sections 205 and 206 of the Federal Power Act grant FERC authority to determine just and reasonable rates for the transmission of electricity subject to the jurisdiction of the Commission, i.e., transmission of electric energy in interstate commerce. Thus, even if a State commission is found to have jurisdiction to order retailing wheeling, it may be argued that the FERC has exclusive jurisdiction over the rates and terms for these services.

The California commission, for example, in its proposal to restructure the electricity industry in California, indicated its intention to impose a charge on all consumers, including direct access customers, to recover stranded investment costs. However, if this charge were to be imposed as part of a transmission rate subject to FERC jurisdiction or as an "access charge" that is found to be a component of a transmission rate (rather than as a condition of receiving a service regulated by the State) it would be outside the California commission's control. This would create an obstacle to implementation of the California commission's proposal.

Should such a stranded investment charge be subject to FERC jurisdiction, it would be subject to FERC's general guidelines regarding the recoupment of stranded costs. These guidelines currently restrict recovery via transmission rates to circumstances in which: (1) the transmitting utility had a reasonable expectation that the customer would continue to purchase power; (2) the stranded cost charge is no more than what the customer would have paid if it had remained a power customer; and (3) the stranded cost charge is subject to reduction to reflect mitigation measures available to the transmitting utility. Massachusetts Electric Company, 66 FERC $61,036, 61,060 (January 13, 1994).

The Energy Policy Act of 1992 amended the Public Utility Holding Company Act of 1935 (PUHCA) to provide that ownership of exempt wholesale generators (EWG's) would not trigger PUHCA regulation. This amendment was intended to eliminate a significant regulatory barrier to entry to the wholesale generation market, and thereby allow greater competition in that market. The amendment did not generally extend the PUHCA exemption to generators making retail sales. As a result, EWG's will not be eligible to participate in this market. Therefore, barring further legislative revisions, State proposals to extend direct access to retail customers could not include EWG's and would likely involve fewer sellers in the retail market than in the wholesale market.

Pursuant to section 210 of PURPA, 16 U.S.C. 824 a-3, and the implementing regulations, an electric utility is required to offer to purchase electricity at its avoided cost from qualifying cogeneration and small power production facilities. The intent

wheeling would presumably be legal and subject to FERC jurisdiction if provided voluntarily by the transmission owner.

[3] The EPAct savings clause precludes any interpretation of EPAct as either enhancing or undercutting any preexisting State authority to order retail wheeling. Congress appears to have intended to avoid opining on whether any such State authority exists.

of the section is to give incentives to alternative supply sources and thus enhance the development of diversified energy supplies. Under retail competition such as proposed by the California commission, investor-owned utilities would operate under the pressures of a competitive retail market but continue to have an obligation to purchase under PURPA. In addition, it appears that new entrants to the California retail market would also be subject to PURPA's obligation to purchase power from qualifying facilities. It is unclear how these requirements would affect development of effective retail competition.

The examples above illustrate that there are important constraints in Federal law pertaining to State commissions' efforts to promote retail competition, and equally important uncertainties related to jurisdiction. It is not known whether or when amendments to Federal statutes or conclusive judicial interpretations of jurisdiction might be obtained. Further, there may be a number of constraints under State law pertaining to reform proposals, thus creating an additional set of legal uncertainties.

DOE Responses to Specific Questions Posed by Chairman Sharp

Question 1: To what degree do you believe the electric utility industry has entered a period of irreversible change, and how widespread do you feel change will be?

Answer: The Department believes that the industry has been undergoing important changes for more than a decade. Some of the changes are reversible and others are not. Competition in wholesale markets is increasing as the Federal Energy Regulatory Commission and the States develop policies to implement the Energy Policy Act of 1992. In addition, technological and institutional changes are taking place that are increasing the efficiency and reducing the cost of new generating capacity and making it easier for alternative resources, such as conservation and renewable resources to compete with traditional supply resources. Moreover, advancements in transmission technology and in metering and information processing are reducing the cost of transmission services and increasing the ability of consumers to receive and respond to more finely-tuned price signals than in the past.

At the same time, increased competition in retail markets sooner or later appears likely as several States, including California, Michigan, Nevada, New York, Massachusetts and others, are in varying stages of addressing retail competition issues.

Question 2. Do you have any recommendations as to how regulators should handle the transition period to a more competitive market?

Answer: The Department believes that regulators should design regulatory approaches that support the basic principles we have outlined in our testimony. In so doing, they need to attend to critical transition questions, such as how to deal with stranded investments. Regulators should establish a basic framework within which competition may take place while also ensuring adequate consumer protection from market imperfections and incorporation of appropriate societal goals (environmental quality, resource diversity, energy efficiency and conservation). Achieving this set of goals will require effective cooperation between State and Federal regulators and agencies. In this regard DOE recently announced that it will hold a National Electricity Forum to enable policymakers and stakeholders from around the country to come together to address issues related to fundamental changes in the industry.

Question 3. How can social and environmental objectives be adapted to a "restructured" industry, in an environment in which services once provided by a single vertically integrated utility are increasingly "unbundled"? To what degree should these objectives be retained, changed, or abandoned?

Answer: In our view, most of these objectives will and should remain important in public policy terms, although methods of achieving them may change. The challenge is for the electricity industry, its consumers, and its policymakers to find ways to achieve these objectives in the context of a more competitive industry.

Responsibility for dealing with this transition will fall on regulators and legislators at the Federal and State levels. For example, several States have authorized or are proceeding to consider requiring their commissions to take environmental impacts into account in their regulatory approaches. In some cases, it appears that new State legislation will be necessary to provide policy guidance or new authority to State agencies to achieve these objectives through alternative means.

Question 4. How should pricing mechanisms be adapted to further carry out Congress' goal, embodied in EPAct, of a more competitive wholesale electricity market facilitated by open access to transmission services?

Answer. Recognizing the need for gradual transitions in rate design from average-cost to marginal-cost methods, the Department believes that pricing mechanisms should move toward pricing approaches that show consumers the marginal costs of service, and toward free and open access to markets and information for all. Generation and transmission services should be unbundled and priced separately. Development of regional transmission groups (RTG's) should be encouraged.

Question 5. What are the most important questions involving State and Federal regulatory authority? Please feel free to offer any opinion you may have on the state of the law, or on the need for changes to, or modification of, the respective roles of States and Federal regulators.

Answer: DOE's views on Federal/State jurisdictional issues are presented in our testimony. While the Department believes that no one knows today how these jurisdictional uncertainties will be resolved, it is our view that development of an efficient, reliable, competitive, and socially responsible electricity industry depends on successful resolution of these issues. The Department will support increased cooperation and coordination between Federal and State agencies to resolve these matters.

Thank you very much. This concludes my prepared remarks. I would be pleased to respond to other questions from the committee.

Mr. SHARP. Thank you very much, Ms. Tierney.

Mr. Moskowitz, we would be very pleased to hear from you now.

STATEMENT OF DAVID MOSKOWITZ

Mr. MOSKOWITZ. I am with the Regulatory Assistance Project. The Regulatory Assistance Project, or RAP for short, is a nonprofit organization funded by the U.S. Department of Energy, the U.S. Environmental Protection Agency, and a number of private foundations for one sole purpose, and that is really to provide State utility regulators with training, education, and advice on anything to do with energy policy, integrated resource planning, energy efficiency, and more recently with respect to competition.

We have already since 1992, since we began, provided workshops in over 30 States, and the most frequently requested topics are integrated resource planning, regulatory reform, and, as I said, competition. So with that, let me turn to competition, the issue of the day.

Let me first say that the current debate over the future of the electric utility in this debate tends to be stylized as a conflict between competition on the one hand and Soviet style central planning on the other. This simply is not the case. All sides, at least from everything I have seen so far, all sides in the debate are in favor of the creation of competitive markets. I'll get into that more soon.

What separates people in the debate are the rules under which that competition will take place: The structure of the market and the rules that will govern that, and particularly the rules that will govern in the integration and how those rules will integrate, integrated resource planning, energy efficiency, and environmental improvement.

Recognizing that it is not really competition versus central planning, it is really the rules under which competition will be structured—how will we structure and create and regulate, in effect, these markets, since all of the free market is regulated in some fashion or other, at least the rules of the game are put into place—recognizing that, it is actually terrific to see that the Department of Energy and NARUC have also sort of formed an activity that is about to get under way to analyze really what the options are, how can these markets really be structured in order to make sure that all of the Nation's important interests are covered.

Now let me be perfectly clear because there are some things that could be incompatible—there are some potential incompatibilities between integrated resource planning and competition. Again, the potential exists depending on how you structure that competitive market.

The area that is most certainly going to end up being inconsistent with integrated resource planning is if you structure competitive markets that focus exclusively on the price per kilowatt hour of electricity without regard to the overall cost of energy services delivered and without regard to the environmental implications of that kind of structure.

Now for the last year and probably, unfortunately, I think, for the next few, there is a near state of paranoia out there with respect to where this industry is going, what the structure will be, what the rules will be of the game, and while that paranoia, while that uncertainty is high out there, we actually have seen substantial retrenchment in the past year with respect to some of the very innovative, some of the very progressive, some of the very successful pursuit of environmental efficiency improvements, and some of the innovations in the environmental improvement area.

I believe that the primary and most immediate role for all of government, all of the public really, is to begin the dialogue of establishing the goals, establishing the principles that will dictate the structure and the rules of the market and then move rapidly toward the implementation of that competitive market. With those goals firmly in hand, it won't be difficult to really fashion competitive structures that give us the type of cost performance and efficiency that we have all been looking for.

With that, let me turn to the five specific questions that you raise, Mr. Chairman.

First, the degree to which we have entered a period of irreversible change. If there is one thing about change, it is never irreversible. This is just a pendulum we are on, and it is just a question of when we happen to take a snapshot of that pendulum.

The change that we are involved in now really is just the next evolution of change that really began in 1978 with the passage of PURPA. Without even knowing it, we really began the competitive market—the push toward competition in a very major way. From there, it moved to integrated resource planning, and then integrated resource planning very quickly moved to the adoption and implementation—perhaps superimposition of competitive bidding integrated with integrated resource planning.

Now I know that this committee played a pivotal role in the Energy Policy Act of 1992, very wisely embraced broad policies of increased efficiency in the use and production of electricity and a steadily improved environment. The tools that you chose to achieve those policy ends were IRP and competition—integrated resource planning and competition.

In fact, I think that you found—at least certainly I find—that the most compelling case, the most compelling case for increased competition, has actually been put forward by those States that have most aggressively pursued integrated resource planning because those States that have most aggressively pursued integrated resource planning also supplemented it with competitive bidding, competitive market forces, and when those two came together we actually saw very substantial benefits that accrued to the Nation and consumers.

The best way to sort of think about the way competition and integrated resource planning fit together is to recognize that competition really gives you more resource options, lower costs for those options. What integrated resource planning gives you is the tools to be able to pick between the options and figure out who wins and who loses the competition. States that have implemented integrated resource planning most successfully know that distinction, know that competition and integrated resource planning fit together hand in glove.

Well, how fast—how widespread will these changes be, and how fast will they occur? Before joining the Regulatory Assistance Project, I was a commissioner in the State of Maine Public Utilities Commission until 1984. Actually while there began competitive bidding—the Nation's first entre into competitive bidding was in the State of Maine in 1984. Now, 10 years later, little more than half of the States have adopted what really has been one of the fastest and most widespread adoption of a new thing in this industry. So while we are moving sort of at lightning pace, you have to recognize that 10 years after competitive bidding started we are about halfway there.

Another good example, if you want to think about California for a moment, in 1986 California adopted competitive bidding. In 1994, 8 years later, they are almost done with their first round. That gives you some sense perhaps of how fast some of these things will actually end up moving. The pendulum is swinging, but it doesn't really swing quite as fast as we tend to think it might.

Recommendations for regulators in the transition period—really, two basic pieces of advice. One is to recognize that regulators have to and they really are beginning to lead this discussion. This is not a question of sitting back and watching where the industry is going; we know where the industry is going; the industry is becoming increasingly competitive. That is not the issue. The issue is the structure of the rules and the structure of the market and the rules under which competition will take place, and that is where regulators are now starting to really focus their effort.

The other real important piece of advice is to understand what competition really can and can't do. What we can actually gain from competitive, particularly in the short run, is really going to be quite limited. As you look out there across the States and recognize that there right now is a fairly substantial gap in many States between the retail price of electricity and what is perceived as being the market price, the marginal cost of electricity, and it is in large part this large gap that is causing the increased emphasis for things like discount rates for large industrial customers—John—most of which have already been given, and things like retail wheeling.

Mr. ANDERSON. Not all yet, David.

Mr. MOSKOWITZ. I'm sure by the end of today the last discount will be given.

You have to look out there and recognize that what competition really can do, and I think can do very effectively and we ought to be pursuing it for that, is to reduce the inefficiencies that are embedded in the current system.

Competition can't eliminate sunk costs. All we can do with sunk costs is decide who is going to pay for them, and I know you will be hearing more about that during the course of the day.

Number three, how can social and environmental objectives be adapted? I am glad to see the question was "how can" as opposed to "should." The answer is, they can. We are not going to have, obviously, the time to go into it, but by deciding what the rules of the game are, if we decide that we want environmental goals, energy efficiency goals, incorporated in our competitive market structures, the task is really not very difficult.

If we want to do that, what we have to do, however, is to articulate those goals. We have to articulate the goals that will guide the formulation of the specific rules that will govern these new markets. The first goal—I have three for you—the first goal is, we have to have an industry structure created, we have to have a market structure created that is designed—not merely a bandaid solution at the end—that is designed in the first instance to encourage increased efficiency in the end use of electricity. That has probably been the single item that has most confounded debates in this area for the last decade or so.

There are really two facts that will always—always have and always will lead us to this conclusion that energy efficiency is important. First, of course, is, we know it saves money. But the other two facts that really lead us there are, we know from the history of the last decade that there are very large and very real and very persistent market barriers to the increased application of energy efficiency and that utilities are very effective at reducing those barriers. We also know that Federal energy and environmental and economic policies are always pushing us in the direction of increased energy efficiency.

The second goal is, we have to be cognizant of the environmental impact of energy production and use. Now here I am just going to relate one additional sort of piece that shocked me at least about a month ago in the State of Maine.

A month ago the State of Maine issued a warning that said if you are a pregnant woman, a woman who plans to get pregnant, or a child under eight, you can not eat any fish, not even one, from any fresh water lake, stream, or river in the State of Maine. It turns out, 30 States have issued similar kinds of warnings. The reason is mercury, and mercury emissions from coal-fired power plants are believed to be one of the major causes.

Obviously, the last goal I will leave you with before turning on to Mr. Anderson will be the goal of making sure that we also have increased efficiency in the regulatory community, increased efficiency of regulation for those elements of the business that end up continuing to be regulated. There we know NARUC had it exactly right when they adopted their resolution that the utilities' successful implementation of a least cost plan should be their most profitable course of action, Congress embracing that both in the Clean Air Act and the Energy Policy Act.

With that, Mr. Chairman, I'll turn it back to you.

[The prepared statement of Mr. Moskowitz and responses to subcommittee questions follow:]

STATEMENT OF DAVID MOSKOWITZ, THE REGULATORY ASSISTANCE PROJECT

My name is David Moskowitz. I am a Director and cofounder of The Regulatory Assistance Project. The Regulatory Assistance Project (RAP) is a non-profit institution funded by the U.S. Department of Energy, U.S. Environmental Protection Agency and several private foundations. Our purpose is to provide State and Federal energy regulators with educational assistance on all matters dealing with integrated resource planning, regulatory reform, energy policy and related matters. We have provided workshops in over 30 States since we began this work in 1992. The agenda for the workshops are governed by each State regulatory commission but the most frequently requested topics are integrated resource planning, regulatory reform and competition.

The current debate over the future of the electric utility industry tends to be styled as a choice between increased reliance on markets and competition on the one hand and Soviet style central planning on the other. This is not the case.

All sides in the debate, in fact, are in favor of the creation of competitive markets. What separates people is deciding what structure and rules should govern new competitive models in general and more specifically, what the roles of Integrated Resource Planning (IRP), energy efficiency and environmental improvement should be.

First, it is important to recognize that IRP is not central planning. The putative conflict between competition and IRP is far more imaginary than real. IRP and competition are wholly compatible. In fact, it has been IRP's insatiable search for new and lower cost supply and demand-side resources that has brought competition to a monopoly business that had little interest in non-utility or non-traditional suppliers.

Competition is a form of economic organization, not a theological imperative. It is an attractive system because it leads to economically efficient production for most goods, most of the time. Even in areas such as environmental protection, where markets generally do a poor job, competitive, market-based regulatory mechanisms can be used to lower the cost of achieving environmental goals.[1]

But, to be perfectly clear what is incompatible with IRP is a model of competition that focuses exclusively on the kilowatt-hour price of electricity without regard for the total cost of energy services or the environment. The national and international stakes in this debate are very high.

For the last year and probably for the next few, uncertainty about the future of the electric utility industry, the structure and rules under which competition might take place and whether some investments may not be fully recovered has caused utilities around the country to cut back on their very impressive achievements in the energy efficiency and environmental areas. This is a substantial concern.

The primary and most immediate role for Congress, the States and the public is to establish the goals and objectives that will dictate the structure and rules of the market. With our goals firmly in hand, fashioning new competitive market structures will not be difficult and the chill of uncertainty should dissipate.

The remainder of my testimony will address the five specific questions that were posed to this panel.

1. To what degree do you believe the electric utility industry has entered a period of irreversible change, and how widespread do you feel change will be?

It is important to put the issue of change in perspective. In most regards, the electric utility has been in a period of dramatic change since the late 1970's. Hardly a single year has gone by when there has not been a major series of conferences focusing on the dramatic "change" in this industry. In years past, the changes fell under the banners of diversification, holding companies, acquisitions and mergers, bankruptcies, integrated resource planning, reduced use of oil and natural gas and environmental compliance. Even the current focus on increased competition started over 16 years ago with PURPA, IRP and competitive bidding.

This committee played a pivotal role in the Energy Policy Act of 1992 (EPAct). There, you wisely embraced a policy of increased efficiency in the production and use of electricity and a steadily improved environmental performance. IRP and competition were the means to achieve these goals.

---

[1] Competition is not a code word for resigning ourselves to an environmentally inferior electric utility industry. While a debate on the environmental standards for the industry falls beyond the range of this testimony, the current level of environmental commitment is assumed as a standard that must, at a minimum, be maintained in any competitive/regulatory approach.

The most compelling case for increased competition has been made by States that have aggressively implemented IRP and supplemented it with competitive market forces. It is the experience of those States that has taught us that competition does two things: it gives us many more resource options from which to choose, and it gives us those options at lower prices than generally expected. IRP complements this by giving us the tools to figure out which resources win in the competition.[2]

There are, however, several areas in which IRP and competition are at odds. One such area is the level of regulatory involvement or process. Some States have adopted IRP and competitive bidding and have, for a variety of reasons, chosen to use a highly prescriptive, highly litigious regulatory process. California and Wisconsin are probably the best examples. Other States such as New York, Vermont and Oregon have adopted IRP and competitive bidding with a much less heavy handed approach.[3]

The trend is clearly to rely increasingly on competitive options and less on regulatory oversight. The key to allowing this to happen is to reform regulation and, where necessary, reform industry structure to better align utility shareholder interest with the successful implementation of IRP. Congress recognized the importance of this simple principle in EPAct (Section III and Title I of the Clean Air Act Amendments).

How widespread will the changes be? The answer depends entirely on how satisfied the public is with the results. Using competitive bidding again as an example, competitive bidding began in the State of Maine in 1984, and today, 10 years later, it has spread to over half of the States. This has been a remarkably fast and widespread adoption of a new reform principally because it has been a largely positive experience, producing obvious consumer benefits.

2. Do you have any recommendations as to how regulators should handle the transition period to a more competitive market?

Yes, I have two pieces of advice. First, regulators should not assume the transition to competitive markets will simply happen and that their job is to sit by and watch it occur and adapt it as necessary. Second, regulators should not expect too much from competition. I will elaborate on each of these points.

The movement to a competitive market is being and should be led by regulators. All competitive markets function within a set of rules, guidelines or framework created by government to control potential abuse and to achieve consistency with other public policy goals. Competition in the electric utilities sector will be no different.

It is not whether we want more competition—of course we want more competition. This is not a conflict between competitive market forces and central planners. The debates that ought to be occurring here and elsewhere are about rules under which that competition takes place. Regulators need to be out front, leading the discussion on the form of competition, the rules of the game and industry structure.

Regulators should also be realistic about what competition can and cannot accomplish. In particular, it is useful to ask to what extent can competition and more market-based regulation lower costs? In answering this question, we first have to recognize what competition is not.

Competition is not simply a mechanism to shift costs away from some electric customers and on to others. To be sure, competitive forces will influence the rates individual customers may be charged. But this does not mean that industry structures should be allowed that encourage uneconomic decisions. Some proponents of retail wheeling[4] advocate that the current rules should be modified to allow large cus-

---

[2] An easy way to understand use of IRP in a competitive environment is to think of competitive bidding as telling you what different options cost. One bidder might bid 4 cents per Kwh for a wind farm, another bids 4.5 cents for a combined cycle gas-fired facility and another bids 12 cents for a photovoltaic facility. These options, though, are so fundamentally different from each other that simply comparing their bid prices is a waste of time. What IRP does is it tells you what each of these resources is worth. Once you know what each option is worth, the job of picking winners and losers is easy; you pick those options that cost less than they are worth, thereby minimizing total costs.

[3] IRP must be understood as a set of planning principles and should not be confused with a particular regulatory process. IRP regulation can be implemented in a "light-handed" fashion, a "heavy-handed" fashion or anywhere in between. "Light-handed" performance based regulation, combined with careful attention to the alignment of utility financial incentives with IRP is the best alternative.

[4] Retail wheeling occurs when a retail customer of a given utility is allowed to shop around for electricity from suppliers other than their host utility. The customer would pay the host a fee for wheeling, that is transmitting, the power to the customer's premises.

tomers to benefit even if everyone else's costs rise by an even greater amount.[5] Competition should not be used to support this type of inequity and economic waste.

Understanding why there is a gap between average prices and current market prices helps to shape successful reforms. It is convenient to divide the sources of this gap into four categories:

1. Inefficiencies associated with the institution of regulated, vertically integrated electricity monopoly.
2. Sunk costs that exceed current market values. These are "good" investment decisions, made in the past, that have turned out to be uneconomical as market conditions changed over time.
3. Some combination of economies of scale and declining marginal cost.
4. The impact on prices of efforts to meet other legitimate goals such as DSM and environmental improvement.

For the average utility, the bulk of the gap is associated with sunk costs that are now uneconomical or just a high proportion of fixed costs. A small fraction is associated with differential State and local taxation of utility property and sales. The smallest fractions are associated with DSM, environmental improvements and actual economic inefficiencies.

Regulatory reform and competition can reduce the size of the gap due to inefficiencies created by prevailing institutional arrangements. Successful regulatory reform can also reduce the likelihood and magnitude of future planning and investment errors. As powerful as regulatory reform and competition might be, neither can make sunk costs go away. All we can do is to decide who is going to pay for them and over what time period.

3. How can social and environmental objectives be adapted to a "restructured" industry, in an environment in which services once provided by a single vertically integrated utility are increasingly "unbundled"? To what degree should these objectives be retained, changed or abandoned?

Congress and regulators have consistently embraced energy efficiency, resource diversity and environmental goals. The current focus on industry reform would advance most rapidly and constructively if it were made clear that these goals must be met.

There are no doubt many goals that can be both readily agreed upon and incorporated into new competitive regimes. I list three goals below which are both important goals and happen to be the items that most separate the sides in utility restructuring debates.

Goal 1. Industry and regulatory structures should be designed to encourage increased efficiency in the use of electricity.

The seemingly endless and repetitive debates about the role of end-use efficiency and the role of electric utilities in advancing energy efficiency goals may, more than anything else, paralyze discussions about the future of the industry. This goal lays the issue to rest by stating that end-use energy efficiency must be a design principle. It is tempting for some to say, and even wish, that utilities should be in the business of delivering kilowatt-hours and pricing the electricity so consumers make rational decisions about how much electricity to use. But saying it or wishing it does not make it so.

Two facts lead to this conclusion:

1. There are very real and persistent barriers to increased energy efficiency, and utilities are very effective at reducing and removing these barriers.[6]
2. Federal energy, environmental and economic policy particularly as they relate to electric utilities place a high priority on increased energy efficiency.[7]

By far the simplest way to meet this goal while creating a more competitive electric industry takes two steps. First, separate generation from the monopoly business. Second, regulate the remaining utility in a fashion that decouples profits from sales and ties earnings to performance.

---

[5] This turns on the question of what wheeling rates or other fees will be charged to retail wheeling customers. If the wheeling rate is appropriately set, then retail wheeling will only be undertaken where it results in lower overall costs.

[6] The recent and exhaustive evaluations of DSM programs in Massachusetts and California convincingly show that utilities are able to squeeze very large levels of energy efficiency at total costs that are far below the direct cost of power supply. Given the relatively high retail prices in these States it is clear the market barriers are substantial.

Customers responding to accurate prices have been and will continue to be drivers for increased energy efficiency. Better pricing clearly helps. But with or without better prices there remains a very important energy efficiency role for electric utilities.

[7] The Clean Air Act, The Energy Policy Act, and the administration's Climate Action Plan are the most recent examples.

Goal 2. Industry and regulatory structures should aim toward reducing the environmental cost and risk of energy production and use.

The public's desire for increased environmental protection and a growing list of new environmental risks associated with energy production are two of the important reasons energy efficiency receives so much attention.

Energy production and consumption are inextricably linked to the environment. Electric utilities account for 69 percent of national sulfur dioxide emissions. Utilities account for 35 percent, 32 percent and 5 percent of carbon dioxide, nitrous oxide and particulate emissions respectively. For these reasons and others, utilities are an early and frequent focus of environmental regulation and public attention.

The focus of the next round of environmental regulation will be on tighter or new controls for nitric oxides, air toxics including mercury, fine particulates, $C0_2$ and other greenhouse gas emissions. For example, prestigious medical journals have pointed to 50,000 deaths per year (more than the number of US casualties in Vietnam) caused by fine particulates. Power plants are a major source of these pollutants. With respect to mercury, the State of Maine has recently warned pregnant woman, women who plan to get pregnant and children under eight not to eat any fish caught in any of Maine's freshwater lakes, steams or rivers. Coal-fired power plants are believed to be a major source of mercury reaching Maine's remote pristine lakes. Similar warnings have been issued in roughly 30 other States.

Sooner or later new regulations will address each of these areas, and electric utilities will be one of the first targets. The bottom line is that electric utilities have been, and will continue to be, expected to make and implement energy decisions that offer energy services without jeopardizing environmental quality.

Utilities already do a reasonable job complying with existing laws and taking these laws into account in their planning. Where utilities generally fall short is in incorporating expectations, probabilities and cost estimates about future environmental constraints into the planning process.

New regulatory systems and new market structures should encourage utilities and others to improve the way that they incorporate uncertainty into their planning and investment decisions, including uncertainty about future environmental compliance costs.

Goal 3. Regulation should increasingly rely on competition and performance based regulation to reduce costs.

There are two important items to keep in mind when designing and evaluating new ratemaking mechanisms.

1. Congress and NARUC were exactly right when they called for regulatory reforms to make a utility's successful implementation of its least cost plan the utility's most profitable course of action. The key is to design regulatory systems that inherently reward superior performance.

2. The distinction between electricity prices and electricity costs must be recognized. DSM or energy efficiency is cost effective and desirable whenever the total costs of saving a kill or KWh is less than the cost of supplying power. The impact of DSM on utility prices is largely indifferent to who pays for DSM. Given today's "rate gap", if we magically removed all DSM market barriers, and customers invested in DSM, electricity prices would increase. (Magic wands, hence, fail the rate impact test.) This means new performance based incentive systems cannot focus exclusively or primarily on electricity price.

4. How should pricing mechanisms be adapted to further carry out Congress' goal, embodied in EPAct, of a more competitive wholesale electricity market facilitated by open access to transmission services?

We do not yet have a competitive wholesale electricity market. The key to creating a fully competitive wholesale electricity market is access and transmission pricing rules. But getting these rules right under the current vertically integrated utility structure is virtually impossible.

The operation of the transmission system and, hence, pricing of the transmission services is so complex that anyone who has an interest in delaying resolution of access and pricing terms has an easy job. The simplest solution to the access and transmission pricing issues is structural. Separating the transmission business and creating a single entity that does nothing but owns and operates transmission systems is the surest way to get a reasonable set of rules in place. This is the most important step in the creation of a competitive wholesale market. I applaud David Freeman's proposal to do this in New York.

5. What are the most important questions involving State and Federal regulatory authority? Please feel free to offer any opinion you may have on the State of the law, or on the need for changes to, or clarification of, the respective roles of State and Federal regulators.

I have three brief suggestions.

1. Susan F. Tierney, Assistant Secretary, Office of Policy Planning and Program Evaluation presented a very comprehensive and thoughtful blueprint for reform in her comments filed June 8, 1994 in a California Public Utilities Commission Proceeding. These comments highlight the national interests and the areas in which coordination between State and Federal jurisdiction is needed.

2. Retail wheeling would more likely than not lead to higher, not lower costs. Yet, as I point out in Attachment 1, it is possible to construct a retail wheeling rate that is entirely consistent with IRP. Doing so however depends entirely on setting the right price for retail wheeling services. Just who has the authority to set these prices is unclear.

3. Federal/State jurisdictional lines stopped making sense long ago. The clearest example is exclusive Federal authority over a one MW wholesale transaction between two utilities within a given State compared to exclusive State jurisdiction over a 1,000 MW power plant proposal by a utility with operations in three States. If it is fair to put every aspect of industry structure and regulation on the table why not at least ask whether it is not time to think about how to best allocate authority between State, Federal and regional bodies.

Mr. SHARP. Thank you very much, Mr. Moskowitz.

Mr. Anderson, we are pleased to welcome you again.

STATEMENT OF JOHN A. ANDERSON

Mr. ANDERSON. Thank you very much, Mr. Chairman.

I would like to first compliment you and the committee on your vision not only in passing the Energy Policy Act but then in holding hearings such as this with insightful questions.

I am John Anderson, executive director——

Mr. SHARP. The State of California deserves credit for the wide attendance, I think. We are happy to have you.

Mr. ANDERSON. But there are many other States besides California that are taking some bold actions also.

I am here representing large industrial electricity customers, as you know, and we have a very large interest in the questions that you put forth. My comments today will summarize my written testimony, and I will direct them around the five questions that you asked.

In the first question: To what degree has the electric industry entered a period of irreversible change? I would like to begin by quoting from an editorial that was in just last week's Wall Street Journal, and I love the opening lines, and it said, "Competition is contagious. Faster than the control freaks in the political process can bottle it up, it spreads to another area of daily life, transmitted all the while by technology. Now it looks to be coming to electricity." We agree with the Journal.

The electric utility industry has been subjected to competitive pressures for more than a decade. Competition will keep increasing, with or without further legislative or regulatory initiatives. The key question in our view at least is, how should a more competitive market be structured to assure fairness and equity for all stakeholders?

Competition in the industry is coming for several reasons. First, both the level and the magnitude of rate disparities have grown significantly over the past two decades, and that is for all customers, not just industrial customers. Indeed, the rate differentials of far more than two-to-one exist throughout the Nation today; and, further, most utilities today charge customers far more than the cost of new supplies.

Second, these conditions give customers strong motivation to seek lower cost alternatives. For example, a typical industrial often pays a penalty of several million dollars a year at a single plant site compared to a competitor located only a few miles away.

Increasingly, industrials have identified options available to them to free themselves from being captive to the high cost local utility. Some of the more obvious include self or cogeneration inside or outside defense EWG's, municipalization, and privatization, and wheeling to political subdivisions of the State.

Third, the Energy Policy Act increased the viability of several of these options. These activities are raising the specter of significant stranded cost. Utilities are responding with a plethora of special rates and deals to help them maintain market share, and I would like to emphasize that these things that I have discussed are all occurring without retail wheeling.

In summary, the push toward deregulation is not reversible. Fine-tuning the existing regulatory regime will not fix the electric industry's problems.

Relating to your second question about the transition period, all customers should be able to competitively source their power needs. Competitive sourcing, or, more nearly, retail wheeling, can produce substantial benefits for all customers. The planning process will be greatly improved, and it needs great improvement. True, least cost planning will be assured, not the kind of least cost planning that is sometimes now accomplished. Those who plan will be made risk takers. This will force planners to consider the broadest range of technological options to control cost and innovate in ways never achieved by monopoly utilities.

Retail competition also gives customers a perfect opportunity to purchase cleaner but more expensive resources if they really want to do so. Retail competition can provide regulators with an important benchmark for those who choose not to shop. Regulators can compare the price charged to nonshopping customers to that obtained by shopping customers. This may be a more efficient means of establishing rates than the highly litigious regulatory hearing room, at least during the transition period.

Some will argue that the potential stranded costs are so large that retail competition should not be allowed. This assertion is incorrect for at least two reasons. First, stranded costs may be significant, but they are not near the $200 or $300 billion that are often claimed. Besides, whatever the level of actual transition costs, they are a real indicator of the value of retail competition. Many costs are stranded because they are uneconomical. Retail competition will make it impossible for uneconomical costs to be foisted on unwilling customers. The elimination of this cost burden constitutes, dollar for dollar, a powerful and a direct economic stimulus to the economy. Several States have taken specific actions to investigate retail competition, California and Michigan perhaps the most visible examples today.

We recognize that the transition to a competitive market will not be easy. We thus compliment those States that have recognized the values of retail competition and are taking bold and courageous steps necessary for a successful transition. In our prepared remarks we offer eight specific principles as a way to minimize the objections in the movements.

Your third question is: How can social and environmental objectives be adapted in a restructured industry? I would like to emphasize that large industrial electricity customers strongly support energy efficiency. They have to. Fierce domestic and international competition require them to implement energy efficiency improvements that are truly cost effective.

However, we have serious concerns how conservation programs and other social objectives are being pursued by regulated electric utilities. We identify six specific points in our written testimony.

In our view, industrial planning is neither as effective nor productive as market forces for cost-effectively achieving the goals of environment protection and other social objectives. This repeatedly is being demonstrated in other sectors of the economy where market mechanisms are proposed as a more workable alternative to traditional forms of government intervention.

It is important to know, though, that there is an inherent conflict between the degree of customer freedom necessary for competitive markets to operate efficiently and attempts by the traditional regulatory apparatus to restrict those freedoms in order to force more market outcomes on predetermined conclusions.

Further, affluence generated by economic growth increases total resources as that society is willing and able to devote to environmental protection and other social objectives. Economic growth will enhance our society's willingness to pay for environmental and other social amenities.

Finally, attempting to achieve public policy objectives by using regulated electric utilities as the primary delivery mechanism for providing these social services by definition significantly reduces the purchasing power of ratepayers. Indeed, when some utilities collect a dollar from ratepayers to fund these kinds of programs, perhaps less than a dime is actually spent on real energy savings.

Within a competitive electricity market, the full accounting costs of social and environmental programs will be visible as a separate line item on a customer's bill. Such costs will not be bundled or hidden within the overall operating costs of traditional utilities. The public will have a chance to vote with their dollars for these programs.

In sum, we strongly believe that competition will not inhibit implementation of legitimate social or environmental objectives. To the contrary, we believe that these objectives will be enhanced as a result of a restructured industry.

In your fourth question, you asked about pricing mechanisms. I simply point out that ELCON, the American Iron and Steel Institute, and the Chemical Manufacturers have filed comments with FERC, and we have included them in our written testimony, and we refer you to them.

But on a broader note, several parties have proposed, and you have even heard today, the creation of regional wholesale power markets in lieu of trying to implement direct access or retail wheeling. This might have been a great idea 20 years ago. However, developments over time render such proposals delay tactics at best.

These proposals generally assert that the benefits of competition can be realized by a scheme reliant on customers' continued acquisition of electric services on a fully bundled basis. Such proposals will not promote greater competition for several reasons.

First, they fail to recognize that electricity rates simply are too high. The SOCAL Edison, Professor Hogan, Professor Joskow proposals may reallocate costs, but they do nothing to lower costs. Under those proposals rates will remain too high.

Second, such proposals cannot be implemented in a timely fashion. The implementation of time-of-use rates for all customers would require the installation of new meters on every household, business, and farm. This would take years and cost billions of dollars.

In addition, their proposals require the establishment of an independent spot market that is open to all suppliers and not subject to manipulation. This requires either years to establish the necessary regulatory apparatus as the painfully slow development of the Western System's power pool has shown, or vertical disaggregation of generation and transmission—again, a rather lengthy process. There is a better way. The California Public Utilities Commission calls it direct access, and we think it is a great idea.

Third, the establishment of a wholesale-only market will not in any way assure economic benefits for ultimate customers. As long as buyers in a wholesale market know that their customers are captive, they have neither the motivation to purchase the lowest cost power nor the incentive to pass through the benefits of competition to ratepayers.

In their reply comments in California, National Power, PLC, of the United Kingdom said, "National Power supports the POOLCO concept put forth by Southern California Edison but not in the absence of simultaneous direct access. The majority of the real benefits from competition come from direct access, not from the pool mechanism itself."

Finally, nothing in the proposals assures that the continued implementation of social programs by regulated utility will be any less efficient or effective. Many, if not most, existing DSM and renewable energy procurement programs are not cost effective because they are shielded from market discipline. The status quo is simply unacceptable.

To us the point is clear. Attempts to establish a wholesale market with customers facing the spot price through new rate designs sounds attractive but won't provide real necessary benefits in a timely fashion, if at all. What they will do is greatly delay action on the one thing that really will produce benefits for all customer, retail competition.

Your fifth question asked about jurisdictional questions. I have attached to my written testimony a legal memorandum saying that we believe very strongly that States do have the authority and the jurisdiction to do the kinds of things that California and other States are now looking at.

But I would like to conclude with a note that a fascinating situation may occur if our analysis is wrong. A ruling that the States have no jurisdiction over transmission may well result in the functional equivalent of disaggregation of the electric industry. This idea was proposed recently by Ashley Brown, both a former Ohio commissioner and chairman of NARUC's Electricity Committee.

Brown proposes that if States do not have jurisdiction over retail wheeling, then they do not have nor have they ever had jurisdiction to approve any bundled rate. This means that States should immediately remove all transmission costs from rate base and request refunds for costs overcollected since these bundled rates became effective, which would probably be 100 years.

This same logic can apply to generation. Therefore, these costs must also be removed from rate base. Indeed, the only things that States would have jurisdiction over are the distribution facilities and the ability to approve prudent market-based power purchase contracts.

Thank you again for the opportunity to be with you today.

[The prepared statement of Mr. Anderson and responses to subcommittee questions follows. Attachments to the prepared statement are retained in the subcommittee files.]

STATEMENT OF JOHN A. ANDERSON, EXECUTIVE DIRECTOR, ELECTRICITY CONSUMERS RESOURCE COUNCIL

Electricity rates are too high in most of the United States. The effects of high cost power reach far beyond the communities where the power is generated and sold. Ultimately, these costs hinder the global competitiveness of U.S. business and industry, placing at risk tens of thousands of jobs throughout the country. Now is the time to restructure the electric utility industry and for the creation of retail competition to preserve American jobs and enhance U.S. competitiveness. The long-term promise is lower cost power for all consumers and a growing, more productive, domestic economy.

Fortunately, the restructuring has begun. Independent power generators are offering competitively priced electricity. Large electricity users are eager to buy it. Power marketers, brokers, energy service companies, and even some utilities are ready to enter a more competitive market. Some utilities are beginning to wheel electric power in and out of their service territories in order to make competition work. Most, however, prefer their monopoly status. They act as bottlenecks, often pricing transmission service so high (if they offer it at all) that they discourage competition, allowing the utility to maintain high rates for their own generation and distribution monopolies.

Lawmakers and regulators must act to accelerate this evolution. All market participants must have access to transmission and distribution services on a nondiscriminatory basis. Any buyer, seller, or trader should have access to both wholesale and retail electricity markets if competition is to work.

No one restructuring model will likely apply throughout the United States. Nor are all parts of the industry equally responsive to competition. Generation and transmission services should be unbundled, i.e., offered individually and priced accordingly. The output of generators can then be priced by market forces. Natural monopolies, such as transmission and distribution facilities, should remain under rate regulation and be priced according to the actual cost incurred.

The threshold issue is forcing traditional utility monopolies to compete. Only when customers can obtain their electricity from competing sources, with retail competition, will utilities be motivated to meet the competition. This will lead to improved service to all consumers at lower prices. The long-term planning process will also improve. Subject to market forces, generators will build only plants actually needed, and only at costs customers are willing to pay. Generators will bear the risks of the technologies they own and operate, including changes in environmental laws. Environmental protection and energy efficiency will be achieved at lower cost to society.

The ease or difficulty of the transition to retail competition will depend on local market conditions. Some utilities will have "stranded" costs, such as uneconomical generation facilities, that cannot be recovered from customers in truly competitive markets. Arrangements need to be made for these facilities; some write-downs or write-offs may be necessary if the assets cannot be sold. No technical barriers stand in the way of an orderly transition to true retail competition in U.S. electricity markets. The only impediment would be the failure on the part of lawmakers and regulators to implement the necessary changes in public policy.

RESPONSES TO SUBCOMMITTEE QUESTIONS

Question 1: To What Degree Has the Electric Utility Industry Entered a Period of Irreversible Change, and How Widespread Will this Change Be?

ELCON Response: The electric utility industry has clearly entered a period of irreversible change. This change is being provoked by powerful domestic and global market forces, new technologies, and rate disparities brought on by a legacy of inefficient regulation.

Like PURPA in 1978, the enactment of the Energy Policy Act of 1992 (EPAct) is also helping to bring about these changes. EPAct finally ended the decade long debate on the desirability and need for wholesale transmission access. Yet, the law does not guarantee that a workably competitive bulk power market will indeed materialize. Actions by the Federal Energy Regulatory Commission (FERC), State authorities, and stakeholders are needed to meet this objective. Proposals addressed to these concerns range from fine-tuning regulations and policies in the context of the existing vertically integrated industry structure to imposing greater market discipline in a restructured industry by replacing regulation with competition where natural monopolies no longer exist. The latter approach would allow alternative pricing regimes that model the way other commodities are priced in established competitive markets.

The current industry structure gives vertically integrated utilities tremendous opportunities to stifle access to the grid. Elimination of this market entry barrier is an essential condition for wholesale competition as envisioned by the EPAct. Utilities' market power is derived from their joint ownership of generation, transmission, and control (or dispatch) centers. The big money is in generation and utilities have always found creative ways to leverage their ownership of transmission and dispatch to maximize their revenues or profits from it. However, this conduct has resulted in a generally inefficient industry—exceptions notwithstanding. Retail rates that greatly exceed the cost of new generation are pervasive and have become a debilitating burden on domestic business, industry, and other end-users.[1]

These inefficiencies are unevenly spread throughout the industry. Unconscionable rate differentials now exist within most States and all regions. Appendix A contains a comparison of the power costs for a representative industrial load as served by 205 investor-owned utilities.[2] Rate differentials within a State or region can result in sizeable bill impacts with potentially severe anticompetitive consequences.[3] Figure 1 consists of two simple bar graphs that illustrate representative rate disparities within the NERC regions. The data show: (1) the highest rate per kilowatt-hour (excluding Hawaii and Alaska) is almost four times the lowest rate, and (2) over 70 percent of the utilities listed charge a retail tariff rate to industrial end-users in excess of typical generation replacement costs. Industrial ratepayers faced with intense domestic and international competition are forced to seek alternatives to their local supplier if the result will lower their costs. These alternatives are listed below. Such actions can directly affect decisions to shut down a manufacturing facility, to expand production at an existing site, or to build a new facility. Relocation or loss of major industrial loads should be of concern to franchise utilities.

---

[1] This point is important to large industrial users of electricity because they generally take service at transmission-level voltages and, in many cases, generate themselves. Thus, they understand very well what it costs to generate electricity.

[2] These rates are based on tariffs effective January 1, 1993. As such, the average annual rate for an individual utility may, in some cases, be substantially different because of seasonal variations in costs. The examples used in these comments are meant to highlight the fact that significant rate differentials are ubiquitous.

[3] For example, in the Western Systems Coordinating Council (WSCC), an industrial facility in one State would pay over $407,453 per month for 10-MW service with a load factor of 68 percent, while a competitor located in an immediately adjacent State might pay only $155,129 for the same power requirements. On an annualized basis, the industrial facility served by the higher-priced investor-owned utility would be forced to pay a yearly premium of $3 million compared to the more fortunate competitor located in a lower-priced utility's franchise area. Even within a single State (e.g., California), the monthly bill differential ranges from $268,627 to $407,453, which has an annual penalty of $1.7 million. In ECAR (East Central Area Reliability Coordination Agreement), monthly bills would vary from $151,429 to $381,810. An industrial facility served by the high-priced investor-owned utility would be forced to pay a yearly premium of between $2 and $3 million compared to a competitor with the same load requirements, but who is served by a lower-price utility. The anticompetitive consequences of rates are not limited to the highest cost utilities or regions with higher than average costs. It is the rate differential between what an industrial must pay and what their competitor pays that counts.

ALTERNATIVE SUPPLY ARRANGEMENTS

Self-Generation/Cogeneration
Inside/Outside-the-Fence Exempt Wholesale Generators (EWG's)
Retail Wheeling
Affiliate Wheeling
Utility-Brokered Power
Joint Ventures with Gas LDC's
Switching Franchises
Municipalization/Privatization
Wholesale Brokerage by "Any Political Subdivision of a State"
Distributed Generation/Dispersed Energy Facilities (DEF's)
Franchise Competition
Franchise Law Loopholes
Reallocation of Production to Other Plants

Rate disparities, new technologies, and global market conditions have rendered the traditional electricity utility industry structure obsolete, and as a result, much—but not all—of that industry should be functionally deregulated. The push toward deregulation is irreversible. Like the California Public Utilities Commission (CPUC), we have reached the conclusion that fine-tuning the existing regulatory regime will not fix the industry's problems. As a trade association representing large electricity consumers, ELCON believes that now is the time to return the market for electricity products and services—in both wholesale and retail markets—back to the consumers.[4] The fundamental premise driving the CPUC's vision is the irrefutable conclusion that government central planning of an industry or a whole economy does not work. The CPUC's proposal recognizes that: "competitive markets, where they exist and function reasonably well, offer a superior regulatory tool when compared to command-and-control regulation and government central planning designed for an era that has passed and will not return."[5]

Thus, generation and other technologies should be planned and operated subject to the discipline of the marketplace—not the vagaries of regulatory processes. End-user technologies and appliances, including demand-side management ("DSM"), should be planned and provided in competitive markets, and not by the dictates of government agencies and special interest groups. The CPUC correctly envisions a competitive, "consumer-driven" electric services industry where, "[a]s in other product and service markets, choice will appropriately decentralize decision-making and give consumers direct influence over the development, delivery, consumption, and price of energy services."[6]

ELCON has long advocated an electricity industry structure conducive to real competition.[7] Historically, the U.S. economy has been an energy-intensive economy because the Nation is blessed with abundant energy resources. The United States has a competitive advantage in many business and industrial activities that require access to low cost energy. However, that competitive advantage can only be sustained if those energy resources are competitively priced and used. The Nation's ability to compete in the global marketplace is contingent on the efficient exploitation of our energy resources.

All energy resources, except electricity, are supplied and priced in largely competitive markets. It is time to deregulate the electricity industry and California has set a noble standard. Only truly competitive energy markets can provide the growth, productivity, and job opportunities necessary to assure the highest possible standard of living for all the citizens of California, other States that follow California's lead,

---

[4] Large industrial end-users are not the only ratepayers that advocate radical restructuring of the electric power industry. For example, in New York, public school districts on Long Island petitioned the New York Public Service Commission on May 12, 1994 for authority to lower their operating costs by purchasing electric power in the wholesale markets and having it replace the power which would otherwise be generated or purchased by the Long Island Lighting Company. In Massachusetts, the Massachusetts Bay Transportation Authority (MBTA) was municipalized by the State in order to allow it to shop for lower priced power.

[5] California Public Utilities Commission, Order Instituting Rulemaking and Order Instituting Investigation, Docket Nos. R.94-04-031 & I.94-04-032, April 20, 1994, at 34.

[6] Id at 13.

[7] In 1987, ELCON published a special report entitled: Electricity's Future. That report advocated increased competition in the electricity industry by expanding access to the transmission grid. The report noted that many utility customers were evaluating alternative suppliers of electric service such as municipalization, self-generation, and cogeneration. ELCON has also been an intervenor in rulemakings and rate cases involving competition at the FERC. These cases involve transmission access and pricing, market-based pricing arrangements, utility mergers, the implementation of PURPA, and independent power producers.

and the Nation as a whole.[8] Only truly competitive energy markets will ensure that resources are employed with efficiency to protect the environment at the lowest possible cost to society. We believe that the CPUC's vision—if applied throughout the United States—will help the Nation achieve the dual goals of job retention/creation and environmental protection, and thus optimally balance those critical public policy objectives in the most efficient and equitable manner.

Question 2: How Should Regulators Handle the Transition Period to a More Competitive Market?

ELCON Response: In recognition of the tremendous market pressures that already exist for restructuring the electric services industry, ELCON proposes eight basic principles for easing the transition to a more competitive, efficient, and equitable retail electricity markets. These principles are explained in a special report entitled: Retail Competition in the U.S. Electricity Industry. A copy of that report is attached as Appendix B and is incorporated by reference into our testimony. We focus on "retail" electricity markets because ELCON's members—like all ratepayers—currently procure their electric services (other than that which is self-generated or cogenerated) at retail from a traditional franchised utility. We also believe that an electric services industry that is limited to wholesale competition will not work. Arguably, "wholesale competition" is an oxymoron. Therefore, "direct access" by end-users to alternative suppliers of electric services must be an option in the new industry structure.[9]

Our "eight principles" generally affirm the fundamental premises espoused by the California PUC in its historic April 20th proposal. These principles represent our recommendations to regulators and other policymakers for achieving a swift and equitable transition to a competitive electric services industry.

Principle No. 1.—Market forces can do a better job than any government or regulatory agency in determining prices for a commodity such as electricity.

Electricity rates throughout most of the country are too high, often exceeding the national average by a wide margin. We also believe that even the national average is too high and that consumers everywhere would benefit from a restructured electric services industry. Traditional regulation was supposed to be a surrogate for competitive markets and produce rates that would reflect prices set by competition. But in most parts of the country, regulation has failed to achieve that goal. Where real competition can exist, there will be no need for the surrogate. The CPUC said, "that the institutional, contractual and financial arrangements, and the technical capabilities required to bring to consumers the benefits of increased choice through direct access already exist and continue to develop."[10] We agree. Therefore, let's do it.

Principle No. 2.—Laws and regulations that restrict the development of competitive electricity markets should be rescinded or amended. The need for burdensome regulation will be reduced where competitive electricity markets are allowed to flourish.

The only barriers to restructuring the electric services industry are laws and regulatory practices that have long outlived their usefulness. Industrial consumers throughout the country are ready to work with their respective PUC's and State legislatures to accomplish these goals.

Principle No. 3.—The benefits from competition will never fully materialize unless and until there is competition in both wholesale and retail electricity markets. But not all retail electric services are natural monopolies, and therefore they should not be regulated as such.

Consumers are not guaranteed any benefits from competition if only wholesale-level competition is allowed in the electric services industry. The potential for wholesale competition has existed at least since the enactment of PURPA in 1978; however, few, if any, real benefits have subsequently accrued to end users. Only retail competition will ensure the materialization of a vibrant, competitive electric services industry with real benefits for all consumers.

Principle No. 4.—The owners and operators of transmission and distribution facilities, and the providers of coordination and system control services, should be required to provide access to those facilities and services to any buyer or seller on a nondiscriminatory, common-carrier basis.

---

[8] It is widely observed that the affluence generated by economic growth increases the resources a society is willing to pay for environmental protection. Hence, the most stringent laws on environmental protection have been established by the most successful industrialized countries. See Frances Cairncross, "An Enemy, and Yet a Friend," The Economist, September 8, 1990.

[9] This point is discussed in greater detail in our response to question 4.

[10] Id. at 24.

Principle No. 5.—Rates for the use of transmission and distribution facilities should reflect the actual cost of providing the service. If the facility is a natural monopoly, those rates should be based on actual costs and the services provided on a nondiscriminatory and comparable basis to all users.

The development of a competitive electric services industry will only happen when potential market entrants have fair and nondiscriminatory access to the use of transmission and distribution facilities as well as the services necessary for coordination and system control. Potential new market entrants include "direct access" customers, power marketers, brokers, exempt wholesale generators ("EWG's"), energy service companies, and end-user power purchasing cooperatives. We strongly support the pro-competitive efforts of the FERC as it implements provisions of the EPAct that require greater access to the Nation's bulk power grid.[11] We urge State commissions to work with the FERC and stakeholders to expeditiously resolve any potential jurisdictional concerns that might impede the establishment of a competitive electric services industry.

Restructuring the industry will deregulate most, but not all, of the business functions associated with the generation and delivery of electric services. There will be residual regulation: transmission and distribution "wires," control or dispatch centers, some substations, and perhaps the use of other facilities that are natural monopolies. Regulators will be able to sort what should and should not be regulated, once existing bundled electric services are adequately unbundled. The rates charged for any unbundled service of a natural monopoly whether provided on an unbundled basis or bundled with other services—should be based on the actual costs incurred in providing the service.

Principle No. 6.—Resource planning is not natural monopoly. The types and market shares of generation and end-user technologies that will be supplied in wholesale and retail markets should be decided in the marketplace.

The resource planning function will not be compromised by creating a more competitive electric services industry. It can only be enhanced for the benefit of all consumers and the environment. The traditional command-and-control type of planning, that is the basis of the monopolistic integrated resource planning ("IRP") approach, aggregates system planning and market planning. It is process-directed, not customer-directed.[12]

In a competitive electric services industry, only the most efficient producers can sell their products and increase their market shares. If a generator's output cannot be sold, that risk is absorbed by the plant's owners, not by captive customers. The market will also decide the least-cost resource mix—not the highly litigious, adversarial environment of the regulators' hearing rooms. The market share of each viable technology will be a market outcome, not a planning variable. Only the most economically efficient generating technologies will be planned and allowed to operate. The new generating technologies employed by the independent power industry are among the cleanest generators from an environmental perspective. Therefore, a competitive industry will meet environmental objectives more efficiently than the current command-and-control model.

In the EPAct, Congress directed the FERC to create a competitive bulk power market. Congress also preserved the right of States to create competitive retail power markets.[13] Competition will no more endanger the efficient planning at the retail level than wholesale competition will harm planning at the wholesale level. The planner in a competitive business "learns their customers" needs and plans accordingly. That planner will also seek ways to modify customer demand in order to achieve a larger market share for their products and services. The firm will only achieve that goal if it can be innovative and control its costs and there is a demand for the product. Thus, the demand for products and services that cost-effectively promote energy efficiency (e.g., DSM programs) should flourish in competitive markets without the need for subsidies or other artificial inducements.

In competitive markets, customer choice will determine the market shares of competing technologies, not government planners or special interest groups. Real competition will force all suppliers to choose and offer only the most efficient, least-cost generation resources. The planning of transmission and distribution facilities, which remain natural monopolies, will be the focus of a top-down regional process that in-

---

[11] Our response to question 4 discusses the FERC's role in greater detail.

[12] Integrated resource planning was an attempt to devise a regulatory solution to a regulatory problem. It has generally failed to achieve its objectives and now a more market-oriented solution is warranted. See George M. Galloway, "A Response to Scott Hempling: The Role of IRP Under a Retail Competition Regime," The Electricity Journal, June 1994 at 2.

[13] See section 722 of the EPAct, which amends section 212 of the Federal Power Act. Also see ELCON's response to question 5 and Appendix C.

cludes regulatory oversight. However, there will be no integration of the monopoly planning functions with market planning.

Clearly opposition to restructuring will emanate from those constituencies who directly benefit—at great cost to society and consumers—from the rampant inefficiencies of current planning practices. Their pleas to maintain the status quo must not be allowed to delay or prevent the formation of a competitive electric services industry that will benefit all consumers.

Principle No. 7.—Legitimate and verifiable transition costs that develop as a result of competition should be recovered by an equitable split among ratepayers, shareholders and taxpayers. The costs of assets that were uneconomical in the existing regulatory regime are not transition costs.

Principle No. 8.—The potential for transition costs should not be used as an excuse to prevent or delay the onset of a competitive electricity market.

The potential exists for certain "stranded" costs during the transition from the current regulatory regime to a competitive electric services industry. Those costs which are legitimately and verifiably associated with the transition should be recovered equitably between current and former utility ratepayers, shareholders, and taxpayers. We commend the FERC for issuing its recent notice of proposed rulemaking ("NOPR") on stranded costs, and like the FERC, believe that the public interest is served by providing "regulatory certainty concerning recovery of these potential significant costs as early as possible." [14]

However, costs "stranded" in the regulatory regime—for whatever reason—are not true transition costs, and therefore are not recoverable from ratepayers. Utility shareholders should not be allowed to recover greater costs under competition than they would reasonably expect to recover under regulation. Forcing the recovery of all costs associated with the current regulatory regime—by the laws of conventional arithmetic—will not change the rates faced by consumers, and therefore they will derive no benefit from the transition to a competitive electric services industry.

Finally, regulators should apply consistent treatment to all assets currently in rate base that may deviate from a fair market value in a competitive electric services industry. If utilities are allowed to recover from current or former customers the difference between the book value of an asset and that asset's lower market value, then customers should receive payments equal to the appreciated value of any asset whose current book value is below market value.

Question 3: How Can Social and Environmental Objectives Be Adapted to a "Restructured" Industry, in an Environment in Which Services Once Provided by a Single Vertically Integrated Utility Are Increasingly "Unbundled"? To What Extent Should These Objectives Be Retained, Changed or Abandoned?

ELCON Response: A variety of social and environmental objectives have been pursued under the traditional regulatory regime. We specifically comment on seven of these objectives: (1) energy efficiency, (2) renewable energy resources, (3) economic development, (4) the regulatory process and public involvement, (5) assistance to low income ratepayers, (6) environmental protection, and (7) the application of environmental externalities.

First, we do not believe that any legitimate social or environmental objective will suffer as a direct consequence of the transition from a regulated industry to a more competitive industry. To the contrary, we strongly believe that those objectives can be more effectively and more efficiently met if the industry is restructured. We again emphasize that electricity rates are too high in most regions of the country and that high rates have a punitive effect on the Nation's economy. Innovative proposals to restructure the electric services industry—such as that offered by the California PUC—will greatly increase the productivity of that important sector. Higher productivity will stimulate greater economic growth and create more jobs.

Second, high electricity rates also unduly burden low income citizens by: (1) forcing them to pay higher electric bills that they can ill afford, (2) increasing the prices they pay for electricity intensive products and services, and (3) limiting job opportunities. We believe that the rate reductions that can be achieved in a restructured industry will provide low income ratepayers with greater relief than what is currently provided by special assistance programs.

Third, the affluence generated by economic growth increases the total resources a society is willing to pay for environmental protection and other social objectives. Higher economic growth can be achieved if the electric utility industry is restructured to include direct access (a.k.a., retail wheeling). Thus, society's willingness to pay for environmental and other social amenities can be supported accordingly.

[14] Federal Energy Regulatory Commission, "Commission Proposes Rule for Recovery of Stranded Costs," News Release, June 29, 1994.

Fourth, "central planning" is simply neither as efficient nor productive as market forces for cost-effectively achieving the goals of environmental protection and other social objectives. This is being repeatedly demonstrated in other sectors of the economy where market mechanisms are proposed as more workable alternatives to traditional forms of government intervention.[15] All too often central planning results in "one-size-fits-all" type programs which rarely fit anyone. Alternatively, market forces rely on consumer choice, i.e., letting consumers make decisions that best fit their specific wants and needs.

Fifth, attempts to achieve public policy objectives, by using regulated utilities as the primary delivery mechanism for providing these social services, by definition, significantly reduce the purchasing power of ratepayers. A dollar given to a utility to fund a social program such as DSM has a very high opportunity cost to society because the utility will ultimately deliver only a small fraction of that dollar in final services. Hence, the argument that a restructured industry will prevent consumers from pursuing cost-effective opportunities to conserve energy are wrong. Within a competitive electricity market, the full accounting cost of social and environmental programs will be visible as a separate line item on customer bills. Such costs will not be bundled—or hidden—within the overall operating costs of the traditional utilities. One reviewer of the California PUC's proposal has warned: "This has both pluses and minuses for all parties. On the one hand, it may prevent special interest groups in favor of specific programs from hiding the costs of these efforts within a rolled-in rate. On the other hand, it will allow for a clear public policy debate on the real cost/benefit of achieving these objectives. In addition, it should be stressed that with or without an environmental or DSM overlay, to the extent that competitive pricing drives resource choices, unambiguous DSM and environmental benefits will be promoted."[16]

Finally, we offer regulators several recommendations which, if adopted, will greatly increase the cost-effectiveness of environmental protection and other social objectives relative to the results typically realized by programs administered under the traditional regulatory regime.

Industrial end-users were strong supporters of the Federal standards for integrated resource planing ("IRP") and demand-side management ("DSM") as codified in Title I of the EPAct. The new standards embody several fundamental principles that are essential if IRP and DSM are to help increase the efficiency of the development, delivery, and consumption of electric services, while also achieving the goals of environmental laws and regulations in a least-cost manner:

—All cost-effective resource options should be considered by regulated utilities in their planning process—just as any unregulated supplier in a competitive market would be forced by market pressure to consider all options if it wants to remain in business.

—The energy savings attributable to utility DSM programs must be accurately measured and verified. Otherwise, the cost-effectiveness of those savings is purely speculative and consumers may be forced to pay twice for the same energy resource or emissions control.

—All resource options should be compared on a consistent basis using life-cycle analysis. The resource value of a utility DSM program must be accurately estimated over time such that it can be directly compared with the metered resource from any alternative supply option it may attempt to displace or avoid.

—The recovery of lost revenues associated with utility DSM programs is not a utility or shareholder entitlement, especially when the net energy savings attributed to the programs are not subject to accurate ex post measurement and verification.

—The IRP process should provide for public participation and comment.

—Finally, the IRP process must not be used to impose environmental controls that go beyond the requirements of existing laws, and hence, evade the constitutionally guaranteed checks and balances of the legislative process. The selective internalization of certain environmental externalities in the IRP process violates the basic precept of IRP to produce efficient, least-cost results.

The most outspoken advocates of IRP practices typically claim that utility DSM programs should be treated in the IRP process as a direct substitute for traditional supply resources. If a utility DSM program can eliminate capacity and energy re-

---

[15] Proposals to use market mechanisms—including the privatization of traditional government functions to more cost-effectively deliver social services are being considered in several major sectors, e.g., health care, insurance, schools, and penal institutions.

[16] See "Reply Comments of National Power PLC on California Electric Industry Reform," California Public Utilities Commission, Docket Nos. R.94-04-031 and I.94-04-032, June 21, 1994, at 4.

quirements, otherwise met with more costly resources, which may also be less friendly to the environment, why shouldn't the utility do so?

While the logic of these arguments has surface appeal, practical demonstration of the real efficacy of the concept has been lacking. There is little or no evidence, despite the best of intentions to prove otherwise, that utility DSM programs actually displace more costly supply resources or result in a net reduction to the utility's revenue requirements.[17] In fact, the opposite effect seems more widespread. Utility DSM programs are more apt to be implemented by utilities that do not need new capacity or energy resources, resulting in higher—not lower—rates for all ratepayers.[18]

Major controversy remains with respect to a utility's ability to cost-effectively measure and verify the level and persistence of any real savings achieved with such programs. If the level of energy savings is uncertain, any claim of environmental benefits resulting from these programs must be viewed with some suspicion. Other important concerns are raised by allowing regulated utilities to be the exclusive suppliers of these services, particularly when the competitive marketplace can more efficiently provide any product or service that is not a natural monopoly. After all, in enacting the EPAct, Congress assumed that generation resources are no longer a natural monopoly. Any competitive alternative to generation should be forced to compete in the same market structure.

There is an inherent conflict between the degree of consumer freedom necessary for competitive markets to operate efficiently and attempts by the traditional regulatory apparatus to restrict those freedoms in order to force market outcomes to predetermined conclusions. That conflict is demonstrated by the two approaches integrated resource planning practices can take:

Command-and-control, or central planning-type IRP within the context of a tightly regulated, vertically-integrated industry structure, or

Market-oriented, or decentralized IRP within the context of a competitive electric services industry in which the industry structure is vertically de-integrated.

Experience with the first approach demonstrates that this type of IRP, at both the generation (bulk power or wholesale) and end-user (retail) levels, is structurally unworkable because it results in inefficient and uneconomical resource selections. The central planning approach sanctions utility and regulatory control of supply resource selection and end-user choices for products and services that are not natural monopolies. Essentially, the IRP process is used to usurp the marketplace and individual choices. The market shares of generation resources, as well as end-user appliances, equipment, and processes, are dictated not by the marketplace, but by the regulators or utilities.

The second approach to IRP limits the utility and regulatory roles to the planning and operation of facilities that are natural monopolies (e.g., distribution substations, transmission lines, and dispatch centers). Technology choices at both retail and wholesale levels are made in each respective market. Competition drives out inefficient suppliers, thereby minimizing or preventing the penetration of uneconomical resources and technologies into the marketplace.[19]

Theoretically, central planning-type IRP might be capable of producing a least-cost mix of supply and end-user resources. However, the merits of this form of IRP, and any perceived need to preserve it, should be based on the evidence from practical experience, not on how it might work in ideal circumstances. Unfortunately, the IRP practices, in States such as New York and California that adopted the central planning approach, have violated some or all of the fundamental principles of IRP enshrined in the EPAct standards.

The IRP process forces utilities to buy resources they do not need, resulting in higher rates than they otherwise would be had the utility been allowed to plan for and operate a truly least-cost resource mix. When utilities need new resources, the

---

[17] We acknowledge recent, if belated, efforts in California, New Jersey, and other States to establish "protocols" for the verification of energy savings attributable to utility DSM programs. Whether the protocols will produce accurate results remains to be seen. Nonetheless, States such as California have required utilities to commit billions of dollars to energy efficiency programs since at least the 1970's and therefore, absent proof to the contrary, anyone can reasonably argue that much of that money may have been wasted. Even with the successful implementation of these protocols, there is no guarantee that any DSM products and services will be deemed cost-effective. The high cost of measurement and verification may doom many DSM options.

[18] See "Comments of the Electricity Consumers Resource Council," U.S. Department of Energy, Notice of Inquiry and Request for Comments on the Impact of Integrated Resource Planning on Utility Systems and Consumers, 59 Fed. Reg. 9481 (February 28, 1994), April 1, 1994.

[19] A more complete elaboration of these principles is presented in ELCON's initial comments to the California PUC filed June 8, 1994.

full range of technically and economically feasible resource alternatives is not considered by the resource selection process. Ratepayers are typically confronted with DSM solicitations on a take-or-pay basis. If ratepayers participate, DSM programs are offered without charge, or heavily subsidized. They pay for the programs even if they don't participate. Renewable resources are selected not on the basis of their true economic merits, but with an ad hoc formula. The full range of end-user options, particularly by large end-users that also generate or cogenerate electricity, are not integrated within the planning process. This shortcoming always compromises the intellectual integrity of the IRP process.

The energy and/or capacity savings attributed to utility DSM programs are not measured and verified. Since the savings resulting from these programs may, in fact, not exist or be highly overstated, ratepayers are forced to pay for a resource that also does not exist or from which they derive no tangible benefit. The "success" of a utility DSM program is typically measured by the speed at which the program's budget is expended. Utilities are rewarded for the dollar volume they commit and spend, not by the amount of any resulting legitimate and verifiable net energy savings or emissions reductions. Utility shareholder rewards routinely exceed the collection of prudently incurred costs plus a fair rate of return.[20] The products and services provided to participants of utility DSM programs are not cost-based. Specifically, those customers responsible for the costs, or who benefit from the programs, are often not the ones who ultimately pay for the programs.

Utility DSM programs and supply resources are not considered on a consistent basis, resulting in the implementation of inefficient and uneconomical DSM programs. The focus, if not obsession, of these IRP processes seems to be the implementation of DSM programs at any cost. Inherently inefficient utility DSM programs are offered to, or forced on, large industrial customers. In fact, industrial DSM programs may make industrial processes less energy efficient. Utilities typically return to DSM participants only a small fraction of every dollar collected from ratepayers for these programs. Large industrial ratepayers (or any ratepayer) could finance greater energy efficiency improvements on their own if they were allowed to keep the money.[21]

Decoupling, and other mechanisms for handing alleged lost revenues, is applied to fix a problem when perhaps a fix is not needed. In particular, utilities are recovering lost revenues for reasons other than those related to DSM programs.

Finally, at great expense to consumers, the IRP process forces the consideration of environmental concerns that go beyond the requirements of environmental law. Often, these additional requirements are predicated on speculative scientific evidence that would warrant any such action. In effect, these practices attempt to bias the resource selection process in favor of uncompetitive technologies in the marketplace.

In summary, any net energy savings achieved with traditional IRP and DSM, are the result of a very expensive and inefficient delivery mechanism. Only a small fraction of every dollar collected from ratepayers to sponsor DSM programs is actually returned to participating ratepayers in the form of electric services, including financial arrangements. The rest of the money is wasted in the regulatory process or is a windfall to utility shareholders. For example, in addition to recovering direct DSM program costs from ratepayers, utilities also collect shareholder incentives and so-called lost revenues. Estimates of lost revenues are typically based on ex ante, engineering estimates of a program's potential energy savings. These estimates almost always grossly exceed actual net savings based on ex post measurement.[22] Thus, many utilities may collect (and keep) more ratepayer money for shareholder incentives and lost revenue recovery than is actually collected for direct program costs.[23]

---

[20] See Michael Foley and Ann Thompson, "Electric and Telephone Utility Stockholder Returns: 1972-1992," National Association of Regulatory Utility Commissioners, September 13, 1993.

[21] See A.C. Price, "Effect of Utility Programs in Encouraging Industrial Efficiency," Presented at the ACEEE Workshop on Demand Side Management Programs, Portland, OR, September 20, 1993.

[22] See Paul L. Joskow and Donald B. Marron, "What Does a Negawatt Really Cost? Evidence from Utility Conservation Programs," The Energy Journal, Volume 13, No. 4, 1992 at 62-64. For a spirited rebuttal of criticisms of the Joskow & Marron paper, see Paul L. Joskow and Donald B. Marron, "What Does a Negawatt Really Cost? Further Thoughts and Evidence," The Electricity Journal, July 1993; and Paul L. Joskow, "More from the Guru of Energy Efficiency: There Must Be A Pony" The Electricity Journal, May 1994.

[23] Very little credible or consistently reported cost accounting data on utility DSM expenditures are readily available in the public domain. However, some data are available for New York. In 1993, New York utilities budgeted $285 million for their DSM programs. In addition, we estimate that those utilities collected at least $443 million for shareholder incentives and lost revenues. If the administrative costs are 20 percent of total direct program costs, than New

There is also no guarantee that the actual amount returned to participants results in net energy savings. Adjustments should be made to account for administrative costs, measurement errors, free riders, and other factors.[24] Such practices are beginning to result in calls for change.[25]

In recognition of the rapidly emerging competitive electric services industry, resource planning must be focused away from central planning-type IRP practices to more market-based, decentralized planning practices. The following five recommendations will refocus that process and help electric services suppliers and consumers posture themselves for the new competitive markets.

If a supplier claims DSM as a substitute for a supply resource, it should be treated like any other supply resource.

The goal of resource planning should be to produce the least-cost resource mix. For suppliers that elect to remain subject to rate regulation, a least-cost resource mix will minimize the net present value of the utility's long-run revenue requirement necessary for an adequate and reliable electricity supply. This criterion will minimize the rate impacts on all customers. If the utility does not need new capacity or energy resources, the only justification for implementing utility DSM programs is that they will lower rates. Additionally, the utility should select the resource that will be least-cost for its customers.

Generation is no longer a natural monopoly. Generation markets are increasingly becoming more competitive and subject to market discipline. Thus, in the short run, regulated utilities should be required to competitively procure all their resources. If one type of resource is procured in competitive markets, other resources offered as substitutes should be selected in the same manner. The planning and operation of the electric utility industry can be made more efficient by increasing the role of competition as a disciplining factor, not by continuing or expanding the central planning model of traditional regulation. Utilities should not offer DSM programs if any independent supplier can do the same at less cost, or more reliably, to end-users. This is analogous to independent power producers who become the supplier of choice if their resources can be provided at less cost to end-users.

If they are used at all, utility DSM programs should encompass the full range of feasible energy efficient improvements that are cost-effective to the utility and its ratepayers. These programs should avoid or defer more costly resources or reduce a utility's other out-of-pocket costs, such as the need to comply with environmental laws. Externalities (which are discussed below) should never be used in the IRP process to deliberately bias the selection of any resource option. DSM programs should not be intended as social welfare programs, customer handouts, or ratepayer entitlement. Ratepayers should not be forced to participate in utility DSM programs by charging them for benefits not actually received or that do not exist.

The supply and marketing of energy efficiency products and services are not natural monopolies and therefore regulated utilities should not be given an exclusive franchise to provide them.

Utilities are regulated because they provide services that are a natural monopoly. Regulatory commissions exist solely to protect consumers and shareholders from potential monopoly abuses (eg., price discrimination or restrictions of supply). Rate regulation allows a utility to charge rates to recover a revenue requirement that is limited to all prudently incurred costs, plus an opportunity to earn a fair rate-of-return. Regulated utilities do not have an exclusive right to market or sell energy efficiency by means of DSM or other end-use programs. If they are implemented at all, utility DSM programs should promote the availability and marketing of such

---

York utilities returned to participating ratepayers only 31 cents out of every dollar collected to support the DSM programs. When that amount is further adjusted for "phantom-watts"—the difference between ex ante engineering estimates and ex post measurements—and free riders, the amount actually spent on real net energy savings may be less than a dime. The other 90 cents is lost in the process. In other words, if utilities are used as the "principal agent" for delivering DSM services, society may be paying a dollar for a dime's worth of energy savings. Some bargain.

[24] Op.Cit. Joskow & Marron (1992).

[25] Consulting firms that supply utilities with various DSM support services have grown tremendously as utility DSM program expenditures have escalated across the country. Some of these firms are not blind to emerging changes in the industry and, in a few cases, have candidly acknowledged the economic distortions created by these programs. A representative of one firm has advanced five ideas for "reengineering DSM programs" in order to make these programs more compatible with competitive reality: (1) reduce or eliminate incentives, (2) return to customer education, (3) require increased customer contribution for the total cost of the DSM service, (4) offer financing and leasing services, and (5) include electro-technologies that increase sales. See Greg Collins, "Utility Marketing: Preserving the Industrial Customer Base," Barakat & Chamberlin, presented to Infocast's Conference on Competitive Power Sourcing for Industrial Companies, June 16, 1994.

products and services in the non-regulated marketplace. However, utility DSM programs must not interfere with competitive markets or impose new market barriers or other impediments to trade.

If utilities wish to market products and services that are not natural monopolies, they should only be allowed to do so on an unregulated basis in competitive markets. Such entities should be separated under a truly arms-length relationship from the regulated affiliate. The same regulatory concerns regarding self-dealing should apply to these affiliate relationships as do situations where affiliate power producers attempt to sell supply resources to their regulated parent or affiliate company. Utilities have shown a willingness to establish unregulated affiliates to take advantage of viable market opportunities.

All utility services should be priced based on cost-of-service principles.

As a pre-condition to the implementation of any utility DSM program, utilities should set rates based on cost-of-service so as to send appropriate price signals to end-users. Rates should be offered on a bundled and unbundled basis as necessary to meet the needs of the utility's customers. This should include time-of-use, real-time, curtailable or interruptible, and other, more advanced rate structures. Utilities have an obligation to develop innovative rates that take maximum advantage of state-of-the-art metering and communications technology.

Every participant and non-participant should know the true cost of each utility DSM offering so they are able to make more informed decisions regarding all their purchases and investments. Customers should always be able to exercise their right to choose to participate or not in utility DSM programs. Customers who choose not to participate should not be obligated to subsidize any costs incurred by the utility or those customers that do participate.

If a utility DSM program is truly cost-effective, the utility should always be able to charge a rate that leaves both participants and non-participants better off. If a utility DSM product or service is properly priced, i.e., not given away or subsidized, utilities will not need special incentives or lost revenue adjustments as an inducement to market these products or services.

Energy efficiency products and services marketed by unregulated suppliers in competitive markets should be priced by those markets and sold for a profit. Firms that supply and market energy efficiency in competitive markets will be price takers. The profit they can earn will be in direct proportion to their abilities to be innovative and to control costs. Examples of entities that might operate in a truly competitive market are power brokers, power marketers, energy service companies, original equipment manufacturers ("OEM's"), and power purchasing cooperatives. Some or all of these entities might market traditional electric services in a bundled package with DSM-type products and services.

DSM products and services provided by regulated utilities should be priced based on the cost of service and the utility should be allowed the opportunity to earn a fair rate-of-return on only actual costs prudently incurred and in rate base. Participants should pay the full cost incurred to provide the service.

—If a utility DSM program is to be deemed a truly viable resource, they should be cost-effectively procured without being subsidized.

Utility DSM program costs, including all costs necessary to accurately measure, evaluate, and verify the program benefits, should be properly allocated and rates should be set to recover those costs including the use of an appropriate discount rate. Utility DSM program costs should be recovered from participants with shared-savings agreements or energy service charges that are assessed through their bins.

Ratepayers should not be held responsible for the costs of utility DSM programs where the claimed benefits have not been accurately measured, evaluated, and verified over the life of the resource. The DSM "resource" provided by a utility should be subject to the same regulatory scrutiny as any supply resource (eg., nuclear plants) for which the utility claims cost recovery.

If rebates are offered as part of a utility DSM program, the maximum payment a utility should make to a participant is the avoided cost minus the participant's bill savings. This is the "no loser test." The no loser test is as much an essential restraint on regulation as it is a binding constraint on utility resource planning. The avoided cost should reflect market-based resource values and not the invention of an administrative process.

—Market barriers to cost-effective energy efficiency improvements are created by the inefficient regulatory regime and can be eliminated by establishing a truly competitive electric serves industry.

Under the existing regulatory regime, barriers discourage investments and expenditures by end-users on energy efficiency improvements. These barriers result from market imperfections created by inefficient regulatory policies.[26] These imperfections include inappropriate rate designs, billing practices that disguise the price signal, abuses associated with fuel adjustment mechanisms, cross-class subsidies, ineffective utility marketing programs, and the corporate culture of a regulated industry. To overcome these barriers, utilities have had to give away or heavily subsidize any products or services they attempt to sell. These barriers exist for both products and services that build load and those that conserve energy.

The solution to these problems is not more DSM—which can only solve a problem if there are no restrictions on cost—but rather the establishment of a competitive electric services industry. A competitive industry, as properly envisioned by the California PUC, will allow a market for energy efficiency products and services to flourish. Instead of one inefficient supplier of these products and services, there will be many efficient suppliers. New market entrants will have guaranteed access to potential customers, and therefore take the risk to develop and supply new innovative products and services. End-users will be offered a greater array of options, including one or more that they win buy because these options meet their needs. The price signal established by a competitive market will allow all buyers and sellers to benefit fairly from each transaction. Competition will force all suppliers to hustle for customers—both to keep their existing customers and to seek new ones. Competition will force all suppliers to be innovative and to control costs. There will be no "barriers" to cost-effective energy efficiency improvements.

The implementation of utility DSM programs and the promotion of renewable energy resources seem to be an obsession with many IRP practitioners. Independent power producers are also given preferential treatment in some IRP's. However, the focus of central planning-type IRP processes has not been "to treat all options—supply, storage, delivery, and demand-side—on a consistent basis using a systems approach without preconceptions or bias for or against any particular option, fuel type, or form of ownership;"[27] rather, the focus seems to force utilities to invest in DSM, renewables or other resources, whether resources are needed or not. We have already articulated our views with respect to DSM. Our suggested treatment for renewable energy resources is not much different.

A vibrant market will develop for renewable energy resources provided the market is allowed to make it happen. Central planners are poor judges of the long-run economic viability of any technology. For example, in the early 1980's, they misread the market for cogeneration and small power producers. Their skills at picking and choosing other technologies and establishing the appropriate "economic" incentive for their development will be no better. Since such practices will discourage further innovation, set-asides and quotas for preferred technologies will do more harm to the nascent industry than help. The long-run viability of a renewables industry would be better served if each technology were forced to compete on its own merits and if it were not shielded from direct competition.

In general, a competitive electric services industry will more efficiently determine the appropriate level of fuel and technology diversity because markets do a better job identifying and managing risks and costs. The private owners of generating resources bear the risk associated with environmental compliance, including new control requirements that will be mandated in the future. They will not be able to shift the costs of these and other risks to captive ratepayers as is often done in the current regulatory regime.

In responding to concerns regarding the compatibility of various restructuring proposals with current regulatory efforts to promote economic development, we offer two comments one at the macroeconomic level and the other at the microeconomic level. However, overall, we strongly agree with the California PUC that restructuring the electric services industry is at the heart of its effort to promote a robust economy.

First, we urge the Congress, State PUC's, and other policymakers to consider two recommendations of the National Research Council ("NRC"):[28]

---

[26] Many of the alleged "market barriers" that limit the implementation of all technically feasible energy efficiency improvements are really the result of transaction costs. Transaction costs are a natural part of any market. In fact, a "market" can be defined as an institution that exists for the sole purpose of minimizing transaction costs. Hence, the argument that transaction costs exist, therefore markets don't work, is fallacious. See R.H. Coase, The Firm, The Market, and The Law, (Chicago, IL.: The University of Chicago Press, 1988), at 6-7.

[27] See U.S. Department of Energy, Integrated Resource Planning Program Plan. Fiscal year 1994—fiscal year 1998, Final Draft, January 24, 1994 at 8. The quote represents DOE's stated approach to IRP.

[28] See National Research Council, Electricity in Economic Growth, (Washington, D.C.: National Academy Press, 1986), at xvi.

The relationship between electricity and productivity is so important that it should be considered in developing Federal and State energy and economic policies.

To foster increased productivity, policy should stimulate increased efficiency of electricity use, promote the implementation of electrotechnologies when they are economically justified, and seek to lower the real costs of electricity supply by removing any regulatory impediments and developing promising technologies to provide electricity.

A restructured industry which gives all customer direct access to alternative suppliers of electricity products and services is the appropriate response. Electricity rates are too high. As a result, U.S. manufacturers face a competitive disadvantage in the use of electricity. Far more electricity is now embedded in finished goods that are imported into the United States than is embedded in products that are exported.[29] The imported products are manufactured in countries with less stringent environmental laws. States should make every effort possible to increase the efficiency by which electricity is produced and delivered in order to lower the real costs of power and still comply with all existing environmental laws and regulations.

Second, we have always been concerned that a regulated utility could use its monopoly power to unfairly undercut private sector suppliers and thus distort the operation of the market. This has the potential to harm the thousands of small businesses and the jobs they provide. As part of the movement toward greater competition in the industry, the supply and marketing of any energy efficiency product or service should be provided exclusivity in unregulated competitive markets and not as a quasi-monopoly service by regulated utilities. We question whether regulatory safeguards can ever be sufficient to protect consumers from all potential monopoly abuses. Since these products and services are not natural monopolies there is no innate need to continue supplying them in a regulatory regime.

The IRP process in some States has become a tremendous regulatory burden. These burdens are particularly created by the expanded scope of regulation when IRP processes are used to simultaneously micro-manage the decisions of utilities and their ratepayers. The IRP process is less open to public involvement, and more importantly, public oversight, because the process has become too complex, jargon-laden, and expensive. Large industrial ratepayers who often reduce their intervention costs by joining intervention groups, increasingly have to stop participating because of high costs and low benefits. All too often, the "public" process is dominated by special interest groups that benefit from the subsidies that many IRP processes lavishly provide and, indeed, are often paid by the utilities or regulators to advocate programs opposed by ratepayers. To the extent that restructuring initiatives result in an overall net reduction in the scope and amount of regulation, all consumers, large industrials and small residentials alike, will benefit by minimizing this costly burden.

The level of assistance provided to low income ratepayers by a variety of State and Federal programs may not adequately compensate for inflated rates that are charged to residential ratepayers. In other words, low income ratepayers are not made whole by these assistance programs; the assistance programs are helpful, but in the end the ratepayers still pay too much.

Low income ratepayers will enjoy greater benefits from rate decreases achieved in a competitive electric services industry, including the opportunity to receive new bundles of products and services by joining power purchasing cooperatives. Requiring regulated utilities to continue being the "principal agent" for such services is inefficient and wasteful for the same reasons that utility DSM programs are so costly. Utilities are simply the wrong delivery mechanism. To the extent that low income consumers need further assistance after the transition to a competitive industry, States might consider an assistance program modeled after the Food Stamp program for providing such aid. For example, a "BTU Stamp" program might be established funded by a State or the Federal Government from general revenues.

The environmental consequences of utility DSM programs are directly related to the reliability of any energy savings achieved by those programs. As stated above, utility DSM programs are routinely implemented without subsequent determination that legitimate and verifiable energy savings were actually achieved. Any claimed net energy savings or emissions reductions are the result of wishful thinking, not a legitimate effort to achieve and verify results. Hence, elimination of the subsidies from these programs, which is the only feature of the programs that is truly at risk as a result of restructuring, does not pose any material risk to the environment. In fact, the opposite will be true. The creation of a competitive electric services indus-

---

[29] See Margaret B. McCarthy, "Foreign Merchandise Trade and the Demand for Electricity," The interindustry Economic Research Fund, Inc., December 1985.

try will greatly increase opportunities to achieve cost-effective environmental protection compared to current regulatory practice.

A competitive electric services industry will expand the opportunities for new environmental technologies—both the technologies for mitigating and controlling harmful emissions as well as new technologies for generating and delivering power more efficiently. However, such new opportunities are only created by guarantying access to the market. That means free access to all end-users and any suppliers on the interconnected grid. Wholesale-only competition, to the extent it is competition at all, limits the size and scope of the market and therefore restricts opportunities for innovation and entry into the market.

Direct access by end-users will provide valuable environmental protection by changing the manner in which power plants and transmission facilities are sited in the future. Many large industrial or commercial end-users will seek to purchase some or all of their power requirements from new or repowered generating units located on or near their facility. This produces several environmental benefits: (1) the need for new greenfield sites in environmentally sensitive areas is minimized; (2) the need for new transmission facilities and transmission corridors is reduced; (3) greater use of cogeneration and hence, more efficient fuel use are promoted; and (4) line losses are reduced. Substantial reductions in power plant emissions and electrical use can be achieved when large users have the flexibility to generate their own power and use it at more than one manufacturing location. For example, a 1993 survey of the chemical industry indicates significant opportunities for energy efficiency and emissions reductions within that industry if affiliate (or "self-service") wheeling were allowed:[30]

Reduction in Total Electrical Energy Use—10 percent
Reduction in SO_2 Emissions—26 percent
Reduction in NO_x Emissions—21 percent
Reduction in CO_2 Emissions—10 percent

A competitive electric services industry will promote more efficient compliance with environmental laws and regulations. New environmental mandates increasingly use market-based approaches for achieving truly least-cost emissions reductions goals. The acid rain provisions of the Clean Air Act Amendments of 1990 ("CAAA") established an allowance trading program to help achieve the CAAA's ambitious goals. Estimates of the savings from the trading system range up to $3 billion annually over what would have occurred with the same SO_2 emissions reductions under command-and-control regulation.[31] However, regulated utilities are not taking full advantage of the cost-saving opportunities created by allowance trading.[32] There is greater compatibility between market-based mechanisms for environmental control and a competitive electric services industry. Therefore, more cost-effective compliance strategies can be achieved in a restructured industry. Competition forces the most cost-effective compliance with environmental laws because shareholders, as opposed to captive ratepayers, assume all risk associated with noncompliance.

A competitive electric services industry will help conserve natural resources and reduce emissions by increasing the capacity utilization of existing facilities. In a restructured industry, the owners of transmission facilities will only make money by increasing, rather than discouraging, the use of their transmission facilities. The market will provide powerful incentives for the development of new technologies that maximize the throughput of the grid, and thus avoid the siting of new transmission corridors. A competitive market win also provide greater incentives to repower existing generating facilities to extend the economic lives of those units.

The environment does not suffer under competition because a more efficient economy can better afford the necessary safeguards. The affluence generated by economic growth increases the resources a society is willing to pay for environmental protection. Hence, the most stringent environmental laws have been established by the most successful industrialized countries—not because they polluted the most, but because they could afford to pollute less. It is no secret that pollution generally declines as per capita income rises.[33]

---

[30] Chemical Manufacturers Association (1993)

[31] See Kenneth Rose, Alan S. Taylor, and Mohammad Harunuzzaman, "Regulatory Treatment of Electric Utility Clean Air Act Compliance Strategies, Costs, and Emission Allowances," NRRI 93-16, The National Regulatory Research Institute, December 1993.

[32] See Douglas R. Bohi, "Utilities and State Regulators Are Failing to Take Advantage of Emission Allowance Trading," The Electricity Journal, March 1994.

[33] SeeJagdish Bhagwati, "The Case for Free Trade," Scientific American, Volume 269, No.5, November 1993.

A frequent criticism of proposals to establish a competitive electric services industry is the claim that markets don't adequately internalize externalities. Therefore government central planners must intervene in the market to force "correct" price signals. The planners—as the argument goes—know the right price when they see it.

Externalities are costs and benefits of production or consumption activities that are not borne by producers or consumers. Externalities are external costs and benefits because they are external to the transaction, and therefore not reflected in the market prices of goods and services. Market prices are said to reflect only internal or private costs and benefits. Economic theory states that resource investment decisions based strictly on market prices may not reflect all costs and benefits that impact society. Therefore the allocation of resources may not be optimal. Implicit in the definition of externalities is the need to strike a balance between the costs (or benefits) of the externality and the costs (or benefits) of the "cure." Also implicit in attempts to directly "internalize" externalities is the need for government agencies to set all applicable prices by administrative fiat. This would follow a process in which all market prices must be periodically re-assessed to determine the extent to which each externality has already been: (1) over-internalized, (2) under-internalized, or (3) precisely internalized.

Examples of negative externalities are visibility problems associated with certain air emissions or the uncompensated wages and benefits lost by workers whose jobs were temporarily or permanently displaced by a plant shutdown including the social costs of diminished economic opportunities (e.g., increased health-related traumas and criminal aggression). Positive externalities include the national security benefits of domestic energy production, survival in cold weather, health benefits of medical technology such as life-support systems, or improvements to the Nation's productivity and competitiveness of industrial innovations. Many of these impacts—both positive and negative—already are fully, partially or overly internalized because of the large number of laws that have been legislated to dealing with these concerns.

Some States require utilities to explicitly consider externalities in their IRP process. Typically, utilities are required to internalize certain negative environmental externalities when evaluating new resource options, while ignoring the impacts of other, usually positive externalities. We strongly doubt the effectiveness of such policies to achieve the goals of energy efficiency or environmental protection.

Regardless of the industry structure that is ultimately adopted, any consideration of externalities in the resource planning process should recognize the following principles:

—Attempts to recognize or internalize any external costs, if they are to be made at all, are the responsibility of legislative, not regulatory bodies.

Any regulatory attempt by a State PUC to internalize selected externalities would usurp the legislative role to make policy. Such actions would also prevent the U.S. Congress and the State legislatures from exercising their right to delegate administrative responsibilities for implementing and enforcing policy. Environmental laws should impose only those costs which society as a whole is willing to pay to address environmental concerns, however uncertain social costs and benefits may be.

—Competition in free markets inherently assures the most efficient use of all resources, including energy resources. Maximizing competition in retail and wholesale energy markets will help all consumers to efficiently use resources consistent with environmental protection.

A regulated utility's acquisition of any new resource should proceed in accordance with the development of its IRP. That resource plan should begin with a determination of need and include consideration of all cost-effective resource options, and compliance with all applicable laws, including environmental laws. All the utility's resources—including equipment and materials used for environmental compliance—should be procured by fully competitive processes. This will ensure that environmental externalities are adequately internalized at least cost to society.

Unregulated suppliers in a competitive electric services industry will develop their resource plan subject to market forces. As stated above, suppliers in competitive markets can outperform their regulated competitors in terms of achieving cost-effective environmental protection. Markets will internalize externalities as required by law.

—Regulatory policies that mandate the consideration of environmental and other social factors that go beyond the requirements of existing laws are misdirected and counterproductive.

Economic theory states that welfare may be maximized only when all externalities are internalized. Internalization requires compensation for both beneficial and harmful effects, up to the point where all marginal social costs equal all marginal social benefits. Those who receive benefits should pay for them; those who are victims of harmful effects should be compensated for harm done, to the extent external impacts are not remediated. Environmental externalities typically are multi-regional and multi-national in origin, and therefore outside the jurisdiction and control of individual States. State regulators cannot collect compensation from entities outside their jurisdiction which receive beneficial effects, nor can they compensate entities outside their jurisdiction who are victims of harmful effects. There are literally an unlimited number of externalities, both positive and negative, and upstream and downstream. Most can neither reasonably be qualified nor quantified. It is impossible to ascertain whether net external costs are positive or negative, and therefore the presumption that net external costs are zero is not unreasonable.

Nothing in economic theory suggests that the internalization of externalities on a piecemeal basis improves overall economic or social welfare. In fact, internalizing some but not all externalities may impose net costs on society, compared to a situation where no externalities are internalized. Attempts to internalize selected negative environmental externalities only in electric utility sector, will arbitrarily raise the relative costs of electricity relative to any substitute energy resources or technologies. This will result in an inefficient allocation of resources. Price signals will also be distorted in retail and wholesale energy markets, resulting in inefficient investment decisions and consumption behavior. This may force a shift in energy consumption and associated economic activities to utilities, States, or countries where costs are lower. Such shifts may actually increase adverse environmental impacts, which is counterproductive to the goals of environmental protection. This policy would also unfairly impose costs on ratepayers who may not be directly responsible for the negative externalities. Finally, attempts to "internalize externalities" will unduly discourage the development of potentially important and more environmentally acceptable technologies such as clean coal technologies and innovative industrial electrotechnologies.

Question 4: How Should Pricing Mechanisms Be Adapted to Further Carry Out Congress' Goal, Embodied in EPAct, of a More Competitive Wholesale Electricity Market Facilitated by Open Access to Transmission Services?

ELCON Response: ELCON and other industrial groups have consistently supported the FERC's efforts to institute transmission pricing reform and recognize that genuine efforts to promote competition in bulk power markets can help achieve the broader objectives of economic growth and stability.

The problem of an inefficient power sector is not just endemic to the United States. Countries all over the world are debating the same or related concerns. These debates all tend to focus on the need for greater competition and the importance of nondiscriminatory open access policies to facilitate such competition. Unfortunately, in countries or regions dominated by large vertically integrated utilities, the commitment to initiate real competition has not yet materialized. The United States and European situations have been described as follows:

You Americans and Europeans are trying to achieve the impossible. How can you seriously expect that the owner of a transmission line will provide comparable access to another supplier who is competing to make the same power sale? Why would the transmission owner act against his own economic self-interest by providing transmission service to a lower-priced competitor? And do you really believe a regulator will be able to police the 101 subtle techniques that the transmission owner can devise to sabotage comparability in practice while claiming to abide by it in principle?[34]

The FERC has been forced to wrestle with this problem as part of its mandates under the EPAct. ELCON, the American Iron and Steel Institute ("AISI"), and the Chemical Manufacturers Association ("CMA") have filed comments with the FERC in response to the Commission's notice of inquiry on transmission pricing.[35] Those comments advanced several recommendations that will promote the objectives of the EPAct. These recommendations complement, at the wholesale level, the "eight principles" we propose for achieving return competition in the U.S. electricity industry.

Recommendation No. 1: The FERC Should Use Market Discipline to Promote Efficiencies.

Real competition only works in real markets where electricity is provided by many competing suppliers and prices are established by the same institutional arrangements that govern the pricing of other tradable commodities. Electricity is a commodity. The FERC's goal to establish greater competition in wholesale markets will

---

[34] See Bernard Tenenbaum, Renier Lock, and James Barket, Jr. "Electricity Privatization: Structural, competitive, and Regulatory Options," Energy Policy, December 1992.

[35] Federal Energy Regulatory Commission, "Inquiry Concerning the Commission's Pricing Policy for Transmission Services Provided by Public Utilities Under the Federal Power Act," Docket NO RM93-19-000.

not be achieved unless competition is also established in retail markets. Native load customers must no longer be forced to be captive to a single monopoly supplier. Thus, the "electric utility industry" needs to be replaced by competitive "electricity markets." Indeed, there should be no entry barriers in the new markets that unduly restrict the opportunity to trade by any potential supplier or buyer. All suppliers should have access to multiple markets. All buyers should have the option to competitively source their power needs. Real competition cannot be administratively emulated nor simulated by regulation.

The "regulatory compact" gave natural monopolists adequate incentive to build and operate the transmission system. Arguments that utilities needed or still need any additional incentive to encourage them to open up their grids to other users are groundless. A utility's motive to withhold access results from the traditional bundled nature of power contracts and wheeling services. Thus, a transmitting utility would insist on a "cut" of the generation savings as part of its "price" to wheel the power. This behavior cannot be changed by giving utilities bribes. Whether the supplier operates in regulated or competitive markets, the appropriate "incentive" should always be the opportunity to retain their customer's business if they are willing to provide an adequate and reliable product that is competitively priced.

The development of competitive electricity markets will require regional solutions on planning and pricing, and we highly commend the FERC for its leadership in promoting the concept of regional transmission groups or RTG's. Premature or rushed attempts to create RTG's before the requisite transmission pricing policies have sufficiently advanced will jeopardize their ability to operate as truly competitive power markets. The perceived gains from the transmission policy reforms mandated under EPAct will be lost if inappropriately structured RTG's are created. Therefore, the Commission must set very high standards for its review of each RTG proposal that comes before it.

Recommendation No. 2: The Primary Goal Should Be Lowest Possible Delivered Prices.

The real objective of on-going changes in Federal policy is lower rates for retail electricity. Open-access and the perfect rate for wheeling services are not ends in themselves. Consumers will not benefit from wholesale competition if those benefits are not fully reflected in retail rates, and as long as customers are captive to a single monopoly supplier, that the supplier does not have to pass through those benefits. The goal must be retail prices for electricity that are lower than they would be otherwise, and that are consistent with an adequate and reliable supply. Competition in wholesale markets alone won't achieve that objective. There must also be retail competition.

Recommendation No. 3:

We Urge the FERC to Apply Its "Or" Policy to the Exception, Not the Rule.

For the time being, rates based on rolled-in embedded costs should be the primary pricing formula for wheeling services. Embedded cost-based rates have proven to adequately compensate utilities and to have necessary transmission facilities built or upgraded. The focus of the FERC's pricing policy debate should not be the various theoretical nuances of the current "or" policy, but instead, on a longer-term pricing policy that helps achieve real competition. Nonetheless, the FERC should remain vigilant to any deviation from cost-based rates. The main—if not sole—criterion for moving away from embedded cost-based rates should be the establishment of greater competition and the creation or mitigation of the market power of vertically integrated monopolies.

The FERC should consider the following evolutionary approach to pricing in order to promote the objectives of the EPAct: (1) start out with simple postage-stamp rates based on rolled-in embedded costs; (2) modify those rates whenever competition can be advanced, that is, market power has been mitigated, first, in generation markets, and second, in unbundled transmission and other non-generation markets; and (3) test other, more advanced pricing approaches as new market arrangements become feasible and industry structures evolve and are accepted to accommodate these pricing regimes. The FERC should act swiftly to encourage the development of standards and protocols for market pricing and trading information. Electronic bulletin boards and power exchanges are now being conceived and the creation of uniform standards and protocols would greatly advance both market development and more efficient pricing mechanisms. New pricing regimes need not, and should not, be so complex that they inhibit rather facilitate the efficient operation of the future electricity marketplace.

What steps should the FERC take right now with respect to its current pricing policy?

—First, we strongly urge the FERC to take the "and" option off the table. Indeed, we believe that the Court's decision in Penelec requires the Commission to do so.

—Second, be skeptical of the "or" policy. The "or" policy sanctions the development of hybrid rates that can wreak all kinds of havoc in nascent competitive markets. Both the "and" and the "or" policies give transmission monopolists too many opportunities to stifle competition at both the wholesale and retail levels.

—Third, reject requests for relaxed regulation until truly competitive markets have been established, that is, until utilities' market power has been fully mitigated. This requires retail wheeling. In the future, many regulations will be stripped from the books, but only after the development of truly competitive markets—not before.

—Finally, require the same standards of market power mitigation in transmission tariffs that are "voluntarily" filed, as for tariffs that are mandated or made conditional for other reasons. The FERC should not presume that any voluntarily-filed tariff is automatically just and reasonable.

Recommendation No. 4: Transmission Pricing Policy Should Evolve With Concomitant Changes in Industry Structure.

Rate designs for transmission services will need to change and evolve as the prevailing industry structure changes and evolves. Tariff rates should be supplanted by market-based prices when: (1) market power has been adequately mitigated, or more importantly, (2) it has been demonstrated that a "natural" monopoly no longer exists. However, pricing flexibility must complement structural changes that facilitate real competition.

Reform of the FERC's existing transmission pricing policy must not become a mechanism to perpetuate the anticompetitive market power of the vertically integrated utilities. For example, the potential for stranded costs must not be an excuse to delay or derail the development of competitive markets. Many of the costs that utilities allege may be stranded are not associated with industry restructuring. To the extent true transition costs are identified, they should not be dealt with on a generic basis—particularly at the Federal level. Certain legitimate and verifiable costs may need to be considered on a case-by-case basis. Those costs are largely associated with generation assets and therefore any reconsideration of cost recovery or cost sharing rightfully belongs—and should remain—with the States, not the FERC. As largely generation-related costs, stranded costs must never become a component of the cost of service for transmission services. The FERC should not invite States to bring these problems to its hearing room. The FERC already has been asked to act in several cases involving alleged stranded costs. If a loose standard is established for what constitutes "legitimate and verifiable" stranded costs, the FERC may become inundated with other requests and efforts to advance competition in the industry will simply cease.

Recommendation No. 5: Any New Federal Transmission Pricing Policy Should Complement Actions Taken or Contemplated by States.

The FERC's current policies should begin to complement actions being taken or contemplated by States to achieve the broader goals of competition at all levels—retail and wholesale. States (eg., California) are aggressively pursuing their own restructuring agendas. These actions are symptomatic of the growing interest in retail, not just wholesale competition. But some actions would preempt the right of States to encourage retail competition. For example:

—Some open-access transmission tariff proposals would restrict or ban the applicability of that tariff for State-ordered retail wheeling.

—Other transmission tariff proposals would limit service availability on the basis of the intended beneficiary or "previous condition of servitude."

—Similar provisions in the bylaws of regional transmission groups or RTG's are also anticompetitive and should not be allowed.

A Federal policy rejecting such provisions in no way violates the EPAct's ban on the FERC's ability to order retan wheeling. In fact, this policy would enable the FERC to better fulfill its duties under EPAct to promote wholesale competition.

Recommendations Nos. 6 and 7:

Pricing Comparability and Unbundling Should Be Instituted Now.

Pricing comparability should be instituted for all wheeling services that apply to the needs of both native load or third-party users. Only absolute pricing comparability will avoid the potential for price discrimination (and cross-subsidation) among the different users of the grid. Pricing comparability cannot be achieved under the "or" policy, and without pricing comparability, there will be no competitive market. But pricing comparability cannot be achieved without adequate unbundling of prices and services.

Therefore, the FERC should encourage all suppliers of generation and transmission services to offer those services on an unbundled basis. Unbundling will promote the most productive use of each asset, allow new services to develop that cannot be anticipated, and encourage new market entrants. Users should be able to procure both bundled and unbundled services that meet their total power and wheeling service requirements. With both pricing comparability and unbundling, utilities will not have to be given "incentives" (or bribes) in order to induce them to operate more efficiently.

Several parties proposed that the California PUC implement a regional wholesale power market before trying to implement direct access.[36] This might have been a great idea 20 years ago. However, developments over time render such proposals delay tactics, at best.

These proposals generally assert that the vision of the CPUC can be realized by a scheme that would rely on customers continuing to acquire electricity products and services on a fully bundled basis. The main difference between the status quo and these new schemes would be that customers would pay a fixed charge that included subsidies for various social and environmental programs and a variable charge based on the spot market price at the time of consumption. In effect, all customers would be served on a time-of-use ("TOU") rate. It is even stated that the implementation of such a wholesale market will move us past the "polarized and false debate about direct access."[37]

Such proposals will not allow the development of a truly competitive electricity market to be realized for at least the following reasons:

First, they fail to recognize that the basic premise upon which restructuring proposals are advanced is that electricity rates are too high. Indeed, the CPUC dearly states: "[f]irst, prices for electric services in California simply are too high."[38] Further: [t]he majority of California businesses . . . depend heavily on reliable, safe and competitively priced electric services."[39] The CPUC even further noted:. "California must shy away from proposals which simply maintain the current level of economic activity, and we must opt instead for a strategy which fosters economic growth, enhancing the State's competitiveness, and increasing business opportunities."[40] Revitalization of the California economy requires real change now. These proposals may reallocate costs among various ratepayers but they do nothing to lower costs.

Second, these proposals cannot be implemented in a timely fashion. The implementation of TOU rates for all customers would require the installation of new meters on every household, business, and farm. This would take years and cost billions of dollars. Even though it may have been a good idea in the past, the time now required for implementation renders these ideas useless as a means to stop the tremendous market forces that are motivating pleas for change. Further, any such attempt would actually prevent the market from developing and adopting more advanced metering technology since, in practice, it would require the immediate use of today's off-the-shelf technology. These proposals also require the establishment of a "transparent, arm's-length spot price that is available to all suppliers and not subject to manipulation."[41] ELCON asserts that this requires either: (1) years to establish under regulatory scrutiny (as the painfully slow development of the Western System Power Pool, or "WSPP," clearly demonstrated), or (2) vertical disaggregation of generation and transmission.

Third, those that propose a wholesale-only market generally recommend the establishment of a power pool similar to that now operating in the U.K. This will not be an easy or a quick task. However, these advocates seem to brush potential problems aside by saying: ". . . but as demonstrated in the U.K. and elsewhere, it can be done."[42] Unfortunately, the U.K. model is far from ideal. There are very serious transition problems with the U.K.'s power pool. Industrials are very much in favor of the establishment of a spot market for electricity and are working hard to achieve that goal. However, in establishing a truly workable spot market, it is essential that

---

[36] See, for example, (1) "Response of Southern California Edison Company (U 338-e) to Order Instituting Rulemaking and Order Instituting investigation Dated April 20,1994," CPUC Docket Nos. R.94-04-031 and I.94-04-032, June 8, 1994; (2) "Comments of William W. Hogan," CPUC Docket Nos. R.94-04-031 and I.94-32, June 15, 1994; and (3) "Comments of Paul L. Joskow," CPUC Docket Nos. R.94-04-031 and I.94-32, July 1, 1994. Hogan is Thornton Bradshaw Professor of Public Policy and Management, John F. Kennedy School of Government, at Harvard University. Joskow is Mitsui Professor of Economics and Management, at the Massachusetts Institute of Technology. Note that Professor Joskow provides consulting services to Southern California Edison regarding issues raised in this proceeding. Both Professors Hogan and Joskow regularly provide consulting services to investor-owned utilities and their trade association, the Edison Electric institute.

[37] Op. Cit. Hogan, at 1.

[38] Op. Cit. California Public Utilities Commission (herein referred to as: OIR), at 35.

[39] Id.

[40] OIR, at 6, emphasis in the original.

[41] Op.Cit. Hogan, at 10.

[42] Id, at 6.

end-users to be able to negotiate bilateral contracts to assure reliability and price.[43] Industrials actually operating in the U.K. report that it is nearly impossible to negotiate bilateral contracts. Suppliers know that they can bid low and still receive the spot or pool price, notwithstanding either the clear need or the theoretical provision in the U.K. pool system allowing any eligible customer to directly contract with alternative suppliers. ELCON's understanding of the proposals for a wholesale-only market suggests that similar problems would be forthcoming through the improper implementation of the U.K. model in the United States. In a broader sense, proposals such as these actually are trying to force the design of a market. But markets can't be created as such. They will evolve only if people and government let them happen.

Fourth, the establishment of a wholesale-only electricity market will not in any way assure economic benefits for ultimate customers. As long as buyers in wholesale markets know that their customers are captive, they have neither the motivation to purchase the lowest cost power, nor the incentive to pass through the benefits of competition to ratepayers. To emphasize, so-called "tight" power pools in the Northeast certainly are not bastions of planning or operational efficiencies. Inefficiencies can prevail over a very long period of time to the detriment of consumers. Direct access is necessary to assure both an efficient and effective wholesale market and benefits flowing through to ultimate customers. In their reply comments to the CPUC, National Power PLC said:

National Power supports the POOLCO concept put forth by Southern California Edison Company, but not in the absence of simultaneous direct access. The majority of the real benefits from competition come from direct access, not the Pool mechanism itself.[44]

National Power PLC continues:

Finally, and most importantly, direct access must be provided on a non-discriminatory basis to avoid competitive distortions and be fair to all participants.[45]

Nothing in the proposals assures that inefficient and ineffective social programs are disciplined by market forces. Indeed, a critical assumption of the proposals is that: ". . . ERAM-type adjustments would continue with the adjustments assigned across all customers to true-up through the average distribution charges."[46] This point is further emphasized by the Natural Resources Defense Council ("NRDC") when it said, "[unlike the Commission's retail-wheeling regime, the Hogan system seeks to preserve incentives for cost-effective energy efficiency and renewable energy investment, while continuing to "decouple" utilities' profits from their energy sales."[47] Yet, many, if not most, existing energy efficiency and renewable energy procurement programs are not really cost-effective because they are shielded from market discipline.[48] The status quo is simply unacceptable.

ELCON has other minor observations about each proposal. However, the point is clear attempts to establish a wholesale market with customers facing the spot price through new rate designs sound attractive, but won't provide real, necessary benefits in a timely fashion. What they will do is greatly delay action on the one thing that really will produce benefits for all consumers—retail competition. ELCON agrees with Professor Hogan and others that the debate over "direct access" is polarized, but it is far from "false." Those who have built very expensive generating units, those who support very ineffective and inefficient social programs, and those who require large subsidies for special programs and resources will in no way like the discipline that competition will require. However, customers want, and deserve, the greater benefits that can be achieved with retail competition.

Question 5: What Are the Most Important Questions Involving State and Federal Regulatory Authority? What Is the State of the Law, or on the Need for Changes To, or Clarification of, the Respective Roles of State and Federal Regulators?

ELCON Response: ELCON and the American Iron and Steel Institute ("AISI") had their counsel review concerns that State initiatives to restructure the electric utility industry might conflict with Federal regulatory authorities. The attached Memorandum (Appendix C) is the result of that review. The Memorandum concludes that, ". . . there is no Federal law impediment to State assertion of jurisdic-

---

[43] See, for example, "Reply Comments of the New York Mercantile Exchange," CPUC Docket Nos. R.94-04-031 and I.94-04-032, June 21, 1994, at 8-9.

[44] See "Reply Comments of National Power PLC on California Electric Industry Reform," CPUC Docket Nos. R.94-=31 and I.94-04-032, June 21, 1994, at 3.

[45] Id, at 5.

[46] OP.Cit. Hogan, at 10.

[47] Se "Reply Comments of the Natural Resources Defense Council," CPUC Docket Nos. R.94-04-031 and I.94-04-032, June 20, 1994, at 2.

[48] See "Comments of the Electricity Consumers Resource Council," CPUC Docket Nos. R.94-31 and I.94-32, June 23, 1994.

tion over retail wheeling where consistent with State law." Further, ". . . a State may properly scrutinize the terms and conditions of proposed retail wheeling to assure that the public interest is satisfied without impermissibly intruding on FERC's jurisdiction over interstate transmission."

A fascinating situation may occur if this analysis is incorrect. A ruling that the States have no jurisdiction over retail transmission access may well result in the functional equivalent of disaggregation of the electric industry. This idea was proposed recently by Ashley Brown, both a former Ohio commissioner and Chairman of NARUC's Electricity Committee. Mr. Brown proposes that if States do not have jurisdiction over retail wheeling, then they do not have—nor have they ever had—jurisdiction to approve any bundled rate. This means that States could immediately remove all transmission costs from rate base and request refunds for costs that were overcollected since those bundled rates became effective. This same logic also can be applied to generation. Therefore, those costs must also be removed from rate base. Indeed, the only things that States would have jurisdiction over are the distribution facilities and the ability to approve prudent—but market priced—power purchase contracts.

Figure 1

RATE DISPARITIES WITHIN NERC REGIONS
For U.S. Investor-Owned Utilities Only

Based on 10-MW Industrial Loads with 68% Load Factor
Tariff Rates in Effect as of January 1, 1993

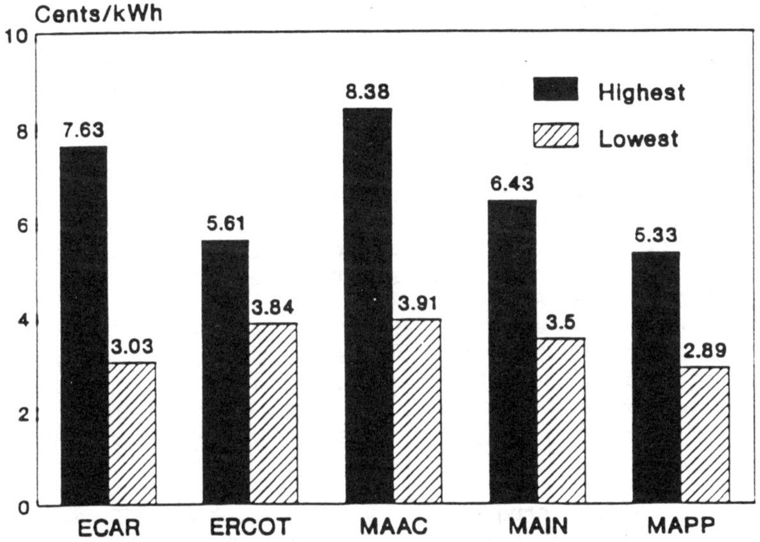

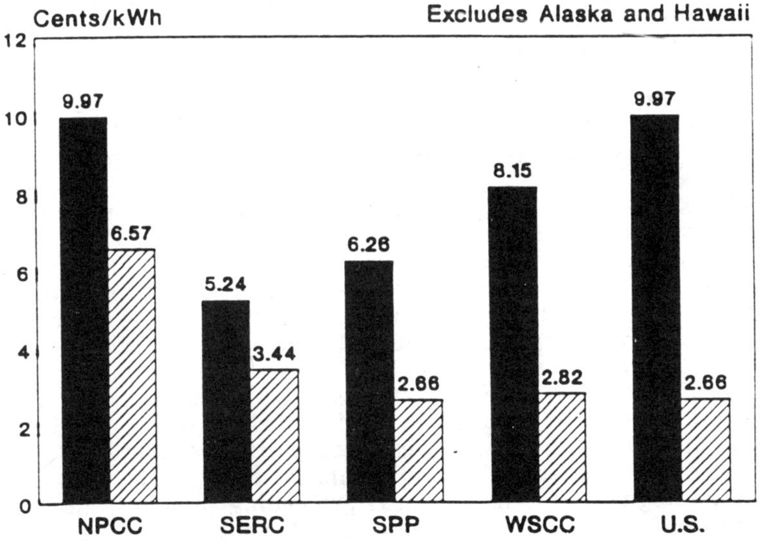

Mr. SHARP. Thank you very much.

Mr. Hempling, we are pleased to hear from you now.

STATEMENT OF SCOTT HEMPLING

Mr. HEMPLING. Thank you, Mr. Chairman.

My name is Scott Hempling, and I appreciate the opportunity to appear on behalf of the Environmental Action Foundation. I wish to discuss first the relationship between structure and competition, then the relationship between competition and planning.

With respect to the relationship between structure and competition, we face today the modern version of a question which began with the industry itself: To capture an efficient electric industry, which activities should be competitive and which should be monopolistic, and what industry structure will maintain reliable boundaries between these two activities?

In 1935 Congress answered this question with the Public Utility Holding Company Act. That statute reduced the mixing of utility and nonutility businesses and limited the size and geographic territory of utilities. The statute presumed that the combination of generation, transmission, and distribution in one corporate family was a natural monopoly. Only those companies which performed all three functions remained in the industry.

In 1992, Congress determined that wholesale generation was not a natural monopoly, but instead of separating generation from the existing natural monopolies, EPAct allowed anyone to perform the wholesale generation function, including existing utilities and including those with an interest in selling or financing generation equipment.

By permitting the mixing of monopolistic and competitive businesses in a single corporate family, EPAct authorized a partial run to the pre-1935 industry structure. If competitive forces materialize, there will be no problem. If competitive forces do not materialize, this industry structure will lead to higher rates, less efficiency, and unnecessary concentration.

What many today are calling competition may turn out to have been a short-term rivalry among established players; the diversity and vibrancy normally associated with a competitive market may not emerge. The jury is out. Defects in the industry structure, unchanged by EPAct, are limiting the establishment of real competition.

There is an unmistakeable consolidation of traditional utilities. Mergers involving Pacific Corp, Kansas Power and Light, Public Service of Indiana, Entergy, Central and Southwest, Sierra Pacific, Northeast Utilities, all have removed established players from the market. Fewer than 2 years into EPAct, most of the model independents have disappeared. The list reads like a witness roster in a 1991 congressional hearing. Long Lake, Hadsen, J. Murkowski, Bonneville Pacific, PSE Energy, all have disappeared, usually through acquisition, usually by utilities. Three of the top 10 independent power firms are, in fact, electric utility affiliates.

Aided by Congress's decision to amend section 211, the Federal Energy Regulatory Commission is proceeding rapidly to implement transmission access. While not without controversy, these efforts reveal a firm resolve to make transmission work. But transmission access alone does not make a market. Last month in the Kansas City Power and Light case, FERC found that there are no inherent barriers to entry in a wholesale markets. The Commission was mistaken and is being challenged. The Nation's wholesale markets contain serious entry barriers. Many utilities resist purchase power with the same vigor with which they resisted transmission access.

There are Government created barriers. Many States authorize only utilities, not nonutilities, to exercise the power of eminent domain. Some States tax utility property at a lower rate than nonutility property. Some States permit only a utility to obtain permission to build a plant. These historic differences create barriers to entry which are hard for independents to hurdle.

This concentration trend is assisted by a FERC policy that fails to distinguish between efficient and inefficient mergers. FERC has stated that a merger is permissible if costs do not exceed benefits. In using this equation, FERC makes three mistakes: (1) it omits acquisition cost, the largest cost, from the calculation; (2) it rewards imprudence by counting as merger benefits to be netted against merger costs those cost reductions that should have been achieved by a prudent company through other means; and (3) it approves less efficient mergers when more efficient mergers are possible.

The result is a form of high-level cream skimming where the acquiring company selects the merger partner which best suits the acquirer's corporate strategy rather than the merger partner most likely to produce the most efficient transaction. The result is a less efficient electric industry.

Other forms of competition are not real competition. At retail, utilities are offering discounts to industrial customers. Discounts by themselves reflect a particular company's competitive concerns but do not necessarily reflect a competitive market structure, particularly if the fixed costs associated with the products sold at a discount are reallocated to captive customers.

Similarly, the efforts of some industrial customers to use unconditioned retail wheeling to have the benefits of natural monopoly service while avoiding the costs of natural monopoly service is not competition, it is cost shifting.

At its core, today's industry structure contains a misalignment between individual incentives and societal efficiency. In the present industry, most utilities maintain virtually permanent control of their exclusive retail franchises plus virtually permanent ownership of their essential transmission facilities plus influence over, if not the right, to determine who should sell generation in their service territories, plus the right to compete with others to sell generation to their service territories, plus substantial influence over who will provide demand side management services in their service territory, plus the right to compete with others to sell demand side management services in their service territory, plus the right to sell generation to other service territories inside and outside the United States using staff and facilities created and paid for by the retail ratepayers within their exclusive service territories.

This structure impedes the development of effective competition because utilities have the incentive and opportunity to favor their own products. It takes a great deal of regulatory resources to make competition work in that structure.

Concerning the relationship between planning and competition, Secretary Tierney is right, competition is a means, not an end. The end is economic efficiency for the benefit of the consumer. Therefore, competition at retail will be efficient only if retail service is not a natural monopoly. That remains to be proven.

Some say that planning is inconsistent with competition. I think Mr. Moskowitz also is right, planning and competition are interdependent, not inconsistent. Planning determines the needs, and competition fulfills those needs. The effective reconciliation of planning and competition will require correction of the present regulatory scheme.

It is not enough to tell States, "Go ahead, be aggressive, but don't step over that jurisdictional line," it requires us all to rewrite that jurisdictional line. Uncertainty over State-Federal relations today is impeding competition.

For example, some States combine integrated resource planning with competitive bidding and then approve a utility's purchase from the winning contestant. That makes sense. But FERC has exclusive authority to determine whether that winning contestant can charge the winning price. To avoid FERC second guessing of State decisions while protecting consumers, States and FERC will have to consider advanced criteria for what makes competition effective.

A key part of State-Federal coordination will be transmission planning. Transmission facilities owned by individual utilities must become components of an integrated multistate transmission highway system. That result will require more cooperation among States and between FERC and States than presently exists.

At present, States and FERC in effect are joint regulators of the same transmission system but they do not work together as joint regulators. They talk about becoming more conversant with each other's needs, having more consultation, more coordination. None of those things truly has materialized.

There is not a system of joint regulation of transmission today. Instead, there is a risk that for every new transaction, the jurisdiction who is rendered worse off can veto it to another jurisdiction's detriment. If that occurs, we will get the suboptimal transaction each and every time. That is not for competition, and that is not proficiency, and we must do better than that.

Lastly on State-Federal, I don't view it as a tension between State jurisdiction and Federal jurisdiction. That is a misnomer. It is normally a tension between customer groups, between States, between groups within the States where there is a need for a neutral party.

Of course, whoever loses says FERC preempted them and there is a problem called FERC-State tensions, but it is a tension among customer groups, among haves and have-nots, among States, and we have to find some way to work those out recognizing that there will be a need for a neutral arbiter.

The ultimate third layer of regulation is the Supreme Court of the United States who says that States cannot erect barriers to interstate commerce, and we have got to find a way to work with that principle before things bubble up that far.

In conclusion, there is much work ahead. Competition brings conflict. The task of regulation is to challenge this conflict toward efficiency. That will not be easy. There is a tendency today to politicize competition, to view it as a battle among interest groups and to turn regulators into legislators. We should resist that tendency. Regulators should focus only on objective rules for efficient production and objective rules for fair competition and let the chips fall. Efforts to preserve a particular competitor's financial integrity from competitor failure have to disappear from regulation if competition is to prevail.

Some have urged that what is least cost for society should be what is most profitable for the utility, or, more bluntly, the rat must smell the cheese. Both propositions are inconsistent with least cost regulation and fair competition. A utility is simply a Government-created monopoly with no particular right to profitability. In a competitive market the cheese goes not to the rat, it goes to the most efficient.

So the proper formulation is not what is least cost for society should be what is most profitable for the utility, the proper formulation is what is least cost for society should be what is most profitable for the entity who can serve society most efficiently. That may or may not be the utility.

Thank you for the opportunity to testify today. Mr. Chairman, it has been a privilege to appear before a subcommittee and to practice law in an industry where you have presided. I look forward to the committee's questions.

[The prepared statement of Mr. Hempling follows. Appendices A and B are retained in the subcommittee files.]

STATEMENT OF SCOTT HEMPLING, ON BEHALF OF ENVIRONMENTAL ACTION FOUNDATION

Mr. Chairman and Members of the Committee: My name is Scott Hempling. I am an attorney in private practice, appearing on behalf of the Energy Project of Environmental Action Foundation ("EAF"). EAF is a nonprofit, research and education organization concerned with energy and environmental issues.

I appreciate the opportunity to participate in this continuation of an historic debate on the changing electric industry. My testimony makes five major points:

I. The future of competition is uncertain.

II. Competition will not be assured unless we correct the structural defects with structural solutions.

III. The Nation can "adapt social and environmental objectives to a "restructured" industry," if it (1) links competition to efficient integrated resource planning and (2) separates political from regulatory decisionmaking.

IV. The uncertainty over State-Federal relations is impeding competition.

V. To ensure that competition prevails, regulators must adhere to objective principles.

I. The Future of Competition is Uncertain

A. Introduction: Temporary Rivalry or Competitive Market Structure?

The subcommittee asks, correctly, whether the electric industry has "entered a period of irreversible change." The present challenges in the industry are a modem version of a debate which began with the industry itself:

Which activities should be competitive and which should be monopolistic, and what industry structure can create reliable boundaries between the two business?

Early in the industry government decisionmakers determined that electric service was a natural monopoly and therefore could be provided most efficiently through a government-regulated, exclusive franchise held by a utility monopoly. Recognizing that electric service was essential to a productive society and that small customers were particularly vulnerable to the decisions of a monopoly provider, these decisiomnakers imposed a public interest, universal service obligation on the utilities. The result was virtual protection of most utilities from competition.

The de-emphasis of competition led to dangerous concentration in the industry. Before enactment of the Public Utility Holding Company Act of 1935 ("PUHCA"), a single corporate system could engage in simultaneous activity in both competitive and monopolistic businesses, without regard for efficiency, the fairness of competition or accountability to local consumers or regulators. The result was a small number of holding companies which controlled all local utilities. A particular holding company might own local utility companies dispersed across regions bearing little or no economic or even electrical connection. The owners of these holding companies often included manufacturers of generating plants or electrical appliances, who benefited from their affiliation with utilities holding legal monopoly franchises.

The result of this industry structure was, among other things, significant overcharges as costs associated with competitive businesses were recovered from monopoly customers, and as the manufacturing and residential infrastructure became overly dependent on the consumption of electricity. Also, nonutility competitors in the equipment businesses had difficulty selling their products to electric companies that were owned by competing sellers of equipment.

The drafters of PUHCA viewed the problems of excess charges and insufficient nonutility competition as a structural problem, flowing from the intracorporate mixing of utility and nonutility business and the absence of size and geographic limits on the ownership of utilities. PUHCA's solution was to confine each holding company to an "integrated public utility system," and either ban (for registered holding companies) or limit (for exempt holding companies) investment by utility holding companies in nonutility businesses. Significant ownership of utilities by nonutility companies also was effectively prohibited.

Although regulators from 1935-1992 still had to address problems of anticompetitive behavior, inefficiency and related problems inevitably associated with regulated monopolies, the structural solution PUHCA substantially reduced the scope and magnitude of this type of risk.

EPAct, again addressing the industry structure, authorized a partial return to the pre-PUHCA picture. In creating the "exempt wholesale generator" exception to PUHCA, EPAct permitted any one utility or nonutility to acquire a wholesale generating company anywhere in the United States and a retail company anywhere in the world. These acquisitions do not have to comply with the "integration" requirement and are largely unreviewed by any Federal regulator.[1]

The Congressional comfort in restoring part of the pre-PUHCA industry structure rested on the assumption that markets would be competitive. To strengthen this assumption, Congress authorized FERC to order access to monopoly transmission highways.

If competitive forces do not materialize, however, EPAct's partial return to a pre-PUHCA structure will produce concentration rather than more competition. As discussed below, defects in the industry's structure, unchanged by EPAct, are limiting the public benefit of EPAct's improvements. Existing utilities, as the chief beneficiaries of the present structure, are resisting the structural changes that will ensure competition. Many are taking actions which, while beneficial to a particular utility, may reduce the vibrancy and diversity in the marketplace.

What appears to be more "competition," therefore, may be a short-term rivalry among established players, which, absent real change in the present structure, will not necessarily evolve not into long-term competition among well-positioned companies.

The remainder of Part I discusses four categories of evidence that competition may not be developing properly:

a. The Trend is Toward Fewer Competitors
b. Wholesale Markets Contain Serious Entry Barriers
c. Retail Markets Remain Legally Insulated from Full Competition
d. FERC's Merger Precedents Invite Inefficient Mergers

B. The Trend is Toward Fewer Competitors[2]

---

[1] EWG status must be certified by FERC, but the certification is ministerial rather than substantive. Also, with respect to registered holding companies, while the acquisition of an EWG is not reviewed by the SEC under PUHCA, financings related to the acquisition, and interaffiliate transactions between EWG's and non-EWG affiliates, are subject to certain SEC regulations under PUHCA.

[2] This section is drawn in large part from D. Penn, "A Critical Look at the Federal Energy Regulatory Commission's Market-Pricing Policies for Wholesale Power" (APPA 1992), and an unpublished article co-authored by the witness and D. Penn.

There is a gradual but unmistakable consolidation of traditional utilities, at a pace and inevitability unanticipated even 5 years ago. For example:

a. In 1988, Pacific Power and Light and Utah Power and Light merged into PacifiCorp, forming a combined utility of $13 billion in assets serving customers in seven States.

b. Kansas Gas & Electric and Kansas Power & Light have merged to form Western Resources.

c. In 1992, Northeast Utilities completed its strategic acquisition of Public Service Company of New Hampshire. The combined company exceeds $10 billion in assets and controls, albeit subject to FERC transmission tariffs, vital transmission corridors in New England.

d. Entergy has acquired Gulf States Utilities for $2.3 billion. The combined company will have assets of $21 billion and serve customers in four States in the middle south region.

e. Public Service of Indiana (PSI) and Cincinnati Gas & Electric have received merger approval, although that plan is being challenged by another Indiana utility hoping to acquire PSI.

f. Central & South West has filed to acquire El Paso.

g. Sierra Pacific and Washington Water Power have announced plans to merge.[3]

The debate over EPAct was dominated by the notion of the "independent power producer," with several IPP's frequently cited as role models. Fewer than 2 years after enactment, most of the independents then cited as role models have disappeared:

a. Bonneville Pacific Corporation has filed for bankruptcy.

b. PSE Energy and Long Lake Energy Corporation were acquired by Dow Chemical subsidiary Destec Energy.

c. Louisville Gas and Electric (LG&E) bought out vocal independent Hadson Power Systems.

d. Just last week, J. Makowski Co. announced its purchase by PG&E/Bechtel.

Only AES, Destec and a few others retain a significant market presence as "true independents."

Independents have had some success penetrating the generation market, contributing over 50 percent of new generating capacity in 1989 and 1990. Their total in service at end of 1992 was 47,597 megawatts (MW) of capacity. In 1991, they added 5,200 MW of generation, and in 1992 they added 3848 MW. Another 93,000 MW are under development or construction, and as much as 78,000 MW could be operating by 2000.[4]

Even the most optimistic projections, however, estimate that independent power's share will no more than double to 16 percent by 2010.[5] More importantly, Jean-Louis Poirer of RCG/Hagler, Bailly, Inc. has predicted that by 1995 the current 170 independent power companies, including about 30 electric utility affiliates, will fall to about 90.[6]

At the same time, utility-affiliated wholesale generators are growing in influence. As early as November 1990, Independent Energy reported acquisitions of independents by utility affiliates and others had left just a handful of truly independent developers not affiliated with a deep pocket player in some way. Three of the top 10 independent power firms now are electric utility affiliates. Southern California Edison's Mission Energy is the Nation's largest independent in terms of ownership in nonutility generating projects. It has 1,432 MW of net ownership and is involved in 25 operating plants totaling over 2,800 MW. Ranked second with 1,346 MW was Enron Power, a subsidiary of Enron Corporation, a major natural gas company.[7] Even a modestly sized utility like LG&E has over 1,000 MW in independent projects operating and under construction in seven States. At least 30 other utilities have affiliated wholesale generating companies.[8]

See also 1992 Ownership Trends in Current Competition (Aug. 1992 at 1) reporting that traditional utilities are increasing their ownership share of the "independent" market.

C. Wholesale Markets Contain Serious Entry Barriers

---

[3] The Electricity Daily (June 29, 1994).

[4] RCG/Hagler, Bailly, Profile VII: A Profile of the Independent Power Market.

[5] National Independent Energy Producers, The Competitive Power Landscape (presented to American Public Power Association, June 15, 1992); 1990 Capacity and Generation of Non Utility Sources of Energy, Edison Electric Institute (December 1991).

[6] See "Utility Independent Power Affiliates Reach Maturity," Electric Light & Power, September, 1992, p. 3.

[7] See report on Independent Power, Electric Utility Week, January 20, 1992, pp. 910.

[8] See also 1992 Ownership Trends in Current Competition (Aug. 1992 at 1) reporting that traditional utilities are increasing their ownership share of the "independent" market.

EPAct made two changes: authorized FERC to order transmission and authorized new entrants into wholesale generating markets, exempt from the traditional requirements of the Public Utility Holding Company Act. Transmission and new entrants, alone, do not make a market. Where there is affiliation between the monopoly and competitive business, regulation is necessary. The present industry structure permits unlimited affiliations between the monopoly and competitive businesses.

Notwithstanding the Federal Energy Regulatory Commission's (FERC) recent finding that there are no entry barriers to generation competition, Kansas City Power & Light, Docket No. ER94-1045-000 (May 13, 1994), entry barriers into the generation market are significant. Examples follow.[9]

1. Utility Resistance to Purchased Power

A utility-as-wholesale-buyer normally has a franchise monopoly over its retail customers. The utility therefore is the sole purchaser of power for its service territory. Where the utility resists purchasing from new entrants (because, for example, the utility does not want to forego the profits available from building, or is concerned about the regulatory risk associated with making a purchase), the prospective entrant may have to incur substantial litigation costs to gain entry, with uncertain results.

A number of financial analysts have given utilities strong reason to offer such resistance, by warning about (and in some situations carrying out) downgrades of bond ratings for utilities that purchase.

Appendix A provides a more detailed discussion about the problem of utility resistance to purchased power.

2. Government-Created Barriers

There are large variances in the States' treatment of nonutility generation. Some States permit only utilities to obtain certificates of convenience or necessity. In that situation, the utility's willingness to forego construction and buy from a competitor, and thereby procure a certificate on behalf of that competitor, is a prerequisite to entry by a nonutility. As discussed above, such utility willingness varies. In some States, an entity may not sell electricity unless it has a "service territory." Since wholesale competitors do not have a "service territory," in these States they are barred from entry.[10]

Most States authorize only utilities, not nonutilities, to exercise the power of eminent domain. A nonutility with no legal right to take property is dependent on the voluntary action of homeowners to give up their homes for a negotiated price. A utility needing the same homes, armed with the power of eminent domain, has a surer and less expensive route to site acquisition. Some States tax "utility property" at a lower rate than other manufacturing property. There likely are other historic differences in the way State laws treat utilities as compared to nonutilities. These cost differences create entry barriers.

3. Economies of Scale

Many economists view "economies of scale" and "economies of scope" as an entry barrier. FERC has found to the contrary:

We also agree with PSI that economies of scale are not considered to be barriers to entry of new electric power supplies. Scale economies in generation have disappeared because both the market has expanded and the minimum efficient firm size is smaller than previously believed.

Public Service of Indiana, 51 F.E.R.C. para. 61,367 n.50 (1990).

This statement does not hold together well. Scale economies for a particular product do not "disappear," although they can shrink. The question is whether the minimum efficient plant size is sufficiently small, relative to total demand in the market, to permit a large number of competitors. The answer necessarily must depend on the particular type of plant demanded by the market and the size of the demand. If, for example, the market demands 500 MW of baseload power, and the minimum efficient plant size is 500 MW, an efficient market will not permit more than one competitor.

Also, there are many sites which already host an existing utility plant, and which can be used for a second plant. The incumbent who controls this type of site can have a cost advantage which can translate into an entry barrier. The question is not only whether there are other sites in the market, but whether there are other sites containing the same advantages. There is no basis for assuming, as FERC im-

---

[9] These arguments have been made by Environmental Action in its Petition for Rehearing in the KCP&L case.

[10] Because the Public Utility Regulatory Policies Act of 1978 mandates utility purchases from PURPA "qualifying facilities," QF's would be exempt from such a rule.

plicitly did, that there is a sufficient number of such sites to ensure the absence of entry barriers.

D. Retail Markets Remain Legally Insulated from Full Competition

At the retail level, competition for all customers is rare. Most retail service is provided by utilities on an exclusive basis. That exclusivity is a result of a governmental decision barring others from competing within that territory.

Most of these exclusive franchises are very long term and rarely if ever contested. Notwithstanding the pervasive "debate" over retail wheeling, these franchises remain exclusive. Efforts to relax this government barrier through retail wheeling are almost uniformly opposed by utilities. The retail service territory is one place where existing utilities do not stress the goal of "getting government out of the market."

There are numerous examples of industrial discounts granted for "competitive" purposes. These discounts, by themselves, are not a sign of a competitive market structure consisting of numerous viable sellers vying for numerous knowledgeable buyers, particularly since the fixed costs associated with the product sold at a discount are frequently reallocated to captive customers.

My point is not to suggest that retail competition is good or bad; but only to suggest that a competitive market structure at retail does not exist.

E. FERC's Merger Precedents Invite Inefficient Mergers

Regulatory policy on mergers should distinguish efficient from inefficient acquisitions. FERC's treatment does not meet this test. FERC's policy instead invites consolidations which do not approve the industry's market structure or consumer welfare.[11]

Section 203 of the Federal Power Act, which applies to most mergers, requires that the transaction be "consistent with the public interest." According to FERC, an applicant meets this standard by showing that the "costs" of a merger do not exceed its "benefits."[12] This test, as applied by FERC, has four defects.

1. Costs are defined incorrectly: FERC finds that mergers meet the test when projected benefits exceed projected implementation costs.[13] FERC thus ignores what commonly is the largest single cost: acquisition cost. One would not buy a rental property merely because the expected rent exceeded the costs necessary to rehabilitate and maintain the space for tenants. One would buy the property only if the expected rent exceeded these implementation costs plus the acquisition cost.

FERC frequently says it will review the acquisition cost, if necessary, in a future rate case. That treatment would make each section 203 proceeding an empty gesture. The proper remedy for an inefficient merger is to reject it, rather than permit it and then require shareholders to absorb the inefficient cost after a lengthy, expensive hearing.

2. Benefits are defined incorrectly: The FERC incorrectly counts as merger benefits, to be netted against costs, cost reductions which should have been achieved through other means.

This policy rewards imprudence. In the section 203 analysis, the purpose of counting merger benefits is to determine whether they sufficiently offset merger costs. Assume the target company has been using quill pens and Roman numerals, and the acquiring company intends to introduce computers, thereby reducing cost by $250 million compared to the nonmerger case. The $250 million represents the elimination of imprudence. Elimination of imprudence is a preexisting obligation of management.

If the "benefit" of eliminating preexisting imprudence can be used to justify merger cost, then the larger preexisting imprudence of the target company, the larger the premium the target company's shareholders can demand without failing the cost-benefit test. That result is not "consistent with the public interest."

The FERC's unstated premise—that the target company never would have improved, nor was obligated to improve its management techniques on its own—is inconsistent with the Commission's efforts to make the industry competitive. In a competitive industry, each utility performs competitively, or fails.

3. "Costs should not exceed benefits" is the wrong signal: The "public interest" is harmed when a merger precludes lower-cost means of achieving benefits. Even if the Commission had counted costs and benefits accurately, its "public interest" equation costs should not exceed benefits is inconsistent with section 203.

---

[11] These arguments were raised by the Arkansas Public Service Commission, which I advise, in the Entergy-Gulf States merger case at FERC, and rejected by the Commission. The case is on appeal.

[12] See, e.g., Utah Power & Light & PacifiCoEp Merger, 45 FERC para. 61,095 at 61,278-79 (1988).

[13] See, L.L.,Entergy-GSU Merger, slip op at 37. By "implementation" cost, FERC means costs incurred after the merger to implement coordination and other possible economies.

Assume two cost reduction proposals, each of which produces $100 in cost reductions. Proposal A, repowering all plants on the system, costs $75. Proposal B, a merger, costs $100 ("acquisition" plus "implementation" costs combined). When the desired result is $100 in cost reductions, it is not "consistent with the public interest" to select a strategy costing $100 over one costing $75.

4. The reliance on "generalities" is improper: The Commission has stated, many times, that merger applicants need make only a "generalized showing of the types of savings and efficiencies which might be achieved through the proposed merger." Entergy-Gulf States, Slip op. at 4. The public is not protected when regulators accept 'regulatees' generalizations about a billion dollar transaction.

In Summary: Given the errors in FERC's policy, the PUHCA's prohibition on acquisitions that fail to produce efficiencies is critical to consumer protection.

F. Conclusion

Many seeking to "relax" regulation, including, for example, the requirements of PUHCA, premise their position on the existence of "competitive markets." In many sectors of the electric industry where competition could be efficient, the markets are not yet competitive and are not necessarily moving in that direction.

My purpose is not to preach pessimism but to remind us of reality. After 70 years of vertical integration and monopoly service, competitive, disaggregated markets will not just "evolve." No competitor gives up its historic advantages voluntarily. As discussed in Part II, the best assurance for competition is a revision of the industry's structure.

II. Competition will not be assured unless we correct the structural defects with structural solutions

Part I explained how real competition was not an inevitable result of the enactment of EPAct. This Part II explains how structural flaws are blocking the natural evolution of competition.

Part II.A explains how today's industry structure creates a misalignment between individual incentives and societal efficiency.

Part II.B describes a special case of misalignment: registered holding companies, whose status appears to be inconsistent with regional efficiencies.

Part II.C explains that a competitive industry structure should have separate ownership of the monopoly and competitive components, with periodic competition for the right to provide monopoly services.

A. Today's Industry Structure Creates a Misalignment Between Individual Incentives and Societal Efficiency

In the present industry, most existing utilities retain:

1. virtually permanent control of their exclusive retail franchises, plus
2. virtually permanent ownership of their essential transmission facilities, plus
3. influence over, if not the right to determine, who should sell generation in their service territories, plus
4. the right to compete with others to sell generation to their service territories, plus
5. substantial influence over who will provide demand-side management services in the service territory; plus
6. the right to compete with others to sell demand-side management services in the service territory; plus
7. the right to sell generation to other service territories inside and outside the United States, using staff and facilities created and paid for by the retail ratepayers within their exclusive retail territories.

In short, the present industry structure, in which large utilities control exclusive service territories and monopoly transmission facilities at the same time they are competing to sell generation and demand-side services in developing markets within their own service territory, impedes the development of effective competition because utilities have the incentive and opportunity to favor their own products.

These structural biases within utilities are exacerbated by traditional methods of regulation, which (a) base profit on size of investment, (b) discourage actions which reduce sales, and (c) frequently force ratepayers to protect shareholders from management errors. Ratemaking does not send clear, consistent signals that efficiency will be rewarded and inefficiency penalized.

Today's industry structure thus creates a misalignment between individual incentives and societal efficiency. When buying from the most efficient producer is against the buyer's self-interest, the economy is bound to underperform. Structural reform requires realignment.

B. Registered Holding Company Status, as a Basis for Federal Preemption, is Inconsistent With Efficient Regional Competition

Many utilities argue that under the Nantahala-Mississippi Power & Light line of cases,[14] FERC and SEC review of interaffiliate power supply transactions within a registered holding company system preempts State review of the prudence of the purchaser or seller in these transactions. The holding companies' position is inconsistent with efficient regional competition.

This section argues that the boundaries of a registered holding company system should not become a bar to State disallowance of utility behavior inconsistent with regional competition. The legal support for this position is set forth in Part IV.

1. In a Real Regional Market, a Holding Company Affiliate Would not be Exempt from Penalty for Declining to Trade with an Efficient Nonaffiliate

The holding companies' premise appears to be that any transaction among affiliates is inherently more efficient than a transaction between an affiliate and a nonaffiliate (which the State regulating utility affiliate might prefer). This premise conflicts with the fact of regional markets envisioned by EPAct; i.e., the path to efficiencies lies in numerous wholesale buyers negotiating with numerous wholesale sellers, with no preconditions on who may transact with whom. Put another way, a "registered holding company system" is not a "natural monopoly" with respect to the generation, transmission and distribution activities within the system boundaries.

It may be that the complex of interaffiliate transactions embodied within each system is more efficient than each company standing alone and transacting with no one. But that is the wrong comparison. In the real world, no company would stand alone and transact with no one. In the real world, a competitive market would force each company to seek those trading partners with whom it could buy and sell electricity at the lowest cost. As the Arkansas Public Service Commission has written:

It is by no means clear that the specific regions served by existing registered holding companies bear any relationship to effective planning regions. Registered holding companies are historical accidents. Congress' intent in enacting PUHCA in 1935 was to break up the 13 holding companies which controlled the industry into local utility systems. In the long process of litigation that accomplished this result, certain entities managed to remain large and spread over several States. In today's industry any individual subsidiary of a multistate company may be able to coordinate and plan with adjacent unaffiliated utilities just as easily as they can coordinate and plan with the affiliated utilities in their registered system. There is some risk that by effectively requiring subsidiaries to coordinate with their affiliates, we forego more efficient coordination with the non affiliates. We should make these coordination and planning decisions on a case-by-case basis, rather than be ruled by system boundaries mapped out 40 or 50 years ago.

Testimony submitted to the Committee on Natural Resources, U.S. Senate, March 1991 (emphasis in original).

2. Regulators Should be Able to Reduce as Well as Expand Registered Holding Company Boundaries

In a world of effective wholesale competition, transmission access, regional transmission groups and interstate cooperation, the possibilities for efficient interutility transactions are numerous. The existing holding company relationship is only one such transaction.

Thus, when some holding company advocates argue for repeal of PUHCA in favor of "the marketplace," consistency requires repeal of any preemption flowing from registered holding company status, so that "the marketplace" rather than decades-old SEC decisions determines the efficient transactions. Just because the boundaries of the Entergy system were efficient in 1940 does not make them efficient in 1994.

In acquisitions involving registered holding companies (e.g., NU's acquisition of PSNH, Entergy's acquisition of Gulf States, CSW's acquisition of El Paso, the PSI-CG&E merger), the applicants have argued for enlarging the boundaries of the existing holding company system in order to create more efficiencies. If the boundaries are not sacred, then they can change in either direction. If a utility can argue for acquisition based on efficiencies, States and consumers should be able to argue for disaffiliation (or at least off-system purchases or sales), again based on efficiencies.

---

[14] Nantahala Power & Light v. Thornburg, 476 U.S. 953 (1986); Mississippi Power & Light v. Mississippi ex rel. Moore, 487 U.S. 354 (1988). Compare Kentucky West Virginia Gas Co. v. Pa. Public Utility Commission, 837 F.2d 600, 609 (3d Cir. 1988) (Nantahala "in no way undermines the long-standing notion [set forth in Pike County, that a State commission may legitimately inquire into whether the retailer prudently chose to pay the FERC-approved wholesale rate of one source, as opposed to the lower rate of another source."); Pike County Light & Power Co. v. Pennsylvania Public Utility Commission, 77 Pa. Commw. 268, 273-74, 465 A.2d 735, 737-38 (1983) (same).

PUHCA nowhere says that efficiencies are available only from enlargement and not from shrinkage.

In short, there cannot be one standard applied to expansion of the system's boundaries and a different standard applied to reduction.

C. A Competitive Industry Structure Should Have Separate Ownership of the Monopoly and Competitive Components, With Periodic Competition for the Right to Provide Monopoly Services

Many industry participants wish to reduce regulation. It is better to reduce the need for regulation, and then reduce regulation as the need declines. The most direct route to reducing the need for regulation is through an industry structure that (1) separates competitive services from monopoly services, and (2) stresses competition for the right to provide the monopoly services.

1. Separation of Ownership

The joint ownership of monopoly and competitive assets is the single largest barrier to competition in our industry. This structural problem requires a structural solution. Generation, transmission and distribution businesses each should be owned independently (truly independently; not through affiliates). This industry structure would align the self-interest of all players (existing and potential) with efficiency and least-cost service.

Restructuring along these lines would require existing vertically integrated utilities to spin off much of their investment to independent companies. The process would require fair compensation to shareholders consistent with their historic risks. Players also would have to devise operational agreements among the newly unaffiliated companies in order to preserve efficiencies associated with vertical integration. An alternative path would be to prohibit the utility prospectively from building generation for its own retail customers. Under this approach, existing utility plants would be replaced eventually by purchased power.

Proponents of expansion by existing utilities often argue that existing utilities have large skills which should not be held back from new markets. These skills can be placed in an independent corporation which can service both the utility and the monopoly business. Such an independent corporation, having no monopoly franchise and being subject to competition in all its efforts, would be unable to behave anticompetitively. A fuel supplier, for example, could supply gas to both a utility monopoly and an independent power producer.

One might argue that a public interest reason for permitting participation by a traditional utility in monopoly and non-monopoly markets is economies of scope and scale: the notion that this combined participation by the same corporation creates efficiencies which otherwise would be lost. No one has supported such an argument with facts.

If utility expansion is permitted, the expansion must be accompanied by the application of interaffiliate pricing rules which ensure consumer protection, and fair competition, including the elimination of entry barriers. These rules are set forth in Appendix B below.

2. Franchise Competition

It is one thing to determine that monopoly service is efficient. It is another thing to determine who should have the monopoly privilege. There should be competition for the right to be the monopoly.

Franchise competition today is not fully effective. Reasons include laws favoring the incumbent, community support for the existing utility, absence of "crisis" reason to change horses, large borrowing costs and other transaction costs for the franchise competitor, and general uncertainty about the process.

In the absence of franchise competition, however, we must use government regulation to measure, reward and penalize utility performance. As problems facing utilities grow more complex, this regulatory responsibility becomes harder to fulfill. In order to move towards a more market-oriented approach to the natural monopoly, policymakers should consider the following principles:

1. Franchise competition should be more regular and less expensive.
2. There must be a predictable mechanism for transferring ownership of the underlying assets to the new franchisee is in place.
3. Competitors for the franchise should be required to make enforceable promises about performance.
4. Transition costs must be assigned fairly.

III. The Nation can "adapt social and environmental objectives to a 'restricted' industry," if it (1) links competition to efficient integrated resource planning and (2) separates political from regulatory decisionmaking

A. Introduction

The subcommittee asks: "How can social and environmental objectives be adapted to a 'restructured' industry, in an environment in which services once provided to a single vertically integrated utility are increasingly "unbundled'"?

Answering this question requires careful distinctions among four steps:

1. determine the social and environmental objectives
2. determine of which objectives should be met by a competitive marketplace and which by a monopoly
3. create a competitive market structure for the competitive services
4. select the most efficient provider of the monopoly services, and apply regulatory principles to ensure continued efficiency

Our industry dialogue, like the industry "transition," has not been this orderly. The question as phrased by the subcommittee accurately reflects our decisionmaking posture. But it also reveals the flaw in that posture. Many debate participants talk of a "restructured" industry without articulating what structure is best. For example, some posit a "restructured" industry in which the purchase of electrons is "unbundled" from all other services. But positing this structure begs the question that should be asked: where is competition efficient and where is it inefficient?

B. A Plan for a "Restructured" Industry Must Distinguish Between Efficient and Inefficient Competition

The goal of a "restructured" industry is not simply "more competition." Competition is a means, not an end. The end is economic efficiency, for the benefit of the consumer. Many proponents of "competition" simply want the type of competition that benefits the proponent. For example:

1. Many utilities argue for the right to expand into the telecommunications business, arguing that "competition" is good. But this same utility may insist on using ratepayer-funded assets, and compensating the ratepayers at only book cost, rather than at a price set by "competition."
2. Some industrial customers argue for retail wheeling on the grounds that "competition" is good. These entities hope to have utility reserves waiting for them when they return from their shopping trips. Those entities articulating this position (successfully) in the Michigan PSC case expect the costs of these reserves to be assigned to shareholders or captive ratepayers, rather than according to the rules of "competition," which would assign the costs squarely to these entities.[15]

To avoid this inefficient opportunism, we must determine where competition is efficient and where it is inefficient. Inefficient competition can occur in at least three ways. (1) disregard of natural monopolies, (2) externalities and (3) discrimination.

1. Disregard of Natural Monopolies

Where the product is a "natural monopoly," the introduction of multiple sellers will be inefficient.[16] A key question in determining whether a particular structure meets our "social and environmental objectives," and does so efficiently, is whether natural monopoly services are being subjected to inefficient competition.

For example, regulation historically has treated retail electric service as a natural monopoly. What is "retail electric service"? If it is merely the physical movement of power from generator to consumer, retail wheeling should cause no loss of efficiency. Individual consumers could shop in the competitive generation market, and then hire the monopoly owner of transmission and distribution to move the purchased power from generator to load (and paying for all ancillary services necessary to complete the transaction reliably). There would be competition for the competitive service (generation), and monopoly provision of the monopoly service (local distribution).

But retail electric service is not mere physical movement of electrons. It is a larger bundle of activities, including long-term planning (i.e., determining the proper level and mix of generation and conservation resources) and resource acquisition. Consider the generation planning principle that given a load curve for a particular service territory, there is a unique mix of baseload, intermediate and peaking plants which will minimize production costs. If each individual customer shopped for generation independently to minimize its own costs, only by sheer coincidence would the sum of all customer purchases replicate this unique generation mix.

2. Externalities

[15] This defect in the Michigan PSC Order is detailed at Part V.D below.

[16] A natural monopoly exists when, due to a combination of economic and technological factors, the cost of service is lower when there is a single provider and higher when there are multiple providers. See, e.g., W. Nicholson, Microeconomic Theory: Basic Principles and Extensions 547 (1989) ("Natural monopolies, by definition, exhibit decreasing average costs over a broad range of output levels."); id. at 535 (in a natural monopoly, "minimum average cost can be achieved only by organizing the industry as a monopoly"); R. Schmalensee, The Control of Natural Monopolies 3 (1979) ("An industry activity is said to be a natural monopoly if production is most efficiently done by a single firm or entity.").

Replacement of vertically integrated monopolies with disaggregated competitors also will require a more comprehensive, rigorous approach to externalities. In real competition, sellers and buyers must bear the full costs of their decisions and victory goes to the least cost, highest quality producer. This rule is violated when some costs associated with the product are borne by third parties.

Recognition of externalities in prices, somehow, therefore is a prerequisite to fair competition. All sellers and buyers must be responsible for the social costs of electric production.

3. Discrimination

Treating some customers as captive and others as competitive can lead to inefficiency. A service is either competitive or monopolistic. It usually is not competitive for some and monopolistic for others.

Individual residential buyers face significant transaction costs. If there is to be retail wheeling (and there should be only if it is determined that retail electric service is not a natural monopoly or if the price is set to avoid any inefficiency associated with breaching the natural monopoly), it will be necessary to design some way to reduce those costs, in order to ensure nondiscriminatory access to this service. For example, it might make sense to require each competing retail seller to sell to a representative group of all customer types. With such a rule, the competition will be won by the entrant who can organize such a group most efficiently.

The notion, suggested in the California proposal, that industrials go first and all others go second, is inconsistent with these principles. The justification—a desire to proceed gradually—is no justification for discriminatory access.

Sometimes the word "competition" is used to cover discriminatory, noncost based discounts (as opposed to volume-based discounts). These discounts are not real competition. Letting a monopoly utility reduce its rates below fully embedded cost, with captive ratepayers picking up those costs, is "retail wheeling without the wheeling," i. e., government authorization to the utility to price below fully embedded cost to particular customers, combined with government prohibition against utility competitors serving those customers.

C. In an Industry Having Both Natural Monopoly and Competitive Services, Efficiency Requires Close Linkage Between the Planning and Competitive Processes

Retail wheeling advocates often say that "planning is inconsistent with competition." "Planning" and "competition" are interdependent, not inconsistent. Planning determines the needs. Competition fulfills the needs.

It is possible that the industry function of "planning and aggregation," presently performed by the utility, may be a natural monopoly. This function is distinct from the traditional functions of generation, transmission and physical distribution. It is the function which has included integrated resource planning and the various so-called "mandates" such as externalities, subsidization of agricultural and low-income rates, and other deviations from conventional competitive market pricing.

The California proposal omits this analysis. The proposal appears to assume, implicitly and with no factual proof, that competition at all levels will be efficient. From this nonexistent factual foundation, it builds a new industry structure. Most participants in the debate now are debating the structure when they should be debating the foundation.

If the planning and aggregation function is a natural monopoly within the presently defined service territory, then there will be a loss of efficiency by allowing some customers to shop, unless (1) a least cost plan is established first, and (2) the price of "retail wheeling service" includes all costs (and benefits) associated with the wheeling customer acting inconsistently with the integrated resource plan.[17] With this approach, retail wheeling becomes a means of implementing, not avoiding, the goals of integrated resource planning.

To make retail competition fair, the State commission should impose the same requirements on all prospective sellers, whether retail or wholesale or utility or nonutility. If the utility must include externalities in prices, so must competing retail sellers. This approach prevents utilities from avoiding retail competition by asserting "we have been forced to carry out public policy agendas from which our competitors are exempt."

These principles logically produce rules for retail wheeling customers as follows:

1. Departure payment to shareholders: Where the departure renders unmarketable any utility investment reasonably made on behalf of the departing customer, the departing customer must make the shareholders whole, unless the shareholders

---

[17] David Moskowitz and his colleagues have designed a retail wheeling pricing proposal that, depending on its application, can be consistent with these principles and warrants serious attention. See D. Moskowitz, et al., Future Utility and Regulatory Structures (Regulatory Assistance Project Dec. 1993).

already have received compensation historically through a return on equity reflecting the risk of unmarketability. This payment should last no longer than necessary to amortize the unmarketable investment; and, the utility would be under a duty to mitigate the loss by selling the surplus to others.

2. Departure payment to remaining customers: Where the departure increases the per-unit cost of utility service under the approved integrated resource plan where the departure is inconsistent with economies of scale), the retail wheeling customer must make pay compensation to the remaining ratepayers.

3. Prospective investment: Before a utility makes new major investments, each customer contemplating future retail wheeling must make an agreement with the utility concerning whether to have the utility plan for that customer's needs. For those customers that "opt in" to the utility planning regime, they must agree in advance to pay off any prudent costs incurred on their behalf before leaving.

4. Consistent application to all forms of "departure": These rules must apply consistently to all forms of departure. Some customers say "if I don't get retail wheeling I'll build my own generation." Self-generation, like retail wheeling, is a breach of franchise exclusivity and must be subject to the same rules. Applying different rules will invite opportunism rather than least cost service.

D. Regulation Should not be Used to Resolve All Societal Frictions

Electricity production involves a host of societal frictions in addition to customer vs. shareholder and competitor vs. competitor. Jobs, air and water quality, land use and other values are affected. Regulation, as a form of government decisiomnaking, is poorly suited to resolving these frictions. Regulation should accomplish two chief goals:

1. Maintain an industry structure that permits competition where competition is efficient, and monopolies where monopolies are efficient.

2. Identify markets where competition is nonexistent or ineffective, and use objective techniques to ensure behavior by participants which produces product price and quality consistent with what a competitive market would provide.

Resolution of conflicts such as environment vs. price, or price vs. jobs, or auto worker unions vs. electrical workers unions, do not belong in the regulatory sphere. Our political branches should resolve these conflicts first, thereby providing regulators with clear values to be implemented. Once the political branches have determined the tradeoffs between, for example, pollution reduction and other values, market mechanisms can be designed to achieve the pollution reduction goal at least cost. Then it becomes the regulators' task to ensure that utilities make prudent purchase and sales decisions in the existing markets.

Examples of incorrect mixing of political and regulatory decisions have appeared at both poles of the present debate. For example:

1. DSM Incentives: Some have argued for DSM "incentives" for utilities on the grounds that "what is least cost for society should be what is most profitable for the utility;" or, more bluntly, "the rat must smell the cheese." Both propositions are inconsistent with least cost regulation and fair competition. A utility is simply a government-created monopoly with no particular right to "profitability." In a competitive market, the cheese goes not to the rat, but to the most efficient.

The proper formulation is "what is least cost for society should be what is most profitable for the entity who can serve society most efficiently." This formulation aligns profit with efficiency rather than profit with utility status. The earlier formulation did value efficiency, but placed a price on it. That price was based not on fair competition, but by a form of political logrolling between government regulator and private monopoly where the latter sought maximum profit and the former guessed at "how much it will take" to get the monopoly to do what a competitive market would do automatically.

The earlier formulation was a logical impossibility because something cannot simultaneously be least cost to society and most profitable to a monopolist. It has no place in the post-EPAct era.

2. Industrial development: At the same time, those who argue that "retail wheeling is necessary to boost industrial development" make the same error. The utility is industry must operate efficiently for everyone, not just selected sectors.

IV. The uncertainty over State-Federal relations is impeding competition

The subcommittee asks for recommendations on how regulators might handle the transition to competitive markets (Question 2), as well as for thoughts on the relationship between State and Federal authority (Question 5). This section responds in combination to those two questions. This Part IV has four subparts:

A. State and Federal Regulators Must Recast Their Relationship to Accommodate Regional Markets

B. A Rational Allocation of Regulatory Responsibility

C. Power Supply Preemption Should Not be Inevitable on a Registered Holding Company System

D. Regulatory Resources are not Sufficient

A. State and Federal Regulators Must Recast Their Relationship to Accommodate Regional Markets

In order to ensure a smooth transition from vertical integration to regional markets, our scheme of State and Federal regulation will have to address at least two key questions.

1. How Can We Avoid Duplicative FERC and State Reviews of the Quality of Competition?

2. What Form of FERC-State Collaboration is Necessary to Produce Transmission Policies Which Realize Regional Efficiencies?

1. How Can We Avoid Duplicative FERC and State Reviews of the Quality of Competition?

Some States combine integrated resource planning with competitive bidding. After hosting (or directing the utility to host) a competition for a generating unit, they pick the contestant whose offering adds the least cost to the systemwide revenue requirement, everything else being equal.

State approval of the results would simultaneously (a) adopt an integrated resource plan, (b) approve a purchase at the price offered by the winning contestant, and (c) make sure the competition was effective.

Because the approved sale was a wholesale sale, jurisdictional problems can arise. FERC is the Agency with exclusive jurisdiction to determine the justness and reasonableness of the seller's price, terms and other conditions. FERC therefore must review the proposal of winning contestant. If this proposal has a market-based price, FERC must determine whether the market was effectively competitive before approving the price as "just and reasonable." But that question—whether the market was fully competitive—would be precisely the same question asked and answered by the State in designing its competition. There is a real potential for differences of opinion between State and Federal views of what constitutes a fully competitive market.

There is no dispute that FERC has exclusive jurisdiction. One way to avoid "second guessing" State decisions is for the States and FERC to agree in advance on the criteria for effective competition which, if existing, will support FERC approval of a winning bid. States then can design their bidding systems accordingly, and contestants can compete without fear of their victories being second-guessed.

2. What Form of FERC-State Collaboration is Necessary to Produce Transmission Policies Which Realize Regional Efficiencies?

In the regional marketplaces envisioned by EPAct, the transmission facilities owned by individual utilities must become components of an integrated, multistate transmission highway system. Efficient competition depends on fair access to the entire integrated grid.

Jurisdiction over the interstate transmission grid is shared among States, and between the States and FERC.[18] If for every new transaction, the jurisdiction rendered worse off can veto the transaction (to another jurisdiction's detriment), then a suboptimal transaction will result each and every time.

"Regional transmission groups," as FERC appears to intend them, would focus on interutility cooperation. Users and owners of the regional transmission system would plan its use jointly, with uniform pricing and access rules. This cooperation would replace existing patchwork of transmission tariffs (where transmission owners have different tariffs created at different times for different purposes, and where some transmission owners have no tariffs at all).

This FERC-created RTG concept, while essential to the success of regional markets, does not substitute for interstate and State-FERC cooperation. FERC's RTG Policy Statement does call for "consultation and coordination" between RTG's and States. This phrase does not itself define how States should work together, and with FERC and the utilities, to ensure efficient use of scare transmission resources.

Interstate and FERC-State cooperation will not "evolve" without some change in present behavior. We must start down that road.

B. Toward a Rational Allocation of Regulatory Responsibility

The previous subsection makes clear that the Federal-State discussion is not about "turf" it is about efficient assignment of regulatory responsibility in a manner that produces the least-cost result for all States. A corollary is that issues which create adversity among States can be left to the States only if the States have a mechanism for resolving conflicts. If they do not, a neutral actor must decide. That

[18] See generally B. Bums and M. Eifert, A Cooperative Approach Toward Resolving Electric Transmission Jurisdictional Disputes (Natl. Reg. Res. Inst. Mar. 1994).

actor cannot be viewed as "anti-State" just because it decides against one State, where its decision necessarily assists other States. At the same time, many aspects of utility regulation do not place States in conflict and can be assigned to individual States acting independently.

A first step is simply to define various regulatory functions and attempt to assign them logically. The scheme below is for illustrative purposes only.

1. Individual State Interests

a. Create and regulate the monopoly utility function

b. Determine the boundaries (geographic and product) of the natural monopoly

c. Determine the standard of behavior expected of holders of the franchises

d. Set all rates charged by the monopoly utility for functions over which the utility has discretion

2. Federal Interests

a. Ensure that economies available from interstate cooperation among utilities are realized

b. Ensure that policies of one State do not affect adversely consumers in another State (both direct harm and reduction of opportunities for efficiencies).

C. Ensure that utilities do not use the monopoly granted by one State to gain unfair competitive advantages in another State.

d. Protect consumers from utility activity which will have interstate consequences

C. Power Supply Preemption Should Not be Inevitable on a Registered Holding Company System

In wholesale electric markets, each utility is usually the sole buyer of electricity for the many captive retail customers within its service territory. It is a truism, therefore, that wholesale competition will work only if these utilities buy only from the most efficient producers. It is up to State commissions to make sure utilities act efficiently.

Substantial uncertainty surrounds the State role in these transactions, as a result of the Nantahala-Mississippi Power & Light decisions. Many utilities argue that these cases apply throughout the industry, not only to transactions among affiliates of registered holding companies. A proper interpretation of these decisions leaves the States with substantial power to protect consumers. The continuing uncertainty, however, serve no useful purpose.

1. The Apparent Reason for Preemption in MP&L: FERC "Ordered" the Buying Utility to Buy from the Selling Utility

Under Nantahala and Mississippi Power & Light, whether State review of a utility's action is preempted by the Federal Power Act appears to depend on whether that utility's action was mandated by FERC. In MP&L, the Court among other things found that (1) FERC had "ordered" the buyer to purchase from its affiliate; yet (2) a State commission set the utility's rates as if the utility had been free to take an entirely different action. As the Mississippi Power & Light Court explained (487 U.S. at 372-73, emphasis added):

Nantahala involved a FERC order fixing the utility's right to acquire low cost power; this case i.e., Mississippi involves a FERC order fixing MP&L's obligation to acquire high cost power—In Nantahala the State court attempted to approve retail rates based on the assumption that Nantahala was entitled to more low cost power than FERC had allocated to it. Here the State court seeks to permit the State to set rates based on an assumption that MP&L is obligated to purchase less Grand Gulf power than FERC has ordered it to purchase.

The Court further stated:

[I]t obviously cannot be unreasonable for MP&L to procure the particular quantity of high-priced Grand Gulf power that FERC has ordered it to pay for. Just as Nantahala had no legal right to obtain any more low-cost TVA power than the amount allocated by FERC, it is equally clear that MP&L may not pay for less Grand Gulf power than the amount allocated by FERC.

Id. at 374 (emphasis added).

2. Preemption is Not Legally Required on Registered Holding Companies

Under Nantahala, MP&L and Pike County, the basis for preemption appears to be a decision by FERC which removes the utility's discretion over whether to enter into the transaction whose prudence is being questioned by the State commission. This statement of the law does not mean that FERC, for all transactions involving registered holding companies, is legally required to remove the utility's discretion.

FERC's "ordering" of purchases on a registered holding company systems should be discretionary with FERC. Whether FERC may decide, as it did in MP&L (but did not in Pike County, to compel a utility subsidiary to purchase from an affiliate should depend on the facts of a particular case. Although the Supreme Court did say that FERC "ordered" MP&L to buy, the Court did not say that FERC was obligated to order MP&L to buy.

In the pre-EPAct era, FERC did come close to making such a statement, although no Court has done so. In AEP Service Corp., 32 F.E.R.C. para. 61,363 at p.61,818 (1985), FERC stated that PUHCA preempted States from questioning the prudence of a utility's "decision" to become a member of a holding company (emphasis added):

[T]he prudence of being a party to the EHV transmission agreement cannot be considered separately from the prudence of being a party to the entire AEP pool relationship. A challenge to the membership in a public utility holding company power pool of a member of the holding company is a Federal matter.

We note that the formation and dissolution of holding companies is regulated by the SEC—Therefore the question of continued membership in a holding company power pool raises in the first instance the question of the composition of the holding company which is within the jurisdiction of the SEC.

The review—by the SEC—preempts State review and determination of the prudence of overall membership and operation of the holding company system.[19]

Under PUHCA, a holding company must operate as an "integrated public-utility system." If FERC or States could treat a utility subsidiary as being free to buy from off-system rather than on-system sources (as in Pike County), the system might not be acting as an "integrated public-utility system." FERC's reasoning appears to be, paraphrased:

Once you have been deemed by the SEC to be an integrated public-utility system, the companies must act like an integrated public-utility system, efficient or not.

This reasoning, ostensibly based on PUHCA, is inconsistent with the Federal Power Act, which demands "just and reasonable" rates and therefore efficient operations. If Arkansas Power & Light (AP&L) buying from a nonaffiliate is more efficient than AP&L buying from an Entergy affiliate, it cannot be consistent with the Federal Power Act for FERC to "order" AP&L to buy from the Entergy affiliate just because the affiliate is an affiliate.

Nothing in the Federal Power Act or PUHCA exempts registered holding companies or their affiliates from the traditional regulatory obligation to operate efficiently. If operating as an integrated public utility system is suboptimal for society (where, for example, more efficiencies are created by having a utility subsidiary buy from a nonaffiliate than from an affiliate), the response should not be to prohibit the transaction in order to preserve the existing "integrated public utility system." Rather, the response should be to further the public interest: by recognizing that the prerequisite for continuing existence of the holding company system—realization of efficiencies for society—no longer exists.

A State's best "defense" against preemptive approval of a capacity sale or purchase on a registered holding company system, therefore, is a showing that there is an alternative action which is less costly for society as a whole (not merely for the particular holding company at issue). PUHCA does not create a permanent right to be a holding company system.

V. TO ensure that competition prevails, regulators must adhere to objective principles

Competition brings numerous interest groups, vying for success. Interest group competition is a healthy part of the legislative process. Regulation, in contrast, should remain indifferent to the success of particular interests. It should apply objective principles rules which allow the efficient to prevail over the inefficient, regardless of who wins and loses.

In several areas of debate, there is a risk that objective principles will become subordinate to interest group analysis. This part recommends treatment for five such situations: "Market Pricing," "Incentives," "Financial Integrity," "Experiments" and "Diversification."

A. "Market Pricing"[20]

The FERC's move to market prices is based on its Commission's confidence that independent wholesale generators will compete, on a sustained, effective basis, with traditional vertically integrated utilities. As discussed in Part I.B above, preliminary evidence says otherwise: a competitive market structure in wholesale generation sales is unlikely to evolve or be sustained.

---

[19] See also W. Nixon and R. Johnston, "Nantahala Affirms Narragansett Whither Pike County?" Energy Law Journal 1, 8-9-(1987).

[20] This section is drawn in large part from D. Penn, "A Critical Look at the Federal Energy Regulatory Commission's Market-Pricing Policies for Wholesale Power" (APPA 1992), and an unpublished article co-authored by the witness and D. Penn.

The Commission has decided several dozen market-based pricing cases.[21] In many situations, the Commission has not linked its approval of market rates with the guarantees of a competitive market structure.

The Commission has approved market pricing when the prospective seller demonstrates absence of market power; i.e., when neither the seller nor any of its affiliates:

1. is a dominant firm in the sale of generation in the relevant market;
2. owns or controls transmission facilities through which the buyer could reach alternative sellers (or, if the seller or any of its affiliates does own such facilities, the ability to block the buyer from reaching other sellers has been mitigated); and
3. can erect or control any other barrier to market entry; or
4. attempts to abuse affiliate relationships.[22]

Focusing narrowly on seller behavior ignores market structure. Absence of seller abuse does not substitute for a competitive process. "Administrative convenience," the only explanation put forward for selecting the "market power" approach over the "competitive process,"[23] does not justify the resulting excess returns.

The theory of market pricing is that the prices will exceed costs, thereby rewarding efficiency and drawing new talent into the industry, which in turn drives costs lower, leads to a new round of rewards and new round of entry, and so on. The theory is sound. But the theory demands that the Commission approve a "market" price only when there is strong evidence of a competitive market structure and competitive market behavior. Put another way, why should we allow seller to charge more than its cost of service? If the price-cost spread results from cost-reducing initiatives that will be emulated and competed away by reactive competitive market forces, then temporary excess profits are worth it.

But opening noncompetitive markets to market pricing will produce permanent excess rates and profits. And that is the Commission's error, as explained below.

In stressing behavior of the seller, the Commission ignores welfare of the buyer. The Commission forgets that market prices could exceed competitive prices for reasons other than seller behavior; e.g., some other utility's market power. The Commission does not condition the extra returns on a competitive process promising lower costs and rates in the future. The Commission does not condition a short term of extra returns on a long term of consumer welfare.

We disagree with this approach. Letting a seller earn extra returns due to entry barriers or other anticompetitive factors, merely because the seller did no wrong, does not advance consumer welfare.

Illustrating the Commission's disregard for the quality of competition is its casual treatment of two key structural issues: the number of sellers and the viability of sellers.

1. Number of Sellers: The extent of a seller's dominance depends in part on the number of alternative sellers. The Commission has insisted that there is no particular number of alternative suppliers which makes a competitive market. Each applicant for market pricing predictably asserts it faces many competitors. The Commission has evaluated these assertions inconsistently. The Commission approved a sale in Cleveland Electric where the buyer had only one alternative to the seller it chose. The Commission rejected the sale in TECO because, in part, the buyer had only two alternative suppliers.[24] Regardless, these numbers are too low to predict competition.

2. Viability of Sellers: A competitive market structure has viable competitors with a reasonable chance of winning the sale. But the Commission has examined the viability of alternative suppliers only rarely. See, e.g., Portland General Exchange, 51 F.E.R.C. para. 61,108. In most other cases, the Commission assumed viability without checking the facts.

B. "Incentives"

The subcommittee asks how "pricing mechanisms [should] be adapted to further carry on Congress' goal, embodied in EPAct, of a more competitive wholesale electricity market facilitated by open access to transmission services?"

---

[21] See, e.g., Tenenbaum and Henderson, "The History of Market Based Pricing," The Electricity Journal (December 1991). The authors are senior staff officials at FERC.

[22] Recently, in KCP&L, the Commission announced it no longer will examine whether an applicant for market-based pricing is a dominant firm in the sale of generation in the relevant market. The Commission has reasoned that there are no entry barriers to new generation; therefore, no one can dominate any generation market. For a critique of this decision, see Part additional discussion of this point, Part I.C above.

[23] Tenenbaum and Henderson, supra at p. 38.

[24] See Cleveland Electric; TECO Power Services Corp., 52 F.E.R.C. para. 61,191; 53 F.E.R.C. para. 61,202.

FERC has made substantial progress toward this goal. There lingers in the transmission pricing debate, however, an emphasis on incentives. The discussion boils down to this fundamental difference:

a. Transmission owners ask: "What terms provide a sufficient 'incentive' to the transmission owner?"

b. Transmission users ask: "What terms ensure nondiscrimination among all users, regardless of ownership?"

Transmission owners want an "incentive" because they see the provision of transmission service as a "business decision." (See, L.L., Comments of the Utility Working Group). The owners lost that battle on Oct. 24, 1992, when EPAct passed. Transmission service is not a "business decision"; it is a legal obligation.

Where a monopoly has a legal obligation, incentive is irrelevant. The appropriate price is the "reasonable" price: the price which protects shareholders from confiscation and protects the customer from exploitation. "Incentive" has nothing to do with it. Adding an "incentive" to a "reasonable" price yields an "unreasonable" price.

C. "Financial Integrity"

There continues to be an ill-defined concern with the "financial integrity" of utilities. Competition must be indifferent to the financial success of the competitors. At the same time, utilities with costs incurred under a pre-competitive era must receive fair compensation for those costs. A utility should not be required by government to make certain investments, and then be told by the government it cannot recover those investments (unless the risk of loss already was reflected in the regulated return on equity).

But even this question is better phrased as "fair compensation for past obligations incurred" than "financial integrity." A utility could receive fair compensation for its past investments, and still suffer a loss of "financial integrity" because it fails in the new competition. Assuming the competition is fair, there should be no cause for alarm. Financial failure is the right result.

D. "Experiments"

Many advocates of retail wheeling and other forms of restructuring call for experiments. Unless designed carefully, an experiment might not prove any usable proposition.

For example, whether retail electric service is a natural monopoly is a question that must be answered by economists studying the economies of scale associated with the various components of retail electric service. A retail wheeling experiment cannot easily answer that question. An experiment can induce utilities to offer discriminatory low rates to industrial customers who qualify for the experiment. Those customers would say the experiment "worked" because it "lowered rates." But if the utility had recovered the fixed costs associated with serving these favored customers from disfavored customers, the experiment would have reproved a proposition known for decades: utilities with market power can charge different rates to different customers depending on their elasticity of demand.

Experiments fail for another reason: they do not test the question of how much the customer will be willing to pay to have the right to return after the experiment. The recent order of the Michigan Public Service Commission illustrates this point well.[25] The order requires the utility, during the 5-year retail wheeling period, to remain ready to take back the departing customer after the 5-year period. The decision states (at 34, emphasis added): [P]articipants should be able to return to full utility service after the program's expiration on the same terms that would be available to any non-participating customer.

"On the same terms" means that the utility has to plan for the return by ensuring that capacity is available. However, the departing customer is not required to pay anything to keep this capacity available.

This approach has one of two possible flaws:

1. If the utility incurs capacity costs during the 5 year period in order to have capacity available at the end of 5 years, and this cost is not recovered from the remaining customers, the government has confiscated shareholder property (by directing that property be maintained but providing no compensation).[26]

2. If the cost of this capacity is recovered from the remaining customers during the 5 years, the result is a cross-subsidy. Not a rhetorical cross-subsidy, but a text-

---

[25] "Opinion and Interim Order Remanding to the Administrative Law Judge for Further Proceedings," Case Nos. U-10143 and U-10176 (Apr. 11, 1994).

[26] This problem will not arise if the utility is able to postpone procuring the capacity necessary to serve the returning customer until the customer actually returns. But given the lumpiness of firm capacity, that "perfect timing" scenario is unlikely.

book cross-subsidy. Some customers will be paying for capacity which is being maintained for the benefit of other customers.

Neither of these results—confiscation or cross-subsidy—is an auspicious beginning for those claiming to want "competition." It is not consistent to argue for competition and simultaneously ask the government to direct a private company to hold capacity available.

The problem gets worse after 5 years. Assume the utility, pursuant to the PSC directive, incurs capacity costs to ensure that capacity is available during the 5 years for the returning customer. At the end of the 5-year period, the PSC decides to make it a permanent program, and the departing customer stays out. Who pays for this capacity which the utility procured pursuant to its legal obligation? The PSC does not require the departing customer to return: the PSC only requires the utility to plan for the return. It is this asymmetry—the utility is obligated to provide universal service to a customer who is permitted to shop the market—which reveals the flaw in retail wheeling.

A customer contemplating retail wheeling, where there is no guarantee that the customer's original utility supplier will accept the customer on return, will hesitate before leaving, or will pay a substantial sum to the utility to ensure future availability. An "experiment" in which the utility must hold capacity available for the returning customer (and any other approach would not be an "experiment") therefore cannot test the price differential which the customer must find in the market in order to make it worthwhile for him to leave.

E. "Diversification"

Utilities faced with "competitive pressures" are seeking to diversify their investments into other industries. The utility industry's experience with diversification has been [H]orrendous in the aggregate and—satisfactory to disastrous for individual utilities.

C. Studness, "Earnings From Utility Diversification Ventures," Public Utility Fortnightly 28-29 (September 1, 1992). Reviewing the results of utilities entry into nonutility businesses in the late 1980's, he found:

In total, the 20 utilities invested about $6.5 billion of common equity in their diversified undertakings and sustained aggregated losses of $387 million during the 6 years from 1986 through 1991. The return on equity on the diversification projects averaged 1.1 percent for the 6 years.

Electricity customers can do without diversification. The record reveals no public interest benefit from it. Utility shareholders are free to diversify by buying stock in independent companies. Utility diversification, i.e., diversifying from a base of operations consisting of a government-granted monopoly, has no recognizable public interest benefits.

The conclusion one draws from these facts is that if we wish utilities to play a significant role in the future of electric service, they should be confined to that service without distraction by other investments.

Taking that route would reduce regulation because it would reduce the need for regulation. Because there is not a consensus for that view yet, however, the need for regulation -will remain. Much has been written on this subject. Appendix B contains a summary of the regulatory principles which should apply to utilities who continue to diversify outside their core local utility businesses.

Mr. SHARP. Thank you very much, Mr. Hempling.

Mr. Popowsky, we are pleased to hear from you now.

STATEMENT OF IRWIN A. POPOWSKY

Mr. POPOWSKY. Thank you, Mr. Chairman, and thank you for allowing me to testify here today.

My name is Irwin Popowsky. I am the Consumer Advocate of Pennsylvania. I am also the secretary of the National Association of State Utility Consumer Advocates, or NASUCA, though I am speaking today on my own behalf.

While the subject of your hearing today is quite broad, I will limit my remarks to a relatively narrow subset of issues that I think are of major concern to retail electric utility ratepayers, particularly the ratepayers who are traditionally considered to be the captive ratepayers of the monopoly utility. Specifically, I will focus on the danger that retail electric competition or even the threat of retail competition may result in nothing more than the shifting of uneconomic utility costs on to those ratepayers who have no competitive alternatives.

First of all, I would note that there is a difference between competition and deregulation. It is relatively easy, I think, to deregulate an industry. It is more difficult, however, to ensure that competitive market forces will be called into play to protect consumers from exploitation. Deregulation in the absence of full and fair competition is the worst of all worlds for consumers.

Second, and of more relevance to your hearing today, I would emphasize that there is a difference between competition and cost shifting. In order for competition in the electric utility industry to serve a valid economic purpose, it must consist of more than the mere shifting of uneconomic costs on to those customers who have no access to competitive alternatives.

We are clearly moving toward a more competitive electric utility industry. To the extent that competition and wholesale bulk power markets and competitive bidding procurement programs can reduce our utilities' costs of obtaining power, that competition should clearly benefit the ratepayers who are served by those utilities. There is little question that the generation of electricity, or, more precisely, the construction and operation of new electric generating units, is no longer a natural monopoly. Nor do I believe should there be much question that utilities and ratepayers can benefit if competitive market forces are brought to bear in the acquisition of new resources at the wholesale level.

The benefits of retail competition among utilities, however, are less clear. When a utility purchases wholesale bulk power at a competitive price, the benefits of that purchase flow to the utility as a whole and to all of its ratepayers. The utility retains the obligation to serve its customers in the most reliable and economical manner, but the utility can turn to a competitive market to best obtain the resources to fulfill that obligation.

With retail wheeling the individual customer seeks a lower cost source of power and then pays the utility a wheeling fee to carry that power from the supplier to the customer. The retail wheeling customer may receive a benefit by obtaining lower cost power but at whose expense? My fear is that it will be at the expense of the remaining ratepayers, particularly residential and small business consumers who may not have access to competitive alternatives but who may be stuck holding the bag of uneconomic costs that are left behind when the retail wheeling customer leaves the system.

Attached to my prepared testimony is a resolution on retail competition which was approved last month by the National Association of State Utility Consumer Advocates, or NASUCA. The resolution concludes that there is no justification to charge remaining captive customers for yet a greater share of uneconomic electric costs that are rendered unrecoverable due to retail competition. The resolution notes that electric utility plant which was added at least in part to serve customers who now have competitive alternatives may no longer be used and useful in either a physical or economic sense and that if a utility is to recover such uneconomic costs at all, such recovery must come from those customers who receive the benefits of competitive alternatives.

I would note that the NASUCA resolution reaches the same conclusion with respect to special industrial discount rates that are designed to keep large customers with competitive alternatives on the utility system. To the extent that such rate discounts simply shift uneconomic costs on to other ratepayers, those rates do not advance the public interest or promote economic efficiency.

I do not believe that retail wheeling is inevitable. I do believe, however, that the threat of retail wheeling is inevitable. My fear is that the response to that threat may be either to provide a retail wheeling option to certain customers or to provide those customers with substantial rate discounts and in either case to pass the costs on to other ratepayers.

If retail competition can be used to encourage or require utilities to lower their costs and to become more efficient, then I, for one, would certainly be willing to examine how to make such competition work for the benefit of all ratepayers. If, on the other hand, retail wheeling is simply used to shift costs from customers with competitive alternatives on to customers who lack such alternatives, then it should be dismissed. In my mind, residential and small business customers are already paying more than a reasonable share of costs for the uneconomic investments, particularly the horrendously expensive nuclear power plant investments of the last decades. We have paid a reasonable share of costs for those investments that have contributed to the great disparity in retail electric rates which in turn has given impetus to the calls for retail wheeling. Captive customers should not be burdened with even more uneconomic costs in the name of retail competition.

The fight over who should pay for the uneconomic investment costs that might be left stranded by retail wheeling should be a fight between the utility which may have to write off some investments that are no longer used and useful in serving remaining ratepayers and those customers who are leaving the system. Arguably, the retail wheeling charge to those departing customers should be sufficient to cover at least some of those stranded costs that were incurred on their behalf.

I would note in this regard that the California Public Utilities Commission in its now famous proposed order of April 20 regarding retail competition essentially commenced its discussion by stating that, "As California confronts the challenges and changes ahead, we will not tolerate cost shifting among customer classes, we will be particularly vigilant in our efforts to prevent costs from being shifted to residential customers."

While the California order has raised a storm of controversy and is under attack from virtually all sides, I believe that at least in setting forth this one principle the California commission got it right. Cost shifting must not be tolerated, and I would urge that this fundamental principle be adopted in any regulatory or legislative determination as this debate goes forward in Congress and throughout the States.

If the only purpose or, more importantly, the only result of retail wheeling is to shift even more utility costs on to captive ratepayers, then retail wheeling is a bad idea and it is a sham that does not promote competitive efficiency.

Again, though, I would submit that if some customers are able to lower their electric bills by forcing their utilities to reduce their costs or perhaps to write off some of the more uneconomical investments, then I would take a serious look at that kind of proposal. But if the game here is to force small customers to eat an even greater share of the uneconomical investments that have previously been shared by all consumers, then that type of competition must be rejected.

While my main concern regarding retail wheeling is the potential for cost shifting, I would like to take this opportunity to raise two other concerns. First, while it is clear that the construction and operation of individual generating units is no longer a natural monopoly function, it is not at all clear that the overall planning for an economic and reliable electric network can be left solely to market forces. Whether it is the utilities, the State public utility commissions, or some regional entities that haven't yet been created, someone must ensure the overall reliability of our electric energy supplies.

Moreover, someone ought to at least think about how our electric demands can be met in the most economically beneficial and least environmentally harmful manner. Perhaps the invisible hand of the market will perform this function for electricity in the same way as it does for other commodities. On the other hand, perhaps the electricity network is unique. It is certainly uniquely important to the American economy and the health and welfare of its citizens. We need to find out how electricity is different from other commodities and how it is similar. As we learned in January 1994, in Pennsylvania and here in the District of Columbia, even a temporary disruption in the electric supply system can have drastic impacts on all aspects of our economic and social welfare.

As we move inevitably toward a more competitive electric power industry, it is imperative that mechanisms remain in place to ensure that the unique economic, social, and environmental attributes of the electricity network are properly considered.

Finally, I would like to comment briefly on the impact of electric utility restructuring on the social objectives that you, Chairman Sharp, have referenced in the letter to the participants in this panel. For better or worse—and I emphasize for better or worse—the historic role of our electric utilities has been much greater than that of a purveyor of kilowatt hours to consumers on a commodity basis.

Leaving aside demand-side management activities which I think should be a part of any utility's efforts to provide least-cost service to its customers, our electric utilities have assumed a number of other roles which are designed to achieve societal objectives which reach beyond the goal of providing electricity to their customers. These roles may range from supporting the local coal industry in States like Pennsylvania to sponsoring research on low-emission vehicles in States like California.

Electric utilities are called on to provide programs for low-income residential customers and economic development programs to attract business to their service territories. I am sure that each of you can think of programs in your district that are sponsored by your local utilities but which are not strictly limited to the utilities' provision of basic electric service.

It seems to me that one of the reasons that we have come to expect electric utilities to serve these additional functions is precisely because they are not competitive. Public utilities may be the only major industries in your respective districts that do not have to face vigorous national and global competition to win or preserve sufficient market share to stay in business.

Again, the California commission's April 20, 1994, order is worth reviewing on this point. The Commission explicitly recognized that electric rates in California have become burdened with the cost of social programs that are not clearly related to the provision of least-cost reliable electric service. But rather than pretend that these costs will no longer exist in a competitive future, the Commission proposed the addition of a special line item to be added to every customer's bill including the bills of retail wheeling customers to cover the costs of these programs in the future.

There are a myriad of objections to this proposal, and it can certainly be argued that electric utilities have no business providing these programs in the first place. Nevertheless, I submit that as we move toward a more competitive electric industry, we must acknowledge that electric utilities do serve a multitude of roles in our society. To the extent we believe those roles are important, we must ensure that someone is in a position, whether it is the utility or someone else, to fulfill those roles in the future.

Thank you again for the opportunity to appear before your subcommittee. I will be happy to answer any questions. Thank you.

[The attachment to the prepared statement of Mr. Popowsky follows:]

NATIONAL ASSOCIATION OF STATE UTILITY CONSUMER ADVOCATES

RESOLUTION

OPPOSING SHIFTING OF UNECONOMIC COSTS TO CAPTIVE CUSTOMERS AS A RESULT OF RETAIL WHEELING AND OTHER FORMS OF RETAIL COMPETITION IN THE ELECTRIC INDUSTRY

WHEREAS, the Federal Energy Policy Act of 1992 increased the potential for competition in wholesale bulk power markets through the creation of exempt wholesale generators and the greater availability of wholesale transmission access; and

WHEREAS, efforts are now being made in a number of States to encourage competition in retail markets through retail wheeling and other measures; and

WHEREAS, some utilities are responding to the potential for competition by implementing special industrial development cogeneration and selfgeneration deferral, "flexible" pricing, and other reduced or incentive rates that are designed to keep large customers with competitive alternatives on the utility system; and

WHERAS, neither retail wheeling nor special industrial rates will promote economic efficiency or advance the public interest if they simply shift uneconomic utility costs to remaining captive customers; and

WHEREAS, regulators already have allocated to captive electric utility consumers, including residential and small business consumers, a substantial share of uneconomic utility plant costs on the ground that such costs were prudently incurred; and

WHEREAS, such captive customers generally have little or no alternative source or supply of electricity at this time other than the monopoly electric utility that serves their territory; and

WHEREAS, retail wheeling and special industrial rates may result in the inability of some utilities to recover previously incurred uneconomic costs from those customers who have competitive alternatives; and

WHEREAS, electric utility plant, which was added at least in part to serve customers who now have competitive alternatives, may no longer be used and useful in either a physical or economic sense; and
WHEREAS, there is no justification to charge remaining captive customers for yet a greater share of uneconomic electric costs that are rendered unrecoverable due to retail competition; and
WHEREAS, if a utility is to recover such uneconomic costs at all, such recovery must come from those customers who receive the benefit of competitive alternatives.
NOW, THEREFORE, BE IT RESOLVED, that the National Association of State Utility Consumer Advocates (NASUCA) urges that all State public utility commissions establish as a fundamental principle that uneconomic utility costs that are rendered unrecoverable as a result of retail wheeling, special industrial rates, or other types of retail competition shall not in any case be shifted to those customers who do not have realistic competitive alternatives; and
BE IT FURTHER RESOLVED, that a copy of this resolution shall be forwarded by the President of NASUCA to the members of the Electricity Committee of the National Association of Regulatory Utility Commissioners (NARUC) and to such other commissioners and public officials as are deemed appropriate; and
BE IT FURTHER RESOLVED, that NASUCA authorizes its Executive Committee to develop specific positions and to take appropriate actions consistent with the terms of this resolution. The Executive Committee shall advise the membership of any proposed action prior to taking such action if possible. In any event, the Executive Committee shall notify the membership of any action taken pursuant to this resolution.

Approved by NASUCA
Submitted by: NASUCA Electricity Committee.

Mr. SHARP. Thank you very much, Mr. Popowsky.

Let me just indicate for our further witnesses and audience that the Chair plans for us to take questions for this panel at this point, then we will take about a 5-minute break unless we have been interrupted by a vote in that process on the House Floor, and then after the second panel we will take a half-hour lunch break and then take the third panel at that point.

With that, let me recognize myself for questions, taking off from Mr. Popowsky's statement that demand side management techniques should be a part of any utility's effort to provide least-cost power to its customers.

Obviously, in PURPA Congress sought to push the States through at least a review of their rate structures, and a portion of what they had to examine were questions about how they could accelerate demand side management. That was the nonmandatory portion of PURPA that gets very little attention given the other QF question.

But a number of States took actions, a number of utilities took additional initiatives, and so we do have in place a plethora of demand side management efforts across the country, and one of the issues that has been raised by the California initiative—the State of California initiative—is that somehow many, most, or all, or a few of those will go by the boards, and I wanted to just quickly run down the panel here and get a sense of your perspective on whether the incentives can be maintained or are there to consider a fairly significant demand side management initiative by a utility system within the context of retail wheeling.

Ms. Tierney, do you want to start? I have to ask you to try to be brief. I know this is sort of the whole enchilada. I mean it is one of many enchiladas.

Ms. TIERNEY. This is the whole enchilada.

No. The tension that is created between a competitive market where you can imagine alternative suppliers would be competing increasingly on short-run costs is posed against a world that we know today in which utilities compare demand side management against a long-run cost of an alternative, a power plant being installed, and decide whether or not it is cheaper to put in place energy efficiency. And a customer looks at its entire price of electricity which includes yesterday's investment as well as today's investment, not just a short-run marginal cost, and the comparison between that long-run approach which brings in cost advantages of demand side management as compared to another world that is more driven on short-run marginal costs. One can imagine energy efficiency would fare more poorly in that world.

Hypothetically, you can imagine bundlers of supply in a competitive market who say, "We have got a power option here and we have got some demand side management that we are going to go put and install in your establishment and create a nice bundled package rate for you." That could happen today, and it is not really happening today.

In a world in which we have shorter-run costs as opposed to today's longer-run bundled costs, it is even harder to imagine—even though it is a nice idea, it is harder to imagine it happening unless we build it in up front.

Mr. SHARP. Mr. Moskowitz.

Mr. MOSKOWITZ. Let me just say, Chairman Sharp, that there are so many different models that you can have to create a competitive environment that the answer to your question really depends on the model, so that is why I really approach it from the other direction.

Mr. SHARP. I mention California only because at the moment that is the one hot on the table.

Mr. MOSKOWITZ. And it is not altogether clear even what the California model is in this respect.

Mr. SHARP. Sure.

Mr. MOSKOWITZ. Let me leave you with this. There is an attachment to my testimony that explains how, for example, demand structure, a competitive market which includes retail wheeling, which includes very vibrantly energy efficiency DSM, it can thrive.

Most models of retail wheeling that are thought of or put forward would absolutely decimate energy efficiency, DSM, as we know it today.

Mr. SHARP. Mr. Anderson.

Mr. ANDERSON. Thank you, Mr. Chairman. I appreciate your question.

Before yesterday I think I would have said that you certainly could have maintained these types of programs with a tax on the wires; you know, it is something that is for residential customers and that kind of thing; but I am even questioning that, given the D.C. Circuit Court's decision yesterday saying there can't be tying contracts which clearly says that generation costs can't be then stuck on the wires. I think that there may be a real legal question as to whether it can done.

But of far more importance to me, I think the question is, is it a good idea? Is it a good idea to continue this thing?

We have brought some charts, and I will ask if we could have our charts for you, and we do have some reduced down hard copy sizes. We believe that implementation of these kinds of programs through an electric utility is really a pretty poor idea. It is very inefficient, and it reduces the purchasing power very substantially.

In the first chart, we looked at some numbers in New York, and, by the way, the numbers are very, very soft. It is real hard to get good numbers on these programs. These are from New York. In the first chart, for every dollar collected from customers supposedly to encourage DSM kinds of programs, 60 cents of it goes to utility incentives and 12 cents of it goes to overhead. That leaves only 28 cents on a dollar to be resulting and supposedly to be recycled and going out to customers. That, in and of itself, isn't very good.

But now the time has moved on, and in the interests of time we will go to the second chart.

Mr. SHARP. The Chair appreciates your speed.

Mr. ANDERSON. Yes. The 28 cents doesn't even really take care of things.

If out of the 28 cents you then negate out what are phantom savings—phantom savings are savings that really didn't occur, that are supposed to have occurred but didn't occur—that gets these numbers down to five cents on the dollar, and then if you take the third chart and it talks about the free riders that could be associated with it—these terms we can get into if you like—it looks to us like you get about two cents on a dollar coming out. That, in our way, is a terrible way to try to provide these things.

We know that if customers are approached directly by energy efficiency providers, if you can bypass the utility and have the dollar from the customer going directly to the people that are providing energy efficiency, it may not be the full amount of money that is spent today but it is going to be a lot more effective than two cents on the dollar.

[The charts referred to follow:]

THE DSM DOLLAR:
What Does It Buy in New York?

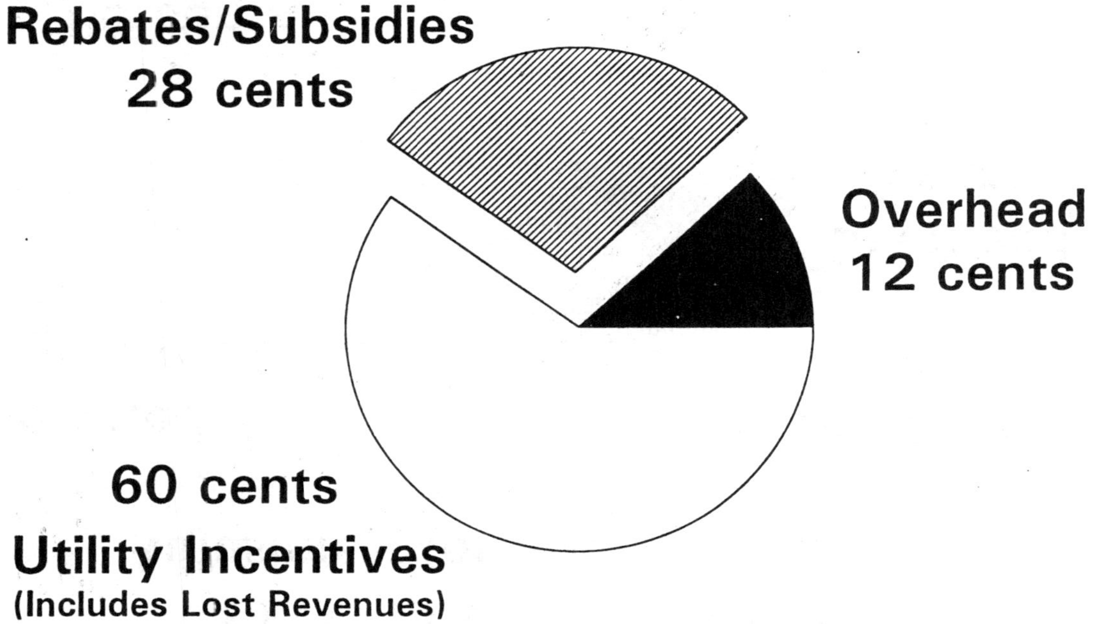

THE DSM DOLLAR:
Adjustment for Measurement Error

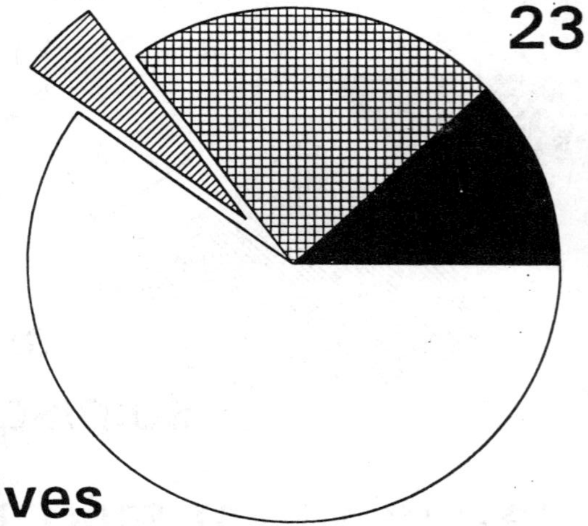

THE DSM DOLLAR:
Adjustment for Free Riders

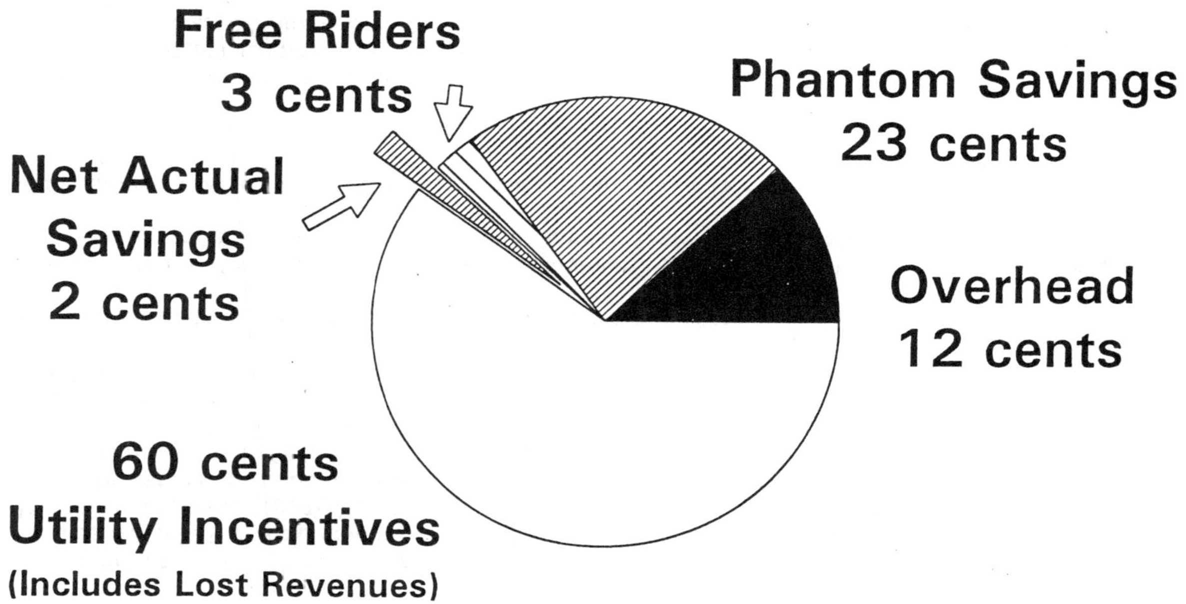

Assumptions:

- The New York Public Service Commission approved 1993 DSM budgets totally $285,000,000. The utilities collected approximately $433,300,000 for shareholder incentives and lost revenues. From Multiple Intervenors (1994), pp. 21-22. Lost revenues for Orange & Rockland and Consolidated Edison had to be estimated based on those companies' claimed energy savings for 1993 and the average rate (revenue per kWh)in 1993. For Con Ed: energy savings, estimated equal to 529,000 MWHs, times their average rate, 16¢, equals lost revenues of $84,640,000. For O&R: energy savings, estimated equal to 90,000 MWHs, times their average rate, 7¢, equals lost revenues of approximately $6 million. These numbers were provided in personal communications with the companies.

- Overhead expenses: 30% of total program costs from Joskow & Marron (1992), p. 60.

- Adjustment for Measurement Error: 50% adjustment of ex ante engineering estimates, plus 33% adjustment for DSM measures' economic lifetimes. From Joskow & Marron (1992), pp. 63 & 66.

- Adjustment for Free Riders: 60%. Rounded average of 29 studies reported in Kreitler (1991).

Sources:

- "Comments of Multiple Intervenors in Response to the January 4, 1994 Notice of Inquiry," Proceeding on Motion of the Commission to Address Competitive Opportunities Available to Customers of Electric and Gas Service and to Develop Criteria for Utility Responses, NYPSC Case N° 93-M-0229, March 8, 1994.

- Paul L. Joskow and Donald B. Marron, "What Does a Negawatt Really Cost? Evidence from Utility Conservation Programs," *The Energy Journal,* Volume 13, N°. 4, 1992.

- Virginia L. Kreitler, "On Customer Choice and Free Ridership in Utility Programs," Proceedings of the 1991 International Energy Program Evaluation Conference, CONF-910807, 1991.

Mr. SHARP. Well, as you said, the figures are soft.

Mr. ANDERSON. We have asked the folks on the other side for better numbers, and we certainly hope that they will give them to us real soon.

Mr. SHARP. I understand. But obviously we have to take your numbers with a little skepticism because they derive from people you get your numbers from.

Mr. ANDERSON. Mr. Chairman, please, we are the ones that are paying these bills.

Mr. SHARP. Well, I can appreciate that you are among the customers paying the bills.

The question I would raise, though, is the argument you make. You represent large users, and at least the rap is that they are wiser and smarter than most of us as customers because they are in competition to get the best deal on whatever they are buying including electricity and so they are going to be very quick to do that.

The issue I would just pose to you, though, is whether, without demand side management programs that try to reach the smaller users who are less capable, less focused, is there a greater need there for a social goal than there would be for a large industrial user who, I have always believed for the last 20 years, is the most likely to take advantage of energy efficiency if they get the chance because they will be sensitive to price? I am not sure it has always worked that way, though.

Mr. ANDERSON. Mr. Chairman, I understand your history over the time.

Let me say first of all that if there were programs that were designed for other customers and not large industrials and the costs were all kept in their customer classes, you would not find my members at least asking me to spend any time trying to oppose them. But the problem obviously is that those costs are not kept in that sort of a class.

Second, I would say that there are some real roles that utilities might take in this area, primarily get price signals right, and there are some atrocious price signals. I mean customers, especially residential customers that get, for instance, levelized bills over a year, they pay exactly the same amount in January and in the middle of the summer as they do in the spring and the fall, have no economic incentive to try to introduce on their own energy efficiency kinds of things. So we would argue very strongly that utilities should try to get price signals correct and that would make all customers, residentials as well as industrials, operate.

Third, information programs make an awful lot of sense. But when you go beyond those kinds of programs, we think at least the central planning kind of an idea just doesn't work. All customers are different, and a central planner can't say what is best for a household relative to another household any more than he can say what is best for one industrial compared to another industrial.

Mr. SHARP. So going back to the initial question, you basically think a lot of the DSM that is in place today would probably go by the boards if we get to retail wheeling.

Mr. ANDERSON. I think, though, that there would be little if any degradation of the environment because of this. What I am saying is, if only two cents on a dollar—or change that to an order of magnitude and make it 20 cents on a dollar, we are not getting anywhere near the kinds of environmental benefits that are claimed from these programs, it is just an ineffective way of doing it.

I think we would reduce substantially subsidization from one set of customers to other sets of customers by them, and you might actually have an improvement in environmental efficiency, because if you get price signals right and have good information, customers might then spend a dollar on a dollar to try to put in programs that really will bring about improvements both in energy efficiency and in environmental benefits.

Mr. SHARP. Mr. Hempling.

Mr. HEMPLING. I agree with the three previous speakers.

Mr. SHARP. That is the great thing about this subject.

Mr. HEMPLING. The question whether DSM programs will go by the boards under a California type regime, those who say yes, I think, make the error of circular reasoning because the question is: What products are going to be subject to competition? If you simply say let's make retail electric service subject to competition on a price basis, then it is a tautology, you have your answer, those who provide electric service at the lowest price will win and anything that is additional will not win. So beginning the question that way which I fear some people in California have is an error.

Again, the question is: What service is a competitive service, what service is a monopoly service, and have competitors provide one and monopolies provide the other.

The other technical question is whether you consider a product like DSM services to be a natural monopoly service. When you ask the question, is it more efficient for society as a whole to determine in advance a level of investment in a DSM product in combination with the other features of electric service and provide that package of services on a monopoly basis, does that produce a lower cost result than separating out these various functions and subjecting them to competition? That is the technical question that I believe is unanswered. In effect, you can read the California proposal four times and you won't see it asked there. That is the error.

At the same time, those who say let's oppose a particular proposal because it will lead to the loss of DSM programs also make the error. They make the error of politicization, of saying we should rely on the regulatory process to make those societal priority decisions about who ought to subsidize whom. I believe that is also an error.

So we have both the technical question at the outset, what is a monopoly service, what is a nonmonopoly service, and a political question at the outset, what group should we have helping other groups, both of which ought to be resolved before we rush into a restructuring proceeding. Then I think the questions would be clearer.

Mr. SHARP. Mr. Popowsky.

Mr. POPOWSKY. Thank you.

Just briefly, to refight yesterday's war which we just finished, some of the nuclear plants in Pennsylvania, if you add in all the costs of those plants and if they actually operate, which some don't, you come out to about 20 cents a kilowatt hour for total costs from these nuclear plants, some of the later ones.

There was a lot of DSM, a lot of fluorescent light bulbs that would have been bought for a lot less than 20 cents a kilowatt hour. If integrated resource planning, if demand side management, had been considered at that time, I think we would have had a more economical system today.

Having said that, my only point is that if the utility is going to continue to do resource planning, that there is no reason why demand side management should not be part of that planning.

I am as anxious as Dr. Anderson to prevent the people I represent from having excessive monopoly profits extracted from them as well. I think it ought to be considered when it is economical and it ought to be priced and adjusted in a reasonable manner to benefit ratepayers.

Mr. SHARP. Thank you.

I have a number of more questions, but I think I should recognize my distinguished ranking member, Mr. Bilirakis from Florida.

Mr. BILIRAKIS. Thank you, Mr. Chairman.

First I would ask unanimous consent, sir, that my opening statement, the opening statement of Mr. Moorhead, one of Mr. Gillmor, one of Mr. Stearns, and one of Mr. Hastert, might be entered into the record.

Mr. SHARP. Without objection.

Mr. BILIRAKIS. Thank you.

[The opening statements of Hon. Michael Bilirakis, Hon. Carlos J. Moorhead, Hon. Paul Gillmor, Hon. Cliff Stearns, and Hon. J. Dennis Hastert follow:]

OPENING STATEMENT OF HON. MICHAEL BILIRAKIS

Mr. Chairman, I want to commend you for calling this timely series of hearings regarding recent market and regulatory developments in the electric utility industry. During these hearings we will probably hear widely differing opinions on how competitive electricity markets will become and what role regulation should play in these developing markets. However, it does seem clear that the electricity industry—the last great regulated industry—has begun to be strongly influenced by rapidly growing competitive pressures at all levels.

The other thing that everyone seems to agree upon is that these changes call for a serious evaluation of Federal energy laws. Some will argue that the Congress should resist, or at least refuse to recognize, the expansion of competition.

However, even if we were to decide that the trend toward increasing competition should be stopped, I am not sure it is within our power to do so. At this point, the best function that the Federal Government can perform is to ensure that our policies, and the mechanisms that are used to achieve those policies, are consistent with the development of fair competition. Some policy tools that worked in a regulated environment may hinder some competitors' ability to do business in a competitive world.

It is not yet clear exactly what the future holds for the electricity industry, nor what role, if any, Congress should play. However, it is not too early to begin to consider these issues, even though it may be only the beginning of a policy debate that will continue into future Congresses.

OPENING STATEMENT OF HON. CARLOS J. MOORHEAD

Mr. Chairman, I would like to thank the chairman for providing this opportunity for the careful examination of the complex issues surrounding recent changes in electricity markets and regulation. I will be very interested to hear how the PUHCA reform and transmission access provisions of the Energy Policy Act are working.

Although the Energy Policy Act only applied to wholesale electricity transactions, competitive forces have reached retail markets. While many predicted that retail competition was inevitable, most have been surprised at the speed at which this appears to be happening. Once again, California is on the vanguard of a new policy trend. Although several States have experimented with retail competition, California's proposal is uniquely comprehensive.

There are difficult transition issues to be addressed in any such dramatic regulatory change. However, the California proposal is an attempt to remove regulatory barriers that restrict customer choice and keep prices artificially high. From comments filed in the California proceeding, it is clear that retail competition involves many Federal issues. Some of these issues will have to be resolved by this committee.

I am sure that this is just the beginning of a long and comprehensive debate regarding the continued evolution of competition in energy markets. As always, once competition is allowed a chance to thrive in one area, it becomes contagious. Our challenge in Congress will be to achieve our policy goals through laws that do not impede the growth of competitive markets. I look forward to working with my colleagues on the committee to that end.

OPENING STATEMENT OF HON. PAUL GILLMOR

I represent a district that has a number of large farms, major industries and a healthy mix of residential areas. In short, my district reflects the entire spectrum of electricity customer.

Certainly, everyone wants reliable and competitively priced electric service. Toward that end, I supported with my colleagues in Congress to pass the National Energy Act of 1992, which provides the framework for wholesale competition among electric utilities. This will allow the benefits of competition to be shared by all electricity users—big and small.

At the same time, most of us in Congress share strong concerns about any measure that would provide competitive advantages to one group of customers while driving up costs for others. For this and other reasons, Congress took the extra step of including a ban on retail wheeling in the national energy act.

My main problem with retail wheeling is this: It has little or nothing to do with actual competition. It would force local utilities to maintain their investments in generation and transmission and would shift costs to small business and residential customers. It would also work against existing efforts across the Nation to promote energy efficiency and environmental compliance.

Contrary to the claims being made by proponents of retail wheeling, Ohio is doing quite well under the existing system. We rank first in the Nation in attracting new manufacturing facilities and expansions. And we're second only to California in the number of manufacturing jobs—more than one million and growing.

In addition, electricity prices in Ohio and throughout the Nation are very competitive with those available in other countries around the world. For example, average industrial rates in Japan—one of our most determined global competitors—are far higher than those available almost anywhere in the United States.

Simple put, I feel that retail wheeling legislation would do more harm than good. We're already on the right track, and the reforms included in the National Energy Policy Act should keep us moving in the right direction for years to come. Thank you.

OPENING STATEMENT OF HON. CLIFF STEARNS

Mr. Chairman, I'd like to commend you on holding this hearing today and also on your decision last month to delay consideration of the Public Utility Regulatory Policies Act (PURPA) reauthorization until after these hearings.

As a number of members brought up at that time, I am concerned that the provisions of PURPA may create unforeseen problems in the competitive markets that Congress has moved to create. While PURPA may have made good sense in a previous, more regulatory era, I am concerned that it may place our larger utilities—investor-owned, rural coops and municipals—in vulnerable financial positions.

As independent power producers who are not subject to PURPA mandatory purchase requirements become larger players in the marketplace, substantial economic pressures may be placed on PURPA-covered utilities. Soon, something will have to give, with either ratepayers paying more or utilities being unable to attract capital competitively.

I hope that as we examine the testimony today of the distinguished group that Chairman Sharp has invited here to testify we can explore this question more fully. Thank you to all of the witnesses who have come here, and I look forward to hearing your testimony.

OPENING STATEMENT OF HON. J. DENNIS HASTERT

Mr. Chairman: I believe that everyone present today would agree with me that the passage of the Energy Policy Act was one of the major accomplishments of the 102d Congress. We all worked diligently in seeing its enactment, and we should be proud of our accomplishment. During my tenure in the Illinois legislature I led the effort which resulted in the adoption of the new Public Utilities Act, reforming the law in my home State to benefit all Illinoisans. The Energy Policy Act of 1992 accomplished that same goal for the Nation. EPAct was passed to reform our electricity laws and to promote more competition in wholesale power markets, to encourage development of new technology, to ensure that fair and full third party access to transmission is made available and that consumers benefit from a more competitive wholesale market.

Competitive forces are alive and well in the generation of electric power. Indeed, in a global marketplace, the price of electricity is a key component of the price of all goods and services. Thus, it is essential that Congress maintain its oversight responsibilities of EPAct. I applaud you, Mr. Chairman for holding this series of hearings so that we can ensure that not only the intent, but also the spirit of EPAct is successfully achieved.

Thank you, Mr. Chairman.

Mr. BILIRAKIS. The minority counsel reminds me that where we have been hearing the term "managed competition" in another—part of another subject these days, it looks like it is working its way into this subject very well too.

When I was in law school, Mr. Chairman, I can't remember the name of the course now, but it was one of those courses where you didn't have an examination basically, you wrote a term paper for it, and my term paper was legalized monopoly, and, boy, that certainly is outdated these days, isn't it?

And I guess that really goes to my question I suppose, and I am wondering, are we on the right path? We see what is happening to telecommunications, the world has completely changed in that regard. Everybody is getting in television and cable and telephones and that sort of thing. We have sort of removed the wraps pretty well. Is that where we are heading as far as electric utilities are concerned, and should we be heading there?

I know, Madam Secretary, well, you also represent consumers; I mean that is certainly part of your job and the rest of you are, I guess, consumer minded, and we all should be. We are all consumers. That is one thing about a subject like this. We all get that electric bill every month at home and sort of mutter and cuss under our breath as we sit down and write that check. So we are all concerned about it all.

But I ask you this not only in terms of rates but also in terms of, you have mentioned things like efficiency, you have mentioned environment, concern about service, concern about other programs in the community. With all this in mind, with all this in the picture, are we on the right track insofar as breaking up legalized monopolies and promoting additional competition and/or managed competition, et cetera? That is the question.

Ms. TIERNEY. That is the enchilada question, actually.

Mr. BILIRAKIS. That is really it.

Ms. TIERNEY. I am going to, of necessity, give you just the start of an answer to your question because that is clearly the jackpot question, and I am going to answer it in two ways.

To harness the benefits of alternative sources of supply, whether in the telecommunications world and the marvelous technological opportunities we have seen there as well as the opportunity to meet different kinds of customer demands, is fantastic, and to the extent that we can tap those kinds of benefits and make them available in the electric market, we are all for that.

I would say that there are some differences that are important in electric supply as opposed to telecommunications, and for me they are clustered in the concept of kind of spillover effects.

When you lay a fiber-optic cable, it is a different thing for local neighbors, it is a different thing with regard to impacts on the environment than when you build a power supply option, when you build a power plant. To make a very long story short, those spillover effects tend to play themselves out in quite vivid local politics, and our system of regulation has been built up over the years, for better or worse, in a way in which we want to enjoy and encourage all of the interstate activity associated with large markets.

But there are inherently some local interest questions in electricity that I think are very different than in other markets that have been traditional monopolies. Those are going to be with us for a long time.

I am oversimplifying a number of very complicated things, but I would say there are some important differences that have weaseled their way into the structure of our regulatory system. We have a very large State presence in electricity that is different than the presence in telecommunications and, frankly, different than that that exists in natural gas.

I will stop there.

Mr. BILIRAKIS. Are you pleased that that larger State presence is there?

Ms. TIERNEY. I am going to say yes, and I look forward to a time in which there are regional—boy, this is going to be very dangerous depending on how I finish this sentence.

Mr. BILIRAKIS. I kind of worry about some of my comments too. Our chairman won't be here next year, but, God willing, I will be, and I know what you mean.

Ms. TIERNEY. Electric markets span State boundaries. Electrons don't care about State lines; markets don't either. Supply options are going to be bought and sold across State boundaries, and that is why Congress has given so much authority to the Federal Energy Regulatory Commission. Of course, electrons pass now between Canada and the United States and Mexico and the United States.

There is, I think, institution building work that needs to be done in the United States to handle those regional issues, which are now crossing international borders as well as clusters of States that represent really natural markets in supply and demand.

So while I do think that there is much of State regulation that is appropriate, and should remain in place—I am a former State public utility commissioner and proud to have been one—we need to figure out innovative ways, institutionally, legally, politically, to deal with the questions that really span beyond the States.

Mr. BILIRAKIS. Thank you.

Mr. ANDERSON. Could I comment?

Mr. BILIRAKIS. Yes, please, by all means.

Mr. ANDERSON. I think your question is a very good one, and obviously we think that we are on the right track. Although it is not going to be a clean and tidy movement from one regime to another, it is a major change.

But to me at least, it should be very obvious that the past track has not been the right one. We have made terrible mistakes over the last decade or two, and the chickens are coming home to roost, so to speak.

The incremental cost of producing power now is about three cents a kilowatt hour. I mean my members can go out, and they are going out, and putting generators down on their plant sites and producing power in the three—maybe even if you don't count them running it 24 hours a day—in the four cent range, and yet the costs that are being charged to customers are 10 cents and 12 cents and 14 cents. It is just orders of magnitude greater than the cost. This is just a mistake.

To me, the mistake in its basic sense has been that the folks that are making the decisions in this industry over the last couple of years haven't really asked the customers what they want. They have made decisions for the customers. They have assumed that they are doing things that are in the best interests of the customers, but they haven't asked the customers: What do you want? Are you willing to pay the 20 cents a kilowatt hour that is going to come out of this nuclear unit? If they had been asked that as the costs rose, those plants would have been canceled a lot sooner and we wouldn't have had the full costs of them in.

In the other areas that you have mentioned, and others, which have been subjected to competition, again, it hasn't been a clean movement, but a couple of things, in my view, have come out. Costs have been reduced for most, not all. Some of my friends that live in wonderful rural areas where I would like to live tell me that airplane costs are much higher going there now than they used to be. But I'll tell you, if you go from hub to hub, prices are way down and the costs are higher going out. So we see again the elimination of subsidies, we see airplanes that are sized to the load a lot better than they used to be.

There are going to be people hurt in any kind of a transition. The same types of things in telecommunication; the subsidies from the long-distance users to the local users are not there any more. It has hurt the local user that is paying more on his monthly bill, but the long-distance user sees the costs that are much more reflective. We think these kinds of things are going to happen.

Finally, I guess I would say the option is not for the continuation of the status quo. There will be change in this industry. My members at least, and if no one else—and I can only speak for them—are going to be—they have been and they are going to continue to make dramatic changes in the way they procure power. If they can't buy it from distant sources, which makes a lot more sense for everybody, then they will have to generate it themselves, and this is going to bring about the kinds of problems that we are all concerned about.

I don't want cost shifting like Sonny Popowsky doesn't want cost shifting, but to me the worst kinds of cost shifts are going to be if industrials go back and self-generate sizable amounts of power; that is where the cost shifting is going to take place.

Finally, I guess I would agree with Scott. The most difficult thing is to try to decide in this industry what portions of it can be competitive and what portions of it cannot be competitive. Encourage competition in those areas where you can have it, and then regulate based on cost those areas where it is going to have to be continued.

Mr. BILIRAKIS. Obviously, my knowledge of the subject is just a small iota as related to yours, and I mean yours in the plural sense. But Mr. Popowsky emphasized community type programs. He, of course, talked about nuclear plants and that sort of thing, but there is so much more that utilities do.

But the point is, we should be taking all that into consideration too, and I guess I am just wondering. I can see where competition—and I am not basically, you know, expressing an opinion as far as all this is concerned, but I can see where competition—and heck, I'm a Republican, we are for competition, free enterprise.

Ms. TIERNEY. So are Democrats.

Mr. BILIRAKIS. In this area maybe.

But you know, I hate to see the rest of it adversely affected. I have mentioned service, I have mentioned the environment, you all have too, but these programs in the community and the part that the utilities play in the community area, would those be adversely affected?

Mr. HEMPLING. Well, Congressman, those are the things that go through utilities' minds too. I can't tell you how many industry structure conversations I have been in where once the talk of competition has died down a bit, some utility says, "But what the about the role of the utility as first citizen? What is going to happen to that?" And, Congressman, that role has no place in a competitive market.

If you are going to ask honestly the question, where should Government be in this industry and where shouldn't it, and you look at the most significant earliest place where Government entered the industry, it was where Government created that exclusive franchise for a utility, which utility now says creates a right to be the first citizen to go to the Kiwanis Club, to take personal responsibility for making charitable contributions that sometimes are made by the ratepayers through their rates.

Mr. BILIRAKIS. Sure.

Mr. HEMPLING. That question of whether a company which has a legitimate monopoly to provide retail service, because that is efficient, whether that should be leveraged into the political gratitude that occurs when somebody says isn't it good to have our local utility doing those volunteer things, that is what is dangerous, and so that when you ask are we on the right track in breaking up these legal monopolies, I am not sure we are on that track. I am not sure the question, which are the monopolies that should be broken up and which are the ones that should be preserved, is being asked and answered very clearly.

So I applaud the question, but I think the answer has to be grounded strictly in principles of competition and efficiency and not on the basis of gratitude.

Mr. POPOWSKY. That is exactly why I raised that issue, Representative.

Mr. BILIRAKIS. You can't pronounce my last thing, you see.

Mr. POPOWSKY. Bilirakis. Yes. Well, my name is Popowsky.

Mr. BILIRAKIS. All right.

Mr. POPOWSKY. I think your first question, are we on the right track, in telecommunications, for example, I think it is greatly to my clients' advantage that they can go to a Radio Shack now and buy a telephone for $19.95 instead of having to pay $4 a month for the rest of their lives to the telephone company. It is to everyone's advantage that the telephone company can now buy switches and telephone equipment from any number of sources rather than always happening to buy it from their own monopoly supplier.

The question, I think, that has been raised by Scott Hempling here and elsewhere is: Are there, nevertheless, some portions of either the electric industry or any industry that are—that do remain natural monopolies and that therefore must remain regulated on a cost basis?

The natural part that most people think is still a natural monopoly in the electric is the wires. To some extent, that is still true in the telecommunications area as well. We really don't have any local exchange competition yet in telecommunications.

So I think that is the fundamental question as to whether any portions of these businesses still remain natural monopolies. If they do, they should be regulated and monopoly profits ought not to be permitted.

Mr. BILIRAKIS. If there is substantial competition existing, should they be as regulated?

Mr. POPOWSKY. Not as regulated, no. Again, that is the difference between competition and deregulation. I don't think you can simply deregulate an industry and assume that competition will flourish.

Mr. BILIRAKIS. Of course.

Mr. POPOWSKY. But I think the next question then is the last question you raise which is: Even if we concluded that competition in every aspect of the electric industry or the telecommunications—let's use telecommunications industry as appropriate—there is still the remaining question, for example, who is going to ensure that universal service is maintained? We have to maintain universal service in the telecommunications industry.

Now again, that doesn't have to be the existing incumbent local exchange. These programs don't have to be provided by the incumbent electric utilities, but I raise the question, and I am glad you raised it. It is just, we need to be sensitive that those roles must be assumed by someone, and I think that is going to be the major debate in both telecommunications and electricity, is to see how those roles develop.

Mr. ANDERSON. Could I just make a brief comment?

I would say that I am very sensitive to the role that you say of a utility. I grew up in a family with my father 40 years with a utility that I won't name, but it is on the west coast of Florida, headquartered in St. Petersburg. I think I know well what you are talking about.

I guess the question that I would simply raise is, is it the most efficient way of providing these kinds of things, and the charts that I showed with DSM I think are the same types of numbers that would show.

I think it is much more efficient if people provide these things directly rather than running it through a regulated utility and try to get the services. It cuts out the middle man, so to speak.

Finally, I agree with Sonny completely that there are better ways of providing these things than through electric utilities. If we need to have assistance in electricity, I think electricity stamps similar to food stamps would be a much more direct way of providing that, clean, direct, you vote on it up and down, you look at it up and down, than to try to do it in the 200 different investor-owned utilities and then all the publicly-owned utilities throughout the United States.

Mr. BILIRAKIS. Thank you. Thank you very much. I know we have a vote.

Mr. SHARP. The Chair would like to request unanimous consent that we include in the record—Mr. Lehman was here and really wished to be here for this hearing, but he is managing the legislation on the House Floor, the California Desert bill right now, for which a vote is underway, so I ask unanimous consent to include his statement and a letter of his to Chairman Fessler of the California PUC into the record at this point.

[The statement of Hon. Richard H. Lehman follows:]

OPENING STATEMENT OF HON. RICHARD H. LEHMAN

Mr. Chairman: I appreciate that the chairman of the subcommittee has scheduled this hearing to help the committee better understand the implications of changes in the electric utility industry following passage of the Energy Policy Act of 1992. I was new to the committee last Congress but my experience on the Banking Committee provided me valuable insights into proposals designed to improve competition. I became actively involved in this committee's debate over changes to the Public Utility Holding Company Act (PUHCA) and transmission access and continue to remain actively interested in these issues.

Less than 2 years have elapsed since passage of the EPAct and this country is already witnessing dramatic changes in the way the electric utility industry functions. In passing the EPAct, the Congress was successful at making long overdue changes to longstanding policy that clearly enunciated a desire to help consumers by allowing independent electricity generators to compete with electric utilities in the wholesale electric market.

In a very deliberate fashion, this subcommittee set into motion a process that allowed independent power producers the opportunity to get its product to market and the Federal Energy Regulatory Commission (FERC) is now on a case-by-case basis determining how to resolve the thorny issues regarding transmission pricing, stranded investment and regional transmission groups. However, due to recent efforts by individual States to push the envelope of competition further, the FERC (and possibly this subcommittee) will possibly have to address additional issues related to authority over aspects of the electrical system.

Most notable among these proposals is the April 1994 proposal by the California Public Utilities Commission to restructure the manner and regulation of the electric utility industry in my home State. This far-reaching document establishes very laudable goals of reducing prices for utility customers and increasing competition in the energy services marketplace. I am, however, concerned that the Commission not jump too quickly into an endeavor which envisions changes in State and possibly Federal law without fully understanding the implications of these changes or contemplating whether the authority to make these changes lies within province of the State or Federal Government.

I am particularly interested in the Commission's proposal to provide electricity customers "direct access" to electricity suppliers. This direct access program would allow customers of existing utilities to shop around for the best price. Traditional customers of existing utilities would be allowed to buy their electricity from other utilities within the State, utilities outside California and presumably even from Canada and Mexico.

My concerns, which I have detailed in a letter to CPUC Chairman Fessler, are that the Commission feels the Congress ceded to them full authority to order retail wheeling. I do not agree with this contention and believe this subcommittee and the FERC will likely be forced to address this issue more forthrightly or expect the courts to resolve this issue in their due haste. The EPAct clearly sought to enhance competition at the wholesale level and did not provide States the clear authority that they now appear to be claiming. In fact, I offered an amendment in this subcommittee to clearly prohibit FERC from ordering retail wheeling through any part of the Federal Power Act or approving sham transactions.

I believe retail wheeling could pose serious problems with respect to demand side management, the energy supply mix, residential consumers and stranded investment. I hope that these hearings help to illuminate these issues and help this subcommittee gain a better understanding of the implications of the California proposal and proposals by other States such as Michigan and Wisconsin. If changes to the EPAct appear necessary, I am prepared to work actively with Chairmen Sharp and Dingell to address these concerns.

I look forward to testimony from members of utility industry, the Federal Energy Regulatory Commission, State regulatory commissions and expert analysts.

Mr. SHARP. I would also indicate our ranking member of the full committee is here, Mr. Moorhead, and Mr. Kreidler.

We could take about 5 more minutes of questions. Let me ask Mr. Kreidler. I would be happy to come back for more with this panel, and Mr. Moorhead, if they have questions for this panel, we can take the break and vote. If not, we have got 5 minutes that certainly we can give Mr. Kreidler who would be next. The gentleman probably wishes to take that time to ask questions. If he wants more, we will bring the panel back.

Mr. KREIDLER. Thank you, Mr. Chairman. Actually, I will pass. I have some interest in questions regarding the Bonneville Power Administration, but I will pass on those with this panel at this time.

Mr. SHARP. OK.

Mr. Moorhead.

Mr. MOORHEAD. I have some questions that I would like to ask.

Mr. SHARP. If you want to ask now, would that work? The gentleman's opening statement has been placed into the record.

Mr. MOORHEAD. Thank you, Mr. Chairman.

I have a question for Secretary Tierney.

What effect would retail competition have on voluntary utility commissions to take certain actions under the global climate change action plan?

Ms. TIERNEY. I hope that you ask the utilities that question. I will try to answer in terms of my thoughts about it, but only they will be able to respond with regard to whether they feel they can meet those commitments in the retail competition or retail wheeling environment.

As you know, I think maybe there are two layers of kinds of effects that you can imagine. Many utility companies are making voluntary commitments with regard to how many emissions reductions they will accomplish over the next couple of years. They plan to do that through a number of different means, such as demand side management programs, renewable resource development, fuel switching. All of those practices are ones that, if their competitors are not making similar kinds of commitments, it will be difficult for utilities to meet. You can imagine utilities feeling they have to pull back on those commitments.

I mean you can imagine that unless there is a more parallel commitment coming forward from competitors in the industry, bundlers of power or independent power producers or international suppliers of power, et cetera, you can imagine that the utilities will feel that it is very hard to meet those commitments.

So we are concerned about that as one of the elements as we look forward to designing a system in which we encourage cost-effective compliance with the Nation's commitments to reduce greenhouse gases.

Mr. MOORHEAD. The Secretary of Energy recently sent a letter to Chairman Sharp stating that DOE will soon release a 5-year plan to provide cost-effective options for the generation of electricity from renewable resources. Will this plan take the existence of increased competition at both the wholesale and retail level into account?

Ms. TIERNEY. I'll make sure it does, but I don't know how we will do that at the moment. As you can guess, it completely complicates the "what if" scenarios, and we are analyzing those kinds of issues.

Mr. MOORHEAD. What, if any, market-based solutions will be recommended?

Ms. TIERNEY. I don't know yet.

Mr. MOORHEAD. Thank you.

Thank you, Mr. Chairman.

Mr. SHARP. Thank you.

Although the Chair would love for us to continue with this panel, we have two other panels today and we will have several panels tomorrow and several panels in the following week. So I do want to thank you for your time and attention to this matter and know that all of you have been intensely engaged in these issues for months, if not years, and will be continuing, and I appreciate that, and I want to thank you all for attending.

We will take about a 10-minute break, and then we will start the second panel.

Thank you very much.

[Brief recess.]

Mr. SHARP. The subcommittee will come to order.

I am very pleased to welcome on our second panel, Mr. Richard Abdoo, the chairman and chief executive officer of Wisconsin Power Company; Mr. James Rogers, Jr., the chairman and chief executive officer of PSI Energy, Incorporated; Mr. Lawrence Hobart, the executive director of the American Public Power Association; and Mr. John B. Howe, chairman of the Energy Policy Committee for the Electric Generation Association.

Gentlemen, you are familiar with our processes, we will put your written testimony in our printed record, and we would be pleased to have your oral summary at this point.

Mr. Abdoo.

STATEMENTS OF RICHARD A. ABDOO, CHAIRMAN, WISCONSIN ENERGY CORP.; JAMES E. ROGERS, JR., CHAIRMAN, PSI RESOURCES, INC.; LARRY HOBART, EXECUTIVE DIRECTOR, AMERICAN PUBLIC POWER ASSOCIATION; AND JOHN B. HOWE, CHAIRMAN, ENERGY POLICY COMMITTEE, ELECTRIC GENERATION ASSOCIATION

Mr. ABDOO. Thank you, Mr. Chairman. It is a real pleasure to be here and share with you some of our ideas from Wisconsin in what we believe to be inevitable in terms of competition in our industry.

I have submitted written comments and testimony, and what I would like to do is just highlight a few points which I think will serve to simplify the issue. I know this issue is complicated, but sometimes the rhetoric makes it far more complicated than it need be.

I believe over a long period of time the regulated electric utility industry has served our Nation well. But we are not looking backwards to bask in a system that is perhaps the most advanced in the free world but to look forward to continue to provide in an economical and efficient manner the lifeblood of the American economy, the electric supply.

All Americans will be affected by whatever Congress does and the local regulators do with respect to this issue. There are a couple of facts, however, that I believe are absolutely essential.

The disintegration of the vertically integrated electric utility system is inevitable. It is not going to remain. We have seen that in so many industries, and we recognize that it simply can't do that. Prices are going to be unbundled, and therein lies the opportunity for efficiency, lower cost, better service, and one major factor that I heard a lot about this morning, and as somebody mentioned, managed competition.

Let me use another overworked term, universal access. I think it is imperative that when you unbundle electricity prices we must provide universal access to all of our customers. That is not negotiable, that is an absolute fact that has to occur, but in doing so I don't believe that we need overly complicate things. We already have, as a result of PURPA on the one hand and Congress's action in 1992 on the other hand, provided incentives for competition on the generation side.

I believe that we should deregulate generation. Let's create an environment where all generators are deregulated. Let customers have universal access, let them pick and choose where they want to buy from. We have part of that today. If you are an IPP or an exempt wholesale generator, you have the ability to be deregulated, but the bulk of the generation in this country continues to be regulated. We should open it up.

We need to do something on the transmission side to assure universal access, and there again we ought to have a common carrier status on the transmission system. Owners shouldn't have rights over and above those of renters. Renters, owners, everybody ought to be able to use the system so that there truly is universal access to all the generation, wherever it may occur.

Now that sounds relatively simple, and I believe it is. However, there is the question of rents: What are we going to pay or charge for universal access on the transmission system? And that is the question that garners much debate, and many will say it is not a question of access, it is price. That may be true, but certainly through the creation of regional transmission grids perhaps within reliability regions we can overcome that and let the participants, all the participants—IPP's, exempt wholesale generators, owners—participate in how they are going to work their little cooperative on the RTG, preserving FERC as a court of last resort, but FERC really need not get into that if the RTG can be formed without any assistance. If we did that, we would have, in essence, deregulation on the generation side and on transmission side with universal access to all.

When we get to the distribution side, I believe there is a role for regulation, and there we don't talk about deregulation, we think in terms of reregulation, so that the local lines that hook all of the customers up to a point where they would have universal access through the transmission to all forms of generation needs to be regulated. But it ought to be regulated on a basis that there is a cap on the price and not so much regulation on the traditional rate base formula but determine what is a reasonable price to charge for those lines perhaps based on historical considerations, and that price would be paid to the owner of those facilities within the price cap whether the customer chose to buy from a thousand miles away or a hundred miles away or somebody else through the system into what—there would be this delivery charge that would be put on there, and that would reflect then the cost of those facilities and be the link that assures this universal access.

In terms of services—and we have heard a lot about conservation and demand side and environmental kinds of things—the services provided should be a deregulated entity, and there will be other competitors. I know in our area, Johnson Controls, for example, is a big competitor of Energy Management Services, and that is wonderful because then customers have choices, they can elect to have the utility provide the service for a fee, they can elect to have some other party provide the service for a fee.

So what you would end up with then is the industrials, the larger customers who pay energy managers to go out and find the least-cost power, can take the risk of saying, "Look, we will buy from wherever it is available, we are going to ship it through this common carrier status, we pay the lineco a fee to get it into our facility, we choose not to buy any conservation services, and that is our bill," and that bill would vary month to month as they decided which supplier, daily perhaps, they are going to buy from.

Our residential customer, the local barber shop or pool room, could decide to do the same thing. They could say, "We have friends in Chicago, for example, who can get us a good deal on power, we are going to buy it from them and ship it up through the system," and their bill, much the way you would in telecommunications, if you want an MCI or an AT&T, you take your pick, and they could choose to buy services or not and pay the lineco to get it there.

On the other hand, they could have the choice and the freedom to say "Look, for my 60 bucks a month it isn't worth the aggravation, I want a contract with the lineco to provide full service, I want the whole kit and caboodle," and the Commission, because the lineco is a regulated entity, has said for full service the price is capped at a nickel or six cents a kilowatt hour, whatever it is, the customer says, "For the next year you just take care of it, I contract with you to provide the full service," and it gets done. But everybody has access to all the generation in the region, transmission, and they are assured, guaranteed by the regulation, that they can get it through the lines to their facility.

We have heard about stranded investment. I prefer to call stranded investment uneconomic assets. I believe that uneconomic assets either exist today or they don't and deregulation will not make an asset economic or uneconomic. The asset is there by virtue of the cost and the product it produces.

Uneconomic assets are a phenomenon that occur in all businesses. Whether it is the big three, the chemical industry, the paper industry, from time to time there are uneconomic assets. It is not a mystery as to how they have been dealt with over the last 30 or 40 years. My view is that regulation protects uneconomic assets, and an immediate benefit out of this process would be to remove that level of protection.

One of the other things that intrigues me about the debate is, many of the people who have argued over the years representing consumers that public service commissions at the local level are coconspirators with the utilities to squeeze them out have now annointed commissions to sainthood in many instances where we must have the commission there to protect us. I fail to understand that. It is an argument which I believe—as I said, on the reregulation, focused on reliability, safety, and assuring universal access is important, setting a price cap, there is a role for the regulator there.

In summary, let me just mention a couple of things that are impediments to this thing that I believe Congress could really help.

You know, one of the things when you have been regulated as heavily as the electric utility industry has been regulated for as many years, nearly 100 years, there is a great reluctance on the part of the regulators to release their embrace, and we get it from the Feds, we get it from the State, and while we love them dearly, the long arm just seems to never end, and it either comes around this shoulder or this shoulder, and when that happens—and I think it is inevitable during the transition period, it will happen—there has to be a recognition that you can't be regulated and deregulated or regulated and reregulated.

Managed competition in the health care for example, when we have seen that occur, what it means is a gravity force pushing bigger and bigger systems. For what purpose? Monopoly status again, I suspect, in a region, but we have to—if we are going to do it, we have to define the role for the regulator, and there is one for them.

Another point, the 1935 Public Utility Holding Company Act was a marvelous piece of legislation in its time. It has long since been outdated. It is an obstacle to open markets and deregulation and reregulation, and with all due respect to Mr. Dingell, I believe it is time that we eliminate PUHCA, repeal it.

PURPA is another piece of legislation which we actively supported as a company, and I personally participated in the rules formulation for that, but it really isn't necessary any longer. We see it used as an excuse. QF's can't get financing unless they get their QF status, and the buyer really doesn't care whether they have got 5 percent steam load or 4 or 7 or 12, it really doesn't make any difference, and I would encourage you to consider repealing it. If you cannot repeal it, certainly modifications are in order.

You know, one issue that has been circulated is the issue of renewable sources, wind, and whether they ought to be capped at

86-340 95 - 4

some size or not. I don't think it really matters, and as far as I am concerned it doesn't matter whether you cap the size or you don't cap the size, and I would argue that these artificial barriers like limiting the size or saying 5 percent here really don't make a whole lot of sense, and besides that, we have got exempt wholesale generators and so we play with words and legalistic kinds of terms that distinguish a little bit here or there, and I think our customers would be far better served if we just got on with it.

Finally, I would say to the committee that competition in a form as I have described will result in better service for our customers, more choices, lower prices. I really don't see a down side in terms of reliability, stability. It ought to be good for everybody, unless you are not efficient and you have high costs, and then I suspect people in those situations will argue for the protectionism.

So in summary, it is a pleasure to be here. I appreciate this opportunity very much. I will take any questions you may have at the appropriate time.

[The prepared statement of Mr. Abdoo follows:]

STATEMENT OF RICHARD A. ABDOO, CHAIRMAN OF THE BOARD AND CHIEF EXECUTIVE OFFICER, WISCONSIN ENERGY CORPORATION

Good morning, and thank you for the opportunity to present my views on the future of the electric utility industry.

My name is Richard Abdoo, and I'm chairman of the board and chief executive officer of Wisconsin Energy Corporation. Wisconsin Energy is the parent company of Wisconsin Electric Power Co., the largest investor-owned electric utility in Wisconsin.

My remarks today will focus primarily on how Wisconsin Electric views the future structure of the industry, future regulatory requirements and role of Federal and State regulators as we make the transition from traditional regulation to open competition. I also will address the appropriateness of incorporating societal and environmental objectives in future energy policy.

Regarding the future of the electric utility industry, I believe the vertical disintegration of the electric utility industry is inevitable and will most likely happen before the turn of the century. These changes will be beneficial to consumers, who, as a result, will have more choices, lower costs and superior service. It is important that utilities and regulators not ignore these business realities. It is not the time to dig our heels in the sand or fight for turf. It is the time to embrace and enable change.

I believe that the generation function of our business will be deregulated and I am comfortable with that scenario. This change may be more disquieting to utilities with uneconomic assets, but I believe the overall benefits for consumers and the majority of producers clearly outweigh the disadvantages for the few.

For some large industrial customers, particularly paper companies, foundries and iron ore mines in our service territory, electricity is perceived as a commodity, just like natural gas.

These companies have experience dealing with different gas producers and suppliers, and some even have the ability to switch to different fuels as seasonal prices dictate. They have tremendous flexibility and control over their natural gas prices. These customers are beginning to expect and demand the same from their electricity supplier. And we want to be ready to give them these options.

But we will not be able to serve them efficiently and cost-effectively without some significant changes that will allow market forces to determine the price of electricity.

Two changes are necessary to enable true competition in the generation market. First, we need to open up the transmission grid so all players in the market have equal access on equal terms. Second, we need to develop some type of dispatching mechanism across entire regions, not just within service territories.

As most of you already know, with the approval of the Federal Energy Regulatory Commission (FERC), each utility that owns transmission facilities sets the price for each transaction to wheel power for others across those facilities. Consequently, long-distance transactions quickly can become cumbersome and cost-prohibitive. I commend FERC for the progress it has made in opening up the transmission system to date and I urge the Commissioners to continue their efforts to give users access to the same terms of service and use that the owners enjoy. FERC also has encouraged the creation of regional transmission groups (RTG) to facilitate open and equitable access to the transmission grid. Wisconsin Electric, along with the 15 utilities and five State regulatory commissions within the Mid-America Interconnected Network (MAIN) reliability council, is developing a proposal for a RTG.

The second need for a truly competitive electric generation market is for the creation of some sort of impartial regional control and dispatch of generation. Today, each utility is responsible for its own dispatching of generation plants, and in recent years, we've seen more brokering among utilities for economic and reliability reasons. In the future, these functions will be expanded across a larger geographic area, resulting in additional long-term contracts and additional day-to-day exchanges based on economics. This will benefit the ultimate consumer by keeping prices as low as possible.

With these two needs—equitable access to transmission and regional dispatch of power—we will find true, wholesale competition in the electric generation market.

On the distribution side of the business, we don't envision deregulation as much as we anticipate re-regulation. Some type of regulation in the form of consumer protection will continue to be appropriate. The distribution activities of today's utilities may be divided in two separate business units: a unit that remains relatively intact as a regulated monopoly to operate and maintain the infrastructure that brings electricity to your home or business. This unit will have an obligation to deliver power from whomever the customer wishes. The second unit will be a deregulated competitive unit and will provide customers with the sales, service and marketing functions. A growing number of companies will likely provide these services as well.

We anticipate that many of today's services will be unbundled and rebundled for our customers. We have more than 900,000 electric customers in the residential and small commercial classes many of whom we expect will prefer the full-service function of today's electric utility. But for those customers who want choices—whether it's real-time pricing of electricity, or whether they want someone else to install, operate and maintain their electrical equipment—our retail unit will be there anticipating and serving their wants and needs.

As stated previously, I expect these changes in generation and distribution will be accomplished by the year 2000.

For true competition in the electric market place to benefit consumers, some change in regulation at both the State and Federal levels will be necessary.

As we make the transition, State regulators need to be catalysts for change rather than roadblocks. The Public Service Commission of Wisconsin has acknowledged that its regulatory role will change and is embracing the change. It initiated a statewide roundtable of regulators, utility personnel, business leaders, consumer representatives and academics to develop a white paper on the future of regulation of the industry within Wisconsin. The commission is also moving ahead with a generic docket to seriously look at the way regulation must change to facilitate the new market. Our commission and a few others are building the regulatory model that the rest will follow.

FERC will play a critical role in the transition as well. Perhaps the most immediate need is for FERC to develop guidelines that set the parameters for future regional transmission companies to establish pricing for transmission service. FERC has pledged to establish those guidelines by this summer, and this would be most helpful for both the RTG's and transmission owning companies. FERC also will need to act expeditiously as the RTG proposals are submitted. We cannot allow these proposals to languish in bureaucratic review while the industry is running a competitive race.

During the transition, regulators will be concerned about the difficulties faced by those utilities with high generation costs. I believe those kinds of uneconomic assets should be treated as they are in any other business—by write downs and amortization of excess costs as determined by the State regulatory commissions. This problem will best be resolved by utilities on their own, on a State-by-State basis. I would not support any kind of Federal mandate to spread that burden across the industry through the use of wire or access fees that do not recognize individual State differences.

There will remain a strong role for State regulation after the industry restructuring occurs. Regulation will concentrate on quality, reliability and other benchmarks of customer service as State regulators will increasingly see their role shift to consumer protection. Price regulation in the form of a price cap for the distribution companies will replace today's cost of service methodology. Regulation in the future truly will refocus on meeting customer needs.

FERC's role after the transition will change. Once RTG's establish open access and fair prices for transmission, FERC's traditional role is diminished. It will need to do less regulating and more enabling of the RTG's. It should remain as a "court of final appeal" for disputes not resolved within the RTG's. It will be a difficult change, human nature being what it is, for commissioners to remain detached and resist the urge to take on a case or get involved in an issue, but it will be imperative for FERC to give deference to the RTG's in order to maintain a truly competitive environment.

Two additional changes are necessary. One is to repeal the Public Utility Holding Company Act of 1935. The Act served a useful purpose in the 1930's to protect against monopoly abuse, but the Act will no longer be relevant if these other regulatory changes take place. The market will promote the discipline that PUHCA currently provides. In fact, PUHCA may serve as an impediment to some of the regulatory changes necessary to facilitate competition. The second change is to repeal the Public Utility Regulatory Policies Act. PURPA may have made sense when it was enacted in 1978 (and we actively supported its passage and the establishment of its rules), but it makes no sense today in a competitive electric generation environment.

Finally, as you know, our company has supported aggressive social and environmental initiatives through energy policy in Wisconsin and at the Federal level. We have one of the best track records in the industry in environmental stewardship and cost-effective demand-side management. We firmly believe these types of social and environmental programs have a place in future energy policy, but the costs need to be absorbed and equalized across all players in the market. These costs should no longer be the sole responsibility of the regulated utility. I would support some type of line charge locally or regionally, or perhaps some sort of access charge mandated federally to pay for these types of programs, as long as it is assessed equally across all competitors.

To conclude, the transition to a competitive environment will not be an easy one. Because electricity is the lifeblood of our economy, everyone in America will be affected by the outcome. As I said earlier, now is not the time for any of us to dig in and resist change. Rather, we must embrace it. Wisconsin Energy intends to be involved in the process. We want to work with you and others to accomplish the appropriate, enabling changes in regulation to facilitate this competitive marketplace. Only then will consumers benefit through lower costs, more choice and better service.

Mr. SHARP. Thank you very much, Mr. Abdoo.

Mr. Rogers, we will be happy to hear from you now.

STATEMENT OF JAMES E. ROGERS, JR.

Mr. ROGERS. Mr. Chairman, members of the subcommittee, my name is Jim Rogers. I am the chairman and CEO of PSI Energy, Indiana's largest electric utility, and I thank the subcommittee for the opportunity to testify today. I really appreciate the opportunity of being on a panel that I am following a witness that I can say, "Me too." I have been in this industry for 6 years, and there have been very few times that I have had an opportunity to be with someone in the industry that shares the same views that I do.

Mr. Chairman, this may be my last opportunity to testify before you. On behalf of the people you have served over the past 20 years, I want to thank you for a job well done in Indiana. Your commitment to the people of Indiana represents the best in our tradition of representative government, and obviously I have not always lived in Indiana and I have been very aware of what you have done these many years in other lives, but it is my belief that you have been a leader in energy and environmental issues, and as a result we have developed policies and laws in this country that will benefit all Americans, not only today but future generations, and for that I want to thank you.

As a Hoosier—at least I think I am a Hoosier; some last year suggested otherwise during the failed hostile takeover—but as a

Hoosier I am proud of your achievements and wish you the best as you start the process next year of reinventing yourself.

Mr. SHARP. Thank you very much.

Mr. ROGERS. You are going to get a good sense of what this industry has been doing the last 5 years in terms of reinventing.

I start out by making one point, and I think this is a point that has been made in all the testimony and made this morning. No one quarrels with the proposition that our industry is undergoing fundamental radical change. The only question is how fast and in what direction it is going.

This change is clearly being driven by the demands of customers, it is being driven by the policies enacted by Congress, PURPA, and the Energy Policy Act, and it is being driven by State and Federal regulatory commissions, and in fact it is being driven by many electric utilities who believe, as our company believes, that injecting competition in a heavily regulated stodgy mature industry like ours is good public policy. That is the state of play.

In generation, we have lost our monopoly on generation as a result of the Energy Policy Act. On transmission, I mean Dick said it best, we have to provide open access. In fact, our company was the first to have an open access tariff. The definition of "open access" is continually being redefined by the Commission. They are moving in the right direction, and certainly the FERC's effort in determining comparability of service is an important step to assuring that all consumers have direct access, equal access to the transmission systems.

With respect to the distribution business, across the country all State commissions are beginning to move from cost of service to performance-based regulation, and even in Indiana we have proposed a performance-based system to our regulators only to find that the consumers in the State are a little reluctant to go in the direction of a performance-based regulation, but I think over time that is the direction that we are going at the State level.

Now let me be a little presumptuous. It is my belief that it is the responsibility of this subcommittee to assure that this industry has the flexibility to evolve toward the most rational, efficient structure possible. I believe that. In this transition, I also believe that we should have one overarching objective, and that is to assure that energy service is provided in the lowest cost, most environmentally benign manner as possible. I think our customers demand it, it is good public policy, and our customers are concerned with energy service, they are not concerned with gas or electricity, they want heating and they want cooling and they want light.

They also want—and I found this—they want an energy-efficient portfolio of services to select from, and certainly DSM has played a key role in the past and is going to play a key role in the future notwithstanding the concerns that many have with respect to the role of DSM in a competitive environment.

It is in light of this overarching objective that I would ask this subcommittee to start the process of reforming the Public Utility Holding Company Act of 1935. It is a serious impediment to the rational evolution of our industry. My testimony addresses in detail these concerns.

Next week I am going to have an opportunity to participate in the SEC's roundtable. My hope is that staff of this committee has an opportunity to participate or at least attend, and as I look at the early drafts of the testimony being submitted there, it is clear that virtually everybody in the industry, to varying degrees, believes that that piece of legislation needs to be reformed. It is outdated, it has many unintended consequences in the new world that we live in, and quite frankly I believe it is an impediment to a competitive market.

Just take a moment. In 1935 we lived in a much different world in the utility business. We believed in the process of electrification of all of American. Today we are at a different point. The emphasis today is on assuring that customers use energy efficiently. Then we were vertically integrated. That is unraveling. I mean we have lost generation, transmission is moving toward regional grids, political boundaries, franchise areas mean nothing with the flow of electrons, and the key in our industry is the ability to think about our business, at least the transmission part, differently.

Back then customers had no choices. That is not true for wholesale customers today. That is not going to be true with respect to retail customers in the future.

With respect to the financial protections in the Act, well, the 1933 and 1934 Acts do a very good job coupled with a very sophisticated financial community in their ability to assess the financial viabilities of utilities today.

The other thing that has changed since 1935 is, the regulators in this country at the State level are far more sophisticated and they are moving from cost of service to more of a performance-based regulation, as I mentioned a moment ago.

For the last 6 years our company has championed the idea of injecting competition into this industry. As I said before, it is good public policy. We started down the road with an open access tariff 4 years ago. We unbundled our generation and transmission and distribution 3 years ago in terms of how we look at our cost and manage our business, and we have the enviable position of having rates lower than 90 percent of the utilities in the United States.

Even in that position, we felt like it was critical 18 months ago to enter into a merger, the primary driver being that we were going to be able to drive our rates down over $1.5 billion, reduce our costs $1.5 billion over the next 10 years, which will translate into reducing our rates roughly 7 percent.

The reason we started down this long road is because we feel like that is the best way for us to create value for our customers, because at the end of the day in our industry customers are going to have choices, and as a utility management we have got to be prepared to give them those choices. We are currently selling power not just in Indiana but all across the country into Canada and the east coast.

We are preparing now a proposal to unilaterally offer retail wheeling to our top 40 customers. We have already offered to our top customers essentially retail wheeling even before the regulators demanded, and the customers are asking for it because our rates are roughly three and a half cents a Kwh, which is very low compared to other utilities. We feel like we ought to give them the ability to choose and select their supplier because doing what the customer wants in a competitive world is the right thing and translates into the right policy.

Let me conclude with one of the questions that you raised and something that I have read a lot about over the last several years and most specifically the last several months.

Somehow people—and I heard a little bit of this on the first panel—believe that by moving to the competitive world, by giving customers choices, by giving customers access to the grid, that somehow utility companies will not aggressively pursue environmental objectives either through the climate change or through DSM programs. I don't believe that is true even though I have heard some people in our industry suggest that much.

My belief is that we are going to continue to pursue our goals environmentally. That is what our customers want so it is good business. But from a demand side management, let me give you one utility's experience. When we proposed that and pushed it forward with our State commission and convinced the consumers in our State to allow us to pursue a demand side management program, they were a little skeptical, but the program has worked and they have become true believers.

But from a utility's perspective and from a shareholder's perspective, the one thing people don't talk, DSM, and DSM gives you an opportunity to build an even better relationship with your customer, and an era when customers have the ability to choose to have a better relationship with them is what you want.

We look at customer satisfaction ratings. In a world where they have choices, customer loyalty is critical, and what we are finding in Indiana is that aggressive demand side management programs demonstrate to customers that you care, and that helps you convert customer satisfaction to customer loyalty. And in a competitive world that is what you want.

So there is a very practical side to demand side management that a lot of people miss out on when they talk about it. My belief is smart utilities in the future will use it as a tool to build relationships as well as to assure that their customers use energy efficiently, and that is one of the important goals of our business.

Thank you very much.

[The prepared statement of Mr. Rogers follows:]

STATEMENT OF JAMES E. ROGERS, CHAIRMAN AND CHIEF EXECUTIVE OFFICER, PSI RESOURCES, INC.

Today, nearly 2 years after the passage of the Energy Policy Act, the electricity industry is in a period of rapid and irreversible change. Utilities who are looking to legislators and regulators to turn back the clock and protect them from competition would do better to prepare to meet the challenge. Customers, and their demands for lower-cost energy supplies, are the primary forces behind this change.

Moreover, policymakers at both the State and Federal level are recognizing that the advent of competition is not only inevitable but beneficial The success of the U.S. economy has been based on a faith in free markets. Regulated monopolies are the exception, not the rule—a response to the peculiar circumstances of a natural monopoly. Regulation in such cases is seen as a second-best alternative, a necessary evil, and the measure of successful regulation is how closely it can approximate the efficiency of a free market.

The recognition that economic and technological forces are undermining natural monopolies presents a great opportunity to harness market discipline to achieve public policy goals. We have already seen the end of natural monopolies in other sectors of the economy, such as transportation, telecommunications, and wellhead production of natural gas, and the EPAct reflects the recognition that the same is true of electric generation. What's more, if competition is possible in markets for local telephone service, it is probably coming to electric distribution, too. And it probably should.

On the one hand, competition in U.S. energy markets is a product of global economic competition. Under pressure to cut costs, American industry is demanding access to low-cost energy, forcing utilities to compete for their business. At the same time, technological innovations—often spurred by competitive opportunities—have made competitive energy markets possible. A key element is the development of a transmission grid that facilitates economic power sales. Advances in "distributed generation," such as on-site fuel cells and photovoltaics, hold similar potential to create competitive opportunities.

Public policy initiatives such as the Public Utility Regulatory Policies Act of 1978 and the Energy Policy Act of 1992 have helped to encourage and shape the growth of competition. A good example of the right approach is embodied in the EPAct. Congress clearly established the goal of a more competitive wholesale market for electricity, then gave the Federal Energy Regulatory Commission the authority to establish policy that would get us there. The FERC is moving forward quickly but deliberately to establish a framework for competition. In my judgment, the Commission is on the right track in its approach to transmission access and pricing, stranded investment, unbundling of services, and comparability of service standards.

The fact is that sound regulatory policy can manage the pace of change, smooth the transition to a more competitive market, and capture the maximum benefits of competition for consumers and the economy as a whole. The current debate is not about choosing between competition and regulation in energy markets, as a matter of public policy. Rather, it is about how to create regulation that is compatible with competition and ensures that competition serves the public good.

One reason that markets are superior to central planning is that they tend to produce solutions that planners could not invent. What that means is that markets are inherently unpredictable, and that fact demands flexibility in the regulatory framework. While regulators must be able to deal with potential abuses, regulation should not preclude unpredictable innovation. What is needed most in this period of transition is the flexibility for this industry to evolve toward the most rational, efficient structure possible.

The most serious impediments to the rational evolution of our industry are presented by the Public Utility Holding Company Act of 1935. I will discuss PUHCA in more detail later in my remarks. One particular problem created by PUHCA—restricting opportunities for U.S. utilities to participate in rapidly growing global energy markets—became apparent during debate on the Energy Policy Act, and the EPAct addressed certain aspects of that problem. But there are many other requirements embodied in PUHCA that no longer serve the law's original purpose and which only serve to prevent companies from doing what would otherwise make the most economic sense.

While I believe that the objectives to which PUHCA was addressed have already been achieved, there are other environmental and social objectives which must be considered as our industry moves toward greater competition. With respect to the former, I firmly believe that a more competitive electricity market will be good for the environment. I take issue with those in our industry who suggest that companies will be forced to abandon environmental stewardship efforts in the face of competition.

Competitive markets are customer-driven, and our customers are demanding environmental responsibility. While it's true that low costs will be the key to winning and keeping customers, utilities will not be able to cut costs by cutting corners on the environment. Instead, we'll have to find the most cost-effective way to meet our environmental responsibilities—because that is what our competitors will be doing.

Similarly, I believe that competition poses no threat to cost-effective demand-side management programs. Providing such energy efficiency services will not only help utilities hold down costs, but will also help utilities differentiate their product and build customer loyalty in a competitive market.

Competitive markets provide both a model and a basis for effective environmental regulation. Competitive markets are not going to replace environmental regulation. Instead, what we're already seeing is the evolution of regulation that harnesses market forces to achieve environmental goals. The creation of sulfur dioxide emission allowances is a perfect example.

The market-based approach to SO_2 reduction provides a model for achieving other environmental goals—a model that gives utilities greater flexibility, but also greater responsibility. In a command-and-control regulatory regime, we could wait until we were told exactly what to do. In the future, competitive advantage will go to companies that pursue innovative environmental approaches.

The market, and market-based regulation, will be ruthless in driving out technologies and companies that cannot compete on both economic and environmental grounds. As particular technologies for power generation or environmental compliance lose out to more cost-effective alternatives, their supporters may seek a return to command-and-control. But the market's rejection of some environmental approaches should be understood as evidence of the success, not the failure, of a market-based approach.

As for other social objectives that have been addressed in utility regulation, I believe that policymakers should begin with the recognition that efficient markets will produce a significant net gain for society. The benefits of efficiency will accrue not only to particular electric customers but to the economy as a whole. With that fact in mind, I would suggest that regulation intended to meet other social objectives be designed, at best, to utilize market forces and, at least, to minimize market distortions.

Focus: The Public Utility Holding Company Act of 1935:

Let me address the remainder of my testimony to the Public Utility Holding Company Act of 1935. PUHCA is an issue of particular importance to the future evolution of a competitive electricity industry, and an issue of particular relevance to this committee.

When PUHCA was enacted in the 1930's, electric utilities were subject to simple definition: a self-contained, vertically integrated provider of electricity to captive customers in a geographically limited area. The physical limitations of the 1930's were literally written into PUHCA, to become a walled fortress against invasion from without and to keep within the culture as it then existed.

In stark contrast, the term that might best be used to describe the future of this industry is "boundaryless." The traditional limits and guideposts no longer apply to an industry that is constantly redefining itself in response to competitive pressure. We are witnessing the erosion of the boundaries defining the industry as we have known it—the industry that was known to the legislators who drafted PUHCA.

The very concept of a "utility" as a vertically integrated natural monopoly is being redefined—if indeed any meaningful definition can be found. Even as electric utilities are unbundling generation, transmission, and distribution, they and others are adding new functions—energy services (such as demand-side management), power marketing, and energy derivatives—that overlap traditional utility activities.

In the future, emerging energy services companies will compete by offering broad energy portfolios—including, for example, electricity from the grid, on-site generation, and natural gas to fuel that electric generation as well as direct applications. A complete energy portfolio will also include energy efficiency services that are as much "information" as they are "energy."

The purpose of PUHCA was to reorganize Rube Goldberg corporate structures based on fictional accounting. PUHCA was both necessary and effective in protecting the interests of, first, utility investors and, second, utility customers. The relevant concern today, however, is that PUHCA has created its own Rube Goldberg regulatory scheme that prevents all stakeholders from realizing the economic benefits of energy industry restructuring.

Consider, for example, PUHCA restrictions on diversification by utility holding companies, and the requirement that registered holding companies maintain a single integrated public-utility system.

Even before major changes began to reshape the industry, PUHCA's provisions on integration and diversification had irrational effects. For example, an exempt holding company can invest in a cattle ranch in a distant State without consequence, but not a transmission line, or a distribution system. An exempt holding company can participate in the growing telecommunications business, but a registered company cannot—despite the potential for economies in utilizing electric utility infrastructure and expertise.

As the industry changes, PUHCA's unintended consequences could be even more counter-productive. Caps on diversification may prove irrational in the future, as unbundling rearranges corporate structures. If, for example, the definition of a utility no longer included generation, that change would not only reduce the utility base for a diversification cap, but would also redefine generation as a form of diversification subject to that cap. In this evolving environment, caps on utility and functionally related investments, and distinctions between registered and exempt holding companies with respect to diversification, will be increasingly difficult to justify on public policy grounds.

In the same way, integration—as required by PUHCA and interpreted by the SEC—has lost its significance for public policy. Integration must be defined by the technology employed to achieve it; PUHCA, however, froze the definition in terms of 1935-vintage technology, notwithstanding significant advances since that time.

As this process continues, restrictive regulation—requiring "integration" defined in terms of 60 year-old technology—will serve only to limit market forces driving toward efficiency and structural rationalization. In 1935, policymakers rightly questioned far-flung operations that appeared to bear "no relation to economy of management and operation or to the integration and coordination of related operating properties." But today, several factors have changed the calculus.

First, the ability to transmit power over long distances and the development of electric grids have increased the potential for achieving integration—in fact, if not in law—of more widely dispersed operations. Second, improvements in communications and transportation have made it possible to achieve economies from coordination of distant operations that are not necessarily integrated.

Third, today's markets are better able to determine what corporate structures are most efficient. As increased competition in energy markets forces energy providers to structure their organizations as efficiently as possible, irrational corporate structures will not withstand the scrutiny of more sophisticated and effectively regulated financial markets. Finally, these corporate structures are also subject to the scrutiny of State regulation that is much more extensive than it was when PUHCA was enacted.

The "integration" requirement is not only anachronistic, it is inimical to the common interest of consumers and investors in more efficient corporate structures. This is particularly true of any restriction on ownership of combination gas and electric utilities by registered holding companies. The fact is that customers want lighting, heating and cooling, not natural gas or electricity. Natural gas has been deregulated at the wellhead, and electric generation also has been effectively deregulated. Given the discipline of the market over production and effective State regulatory oversight, it makes sense to bring together delivery of energy services in one company that can provide customers the optimal combination of energy-efficient options.

PUHCA is no longer needed to protect investors. The SEC has long since completed the industry reform embodied in that legislation. Its enforcement of the '33 Act and the '34 Act, as they apply to all public corporations, along with the work of competent investment analysts, will adequately protect investors from potential corporate abuses.

As utilities emerge from protection into competitive markets, however, what utility investors do need are opportunities for growth. The danger is that utility investors will be caught in the worst of all possible worlds, facing the pressure on margins that is inherent in competitive markets but, unlike other competitive firms, restricted in their ability to develop new earnings opportunities. It is crucial that the Federal regulatory framework remove barriers to such opportunities and provide a level playing field for all companies to enter energy related businesses, both here and abroad.

As for consumers, the advent of competition will go a long way toward protecting their interests. Because abusive practices such as self-dealing and cross-subsidization ultimately must be reflected in higher prices, such practices cannot be sustained in a competitive market. Where utilities continue to exercise monopoly power, consumers will continue to be protected by State and Federal regulation, which has grown in scope and sophistication since PUHCA was enacted.

Yet, as we consider changes in the regulatory sphere, we face a higher standard. Today, it is not enough to protect consumers from abuses. Regulation must be designed to allow consumers to benefit as much as possible from the workings of efficient markets and technological innovation.

If PUHCA were merely irrelevant, there would be no urgent need of reform. What is clear, however, is that PUHCA is a significant impediment to an industry in the process of redefining itself. Incremental reform is not the answer; PUHCA must be replaced with a new regulatory approach that works in concert with the market forces that are reshaping the industry.

Mr. SHARP. Thank you, Mr. Rogers.

Mr. Hobart, we will be happy to hear from you.

STATEMENT OF LARRY HOBART

Mr. HOBART. Thank you, Mr. Chairman.

I want to echo the eloquent statement of Jim Rogers relative to your service on this committee and in the Congress, Mr. Chairman.

I extend the scope of your venue to the United States, not just Indiana.

Mr. SHARP. Thank you.

Mr. HOBART. I would like to give my answers to the three questions that you raised in your letter of invitation to testify at this committee.

The first question you posed was: To what degree do you believe the electric utility industry has entered a period of irreversible change, and how widespread do you feel change will be?

I think the real test of reversibility is results. If consumers find that power costs are lowered and service levels are raised, the present process will likely continue, but if bills go up and service deteriorates, then consumers are going to tell policymakers to try again.

Great Britain is sometimes cited as a guide for changes in the structure of the U.S. electric utility industry. However, a recent report by Morgan Stanley shows that the weekly price averages for baseload power have increased every year since 1990, the year the transition to an electricity market economy took place. A 1992 Oxford University study concluded, and I quote, "Whatever the merits of such a policy, it cannot reasonably be claimed, as so often is claimed in political debate, that privatization was an exercise whose principal purpose was promotion of the interests of electricity consumers. Indeed, thus far consumers have fared poorly from the venture."

That Oxford study showed that residential electricity prices stood at approximately 25 percent above the level that would have been expected on the basis of pro-privatization trends. Nevertheless, as the previous two witnesses have testified, there are widespread changes under way in the U.S. electricity economy.

Electric utilities of all ownerships currently face the challenge of retail competition. There is increased scrutiny of rates and service by communities that are concerned about family budgets, time demands, and local jobs. We have seen over the past two decades a stagnation in real wages of many Americans, and that fact has spread now to the middle class, and people look at their electric bills and they question whether they are paying a fair price, and if electric utilities, regardless of ownership, fail to pass that test of service and price, it can mean not only the loss of industrial customers such as those represented by John Anderson but the inability to attract new business development, or it can actually mean the transfer of the franchise to another system.

As we look at the structure of the electric utility industry and the kind of political and policy formulations that we have built up over the past, I think we need to think clearly about the interests of consumers and the results of changes that we might make. Jim Rogers talks about his belief in the desirability of changes in the Public Utility Holding Company Act, but I think it is important to remember that that act came into being because there were gross abuses that existed in the 1930's and were discovered and documented in the 1920's and resulted in the legislation that we live with in the form of a statute today.

There are certainly differences in the environment in which we operate currently as compared to that bygone period, but I think there is still a need to examine the question of protecting consumers against complicated industrial structures which may be designed and actually achieve hiding of costs in a fashion that is detrimental to the interests of consumers.

Your second question was: Do you have any recommendations as to how regulators might handle the transition period to a more competitive market?

The Federal Energy Regulatory Commission has given official endorsement to the concept that competition exists in the generation of electricity and that markets can effectively regulate the price of wholesale power from future power production units. At the same time, the FERC has repeatedly restated the point that transmission is a natural monopoly and must continue to be scrutinized to ensure it is not used for anti-competitive purposes.

In unanimous decisions, it has ordered network access and required prices that are comparable to what the transmitting utility charges itself. It has urged the creation of regional transmission groups to effectuate this policy on a voluntary basis.

I think it is gratifying to those who supported the 1992 and certainly it must also be gratifying to this subcommittee that the FERC positions are actually getting results. Small wholesale requirements, municipal utilities in States like New Jersey, Maine, Pennsylvania, Illinois, and elsewhere have used the new transmission provisions of the Federal Power Act to bargain with suppliers, and they have received bulk power bill reductions in some cases reaching over 40 percent, a very dramatic change. Regional transmission groups are also advancing in the West, the Southwest, and the Midwest, and again I think this is a direct result of not only the 1992 law but the fashion in which FERC has implemented it.

The key to creation of a continuing competitive market is to move forward in the same direction, and it appears that stranded investment is the next issue up, and there are questions here that are different at wholesale and at retail.

Among the investor-owned utilities regulated by FERC wholesale transactions on average represent 10.5 percent of revenues and 19 percent of kilowatt hour sales. The rest of the revenues and the sales, a major portion, are realized from retail transactions.

The investor-owned electric utilities have far more protection in dealing with wholesale customers than they do with retail. Unlike the usual retail situation, most wholesale transactions involve contractual arrangements where two parties, a seller and a buyer, agree. You have all the usual elements present—competent parties, proper subject matter, legal consideration, and mutuality of agreement and obligations—and these contracts have duration and notice provisions. They employ utility terms and conditions. They do not contain surprises, mysteries, or secrets because they are negotiated by individuals who were knowledgeable, interested, and involved, and it would be ludicrous to label as stranded investment the administration of these contracts by their very provisions and to demand that one party pay a fee or a penalty because the other party didn't like that to which it had already agreed.

This same logic about contracts neutralizing the so-called stranded investment for existing municipal wholesale customers applies to franchise agreements which expire for communities which are considering setting up their own municipal electric system.

Yesterday the U.S. Court of Appeals for the District of Columbia Circuit said that where contracts do not govern, a requirement to pay stranded investment for transmission services is illegal because it is a tie-in sale that violates the Sherman Antitrust Act. The court remanded the decision in the energy case for reconsideration consistent with its opinion.

This decision raises serious questions about the viability of FERC's outstanding notice of proposed rulemaking on stranded investment and the California Public Utility Commission's retail wheeling rulemaking. It would seem appropriate for both commissions to withdraw and rethink these proposals before forcing parties through a lawyer, labor-intensive paper shuffling process. In view of the fact that the court observed that the emperor has no clothes, it seems a suitable time to look for a new tailor.

The third question that you posed, Mr. Chairman, was: How can social and environmental objectives be adapted to a restructured industry in which an environment in which services once provided by a single vertically integrated utility are increasingly unbundled? To what degree should these objectives be retained, changed or abandoned?

Public power systems are committed to the process of integrated resource planning. But it is incorrect to believe that IRP processes must be protected from competition in bulk power supply. Utility planners cannot ignore their customers and market forces to plow through a deterministic evaluation process that uses theoretical ideals to model the optimum energy service machine.

The country is already engaged in real retail competition that tests rates and service. It is likely that customers will demand rates that are comparable to regional and national market rates or they will seek to bypass or relocate. Power supply costs will be a crucial factor. But the fact is that demand side management can help lower customer costs and simultaneously improve service, and Jim Rogers is absolutely correct that this is an important consumer function as far as all utilities are concerned.

Public power systems seek to employ demand side solutions where they are cost effective. A recent study by Oak Ridge National Laboratory reported that for the third year in a row public power utilities spent less on demand side management programs than investor-owned electric utilities yet saved more energy and cut more peak demand, and I have included figures in my statement to document that study.

In public power communities, decisions may be made to carry out plans for the use of low-income assistance, energy conservation, or renewable resources at an extra cost in order to satisfy social objectives which are arrived at by the community itself, and a threat to carrying out those kinds of activities is posed by retail wheeling.

It would appear, for instance, that California's version of retail wheeling would allow a few large direct access customers to undercut conclusions arrived at by the majority of electricity users. For instance, agreements to pay a little extra for green resources could be canceled by a loss of revenues to pay for them.

Retail wheeling is fraught with difficult workability and equity problems, and a number of those problems have been testified to before this subcommittee this morning. It seems that a more realistic way to approach the application of competition in the electric utility industry is to make wholesale wheeling work. If transmission access opens markets further, we can expect to see an homogenization of bulk power supply prices as bidding drives down charges to a common level.

Bulk power supply is the largest part of the cost of electricity at retail, roughly three-quarters of the total. The benefits of competition could be realized by dealing with 3,200 electric utilities in the United States rather than 500,000 industrial customers, 12 million commercial customers, and 100 million residential customers.

Thank you, Mr. Chairman.

[The prepared statement of Mr. Hobart and responses to subcommittee questions follow:]

STATEMENT OF LARRY HOBART, EXECUTIVE DIRECTOR, AMERICAN PUBLIC POWER ASSOCIATION

The American Public Power Association appreciates the opportunity to participate in today's hearings on changes in the electricity industry since the Energy Policy Act of 1992. APPA is the national trade association representing the Nation's approximately 2,000 public power entities, mostly individual municipal electric utilities serving communities from the size of Los Angeles to the more numerous small systems throughout the country,

Since the beginning of the industry, public power utilities have been at the very center of competition-street by street, competition to locate, franchise and yardstick competition. As electricity buyers and often captive distribution-only utilities, they have dealt with limited generation options and even more limited access to transmission. They represent consumers' interests as they are frequently buyers at wholesale and are owned and managed by their customers.

Accordingly, over the last 25 years APPA and its members have been involved with events that finally led to EPAct92, its key Title 7 provisions dealing with competition and transmission access, and their implementation. Related public power involvement includes:

—Development and use of the NRC license conditions;

—The 1973 Otter Tail Supreme Court case;

—The formation of joint action agencies;

—Wholesale and merger case interventions at FERC, including the breakthrough conditions of the PaciftCorp merger case;

—Legislative efforts regarding transmission access language both in PURPA and EPAct92;

—A central role in the follow-on rulemakings and cases that FERC is using to implement EPAct92-use of new section 211 authority, network access, comparability;[1] and

—Participation in the negotiations of regional transmission groups (RTG's) from the pre-EPAct92 "consensus" proposal to the ones now being filed or readied for filing at FERC.

APPA's members have been at the center of the industry's drive toward more competition and fairness to customers. They care where the industry is going and how it gets there. And, in fact, their very survival and the well-being of their customers depends on the proper outcomes. Because of this, I want to concentrate today on what it will take in our view to arrive at a world of effective electricity wholesale market competition. As such, I will concentrate on the question in Chairman Sharp's June 30, 1994, letter of invitation, to wit, "do you have any recommendations as to how regulators should handle the transition period to a more competitive

---

[1] See especially Docket TX 93-2 for the first successful section 211 case mandating AEP to provide transmission access to Blue Ridge Power Agency, Docket TX 934 for the FERC mandate to Florida Power and Light to provide "network" access to the Florida Municipal Power Agency, and Docket ER 93-540-001 for the Commission's clarification that AEP and others will have to provide "access on the same or comparable basis and under the same or comparable terms and conditions as the transmission provider's uses of the system."

market?" At the end of my remarks I will return briefly to answer the other two questions posed in that letter.

In a succinct statement (June 29, 1994) introducing FERC's stranded cost NOPR, Chair Moler stated the goal: "This Commission is committed to developing a competitive, open transmission access, wholesale bulk power market. Our goal is to facilitate the development of competitively priced generation supply options, and to ensure that wholesale purchasers of electric energy can reach alternative power suppliers and vice versa. This vision of the industry is one we all share."

Embedded within the stated goal are several important truths. First, the focus must be on generation competition. That is where the big bucks are in terms of investment and potential consumer savings. Effective wholesale competition can most likely preclude the need for retail wheeling initiatives. Second, open transmission access is a necessary condition for the attainment of generation competition, the very reason it had to be included in title 7 in the first place. Transmission remains a natural monopoly and FERC must regulate aggressively if it is to reach its overall goals. Third, both buyers and sellers are necessary. Wholesale purchasers or buyers of electric energy must be able to reach alternative suppliers, as well as vice versa. And, fourth, for competition to be sustaining, market power throughout the market must be absent.

This vision or goal—a competitive, open-transmission-access, wholesale bulk power market—is the goal of the Title 7 provisions of the 1992 Energy Policy Act. FERC is implementing the Act's provisions and generally doing a very good job of following your legislative vision. However, at this point several things can be said about the transition period toward more competition. They are: we are not there yet; each new related FERC decision replaces the last one as "the most important;" and there are several objectives that must be met before wholesale electricity competition is a reality.

We are not there yet. As rapid as events seem to be occurring to many, we are nowhere near arriving at a point of effective wholesale competition in this country. Despite the FERC activity level and media coverage of these cases, EPAct92 is still relatively new and there have been only a handful of cases so far mandating transmission access in a handful of places in the country. Even these are not necessarily fully litigated and resolved. Also, despite FERC's clear signals, many transmission providers have not reopened their restrictive transmission tariff terms currently on the books. Other owners persist in seeking to offer non-network or noncomparable service. Other situations are nonmaliciously just caught in the pipeline, and still others are in what is probably the most common situation at this time: with all the change and uncertainty, negotiators are hedging their bets and stalling until they get a clearer picture of the final regulatory outcomes.

FERC's recently released NOPR on stranded investment is a good case in point. While the Commission is moving as expeditiously as it can, it has talked with industry sources and struggled with this release for months. Its release now promises the potential for some stranded cost payoff to inefficient generators. This is bound to slow negotiations and delay decisions on write-off or retirement of assets.

Each new FERC decision is the most important. FERC's new power flow information reporting rules implementing section 213(b) requirements on transmission capacity and constraints were revolutionary. Mandatory network as opposed to only point-to-point access stunned the transmission owners last fall. Clarifying utilities must provide access and terms and conditions of service on the "same or comparable basis as they do for themselves reversed decades-old definitions of undue discrimination." It will be transmission pricing this summer, stranded cost compensation and RTG's this fall, and who knows what next. The point is: each new decision is a foundation block just as important to the final structure of competition, and its continuing stability, as the last. Put in another, more worrisome way, as one of the FERC Commissioners has characterized it, if we blow the next one [on stranded investment or whichever], we've blown all of the good work up to now. This could not be more true because of the many conditions that must be satisfied for competition to prevail.

There are three important general requirements that must yet be met before wholesale bulk power markets can be declared competitive in any meaningful way. This is true despite, and in contrast with, FERC's pronouncement in the recent Kansas City Power and Light case (Docket ER 94-1045) that competition in generation has arrived.[2] The three requirements are:

1. Nondiscriminatory open transmission access available to all electric utilities and other sellers of power and energy at wholesale that eliminates all advantages

---

[2] Directly and persuasively on this point, see "Petition for Rehearing of the Missouri Public Service Commission and Kansas Corporation Commission," June 13, 1994 (Docket ER 94-1045).

for transmission owners in-bulk power markets derived from their ownership and control of transmission;

2. The development of all-inclusive regional bulk power markets that yield efficient utilization of regional resources to meet regional load through pooling, market mechanisms, or both; and

3. The nondiscriminatory availability of control area and dispatch services, interchange agreements, and all necessary backup arrangements within any of those regional markets.

Our recommendation is for regulators to accept nothing less than fully meeting these three requirements for competition, and for them to make their individual decisions with these objectives in mind as they proceed through the transition.

As discussed above, we are not there yet. Nondiscriminatory transmission access is just beginning to spread, as is the development of regional bulk power markets. Even where regional markets and/or pools are taking shape, exclusion or choosing not to participate is not uncommon. The subject of nondiscriminatory availability of control area services still remains on the horizon and is seen by many as the last bunker for those who would just say "no" to the forces of competition and change.

Among many other recent positive developments, we would point to Southern California Edison's idea that transmission-owning utilities throughout a western region of 36 individual control areas give control over their transmission facilities to a centrally operated independent pooling agent.See company Response, June 8, 1994, to CPUC Order (R 94-04-031 and I 94-04-032).) This would need to be implemented so it was truly all inclusive and independent and did not become just another transmission owner's club.

Also, we would point to a June 30 advisory committee Report recently approved by the Management Committee of the MAPP reliability council region. The Report recommends a complete reorganization of MAPP to open up membership in accordance with NERC's Report 2000 recommendations and to become an RTG. MAPP would include a reliability council, a generation reserve-sharing pool, a power and energy market with market-based pricing, and an RTG. The RTG would deal with pricing and compensation and may treat the MAPP transmission grid as one system.

Whether or not we finally arrive at a new world of wholesale electricity competition is a function of how the regulatory agencies proceed through the transition. It almost goes without saying that the regulators must be clear and decisive. They must consider administrative and regulatory costs, and adopt the most simple constructs and rules possible to get the job done. They must be aggressive in eliminating transmission access restrictions, other barriers to entry, and continuing market power.

They must not erect new barriers, or invite litigation delays and market uncertainty. This is precisely the risk with seeking a politic compromise on the issue of stranded investment.

The concept of paying for stranded investment is inconsistent with a competitive market. As quoted in the April, 1994, Electricity journal from the new book Electric Utilities Moving Into the 21st Century, private utility Wisconsin Power and Light Chairman and CEO Erroll Davis says:

"In a competitive environment, utilities cannot expect to be held harmless for stranded investment, for example. . . . Private industry routinely builds plant and capacity that is at risk. If a company's product isn't purchased, the capacity goes unused or they convert it to another use. That is the business risk all industries face."

Stranded investment payments have anticompetitive effects in the marketplace and will prevent arrival at the desired new destination of competition:

—by erecting a barrier to entry for alternative suppliers and trades;

—by discriminatorily favoring and shielding certain individual competitors;

—by artificially giving an entrenched competitor a paid-off asset with which to compete with rivals;

by distorting transmission prices if charges are placed here.

Stranded investment payments are antienvironment. They hold in place existing, less environmentally sensitive generation rather than let retirement help convert the generating stock.

Stranded investment is not a surprise.

—In 1988 Electric Utility Week carried 231 articles on competition, IPP's, and mergers and restructuring. There have been between 100-200 articles each year since.

—The February 25, 1994, EEI newsletter reports on its President Thomas Kuhn going to New York at that time to tell Wall Street that it as a "myth" that "increased competition caught the industry by surprise."

—PG&E, for example, says it has been preparing for 9 years (see New York Times, May 11, 1994) and its senior vice president Robert Glynn says "we are not at all surprised . . . we have been preparing for the day" (see Sacramento Bee, May 16, 1994).

—Stranded investment is not a new or unforeseeable (therefore not a risk for which investors have not been compensated) phenomenon. Plant closings, relocations, switches to self generation and prudency review reversals by commissions are common. Court decisions like Otter Tail (1973) are instructive.

Stranded investment is a generation, not a transmission, concept. In almost all cases transmission is not stranded as a result of supplier changes, and transmission prices should not bear any such charges.

The potential problem has been vastly overblown numerically. In fact, IOU total generation assets, net depreciated, equal about $190 billion. U.S. reserves (capacity above peak) are 20-25 percent, depending on definition, while widely recognized good utility practices require about 15 percent reserves. Thus, all told, only 5-10 percent of the $190 billion can be regarded in any sense as excess or stranded without a market. Further, a departing a customer should not have to pay selectively only for the high cost generation as it has paid all along in its rates for a slice of the system.

Since utility shares more often than not sell at a multiple of book value (shareholders put a higher value on the earnings potential of the utility's assets than is indicated by the book value of the depreciated investment), investors have received ample compensation especially as the risks of regulatory changes were factored in share purchase price.

Any stranding of investment must be mitigated as much as possible. All opportunities to sell, and at market rice, must be seized. How much will be grown into and when? When are other units to be retired? One person's stranded investment is another's economy or reliability purchase.

In further point of fact, any remaining stranded investment of generation is very transaction and utility specific. Utilities and State commissions that have been forward looking and had assets written down or off should not be penalized. The ostriches should not be rewarded.

Overall, this is a State, not a Federal, issue.

—The majority of transactions are at retail and not subject to FERC regulation.

—Among IOU's, the most recent EIA statistics show that wholesale transactions on average represent 10 percent of revenues and 19 percent of kilowatt hour sales.

—Wholesale transactions are, in almost all cases, governed by contracts with expiration dates and termination clauses. Especially since the obligation to serve does not exist at wholesale as it does at retail, the supplier has no reasonable expectation to the point of building dedicated capacity that the wholesale customer will be on its system after the contract expiration terms can be met.

On another part of the "how" subject, regulators seeking a competitive outcome must be as concerned with buyers as with sellers. Economists sometimes think of supply and demand each as a blade of a pair of scissors, virtually meaningless alone but very effective when working together. As Chair Moler said, both buyers and sellers must be able to reach each other. Here the temptation to react to the exaggerated fears of municipalization and calls for seizing regulatory jurisdiction should be resisted.

The threat and the actual formation of new municipal electric utilities is precompetitive, not anticompetitive, and is consistent with the competition goals of the Energy Policy Act of 1992.

—Franchise competition, in both directions, keeps utilities on their toes and responsive to consumers. This is a dynamic situation with 56 municipal electrics having been sold since 1980, and 31 having been formed.

—Formation of new municipals (which would only be done to lower costs) is good for the economy and for all customers, residential as well as larger industrials.

—Just the threat of formation of new municipals has resulted time after time in forcing an investor-owned utility response that has greatly benefited customers with lower electricity prices and other community programs (New Orleans; Brook Park, Ohio, for example).

—The precompetition abilities of municipals can blossom now that EPAct92 has helped remove the transmission access restrictions that were a barrier to forming new municipals.

—In the Energy Policy Act of 1992, Congress evidenced a concern about the possibility that corporations might try to game the system by pretending to be a municipal, but Congress did not do anything to prevent the creation of new municipal utilities or restrict their rights to obtain transmission access on terms equal to those already in existence.

—Congress recognized and did not want to interfere with the creation of municipal utilities—the expansion of the number of buyers (new municipals) as well as the number of sellers, and linkage of buyers and sellers together in newly possibly markets.

Municipalization is a local rights issue.

—A community should not be interfered with in its right to determine what public services it will provide its citizens. When the franchise expires, a community should have the freedom to choose its electric provider just as it does for water, sewer, gas, and cable, for example.

—If all of the legal hurdles and referenda and other requirements are fully met, a community's forming a new municipal electric utility is not a sham and should not be discouraged by Federal or State law.

—A community's choice of electric supplier is a local decision, best made in an open public forum, not inside board rooms of large private companies.

—A city like Las Cruces, New Mexico, should not be limited in its search for lower electricity costs for its citizens just to insulate an entrenched competitor and its core customers.

Franchise agreements have expiration dates and termination provisions.

—Thus there is no stranded investment if a city chooses to municipalize.

—The very existence of these franchise agreements with expiration provisions implies that the city has and can exercise the choice to municipalize, and has a fiduciary responsibility to exercise that choice if it is in the best interests of its citizens.

Finally, the most important recommendation to regulators on how to proceed is to keep monitoring closely progress and the market power/competition balance. Competition in generation cannot be assumed. To paraphrase, the price of competition is vigilance.

Depending on economies of scale in generation, financing and capital lumpiness barriers can remain. In addition, by all accounts the U.S. independent power producer segment is shaking out and consolidating. U.S. Generating Co., a partnership of PG&E and Bechtel, just announced it was acquiring the widely recognized fellow IPP, J. Makowski. Not only this consolidation, but also it is another example of the trend toward traditional utility affiliates assuming dominant positions in the IPP sector. Finally, the new and obscure corporate structures being designed around EPAct's exempt wholesale generators (EWG's) would make Samuel Insull and his pre-PUHCA holding company cronies blush.

Regulators should emphasize policies that collect and publish timely information about the marketplace that is needed to eliminate information barriers. Regulators should collect and use the right information to evaluate market power. FERC persists in analyzing whether or not market power exists on the basis of whether or not one utility company has market power. This is insufficient—the entire market must be analyzed. For competition to exist, all of the significant participants in the entire market must be lacking market power.[3] Regulators should set performance criteria for themselves and for competitors and conduct periodic investigations that result in publications to answer former Mayor Koch's question, "how are we doing?"

Before retail wheeling is embraced, for example, such an analysis should be done that demonstrates that significant costs and efficiency gains are likely to occur over and above the benefits that are achieved from wholesale electricity competition. This only points to the tradeoff that pressures for retail wheeling will continue to build until the benefits of open, nondiscriminatory transmission access, regional markets, and competitive generation are obtained.

1. To what degree do you believe the electric utility industry has entered a period of irreversible change, and how widespread do you feel change will be?

We are in a period of change that will not be reversed. However, as the discussion above stated, vigilance will be required to ensure that certain dimensions of change—the level of market power, for example—do not reverse and leave our society with less regulatory protection and more economic and political concentration of control in the electric utility industry. Change will be widespread, but underlying economic conditions will dictate. For example, retail wheeling is not a necessary logical extension of wholesale wheeling if we succeed in obtaining full, wholesale competition.

3. How can social and environmental objectives be adapted to a "restructured" industry, in an environment in which services once provided by a single vertically integrated utility are increasingly "unbundled"? To what degree should these objectives be retained, changed, or abandoned?

[3] See A Critical Look at the FERC Market Pricing Policies, APPA staff report, November, 1992.

One of the advantages of moving to an unbundled competitive market discipline is that social and environmental objectives will need to have accurate benefits and costs attached to them. Cross-subsidies and overriding fiats will not confuse. In this world, valuable and desirable programs will be strengthened and weak ones jettisoned. This is the discipline that already exists in the decisionmaking process for nonprofit local municipal utilities. Community leaders make the decisions and are held accountable for the attendant economic, social, and environmental consequences.

Thank you.

Mr. SHARP. Thank you very much, Mr. Hobart.

Mr. Howe, we will be happy to hear from you now.

STATEMENT OF JOHN B. HOWE

Mr. HOWE. Thank you, Mr. Chairman. EGA thanks the committee for the opportunity to speak today.

My name is John Howe, vice president, regulatory and Government affairs for J. Murkowski Associates, Incorporated, still here and active around the world as an energy project developer.

I am speaking in my role as chairman of the Energy Policy Committee of the Electric Generation Association. EGA is a national association that represents independent power producers and suppliers of goods and services to the competitive wholesale power generation industry.

It is appropriate and very timely for the committee to pose these questions concerning the future structure of the electric industry, and I am going to depart somewhat from my prefiled testimony in order to make a few points of special emphasis in response to other witnesses.

As is reflected in EGA's prepared testimony, we consider the logical direction of reform to be a more discerning approach to regulation, a realignment of functions within the industry, and most notably the functional unbundling of generation. In short—and this has been said by several other speakers today—those services that are naturally competitive should be provided by the competitive marketplace. This is the American way.

Those services that retain monopoly characteristics should continue under an appropriate form of regulation. We must preserve the unmatched reliability of the Nation's distribution system and avoid destructive forms of competition between distribution utilities.

There are close analogies in other industries to the path of reform EGA suggests which is described more fully in a discussion paper appended to our testimony. It is within our reach to unleash huge efficiencies from regional power markets that are constrained not by artificial political boundaries and tariff filing requirements but by nothing less than the great Continental Divide.

But all is not proceeding smoothly and logically in this time of evolution. Unfortunately, the structure of regulation in today's market has led to an unnatural adversity of interests between utilities and suppliers of competitive power. Utilities continue to possess, and in many instances to use, tremendous market power in the conduct of system planning and power procurement.

It is a very timely illustration of EGA's concern that just yesterday Georgia Power Company announced its intention to cancel a major 800 megawatt RFP for power resources in 1997, this despite an acknowledged need for resources in the latter part of this decade. This after the independent power industry spent millions of dollars and man years of effort in time and responding to the RFP in a highly competitive environment that yielded unprecedented low cost for power. This even though many bidders offered the company the opportunity to defer the in-service date of their resources.

This company's regrettable actions, which will most certainly be challenged, are adverse to ratepayer interests, yet they are far from unique, and it is really not my intention to single out one utility. Several companies have taken similar actions in the past 2 or 3 years across the country.

Tactics to thwart the growth of the independent power sector may be characterized, furthermore, as a rational response from the standpoint of shareholders given a regulatory system that rewards building and implicitly penalizes buying least-cost power from the competitive marketplace. The result is a world in which many utilities are fighting tooth and nail not to purchase from independents and at the same time to prevent independents from selling directly to their customers.

What is the upshot? A growing number in our industry find we are simply being locked out of markets in the United States, this despite the Nation's need for clean, efficient generation to meet the mandates of the Clean Air Act, to facilitate the orderly retirement of older, inefficient, and dirty units, and to meet growth and demand which continues to be strongly correlated to economic growth.

Another strategic response by some utilities in lieu of unbundling generation and embracing competition has been described by several previous witnesses. This is to retain the vertically integrated structure and to offer price discounts to competitive customers, offset by cost shifts to captive customers. Mr. Chairman, this is not market-based competition, this is mercantilism. At one time we fought a national war of liberation over that type of economic structure. It will diminish and delay the benefits of true competition.

In today's bulk power markets, utilities are able to sell power for resale at steep discounts because their own native ratepayers subsidize the fixed costs of their generating units in base rates. IPP's, of course, sell based on the total cost of their facilities pursuant to contracts.

Our power markets are by no means competitive in the classical reasons in terms of openness of entry and access to information. As Mr. Hempling noted, we see immense disparities in market power. This year it appears the time has come for a little truth telling. What is occurring in this industry is a struggle over market share and financial environment in a world in which utilities candidly concede that exposure to true competition would subject them to potentially huge amounts of stranded investment.

For this reason, EGA advocates dealing with the stranded investment issue frontally and fairly. We believe there are circumstances in which utilities should be eligible for stranded investment recovery, and we delineate them in our prefiled response, but we must recognize that this is a transitional problem, we must deal with it accordingly in a fashion consistent with the ultimate and overarching goal of a truly competitive generation market. The tail must not wag the dog.

We also firmly believe that increased competition in generation markets properly structured is quite consistent with the attainment of social policy and environmental goals through careful retail rate design. In particular, competition and environmental protection can be mutually supportive. The fierce competition among IPP's to develop cleaner sources to meet new and existing market needs has dramatically pushed the envelope in environmental control technology over the past decade.

We should actively encourage the replacement of older, dirtier, less efficient units with state-of-the-art technology to maintain our international competitive position. We risk being leapfrogged by other nations that are going to step up and make these investments.

While the EGA did not address the status of PURPA in our testimony, our association has viewed PURPA as a necessary and important step in the evolution of the electric industry. As long as utilities maintain a market dominance based on rate-based treatment of their own generating units, PURPA provides an important competitive check. Ultimately, however, as we fashion a truly competitive generation industry, we should not be constrained by outdated constructs.

Our Association's core philosophy is conveyed in our name. We look beyond an era marked, as Mr. Hempling and others have noted, by conflict and litigation between utility generators and nonutility generators. We look to a future in which all are treated simply as generators without regard to ownership or artificial legal categories.

In our testimony we have also outlined our views to you on transmission pricing reform. In this matter as well, we advocate nondiscrimination as to ownership as a central guiding principle as well as efficiency, transparency, and ease of administration in pricing rules.

We have done our best to respond to all of the committee's questions directly and succinctly in our prefiled testimony.

In closing, Mr. Sharp, EGA wishes to thank you profoundly and to congratulate you for an outstanding congressional career as an advocate of competition and consumer interests in the electric industry. We are going to miss you after this year, and we wish you the best of luck in your future endeavors. We stand ready to assist the committee in any way we can, and we look forward to your questions.

[The prepared statement of Mr. Howe and responses to subcommittee questions follow:]

STATEMENT OF JOHN B. HOWE

Mr. Chairman and Members of the Subcommittee: My name is John B. Howe and I am Vice President of J. Makowski Associates, Inc. I also serve as Chairman of the Energy Policy Committee of the Electric Generation Association. I am very pleased to have the opportunity to testify today on the issues of transition and restructuring in the electricity industry since the enactment of the Energy Policy Act of 1992 EPAct. The Electric Generation Association (EGA) greatly appreciates the subcommittee's invitation to address you today. The subcommittee's letter of invitation has presented speakers with a timely and critical set of issues for discussion.

EGA is a national trade asssociation that was formed in 1992 to represent the interests of independent power producers and providers of goods and services to the competitive wholesale power industry. EGA was an active participant in the legislative process leading to passage of the Energy Policy Act of 1992. EGA's name reflects our organization's philosophical core. We believe that the distinction between utility generators and non-utility generators, one that has given rise to so much contention and litigation in the electric industry structure in which all are treated simply as generators, without discrimination as to ownership. Such a structure is necessary if we are, as a Nation, to harvest the full benefits of competition in generation on behalf of all electric consumers, as envisioned by EPAct.

As this industry continues to evolve, EGA aimes to represent the interests of competitive power generators and promote policies that provide the broadest possible benefits from reform. EGA's vision of the logical path of reform is presented in a recent discussion paper, "Electrifying Change—Strategies for Structural Reform in the Electric Industry." A copy of this paper is attached to the association's written comments for the record.

Responses to Committee Questions:

(1). The extent and Irreversibility of Change.—The transition in which we are involved toward greater reliance on market-based competition is inexorable and irreversible, and this is a very good thing for consumers. This transition is directly attributable to the growth of robust competitive conditions in the generation market. Competition has driven down the cost and driven up the efficiency of new generation relative to the existing stock which would otherwise have been built under old regulatory regimes.

Traditionally, the electricity sector has been dominated by vertically-integrated monopolies. Within a model that dated to the era of the New Deal, utilities provided generation, transmission and distribution functions within exclusive franchise areas. This form of organization was clearly one of the exceptions to the rule of open markets that characterizes almost the entire American economy. It was adopted not as a matter of preference but as a second-best substitute for competition. A market-based competitive structure simply wasn't feasible given the immature state of technology and markets.

But times changed, technology progressed, markets matured, certain economies of scale were exhausted and this traditional structure became outmoded. The continued adherence of regulation to the vertically-integrated monopoly model has become, in some ways, actually harmful to the public interest. It is a costly, litigious and more importantly an inefficient system. We must be prepared to dispense with traditional assumptions and modes of regulation. In their place, EGA has called for a more discerning model of regulation that treats each distinct segment of this complex industry in accordance with its competitive characteristics.

As competitive conditions have matured in the generation industry, it is possible to bring about more efficient regional power markets. In EGA's view, the logical evolution of the industry is based on a realignment of functions. Specifically, EGA has advocated the benefits of functionally unbundling generation. This path of reform has close analogies in other industries. The unbundling of wellhead supply from transportation in natural gas markets, as well as long-distance service and equipment manufacture in the telecommunications markets, have made those industries stronger and more efficient. This approach to reform has also been pursued successfully in the privatization of electric power industries in several foreign nations.

While the change underway in electricity markets is pervasive, there is much to be retained in our current system. America's distribution utilities offer a degree of reliability that is unmatched throughout the world. One of the key challenges facing regulators and legislators is how to gamer the benefits of competition in generation while preserving a reliable and financially healthy transmission and distribution industry. Unlike generation, transmission and distribution remain natural monopolies that are more amenable to traditional regulatory oversight. This fact was recognized by the Congress in the transmission provisions of EPAct. EGA believes that unbundling of generation can, in fact, produce the best of both worlds: market-based discipline in generation, along with a reliable regulatory bargain and continued unmatched reliability of service in transmission and distribution.

There is a risk, however, that structural reform efforts may be sidetracked in a way that directly undermines the goals of EPAct. At this juncture the industry is plagued by uncertainty, and an orderly transition is by no means assured. Many utilities, in adapting to increased competition, are inclined to retain their vertically-integrated structure and seek to capitalize on their position of market power. Instead of unbundling, they are seeking greater latitude to engage in price discrimination to retain customers and maximize revenues. EGA respectfully submits that a response to evolving market conditions based on state-sanctioned discounts to competitive customers, offset by cross-subsidies from other customer classes, is not market-based competition. It is a form of mercantilism. It will diminish the benefits of competition and is likely to incite continued litigation. If we are to replicate the successful efforts at reform in other industries that have undergone competitive transformation, EGA urges that reform efforts focus on fostering meaningful competition and choice. Reformers should strive to bring about conditions in which suppliers are similarly situated, rather than play to the unique status of utilities as regulated monopolies.

(2). Methods of Dealing with Transition.—The process of reform toward a truly competitive market structure threatens to be arduous. Utilities are threatened by uncertainty about transition policies, particularly regarding the treatment of costs that may be "stranded" by a rapid change in market structure. Utilities have always been subject to many kinds of risks, and the regulatory compact has tried to compensate them accordingly by affording them the opportunity to earn appropriate returns.

If we are to maximize and hasten the benefits of competition, we must manage this process of transition proactively. It is essential that the electric industry avoid a repeat of the delays and litigation that marked the natural gas pipeline take-or-pay crisis. EGA supports a frontal approach to the issue that provides utilities with appropriate compensation within certain circumstances.

We agree that utilities should be allowed to seek recovery of costs prudently incurred in connection with providing service where (1) the utility had a reasonable expectation of continuing in the role of supplier to a particular customer; (2) these costs would not have been incurred but for the regulatory or market action that could not reasonably have been anticipated; (3) where attempts to mitigate the so-called stranded costs were unsuccessful; and (4) a clear connection exists between claims for stranded asset cost recovery and actual lost customers. Furthermore, in the competitive transformation of the industry, it is not appropriate to assume that all uneconomic costs are stranded assets. Additionally, any stranded asset recovery should be accomplished through procedures that do not distort competition in generation markets on a going-forward basis. Finally, if certain unanticipated market and regulatory changes resulted in gains by utilities who also experienced losses due to other unanticipated changes, then it would seem to be appropriate to determine a net cost. Indeed, taking advantage of the new environment is a form of mitigation.

The issues in play are complex and cannot be easily disposed of without reference to the facts of particular situations. The dollars involved are significant. Nevertheless, EGA urges regulators, legislators and others to recognize the stranded cost dilemma as a transitional issue and to deal with it as such. We must not allow transitional obstacles to cloud the long-run vision of a more fully competitive electric industry. The outline of any transition policy should be determined in light of this ultimate goal, and not vice versa. The tail must not be allowed to wag the dog.

(3). Social and Environmental Objectives.—EGA believes that the social and environmental objectives that have been sought within the framework of monopoly-style utility regulation are, in many instances, worthy goals. They must not be abandoned or neglected. However, they must be pursued in a fashion that does not distort competitive market outcomes in the generation segment of the business. Thus it may be useful to consider converting costs incurred in connection with social and environmental policy goals into a universal access fee or "wires charge" that applies to all utility customers. Such a mechanism could enable costs to be collected equitably from all customers save those that literally pull the plug from the interconnected grid. It may also be appropriate to consider programs in lieu of the utility rate structure since the utility's "taxing power" is diminishing in the face of competition.

While these alternatives are available, there is a growing recognition that regulators are ultimately constrained in the extent to which they can use the ratemaking system to impose costs on consumers, no matter how worthy the social or environmental objective. Retail electric rates have become a factor in inter-regional as well as international competition.

EGA wishes to comment specifically on the issue of environmental protection. In this area, heightened competition in power supply can truly provide a double win. Clean, efficient state-of-the-art new generating facilities are not part of the pollution problem. They are part of the solution.

Many of our pressing environmental objectives will best be attained by subjecting all existing generating facilities to a more rigorous market test with due recognition to environmental factors. With the Clean Air Act Amendments of 1990, Congress has taken the lead in developing more flexible, market-based forms of environmental regulation. EGA hopes to see this approach increasingly adopted at the state level. Our state and national environmental and economic regulatory policies should encourage the orderly replacement of inefficient and dirtier sources of emissions with newer, cleaner, state-of-the-art generating facilities. This is often what occurs with production facilities in industries not subject to economic regulation. As competition becomes pervasive in the generation industry, environmental regulation should encourage such an outcome as well.

(4). Pricing Reforms.—EGA has been a vigorous participant in the debate over transmission pricing reform before the Federal Energy Regulatory Commission (FERC). We consider the issuance of the FERC's upcoming policy statement to be an event of great importance to the continued orderly development of the industry. Clarity in FERC's pricing policies will promote the development of regional transmission groups, which are considered by many to be crucial to the development in tum of fully functional regional power markets.

While EGA has not advocated a specific transmission pricing methodology in that proceeding, we have supported the adoption of a set of principles that would be compatible with the kind of competitive generation market that the Congress sought to advance through EPAct. We are pleased to see many of the principles EGA espouses reflected in a recent string of FERC decisions. Perhaps most notably, we were very excited to see FERC adopt the "Golden Rule" of comparability for services provided for a utility's own use of its system and services to third parties. This principle was adopted in a case involving the American Electric Power Company in May, and is a crucial step in eliminating undue discrimination in generation markets.

To summarize EGA's principles in brief, we believe that transmission rates should be nondiscriminatory, providing comparable treatment of all generators regardless of ownership. Transmission pricing should also mirror as closely as possible what would occur if transmission markets were competitive thereby promoting efficient investment in both generation and transmission facilities. Another essential principle is that transmission rates must be transparent, predictable and stable, since they are a principal factor undergirding long-term investment decisions. Finally, systems for determining prices should be simple to administer. The rules of the road for the power grid must be compatible with a generation market that, in the future, will operate in real time. Competitive markets cannot wait for lengthy reviews and formal hearings. It is our hope and belief that FERC is moving full speed ahead in the proper direction in accordance with the goals established in EPAct. We hope that reforms in state regulation will be similarly farsighted.

(5). Federal-State Jurisdictional Issues.—Debates overjurisdiction center around two particularly contentious issues: the siting of new high-voltage transmission lines and the possibility of retail wheeling. Electricity originally developed as a local service subject to local regulation that was constrained by artificial political boundaries and formal tariff filing requirements. With the progressive interconnection of the national grid, we now have the possibility of power markets that operate in real time, that span half a continent and are bounded by nothing less than the great continental divide.

The regionalization of power markets has the potential to create huge efficiencies for ratepayers. But the historic division of jurisdiction between federal and state regulators threatens to stand in the way of realizing these benefits. If the full promise of the Energy Policy Act is to be realized, a new apportionment of responsibilities is necessary. This new framework must recognize states' vital interest in the land-use impacts of transmission infrastructure, yet facilitate the development of regional markets.

The other critical flashpoint in jurisdictional disputes between state and federal regulators centers around the debate over retail wheeling. EGA has neither endorsed nor opposed retail wheeling in concept; we will eagerly pursue whatever markets regulators and legislators deem that we should serve. However, if a competitive market framework based on direct access or retail wheeling is to be established, then customers must be offered meaningful choices of supply. This will require reforms in federal legislation to permit Exempt Wholesale Generators to compete in retail markets, among numerous other changes. We would urge the Subcommittee to monitor the development of the retail wheeling debate in many states and to ensure that, if the retail franchise is to be opened in some areas, all potential competitors will be given meaningful access to supply those markets. Failure to pass such reforms prior to the opening of retail markets would not lead to market-based competition, but to a struggle among oligopolies to maintain market dominance.

In conclusion, EGA wishes to offer itself as a resource to the Subcommittee in its future deliberations. Finally, EGA wishes to use this opportunity to offer its most profound thanks to Chairman Sharp. Mr. Sharp's efforts on behalf of the interests of competition and electric consumers have yielded tremendous benefits to our nation. Our entire Association and most particularly those individuals in EGA who have known him and worked with him closely over the years extend best wishes as Congressman Sharp pursues the next phase of his career. Mr. Sharp's presence in Washington will be missed by our industry. EGA will continue to advance the agenda of full and fair competition in generation markets at every opportunity, and we appreciate this opportunity to share these comments with the Subcommittee today.

Mr. SHARP. Well, thank you very much. It is not really necessary to testify every time as to what I have ever done, but I appreciate your remarks. There are others who might want to wish equal time on the record after all.

But let me turn to the subject of the sunk costs or stranded costs or, as Mr. Abdoo suggested, uneconomic assets which did come up when we considered EPAct, and it was raised by some of those who did not wish to see us change PUHCA. It is raised by lots of folks, but that was one of the arguments against it, and some of us took the view there have always been in all industries and there is in the electric industry a certain amount—that is inevitable change, it is there.

But obviously when we change the rules of the game, that also changes what investors, managers, and others thought they were banking on, and that does create—at least the Government has created a new situation.

To what degree do you folks put an emphasis on stranded and sunk costs, and what kinds of options do you see regulators taking? I guess we have to distinguish between the world of the wholesale where we have wholesale competition and then also the potential for the retail issue.

Mr. Abdoo, you obviously addressed that part in your testimony.

Mr. ABDOO. Yes, I see it as an unfortunate situation but one in which we should not deprive the majority of the consumers of electricity in this country from universal access to buy from wherever and whoever they want, to buy full service if they want with DSM and all the rest of it, or to buy economy service, self-service, if you will, simply because we have a few situations where there are assets on the books of companies that are on those books at above the economic value of those assets.

Our company, when we have had that situation, we write it off, and our stockholders end up paying for it, and our commission does not look very favorably upon having a lot of regulatory assets on the balance sheet and then allowing you to earn on those kinds of things.

So it in effect gives us in a competitive environment a much better position, but it is because we paid once, and it is like with the 1990 Clean Air Act Amendments which we strongly supported. The concern was that those of us that had been aggressive and had gone out and spent millions of dollars to clean up would somehow now be penalized as a result of others not doing it, and it is not to criticize others, it is not to see that the decision was avoidable, but I think it is a local problem left to the local utility and their State commission.

Mr. SHARP. Let me just push a little further on that, and then others can respond, because obviously I think at this table we don't have folks representing some of the most alarmed on this issue since you tend to be lower-cost operators in both cases of the IOU's here, but some of your colleagues who might not be feeling so collegial at those comments, what if they have a couple or two or three very high-cost nuclear power plants that appear to be running fine, and they have a concentration of assets that they find it difficult to write off? I heard one of them describe what the stockholders may feel like. It sounds like some of our public town meetings. But that—it didn't sound as strange to me as it seemed to sound to them.

But the hostility that would be represented in them and the loss as they described—I am trying to put one of the stronger cases and ask you if you can give some scenarios of how that might be handled or if that is the luck of the draw and that is the way—the truth is, that is the way our society moves forward and people just have to be mature about that.

Mr. ABDOO. I certainly am fortunate in that our company does not own a lot of those things which are overvalued, perhaps. And if faced with it, as I say, I think there may be some justification, depending on how they got into the situation, that they end up with a plant that cost twice as much as its economic value.

But that is a situation that ought to be able to be worked out between their regulator and themselves and perhaps, depending on the complexity and how that decision was made, there may well be a reason to over a period of time allow recovery of part of those costs, particularly during the transition period.

But, in a world where there is truly competition, the thing that distinguishes those who will make handsome profits and those who will have lousy shareholder meetings will be how well you manage your capital, how well you utilize those assets, how efficient you are in delivering a service that is not what you want to sell the customers but what they want to buy, and how well you can meet those markets, and that is the distinguishing factor in American industry.

And I know during the transition, I have described it to our employees. It is hell. It really is, going from being so much regulated to the potential freedom for the customer can—and we pay a price in that transition, and I am not—fortunately, I don't have to deal with it at this point.

Mr. SHARP. I will ask the next panel as well.

But go ahead, Mr. Rogers.

Mr. ROGERS. Mr. Chairman, I vote for maturity, less so as I get older.

But let me make this observation. As you know, I came out of the natural gas pipeline business, and our version of stranded investment was take or pay. And the thing that was unfortunate about the eighties, it took so long for the regulators to finally come to grip with the issue of take or pay and determine how those costs ought to be allocated between shareholders and customers.

I think it is unrealistic for anybody in the utility industry today to think they are going to get off scott-free and the investors are not going to pay some price for these so-called stranded investments. I think that is unrealistic as a matter of public policy because these costs are not going to be put on the backs of consumers. It is just not going to happen.

It didn't happen in the gas industry with take or pay. It is not going to happen here.

I am a low cost supplier and I have some sympathy for, having come out of the pipeline industry with take or pay problems, I have some sympathy for the notion that it should be shared between customers and shareholders.

But I look at the history of my own company, who had a nuclear plant that was 53 percent completed, it was a $2.7 billion write-off, and as a result of that equity fell to 9 percent, and nobody was talking about stranded investment then.

So, I think that maybe one of the realities is, as we go forward is that, to use your phrase, people may have to be a little more mature about this. The reality is if you look at Dow Chemical, if you look at the major corporations across this country, they have made major investments and have written them off.

Our company made a major investment in nuclear plant and wrote it off. Nobody asked the customers to pick it up as stranded investment.

So, I think that at the end of the day, given my natural gas experience and the experience of my own company, my belief is there may be some sharing, but at the end of the day the investors are going to carry, and the shareholders, a portion—have to—a portion of the burden of any stranded investments that we move.

I will make one further comment, and that is, the stranded investment argument should not be a roadblock to moving to a more competitive world in our industry, and it shouldn't be because that is just sort of the natural process.

And, if you think back about the gas industry in 1979 and the bleak future it had, and look at the future that it has today, there has been a dramatic turnaround in the industry, and I think if we can work our way through this over the next 5 to 10 years our industry is going to have an even brighter future as a result of new players, new competition, more innovation and great ideas.

I heard somebody last week at the Aspen Institute say something great, and it really sort of picks up on Dick's point, and that is this. Stranded investment is an issue. But as we move into a competitive environment, the bigger issue this committee ought to worry about is the stranded regulator issue.

Because quite frankly, the role of the regulators are going to have to change as we move into this new world.

Mr. SHARP. Thank you.

Mr. Hobart?

Mr. HOBART. As I indicated a minute ago, Mr. Chairman, I believe it is the sharp distinction between wholesale and retail on this question. But I should point out that for publicly owned electric systems where the user and the owner are one and the same there is no place to hide.

If we make a mistake in acquiring an asset which turns out not to be productive, we have to eat it, and there is no way to spin it off to somebody who holds stock and lives in New York and you are operating on the West Coast. It is just a mistake that we have to live with ourselves.

But in the overall concept of stranded investment I think there are a couple of things to consider. One is that, as I think Scott Hempling and others pointed out this morning, some of the claims are grossly overblown in terms of how much stranded investment there might actually be. They exceed the actual depreciated value of the assets that exist in America currently in some instances.

Second, because of the thrust of competition that has entered the electric utility industry, pushed by this subcommittee in part, it doesn't appear to be desirable to protect people from some of the consequences of competition, and to figure out payments for stranded investment that leave people as they were kind of defeats the purpose of competition, which is to test products and services in a market.

There is a third point which is worth making, and that is that doing such, that is, making payments for stranded investment, could actually have an anti-environmental impact in that it would encourage people to retain facilities that might be old and not be up to the best standards of the business. And that again is not a public policy objective.

Another point I think to make about stranded investment is that I don't think it is any surprise that we are now in the situation that we are. Going all the way back to 1978 and perhaps even earlier, I think discerning individuals in the electric utility industry could determine that there was a trend that was moving in this direction, and certainly actions of Congress were bellwethers as far as the ultimate form that the industry is going to take might be concerned.

The last thing, I think, to say about stranded investment from a public power point of view is that we see no equity at all in a suggestion that if the States should not deal in a satisfactory manner with stranded investment as determined by an investor-owned electric utility or the Federal Energy Regulatory Commission that somehow a portion of those costs should be transferred to wholesale transactions including transmission.

Generation and transmission are two separate facets of the business, and it would be totally unfair for the FERC, for instance, to second guess States, determine that they acted in an adequate way, and penalize public power systems by putting additional charges of transmission services that are totally unwarranted.

Mr. HOWE. Mr. Chairman, I will make just a few observations. First off, and I think this has been said, let's remember that competition is not to be viewed as a problem. It is an opportunity. Competition is what can drive this whole market toward a lower cost structure in the future. So, let's think of this as an opportunity.

I did say that EGA supports the concept that stranded costs should be recoverable in certain circumstances. I have heard John Anderson and others from time to time say, "Well, you know, Ford Motor Company, if they have an automobile factory that doesn't sell cars they don't have an opportunity to claim stranded cost recovery."

I think I am uncomfortable with that analogy because the Ford Motor Company did not have an obligation to serve up Edsels. That is a business venture that they entered voluntarily.

I think we have to recognize the regulatory framework that existed in this country for most of a century and that has imposed an open-ended obligation to serve on utilities. We have allowed to a certain extent stranded cost recovery for telecom and gas utilities, gas pipelines as they have gone through their restructuring processes, and I do think it is worth emphasizing that as a result of restructuring both of those industries in the view of most analysts are healthier today than they have been in some time.

The problem is also to a certain extent self-liquidating over time. I think this is a difference between the take or pay crisis where we allowed it to fester and balloon and grow. This is a problem that to a certain extent is self-liquidating over time, and time may help to cure the problem.

But there probably will be some write-downs and there are some plants that simply should be shut. Those that are truly uneconomic on a going forward basis I think best illustrate the aphorism "If you are in a hole stop digging," and we should encourage the retirement of truly inefficient units.

I think one of the most perceptive comments I have heard on this was offered by a member of your staff who is sitting behind you who I heard say some months ago that he was willing to contemplate stranded cost recovery but "I want to know what I am buying," I think is what you said, Leon, and I thought that that was a very important way of approaching the problem.

If all we are getting is more of the same, if we are paying off utilities for bad previous investments and perpetuating a vertically integrated anticompetitive structure, then we are not interested. But, if it is part and parcel of a comprehensive program of reform that will bring about a dynamic and robust competitive generation industry in the future, then I think it is appropriate for all customers, and inevitably to a certain extent shareholders, to share in the burden.

I think we should—because everybody is going to be benefiting, we should consider net benefits. We should consider the fact that utilities are going to be deriving incremental revenues from this transition to a more competitive structure because they will be providing new services that were not possible under the old regulatory framework.

I think those largely summarize. We also address some other criteria in our prefiled comments.

Mr. SHARP. Our colleague from Connecticut, Mr. Franks.

Mr. FRANKS. Thank you, Mr. Chairman. And I thank our panel for their testimony today.

My first question I believe I know the answer to but I will still ask the question. Do you believe that there should be a requirement to purchase blocks of power from PURPA-qualifying facilities when they are not competitively priced?

Mr. HOWE. Let me be the first to tee that one up. A lot of the myths about PURPA being circulated today are based on the experience of contracts that were entered into 10 to 15 years ago in the very early days of the IPP industry.

I think you have to look at the significant progress that has been made and the greatly increased sophistication in contracting, the fact that independent power producers are now developing dispatchable projects with capacity, energy-type payments, the fact that the pricing is tied to an index instead of being simply sort of a predetermined curve.

The sophistication of this industry has grown to meet every challenge, and I think it can fairly be claimed that the competitive market is the place to which utilities should be turning for new power resources.

Rate-based regulation is, candidly, a second best approach. Competition in every other segment of our economy we acknowledge as the best approach. So, as long as the procurement is through a competitive framework, then I think it is appropriate to require utilities not to go forward with their own construction if they can acquire resources at lower cost from the competitive marketplace.

Mr. FRANKS. So, would you be in favor of forcing utility companies to purchase?

Mr. HOWE. Are you talking about from existing projects under existing contracts?

Mr. FRANKS. Either/or.

Mr. HOWE. OK. Well, I think I have addressed the issue of new contracts.

As for existing contracts, I don't see a difference in principle between a commitment for ratepayers to continue to pay for past prudent investments in generating plants as opposed to a commitment to follow through on the obligations entered into with a contract under a previously approved power purchase arrangement.

Mr. FRANKS. OK. I am going to ask that of Mr. Abdoo and Mr. Rogers as well, because as the electricity markets have changed most people seem to agree that PURPA's policies, particularly the mandatory obligation to buy power would also need to be revisited.

And it is, however, very unlikely that we could get a consensus this year to reopen PURPA, but in my opinion we are faced with the prospect of sending a bad policy signal to renewables when the size limitation goes back into place. And I would like to have Mr. Abdoo and Mr. Rogers address that concern.

And also I would like to ask how would you view extending the removal of the size limitation for a limited number of years, let's say 4 years, so that in coming years we as a Congress would again be forced to examine PURPA's policies?

Mr. ABDOO. On that latter question, Mr. Franks, as I mentioned in my remarks, I have no problem with eliminating the cap. I don't see where—I don't think PURPA needs to be there anymore at all, so anything you did to it that made it more open would not trouble me in the least.

On your other question with respect to buying uneconomic power, if it is from an existing contract, I think that is the IPP version of stranded investment and I take the same view with that. I mean it doesn't matter to me who owns the thing. If it is uneconomic, it is an uneconomic asset. It has no place in the competitive environment.

If it is a new contract and somebody enters into a contract that is uneconomic on its face, sooner or later we will hear they have stranded investment and we will be back to the same question. So, I would not be in favor of doing something uneconomic whether it is for power or poles or wire or anything else. It doesn't make a lot of sense.

Mr. FRANKS. OK. Thank you.

Mr. Rogers?

Mr. ROGERS. I think Mr. Abdoo has done a good job of expressing my point of view on that. The point simply is that for the IPP or cogenerator that is their stranded investment like the utility has a stranded investment, and if it is uneconomical the same rules should apply in dealing with the issue.

Mr. FRANKS. OK. I yield back, Mr. Chairman.

Mr. SHARP. Thank you very much.

Gentlemen, we thank you very much for your time and attention to the hearing today and appreciate your thoughts and comments. Thank you very much. Appreciate it.

The Chair would indicate we have one more panel and Small Business Committee, which has the jurisdiction over this room—and some of you know about jurisdictional competition—has indicated that they would like for us to vacate the room by 3 p.m., if we can.

Therefore, what the Chair would like to do is shorten our break—I had indicated we would have a half hour break for lunch—to about 20 minutes, and everybody can scramble to the carryout, and try to begin again at 2:50 with our final panel and try to complete it by 1:50 p.m.

Oh, pardon me. Yes, 1:50, and complete by 3 p.m. There may be those who wish it were the other. But 1:50.

Thank you.

[Brief recess].

Mr. SHARP. The subcommittee will come to order.

We now call our third panel of witnesses, Mr. Paul DeNicola, president and chief executive officer, of Southern Company Services, representing the Edison Electric Institute; Mr. Robert D. Glynn, executive vice president of Pacific Gas and Electric Company; Mr. Doug Divine, Chairman of the Task Force on Markets and Competition for the National Independent Energy Producers; and Mr. Robert G. Foster, Vice President for Public Affairs with Southern California Edison Company.

Gentlemen, I think you are familiar with our processes. We will be glad to make your written testimony a part of our printed record, and at this point we would be glad to hear your oral comments.

STATEMENTS OF PAUL DeNICOLA, PRESIDENT, SOUTHERN COMPANY SERVICES, ON BEHALF OF EDISON ELECTRIC INSTITUTE; ROBERT D. GLYNN, JR., VICE PRESIDENT, PACIFIC GAS AND ELECTRIC CO.; DOUG DIVINE, CHAIRMAN, TASK FORCE ON MARKETS AND COMPETITION, NATIONAL INDEPENDENT ENERGY PRODUCERS; AND ROBERT G. FOSTER, VICE PRESIDENT, PUBLIC AFFAIRS, SOUTHERN CALIFORNIA EDISON CO.

Mr. DENICOLA. Mr. Chairman, I am Paul DeNicola, president and CEO of Southern Company Services, Incorporated, and I am pleased to appear before you on behalf of the Edison Electric Institute to discuss our competitive electric utility industry.

Congress' goal of promoting competition in wholesale electricity markets is being achieved. Non-utility generators are competing successfully as new supply options for utilities. FERC is promoting a competitive wholesale bulk power market through transmission access initiatives. State utility commissions are reexamining traditional State utility regulation. Many are giving utilities flexibility to address competitive forces.

The Michigan and California Commissions have proposed forms of retail competition. Utilities are reducing costs, streamlining operations, finding better ways to serve their customers, reducing rates and proposing innovative approaches to price regulation. A successful transition to competitive markets requires a fundamental change between electricity providers and regulators to assure that a competitive electric industry operates efficiently and benefits all electricity customers.

These goals can best be met with a vigorous and efficient wholesale electricity market and retail ratemaking innovations such as performance-based ratemaking. I would like to discuss the most important elements of this new scheme outlined in my written testimony.

First, there must be a consistent set of rules for all competitors. Utilities are working hard to be effective competitors. Now that Congress has promoted wholesale competition in the Energy Policy Act we need Congress to take additional steps to make the electricity market truly efficient.

Federal laws provide competitive advantages to certain electricity providers. This undermines a fair, efficient and competitive electric power market.

For example, PURPA accords special treatment for qualifying facilities, or QF's, through the mandatory purchase obligation. It makes no sense to allocate market share in a competitive share where QF's can sell power themselves to any wholesale utility. Either we have a truly competitive wholesale market or we don't.

We support PURPA's goals of increasing energy efficiency and the use of renewables, but not through a mandatory purchase obligation which distorts markets. QF should compete directly with all other wholesale generators.

Also, the restriction on utility ownerships of QF's is counterproductive because it reduces the number of potential competitors in the market. State commissions' administrative determinations of avoided cost under PURPA have imposed billions of dollars in excessive costs on utilities and their customers in at least 15 States.

We strongly encourage this subcommittee to reexamine PURPA in today's competitive power market with particular scrutiny on the need for the mandatory purchase obligation, the administrative determination of avoided cost and the limitation on utility ownership of QF's.

Federal law also grants many subsidies to municipally owned utilities and rural electric cooperatives. These include exemption from many Federal and State taxes, capital formation subsidies, access to Federal hydropower at rates below competitive market levels, and access to low interest Federal, rural, economic development funds.

For competitive markets to operate efficiently, all electricity providers should be able to compete on an equal basis regardless of ownership. Otherwise competition will be skewed.

Second, price transmission realistically. Economically accurate transmission pricing is key to an efficient wholesale electricity market. FERC has acted to restructure transmission access on a competitive basis, but has not yet applied competitive market principles to transmission pricing.

Appropriate pricing is critical to send the right signals to market participants. Prices that exceed the transmission facility's value would discourage competition, but artificially low prices would promote inefficient use of generation, cause poor allocation of the system, and discourage its maintenance and expansion. With the changes FERC is adopting, it makes no economic sense to retain embedded cost pricing for transmission while applying market-based pricing for generation.

Third, provide for an orderly, uniform and equitable transition to increase competition. The objectives of direct retail access can be achieved without all the risk by ensuring that the wholesale markets work properly. However, this will not happen if utilities are unable to recover transition costs. These include costs of uneconomic generation assets, nuclear decommissioning, environmental and social programs, and power purchased under QF contracts.

The failure to recover transition costs will have significant financial impact upon utilities, potentially affecting their ability to raise new capital on reasonable terms.

The financial markets need to know that the transition to a competitive market will be orderly. Last fall the bond rating services downgraded their outlook for many utilities, citing concern about the changes affecting our entry. This helped contribute to a 25 percent decline in average electric utility stocks from September 1993 through May of 1994, while the S&P 500 declined only about 5 percent.

This sharp decline has resulted in the loss of roughly $70 billion of wealth in the U.S. economy. Utility costs are a favored investment for mutual funds, defined contribution plans and individuals seeking current yield, low volatility and the preservation of capital.

The majority of utility shareholders are at or approaching retirement age and are looking for attractive dividends to supplement their retirement income.

FERC's recent stranded cost recovery NOPR fails to address the full impact of the transition caused by the Energy Policy Act. We are particularly concerned about FERC's reluctance to address impacts of its policies on retail transactions and its narrow approach to wholesale transaction issues.

Fourth, equalized responsibility for social and environmental programs. Utilities have spent billions of dollars on environmental and social programs which include improving energy efficiency, promoting renewable energy sources and helping low income customers pay their energy bills. These utility investments would be jeopardized if proper price signals or provision for utility recovery of investments are not established.

The cost of social and environmental programs should be borne equally by all electricity customers. Otherwise, competition will restrict the ability of regulated utilities to participate in these programs. There must be regulatory certainty.

The Federal Power Act was enacted to eliminate the regulatory gap caused by constitutional limitation on State jurisdiction over the flow of electrons in interstate commerce. As the State commis-

86-340 95 - 5

sions consider retail competition and retail access substantial, questions arise about how the Federal Power Act would be applied in a retail access context. The most pressing question is who has the authority to establish rights, terms and conditions for transmission services in interstate commerce to retail customers.

The Michigan commission in its experimental retail wheeling proposal has asserted jurisdiction to establish the rights for wheeling to retail customers under that program. However, FERC has stated that it maintains exclusive jurisdiction over such rights. There is a clear conflict between the two commissions.

Other principles outlined in my testimony are also important. The obligation to serve and the obligation to purchase has got to be symmetrical, and utilities must have pricing flexibility and freedom to innovate in order to compete and serve their customers effectively.

In conclusion, we look forward to continuing to work with you and other policymakers towards the goals of lower electric prices to customers and efficiency in the production and delivery of electricity.

Thank you again for this opportunity.

Mr. SHARP. Thank you very much.

[The prepared statement of Mr. DeNicola follows:]

STATEMENT OF PAUL DENICOLA, PRESIDENT, SOUTHERN COMPANY SERVICES, ON BEHALF OF THE EDISON ELECTRIC INSTITUTE

Mr. Chairman and Members of the Subcommittee: My name is Paul DeNicola, and I am President and CEO of Southern Company Services. I am testifying on behalf of EEI, the national association of the investor-owned electric utility industry. EEI's members generate approximately 79 percent of all the electricity in this country and serve 76 percent of the Nation's ultimate electricity customers.

I thank the subcommittee for the opportunity to testify on the dramatic and fundamental changes occurring in the electric utility industry today. Clearly, competition has come to the electric utility industry.

The Energy Policy Act of 1992 (EPAct) substantially accelerated and broadened the level of competition affecting our industry. EPAct encouraged new participants in the wholesale generation market by creating a new class of wholesale power supply entities, known as exempt wholesale generators (EWG's), which are exempted from the entry and other restrictions imposed by the Public Utility Holding Company Act (PUHCA). In order to help assure the efficient development of a more competitive wholesale power market, EPAct also expanded the authority of the Federal Energy Regulatory Commission (FERC) to order transmission access so long as system reliability and the public interest were not harmed. Congress, however, specifically prohibited FERC from ordering a utility to wheel power to retail customers and, in a savings clause, preserved whatever authority, if any, the States may have had with respect to retail wheeling prior to enactment of EPAct. EPAct also initiated a host of important energy efficiency measures and directed State commissions to consider incentives to encourage utilities to promote efficiency goals.

A post-EPAct snapshot of the industry reveals that Congress's goal of promoting competition in wholesale electricity markets is being achieved. Over 150 applications have been filed at FERC for EWG status. New non-utility generators are competing aggressively and successfully as new power supply options for utilities. New power marketers and others with innovative concepts for conducting wholesale transactions are emerging as well.

Electric utilities are responding to competition by streamlining operations, reducing layers of management, trimming capital budgets, reducing purchased power costs, finding new and better ways to serve their customers, reducing rates and proposing innovative approaches to price regulation.

FERC is actively promoting a competitive wholesale bulk power market through various market-pricing, comparability, network service and transmission access initiatives while preparing to establish overall guidelines for transmission pricing. In addition, FERC recently published a Notice of Proposed Rulemaking (NOPR) which is intended to deal with the costs of the transition to a more competitive wholesale power market. FERC has completed efforts to require the submission of, information relating to transmission planning, use and operations, articulate the content of good faith requests for transmission service and encourage the formation of regional transmission groups (RTG's). Applications for approval of two RTG's are now pending before FERC, and other RTG's are in various stages of formation.

A number of State utility regulatory commissions are reexamining traditional State utility regulation. Many are moving forward with initiatives to afford utilities the flexibility within the existing regulatory structure to address the competitive forces affecting them and their customers, who are seeking lower rates to enhance their own competitive positions. Two State commissions, Michigan and California, have proposed forms of retail competition. The issues raised in these proceedings are provocative and far-reaching and raise important Federal-State jurisdictional questions.

Under these circumstances, the traditional embedded-cost regulation that served the industry and its customers well when utilities were considered "natural monopolies" may be too cumbersome to provide utilities sufficient flexibility to guide efficiently the transition to increasing competition.

EEI believes that there is a need for a new vision of the economic and regulatory principles to assure that a competitive electric industry operates efficiently and in the public interest to the benefit of all electricity customers. We believe that the goals of lower electricity prices to all customers and efficiency In the production and delivery of electricity can best be met by assuring a vigorous and efficient wholesale electricity market and by instituting performance-based ratemaking and other retail ratemaking innovations.

The key to realizing these goals will be a comprehensive new scheme of regulation based on the principles of economic efficiency. The scheme must be founded on the recognition that the transition from monopoly to competitive markets requires a fundamental reshaping of the relationship among the participants in markets. It also requires a basic change in the relationship between those participants and the regulators. We cannot assume that change can be managed through incremental additions to and deletions from the old regulated monopoly scheme.

We are not here today to prescribe all of the elements of that new scheme: all of us have some more thinking to do before we fully understand the implications of EPAct. We do, however, want this committee to be aware that a new scheme will be required if the benefits of competition are not to be outweighed by the costs of piecemeal regulations. To that end, we outline below some of the important elements that should be included in any comprehensive plan to assure that a competitive industry operates efficiently and in the public interest for the benefit of all electricity consumers.

First, there must be a consistent set of rules for all competitors. Without consistent rules and obligations for all electricity providers, competition will be skewed, inefficiency will reign and a few customers may benefit at the expense of all the rest. However, the electric utility industry faces a complex patchwork quilt of legislation and regulation at the Federal and State levels that provides competitive advantages to certain participants and undermines the ability to achieve a fair, efficient competitive electric power market.

First and foremost, the Public Utility Regulatory Policies Act (PURPA) accords special treatment for qualifying facilities (QF's), particularly the act's mandatory purchase obligation, which makes no sense in a competitive market. While its initial goals of increasing energy efficiency and use of renewable energy resources are worthy of support, with open transmission access to wholesale customers for all electric generators (including QF's), there is no longer any policy rationale for the mandatory purchase obligation. QF's should be required to compete equally (under the same set of rules) with all other wholesale power generators. Further, EPAct's provisions authorizing the creation of EWG's and FERC's administrative decisions allowing new generators to sell power at market-based rates eliminate any rationale for giving QF's special exemptions from PUHCA or the Federal Power Act. In the current competitive context, the restriction on utility ownership of QF's is unnecessary and counterproductive in that it serves to reduce the number of potential competitors. Removal of the limitation will enable the power supply market to benefit from the significant experience and expertise of the electric utility industry.

Moreover, the administrative determination of avoided costs under PURPA has been implemented in a manner which imposes billions of dollars in excessive, uneconomic costs on utilities and their customers. In California alone, the State utility commission's administratively determined avoided costs require utilities and California's electricity consumers to pay at least $6 billion to QF's above actual avoided costs as determined by the market. Electric utilities and their consumers in at least fourteen other States also are saddled with costs of QF payments above avoided costs. Utilities are attempting to buyout uneconomic contracts for purchases mandated from QF's in order to reduce their costs and restore their competitive positions.

These excessive costs harm consumers and create substantial competitive disadvantages for utilities competing with power generators that are not saddled with uneconomic purchases from QF's. Now that there is a vigorous competitive wholesale power market which establishes benchmark prices for wholesale power, there is no need for the avoided cost rate standard, or any administrative determination of avoided cost.

EEI and its member companies strongly encourage a reexamination of the policy rationales for PURPA in today's competitive power market with particular scrutiny on the need for the mandatory purchase obligation, the administrative determination of avoided costs and the limitation on utility ownership of QF's.

In addition, Federal law grants many preferences and subsidies to Federal power providers, municipally owned utilities and rural electric cooperatives that are not available to Investor-owned utilities. These include exemption from many Federal and State taxes, capital formation subsidies, access to Federal hydropower at rates below competitive market levels, and access to low-interest Federal rural economic development funds. These Federal subsidies are particularly inequitable given that investor-owned utilities serve 76 percent of all electricity customers in the linked States, including the majority of consumers who live in rural areas or municipalities.

Although we realize that tax issues are not under the jurisdiction of this subcommittee, the various tax advantages enjoyed by certain other electricity providers over investor-owned utilities significantly distort the economic efficiencies of the competitive marketplace. For example, in addition to being exempt from most Federal, State and local income taxes, public power entities are also exempt from numerous other Federal excise taxes, including telephone, gasoline, diesel fuel and highway use taxes. These entities also enjoy exemptions from many State and local taxes, including property taxes, gross receipts taxes and sales taxes.

The tax advantage enjoyed by municipal utilities is clearly demonstrated by the following facts: Between 1987 and 1991, investor-owned electric utilities paid nearly $90 billion in taxes, while during this same period municipal utilities paid roughly $3 billion. While Investor-owned electric utilities clearly have a larger market share (76 percent vs. 14 percent), the tax differential is dramatically disproportionate. For instance, if municipal utilities paid taxes in the same proportion as investor-owned utilities, the tax for this 5-year period would have been over $16 billion.[1]

These tax advantages do not include the added advantage of being able to finance utility property on a tax-exempt basis.[2] Currently, municipal utilities may not use tax-exempt financing for private use (i.e., to compete against investor-owned utilities for sale of power outside the municipal boundaries) beyond a $15 million per project maximum cap. There are two bills pending in the House which seek to repeal this $15 million limitation (H.R. 1938 and H.R. 3630). EEI's member companies oppose these bills because they would have the Federal Government further subsidize public power in this competitive market.

Rural electric cooperatives enjoy many of these same competitive advantages. Many of these cooperatives are exempt from Federal, State and local income taxes.,

EEI also believes that power generated at Federal facilities should be available to all Americans. As a first step, EEI's members support the administration's Climate Change Action Plan Initiative to remove barriers to allow private investment in efficiency upgrades at Federal water facilities and to sell the Incremental power at market-based rates. A partnership between the Federal Government and private developers to improve the efficiency of existing Federal hydropower facilities in an environmentally sound manner will enable our country to maximize the benefits from existing energy resources.

As I mentioned earlier, EEI's member companies are a strong presence in rural America. Of the 55.5 million Americans living in rural areas with fewer than 1,500 people or in unincorporated rural areas, almost 60 percent receive their electric service from investor-owned utilities. We have invested in these areas—not only in the infrastructure necessary to provide electric service, but also in economic development to help promote economic growth and a higher standard of living for rural Americans.

[1] Alternatively, if investor-owned utilities were provided with similar subsidies, the cost to Federal, State and local revenues would be almost $70 billion over this 5-year period.

[2] Tax-exempt financing generally reduces the financing cost by approximately 200 basis points. A financing that would generally require a 9 percent interest rate could be financed at 7 percent.

Federal rural economic development programs are intended to help all rural communities and provide rural Americans an equal chance to benefit from new jobs and economic prosperity. However, unlike rural electric cooperatives or municipal utilities, investor-owned utilities are not eligible to receive Federal rural economic development funds. We believe that all rural Americans, including the majority served by investor-owned utilities, should have equal access to these rural development funds.

These laws, preferences and subsidies were adopted in the context of the traditional utility regulatory model. Yet for competitive markets to operate efficiently, all electricity providers should be able to compete on an equal basis with other generators, regardless of the type of ownership. Unless these fundamentals are consistently satisfied, competition will be skewed. We believe that the changing competitive marketplace is an opportunity for congress and the State governments to reevaluate these subsidies.

Second, price transmission realistically. Economically accurate transmission pricing is key to an efficient competitive wholesale market for electricity. FERC received extensive comments on transmission pricing during this past winter and held very informative hearings on this issue in April. In addition, it has taken actions in individual cases to restructure the provision of transmission access on a competitive basis: it has introduced the concepts of network pricing (a customer may obtain flexible service to all points on a utility's transmission network) and comparability (a customer may receive service of comparable quality to that which the utility provides to itself. FERC has not yet applied competitive market principles to transmission pricing. Moreover, the applicability of the network service and the comparability criteria and their relationship to transmission pricing is being fleshed out in individual cases.

Transmission pricing remains a controversial issue. Prices which exceed the value of the transmission facility would discourage competition. Artificially low prices would promote inefficient use of generation resources, cause poor allocation of the system and discourage its maintenance and expansion. FERC received substantial testimony that its current pricing approach deters transmission system expansion. However, FERC has not yet reexamined the basis for its current pricing policy in light of its comparability and other new initiatives. FERC must be encouraged to, find the right balance by pricing transmission services at the current value of such services based on competitive market principles.

Third, provide for an orderly, uniform and equitable transition to increased competition. The transition to a more competitive market should be accomplished in an economic and efficient manner by providing the proper price signals for development of now generation alternatives at the lowest marginal cost. The most important obtain for all customers the benefits of efficient competitive wholesale markets as Congress contemplated in EPAct. By ensuring that the wholesale markets work properly, most, if not all, of the objectives of direct retail access can be achieved. Thus, transition costs cannot be viewed in isolation; all of FERC's initiatives implementing EPAct influence the nature of the transition and thereby affect transition costs.

There are substantial economic, financial and legal reasons to avoid making shareholders bear the full costs of actions and investments that regulators previously found to be prudently incurred. These transition costs include uneconomic generation assets, take-or-pay fuel supply contracts, regulatory assets (costs incurred for which there has been a regulatory promise of future cost recovery), nuclear decommissioning costs, the costs of demand-side management programs, the costs of low-income and other social programs, and the costs of purchased power under QF contracts. The failure to recognize such costs in prices could lead to inefficient choices for future electric supplies by distorting price signals to new generators, which Is not in the public interest.

Imposing such costs on shareholders also would have significant financial impacts upon utilities, potentially affecting the viability of mortgage indentures and other financial instruments. The financial risk profile of utilities would change substantially. Even the mere presentation of the California State utility commission's restructuring proposal has resulted in an approximately $3 to $4 billion reduction in the market value of California's utilities. How will utilities be able to raise new capital on reasonable terms of new regulatory rules preclude recovery of prudently Incurred costs? Will investors have sufficient confidence in the financial integrity of utilities to enable them to maintain credit and attract capital on reasonable terms?

The financial markets need to know that the transition to a competitive market will be orderly. Last fall, the major bond rating services, expressing concern and uncertainty about the changes affecting the electric utility industry, downgraded their outlook for many utilities and downgraded the bond ratings of some companies. This factor, combined with higher interest rates, contributed to a sharp decline in average utility stock prices. From September 1993 through May 1994, average electric utility stocks declined about 25 percent, while the S&P 500 declined about 5 percent. The threat of direct retail access, particularly as proposed by the California utility commission on April 20, ended a weak rally in utility stock prices; between that date and May 9, the industry's average stock price declined 6.2 percent.

The sharp decline in the market value of utility stocks since last fall has resulted in the loss of roughly $70 billion of wealth in the U.S. economy. Utility stocks have been a favored investment for mutual funds, defined contribution plans and individual investors who are seeking current yield, low volatility and preservation of capital. The majority of electric utility shareholders are over age sixty-five or approaching retirement age and are looking for attractive dividends to help contribute to their retirement income.

Our preliminary analysis of FERC's Stranded Cost NOPR indicates that FERC has failed to address the full impact of the transition caused by FERC's implementation of EPAct. We are particularly disturbed by FERC's reluctance to acknowledge that its policies do affect retail transactions and its insistence on constraining its analysis of wholesale issues to a narrow set of contractual issues involving wholesale requirements customers (those customers which purchase all their electricity requirements from another utility).

FERC's stranded cost proposal assumes that the rules and requirements which FERC is establishing for a competitive wholesale electricity market, particularly its pricing, access and comparability initiatives, will have little effect on prices or competition for electricity sales to retail customers. We disagree. So do all our member utilities which are streamlining their operations and reducing costs and rates to better serve their customers. Retail customers will benefit directly from price reductions utilities can gain from more effective wholesale competition. However, such competition may strand costs and investments which regulators previously found to be prudent.

Of course, the specific manner in which FERC implements its transmission pricing, access and comparability initiatives will also directly affect the nature of the competitive market which evolves. Because FERC's rules for wholesale competition are still evolving, it may be too soon to determine precisely the full economic impact and thus the transition costs they will impose on utilities.

FERC's reluctance to address transition costs caused by a new municipalization is extremely troublesome. Although EPAct prohibits FERC from mandating wheeling to a retail customer, it authorizes transmission access for municipalities when In the public interest, which coupled with under-priced transmission, creates the economic incentive for at least some retail customers to municipalize. FERC has not otherwise been so reticent to regulate transactions involving costs which had once been in a retail ratebase. In fact, the NOPR asserts FERC jurisdiction over transmission services in interstate commerce to retail customers. This means that if a utility provides transmission access to a former retail customer, FERC will regulate the rates for use of transmission (and distribution) facilities which had previously been in retail ratebase.

We are also concerned by FERC's narrow, contractual approach to wholesale requirements transactions. FERC has established a rebuttable presumption that all notices of termination clauses in contracts were intended to avoid stranded investment without making any findings establishing what those contracts provide or why those clauses were included. More fundamentally, FERC has failed to address the significance of its own regulatory history and actions. Does FERC's "exit barrier" regulation, which prevents a utility from terminating service until FERC gives specific approval of such termination, establish a presumption that utilities had an obligation to continue to serve wholesale customers even after contracts terminated? Does the fact that contracts were subject to FERC approval and regulation on an embedded cost basis suggest that rates, terms and conditions of such contracts did not reflect competitive market prices or risks? Why has FERC excluded contracts with other, non-requirements customers from stranded Investment recovery? EEI will address these and many other important Issues in our comments to FERC on the stranded cost NOPR.

Fourth, equalize responsibility for social and environmental programs. Electric utilities are major partners with the Federal Government in furthering national environmental and social priorities to which Congress has consistently demonstrated its commitment. Utilities have spent billions of dollars on these objectives, which include improving energy efficiency, promoting the development of renewable energy sources and providing assistance to help low-income customers pay their energy bills. However, these utility investments will be seriously jeopardized if proper price signals or provision for utility recovery of investments are not established.

Clearly, one of the centerpieces of EPAct is the energy efficiency title. Congress's strong commitment to improving energy efficiency was a driving force behind passage of the comprehensive energy bill. Several key provisions in that title are designed to encourage utilities to engage in integrated resource planning (IRP) and to make investments in conservation and demand-side management programs, as well as energy efficiency improvements in power generation.

Utility energy efficiency programs have grown rapidly in numbers and scope since their inception in the late 1970's. The electric utility industry is currently spending roughly $2.5 billion per year on over 2,300 of these programs. Utility-sponsored energy efficiency programs have deferred the need for almost 25,000 megawatts of new generating capacity. By the year 2000, the amount of deferred capacity is expected to reach 45,000 megawatts.

Congress also established greater utilization of renewable energy resources as a goal in EPAct. In addition to promoting research and development, EPAct authorized Federal incentive payments to qualified renewable energy facilities and extended tax incentives for energy production from renewable energy resources.

Just as with energy efficiency programs, electric utilities have been major stakeholders In research, development and deployment of renewable energy resources. In the case of photovoltaics (pv), over 80 utilities are members of the Utility Photovoltaic Group (UPVG). Utilities are planning or developing several hundred megawatts of wind generation across the country. In addition, the Utility Biomass Energy Commercialization Association has been formed to evaluate the variety of biomass technologies. Finally, electric utilities are active players in the Federal-private collaboratives on commercialization of photovoltaics, wind, biomass and geothermal energy proposed in the administration's Climate Change Action Plan.

Most recently, almost 100 investor-owned utilities have responded to the Climate Challenge, a voluntary partnership program initiated by EEI and the electric utility industry with the Department of Energy (DOE), to take measures to reduce, avoid or sequester the growth of greenhouse gas emissions. In combination with the municipal utilities and rural electric cooperatives participating in the program, DOE estimates that the Climate Challenge utilities may represent as much as 80 percent of the total electric generation in the United States.

Utility measures to reduce greenhouse gas emissions will include both demand-side and supply-side options including greater emphasis on increasing energy efficiency, the use of cost-effective, environmentally sound electric technologies, and greater deployment of renewable energy resources. Obviously, utility activities must take into consideration impacts on ratepayers, competitive situations, resource planning and the mandates and positions of regulatory commissions.

One of the most important social programs in which utilities have been involved is the Low-Income Home Energy Assistance Program, which helps low-income households moot their basic energy needs. In addition to being vigorous proponents of this Federal program, utility companies also provide 48 percent of the funds for the country's fuel funds, which operate in 44 States and the District of Columbia. Private fuel fund resources have increased steadily over the years. In 1991, fuel funds provided over $46 million to more than 396,000 households.

However, investments in environmental and social programs exert upward pressures on electric rates. To promote efficiency and fairness, the costs of such social and environmental programs should be borne equally by all electricity customers. Otherwise, competition could have the potential to restrict severely or to negate entirely the ability of regulated utilities to participate in such programs and initiatives in an effective manner.

The situation could be exacerbated if utilities are not allowed to recover the full costs of past expenses in such programs that were previously found to be prudently incurred and that were capitalized for recovery over a number of years, rather than expensed in the year in which they were incurred. The costs of such programs constitute a large amount of the costs that could be stranded by the transition to competition. If investments in such programs are placed at risk, now utility investments of all kinds, particularly investments in demand-side management and renewable energy resources, which may lower long-term costs of energy use but raise customer costs in the short term, would be substantially discouraged.

Fifth, make adequate provision for system maintenance and control. All generators and customers using the integrated electricity grid must bear appropriate responsibility for assuring its safe and reliable operation. The rules allocating responsibility for matters such as spinning reserves, generating reserves, need for new transmission capacity, and other system issues (control of voltage, frequency and reactive power) and the provisions for recovering the costs for ancillary services must be clear and explicit. The cost accounting and analysis methods for unbundling costs associated with these services will take time to develop properly.

Sixth, make the obligation to serve and the obligation to purchase symmetrical. The obligation to serve is a fundamental component of the traditional regulatory compact and provides important assurances to utilities and customers alike. Many customer groups which advocate competition still insist on the right to return to their original utility at regulated rates. Regulatory commissions are also reluctant to waive the obligation to serve. For example, while FERC's NOPR on stranded investment requires termination clauses in contracts, FERC does not propose to eliminate its regulatory authority to approve such terminations before they take place. Similarly, the retail access proposals of the California and Michigan commissions require traditional utilities to continue to offer service to ex-customers if they choose to return. The duty to serve returning customers will require utilities to maintain extra generating capacity, at significant costs, and significantly disrupt efficient planning.

In effect, customers and commissions are currently unwilling to accept the full risks of competition. A utility's obligation to customers who have competitive alternatives should be based primarily upon competitive market principles—not regulatory rules. In competitive markets, competitors have the right not only to enter markets, but to exit as well. Conversely, customers choosing to participate in and benefit from competitive markets should accept all risks and responsibilities of participating in such markets. They should not be able to play the market while forcing the traditional utility to be the supplier of lest resort. Any regulatory structure which allows customers to play off competitive and regulated rates will guarantee inefficiency and is guaranteed to fail.

EEI believes that these positions of customers and commissions demonstrates the continuing importance of the obligation to serve and the importance of preserving—and reforming—the traditional regulatory model.

Seventh, allow pricing flexibility and freedom to innovate within the present regulatory model. Lower electricity prices to all customers and efficiency in the production and delivery of electricity can best be met by assuring a vigorous and efficient wholesale electricity market and by instituting retail ratemaking innovations. EPAct's prohibition on mandatory retail access, which we believe is essential, properly focuses attention on these goals and avoids the need to address the many complex questions inherent in implementing direct retail access and the serious risks to customers and utilities which could follow from incomplete or improper resolution of these issues.

Traditional utility regulation—that is, cost based rate of return regulation—was never designed for competitive markets. As practiced by FERC and State commissions, traditional regulation tends to result in rigid prices, regulated capital structures, an obligation to serve all customer segments (whether they are profitable or not), slow decisionmaking processes, high administrative costs, limited upside potentials and substantial downside potentials for utilities to take risks, and use of utilities as tax collectors and as agents to forward a government's social agenda. In today's competitive atmosphere, traditional rate regulation can put utilities at a competitive disadvantage, for it is largely incompatible with market forces and can squelch the development of competition. The management challenge for utilities and regulators is to develop regulatory policies that work efficiently with markets and not against them.

Attributes of policies that are consistent with market forces include pricing flexibility, market responsiveness, regulatory certainty, confidentiality of data, and risk-return symmetry. A number of alternative regulatory pricing models are being developed by utilities and tested in the electric utility industry that incorporate these attributes in one form or another. Some alternatives represent variations on traditional regulation, while others are outside of the rate base/rate of return regulatory framework.

These policies should stimulate the development of new products and services. A utility should be free to offer new services to its existing customers and to enter new markets—either directly or through affiliates. The international electricity market and telecommunications offer two good examples. Safeguards can be provided to assure that customers obtaining traditional utility services do not subsidize competitive ventures.

Eighth and lastly, there must be regulatory certainty. The Federal Power Act (FPA) is the source of authority for FERC regulation of the wholesale electric power market and transmission In interstate commerce while PUHCA is the authority for Securities and Exchange Commission (SEC) regulation of utility holding companies. The FPA was enacted to eliminate the "regulatory gap" caused by constitutional limitations on State jurisdiction over the flow of electrons in interstate commerce. It was written and, up to now, interpreted in the context of the traditional regulatory compact in which a utility received a secure franchise in return for a firm duty to serve under regulated rates, terms and conditions. In this context, the rules for the division of jurisdiction between the SEC, FERC and State regulatory authorities were relatively clearly defined, although some important disputes have been litigated all the way to the Supreme Court in the past decade. The introduction of vigorous wholesale competition has raised issues concerning the respective roles of the SEC, FERC and the State commissions.[3]

As State commissions begin to consider retail competition and retail access, substantial legal questions arise about how the Federal Power Act would be applied in a retail access context, and such questions make it difficult to avoid regulatory inconsistencies or regulatory gaps. The most pressing question is who has authority to establish rates, terms and conditions for transmission services in interstate commerce to retail customers.

The Michigan commission proposes to adopt an experimental retail wheeling program and asserts jurisdiction to establish the rates, terms and conditions for wheeling to retail customers under that program. Michigan utilities must shortly begin a proceeding before the Michigan commission to establish such rates, terms and conditions. However, FERC has stated that because the Federal Power Act gives FERC exclusive jurisdiction over the rates, terms and conditions of transmission in interstate commerce, it maintains exclusive jurisdiction over the rates, terms and conditions of such transmission to retail customers. Since Michigan utilities are connected to the interstate grid where electrons flow in interstate commerce, there is a clear conflict between the two commissions over which would regulate the rates, terms and conditions of transmission to retail customers.[4]

A Michigan court recently declined to resolve this dispute because the Michigan commission's decision is still a proposal and therefore not ripe for review, notwithstanding the time and expense that will be incurred to address transmission rate issues before the Michigan commission.

If a State requires retail access, many other questions regarding the division of authority between FERC and the State arise. The Federal Power Act reserves to the States the regulation of distribution facilities, but FERC has already asserted that its authority to regulate transmission services in interstate commerce extends to transmission which takes place over distribution facilities. How will the authority over distribution facilities be divided? Who will have the responsibility to assure that all customers pay their fair share for system maintenance and control, and environmental and social policy costs? And who will have the authority to address the costs of the transition to competition? FERC's Stranded Investment NOPR states that while it has jurisdiction to allow recovery of stranded costs incurred to serve retail customers, it may decline to do so. Is FERC's definition of its authority correct, and if so, would the States retain such authority if FERC declined to act?

Electric utilities and regulatory commissions must have a clear articulation of the boundaries of regulatory authority over their business activities. Many believe that it will take litigation all the way to the U.S. Supreme Court to resolve the myriad of jurisdictional issues that will ensue from any comprehensive actions aimed at mandating direct retail access. This would take years to complete. We would prefer that regulatory actions to achieve any new regulatory and economic structure for utilities be carefully coordinated by FERC and the State commissions and clearly authorized by law before they are taken. However, if conflicts or gaps arise which are not so resolved, legislation could be necessary.

For the foregoing reasons, EEI believes that the goals of lower electricity prices to all customers, efficiency in the production and delivery of electricity, and implementation of competitive electric service business principles can best be met by (1) assuring a vigorous and efficient wholesale electricity market and (2) instituting performance-based ratemaking and other retail ratemaking innovations at the State level. Congress can play a substantial role to play in achieving these goals.

Thank you again for the opportunity to testify on the significant changes affecting the electric utility industry.

Mr. SHARP. Mr. Glynn.

---

[3] Because this committee recently completed oversight hearings on these issues, this statement does not go into detail about these issues.

[4] The savings clause in section 212(h) of the Federal Power Act, enacted as part of EPAct, has no effect on this analysis because it simply preserves whatever authority the States may have had prior to the enactment of EPAct. State authority over retail marketing areas of electric utilities does not after these conclusions. It is not the unbundling of transmission, but the sale of transmission services to others, that affects the regulatory status of the transaction. The sale of transmission services in interstate commerce triggers section 205 of the Federal Power Act, which requires that FERC regulate the rates, terms and conditions of the transmission transaction.

STATEMENT OF ROBERT D. GLYNN

Mr. GLYNN. I thank you, Mr. Chairman, for inviting Pacific Gas and Electric Company to share our thoughts with you on this subject. I am Bob Glynn, executive vice president of PG&E.

Our thoughts on this are based on work we have done since the mid-1980's preparing our business for the onset of significant competition in its electricity part. It is no surprise that these comments are based on things that are going on in California right now, and I recognize the framework for this conversation. However, we can't and won't ignore some of the California issues.

We believe that competition in the supply of electricity is good for customers, is reasonable to begin implementing now, and under the proposal that we have advanced to the California PUC is implementable without a major down side to the many interests that are involved in this industry.

By competition, we mean customers being able to directly access the electricity supplier of their choice, as they do now when they choose their long distance supplier or their natural gas supplier.

In some contexts, people speak as if restructuring of the electric industry is a future event to be contemplated, analyzed, debated. That is pretty far from the truth.

The restructuring began in earnest in 1978 with PURPA. It was facilitated in 1992 by the Energy Policy Act, and today in many places electric generation is simply not the sole province of the former vertically integrated utility business. PG&E today, almost a quarter of the energy that we provide our customers comes from independent suppliers and not from PG&E generators.

The California Public Utilities Commission in its proposal outlined five principles which it felt needed to be addressed in order to see direct access through. Most of us have read those, memorized them, perhaps. But they are short, so we will repeat them: that the benefits of direct access flow to customers; that the financial integrity of utilities be preserved; the reliability of the system be maintained; costs not be shifted between customer classes; and finally, environmental and social programs be maintained.

We agree with those principles. We believe that a transition to direct access can, in fact, be crafted to achieve all of them, and we have proposed such a transition. It would phase in direct access in an orderly manner starting with the largest customers, who are, frankly clamoring the loudest, to have such access. We would seek no special transition charges for PG&E-owned—for our own generation. And we would accomplish the objectives of direct access and customer choice, including the right of customers to choose or not to choose direct access, and fully observe the principles which the CPUC articulated and which we just reviewed.

It turns out that in California since April 20th of this year most of the debate has focused not on what the CPUC itself proposed but on how not to achieve what they proposed. And five red herrings have been identified and are raised in the direct access debate, red herrings that allege that because of them direct access can't and shouldn't be implemented. And those red herrings are the incompatibility of direct access with environmental and social programs, reliability, power pools, transition costs, and the jurisdiction issue.

Each of these is important. Resolving each of them will be difficult. But they cannot and should not become a reason to simply throw up our hands and say customers shouldn't have choice. It isn't worth the effort to let them have that choice.

Very briefly, each of those five red herrings can be discussed.

Environmental and social programs can be funded at whatever level policymakers deem appropriate. The vehicle for that will transition as the industry transitions, but if policymakers want to see environmental and social programs linked to energy businesses continue they can continue.

Renewables can continue. Perhaps not in precisely the same way that they do right now. But there are already mechanisms in place for some components of the renewable industry through tax credits and the potential of production credits that can put them on a more economically competitive playing field. So there is no roadblock there.

Some raise reliability as a roadblock. The PG&E system today operates, as I mentioned, with about a quarter of its daily energy received from generators owned by others, and we also manage on behalf of many other entities, principally municipalities, that supply that comes into our service territory for their use.

Reliability is adequate. The lights are on every day. We get through natural disasters, and hundreds and hundreds of transactions and contracts are observed every day every hour. We can get through the reliability issue.

Power pools. A number of organizations and individuals have advocated enhancing wholesale power pools as an alternative to direct access. Power pools are worth looking at if you believe that customers really shouldn't have the choices that they are asking for. We believe that they should have those choices.

They are worth looking at if you believe that the regulatory structure will deliver lower costs to consumers than an active competitive marketplace will deliver. We believe the marketplace in the end will deliver lower costs.

Enhancing the pools that we have now seems to make sense if we ignore for a minute the working efficiencies of existing wholesale pools like the Western Systems Power Pool, or if we ignore the existing reliability and planning coordination from the Regional Reliability Councils that exist around the country, or if we ignore the promise of regional transmission groups that haven't waited for this debate, but have already filed before FERC and are ready to go. Groups include not only the traditional investor-owned utilities, but also transmission dependent entities, brokers, independent power producers—the whole range of players in a more competitive business.

And, at the end of the day, depending on how far one looks, enhancing wholesale pools make sense if what one is interested in is pools and not customer choice. We think the customer choice makes more sense at the end of the day.

Transition costs is the fourth herring. Transition costs have to be dealt with. The CPUC has a principle which sets out their intention to do so.

PG&E's proposal includes the use of both explicit transition charges to cover the investments made by third parties that we are responsible for delivering to customers, the contracts that we have been required to enter into, and proposes the use of time in lieu of transition charges for investments that we have made.

And there is a couple of key things worth remembering about transition charges. First of all, they arise from a utility's obligation to serve, which does differentiate us from many of our customers, although when I meet with customers they like to say, "Hey, when we make a bad investment we have to live with it."

But they do have the choice as to whether or not to make a particular investment that might or might not turn out to be bad. At the end of the day, utilities have had up to now the obligation to serve customers as they arise.

Maybe more significantly, about transition charges, they don't represent a new cost entered into rates. They are already embedded in rates right now, so they are not a force that would move rates upward. They are already there.

The final herring is jurisdiction. The multiplicity of separate jurisdictions which will probably have some role in approving or facilitating or simply allowing direct access to go forward. Of course, FERC is going to have a role in this. And of course the States are going to have a role in this. And for transactions that cross national borders there are going to be other regulatory entities that have a role in it.

We are confident that if California designs a direct access program which meets the principles that have been identified for it, and one which our proposal addresses directly, it will serve to achieve substantial customer and public benefit, and because of that the issue of jurisdiction will become an issue of implementation.

I don't think it will be tidy but it will be an issue more of how do we get these benefits to consumers and not how can we stop the progress that otherwise would go forward.

The concept of the five red herrings comes from a Dorothy Sayers novel of that same name. Her novel is a murder mystery. The red herrings in the direct access debate shouldn't be permitted to murder consumers' opportunity for choice. Ultimately, as we have seen in so many other industries, competition will replace monopolies and customers will have choice.

Thank you.

[The prepared statement of Mr. Glynn follows. Attachments are retained in the subcommittee files.]

STATEMENT OF ROBERT D. GLYNN, JR., EXECUTIVE VICE PRESIDENT, PACIFIC GAS AND ELECTRIC COMPANY

Over the past century, the Nation's electric utilities have built what is without doubt the world's finest electric supply system, a system which provides light, heat, and power to virtually every American at an exceptional level of reliability, safety, and performance.

This system was developed largely as an integrated business, with local utilities building and owning the generation, transmission, and distribution systems required to meet their customers' needs. Today, that traditional, vertically-integrated system is changing. This Congress gave birth to that change in 1978 when, with passage of the Public Utility Regulatory Policies Act, it encouraged the development of power plants fueled by renewable resources or by using more efficient cogeneration technology. In the ensuing 15 years, a robust and competitive independent power production industry has emerged, capable of supplying electricity to millions of consumers. That industry received further impetus in 1992, when Congress passed the Energy Policy Act of 1992, which established electric wholesale generators as a separate entity under the law.

The direction should be clear to everyone. The generation of electricity is no longer a monopoly business, the exclusive and inevitable territory of the distribution utilities. Today, this is a competitive business, with major players vying for the opportunity to build power plants to serve growing consumer needs.

There are those who still talk of "the future restructuring of the electric industry" as if it is coming someday, but not yet here. I submit that the restructuring, in fact, already is largely here in the true economic sense, that competitive electric generation supply already is a reality. The only question is how soon the legal framework of regulation will recognize and accommodate this reality of multiple suppliers to allow them to become directly associated with multiple, ready, willing retail buyers.

There also are those who fear this change, who believe that allowing customers to choose their own suppliers of electricity will destroy the many positive attributes of the current system, and that consumers will suffer, not benefit These are legitimate concerns, but they are concerns which should be focused on managing the transition to a more competitive world. They should not be allowed to stifle or unreasonably delay that transition.

At Pacific Gas and Electric Company, we believe that competition in the supply of electricity is desirable. More than 5 years ago, we joined with Bechtel Corporation to enter the independent power production business through our partnership, U.S. Generating Company. We believed then, and we believe today, that the development of future electric generation will rest more in the hands of independent entrepreneurs, and less in the hands of vertically integrated, regulated monopolies.

We also believe that, as this competitive electric generation business matures, consumers should obtain the full benefit of the lower prices it will produce. To a large extent, the benefit is being realized through the wholesale market that has been facilitated by the Energy Policy Act and by various regional pools. But, ultimately, the full benefit will accrue to consumers only if they have the ability to choose among competitive suppliers, to have direct access to electricity suppliers in the same way that, today, they can choose among providers of long-distance telephone service or of natural gas supply.

In April of this year, the California Public Utilities Commission proposed a dramatic step in this direction. It suggested that California move quite rapidly to a world in which all consumers would have direct access service so they can directly participate in the competitive electric supply marketplace. The clamor that followed this proposal has been intense, with numerous parties expressing approval of the Commission's proposal and others registering deep concerns about the future which the Commission described in its proposal. Rarely has a piece of regulatory policy received such nationwide, and even worldwide, attention from utilities, legislators, environmentalists, consumerists, and academic economists, not to mention the legion of lawyers and consultants who inevitably surround any major change in regulatory policy.

Obviously, my company is involved in this process. We are the largest gas and electric utility in the Nation, and the largest supplier of electricity in California. We have significant utility generation assets, and we have a direct and pressing interest in ensuring that the California commission's proceedings yield a rational and positive result.

In this process, we have made it very clear—and I want to reiterate for this committee that we do not disagree with the end point which the California commission has described. We share the Commission's vision of the future: we believe that all consumers can and should have the opportunity to have direct access to the competitive electric generation market, and that distribution utilities should provide that access in a manner which ensures continued equity and reliability.

We have filed with the Commission a proposal which would phase in direct access in an orderly fashion, starting with the largest customers who are most likely to benefit from it in the near term. We have suggested that, under this proposal, PG&E would seek no special treatment or "transition charges" to recover any stranded costs associated with our company-owned generation, including our nuclear power plant, Diablo Canyon. I have included our submittal to the Commission, which outlines our proposal in more detail, as an attachment to this testimony.

Much of the debate in California over direct access, or retail wheeling as it is more commonly called in the industry, has properly focused not on whether it should occur, but how we can best achieve it. In its original proposal, the Commission outlined five principles which would guide its efforts. We agree with those principles and believe that it is possible to craft a transition to direct access—along the lines of our proposal to the Commission—which will achieve them all.

The first principle is that, to the extent that competition creates lower prices, those benefits should flow to consumers. This may not occur overnight. But it is essential that, if prices come down through the working of an open and competitive marketplace, the savings should be felt by consumers in their pocketbooks. We believe allowing customers the ability to participate directly in that marketplace, through direct retail access, is the only way to truly ensure that consumers receive the full amount of those benefits.

The second principle is that consumers should be assured that the financial integrity of the utilities will not be impaired. The utility business is a highly capital-intensive business. It is essential that the utilities continue to have the financial strength to raise the capital they need, at a reasonable cost, so they can build and maintain the extensive infrastructure essential to ensure reliable utility service. It is equally important that shareholders not be penalized for investing their savings in the utility industry. Hundreds of thousands of individual shareholders have invested in California's utilities, with the faith and confidence that, under the traditional regulatory compact, their investments would not be confiscated by regulatory actions. The transition to direct access must not betray that trust and reliance on sound public policy.

The third principle is that consumers must continue to receive reliable energy supplies, and that the new regulatory system must continue to assure that all consumers have access to electricity. We will need to look closely at the traditional utility obligation to provide universal service. We will need to recognize that, in a competitive marketplace, our obligation must over time, become an obligation to provide access to electricity—an obligation to connect customers to the distribution grid so they can buy power from whomever they choose—and not an obligation to plan for and to build the actual facilities which produce that power.

The fourth principle is that one group of consumers must not unfairly benefit at the expense of another group. We must be vigilant in assuring that, in implementing direct access, the rates charged for utility service continue to reflect the cost of serving each group of customers, and that we do not have the rates paid by one customer group subsidize low-cost power for others.

The fifth principle is that consumers must continue to receive the benefits of environmental and social assistance programs which currently are delivered through the vertically integrated utility system. We must be certain that customer energy efficiency, low-income energy assistance, and other similar programs—which directly or indirectly benefit everyone—continue to be borne by all consumers, and that those who opt for direct access cannot escape their fair share of these costs. We believe that this can be achieved in the direct access future.

In commenting on the California commission's proposal, some have noted that this list of five principles reads like motherhood and apple pie, but that it simply is not possible to achieve them all. We do not agree. While it may take time, we believe they can be achieved through a carefully designed transition to direct access along the lines we have proposed to the Commission. And we believe that such an orderly transition can be implemented in California if the parties focus on that objective rather than on all of the obstacles to achieving it unfortunately, that has not been the case. Much of the debate about direct access has dealt with the reasons why it cannot be achieved, not why it can and should be. Dorothy Sayers, the noted British mystery writer, once wrote a book called, "The Five Red Herrings." In many respects, that might be an appropriate title for the dialog on direct access which has occurred in California so far. In no way do I mean to minimize the concerns raised by many of the parties. They are legitimate. But they should not become reasons to throw up our hands and decide that providing consumers greater choice simply isn't possible or isn't worth the effort.

What are the five red herrings in the direct access debate? They are (1) environmental and social programs, (2) rehability, (3) power pools, (4) transition costs, and (5) jurisdiction.

The first two on the list, you will note, are issues which I just mentioned in connection with the fundamental principles for restructuring. They are critical. They must be addressed. But they are not "show stoppers."

Environmental and social programs can be continued, at whatever levels policymakers deem appropriate. The key is to make sure they are funded through a separate charge on all electricity customers' bills, and not be imbedded in "bundled" utility rates. They should be paid by all customers who are connected to the distribution utility system—whether they buy power from the utility or from other suppliers.

Direct access should not be withheld from electricity consumers simply to provide a convenient vehicle to fund and deliver environmental and social programs, rather those desirable programs should be structured for success in a direct access world.

The development of renewables also can continue as we make the transition to direct access. It has been the policy of both Congress and the State of California to encourage the development of renewable resources in two ways: by providing taxpayer subsidies through tax credits and similar devices, and by providing consumer subsidies through requirements that utilities purchase power from renewables at above-market costs. In the future, as utilities no longer are the exclusive suppliers of power, it is essential that they no longer be required to buy power at uneconomic prices. Some renewable technologies are approaching commercial viability and soon will be able to compete with other technologies, without subsidies. To the extent that others still require special supports to make them price competitive, those subsidies can and should be provided at the production level, through tax credits and similar supports, and not at the end-user marketplace.

Reliability also is critical, and it will be assured if direct access is implemented in an orderly manner. We already manage multiple transactions through our power grid, and opening those transactions to retail access should not impose a severe hardship. Currently, about 400 independent generators are connected to PG&E's grid, dwarfing in number the about 100 PG&E-owned generators. Rehability is adequate, the lights are on each day, a multiplicity of contracts are observed. As the number of transactions increases, we will need to build the billing, metering, and other infrastructure to manage them. This is one of the reasons why we propose phasing in direct access over a multi-year period, starting with the largest customers whose transactions can be managed today, and leading ultimately to the millions of residential customers for whom new systems will be essential.

The challenge here is not just with the hardware. We also need time for a robust marketing/brokering infrastructure to develop. As we have learned from our experience in restructuring the gas industry, the role of marketers and brokers, and the development of a futures market with financial hedging mechanisms, are important elements for a competitive marketplace. The market must be allowed to develop with minimal regulatory prescription. This will occur once direct access becomes a reality.

This leads me to the third "red herring"—Power Pools. Several of my colleagues in the utility business and some very persuasive members of the academic community have proposed that, instead of direct access, California's commission should create a state-mandated wholesale power pool. They argue that we can achieve many of the benefits of direct access through such a mechanism, and can avoid the many difficulties posed by retail competition.

We agree that wholesale competition is a desirable first step and, in fact, we believe that such a market already exists in the West. We are already blessed with one of the most fully functioning and open wholesale pooling infrastructures in America—the Western Systems Power Pool. PG&E and about thirty other entities, including brokers, independent power producers, and municipal utilities, recently filed to create the Western Regional Transmission Association, which will further enhance system planning, coordination and operation.

Because of the high level of efficiency and coordination which already exists, we believe that imposing a government-mandated power pool would not provide additional efficiency benefits for consumers. Under the most advanced and complex proposal, even if new benefits could exist, real, widespread consumer benefits from the creation of a wholesale pool would start to accrue only after a major investment is made in time-of-use metering and other hardware. It would seem to require a suspension of reality to believe that a regulator mandated, designed and approved wholesale pool would be more economically efficient than customer choice under direct access. In short, a new wholesale pool may simply have the effect of delaying for many years consumer choice and the consumer benefits which can result from beginning an orderly transition to direct access today.

It should be emphasized that a fully functioning power pool probably will be one of the essential pieces of the infrastructure we will need before direct access is extended to large numbers of customers, such as residential and commercial customers. However, we do not believe that it is a prerequisite to beginning direct access. The schedule we have proposed to phase in direct access, starting with the largest customers in 1996, would permit the development of a mature retail power pool over the same time as an increasing number of customers become eligible for direct access. Such a retail pool would form naturally, based on the needs of buyers and sellers, and need not and should not be mandated by the Commission.

The fourth issue we hear discussed frequently is transition costs. This is a very serious issue and one which must be resolved. But it, too, need not be an insurmountable barrier. The problem, of course, is that utilities have made investments and entered into contracts and supply arrangements in the "old world" of regulation. some of these investments may become uneconomic, or "stranded," once there is direct access. Leaving the utilities exclusively liable for these costs would be crippling to the utility industry. At the same time, imposing huge "transition charges" on customers seeking to take advantage of direct access would materially diminish the benefits which providing customers greater choice are designed to achieve.

I cannot speak for other States, but at least in California, this, too, can be resolved by a combination of time and transition charges. The implementation schedule we have proposed allows sufficient time for us to "work off" the bulk of the above-market costs associated with our utility-owned generation. As customers became eligible for direct access, we would be permitted to offer services and prices designed to compete with other suppliers, with minimal regulatory intervention. The amount of any discounts would be borne by our shareholders, not shifted to other customers in higher prices. Over time, as more and more customers had the option to choose other suppliers, we would be more and more at market risk. In the meantime, all other customers would continue to pay rates which reflect the generation costs which regulators already have found reasonable and which we are allowed to include in rates.

We do propose transition charges for above-market costs we incur to purchase power under mandated contracts with qualifying facilities. We also propose an ongoing surcharge on all customers bills to pay for the various environmental and social benefit programs currently funded through rates. These are costs which were imposed on utilities to achieve various public policies, costs which already are included in utility rates, and, therefore, are properly borne in transition charges.

The final "red herring" in the debate is jurisdiction. Again, this is an important issue and undoubtedly a major focus of discussion in Washington as well as in California. California cannot and should not act in a vacuum and, in fact, may not have the power to do so, even if it wanted to. Whatever the California commission does must fit with Federal and State legislation, and with U.S. Department of Energy and Federal Energy Regulatory Commission policy.

This is a serious matter. But it can be dealt with. We are confident in the good motives of the Congress, the FERC, and the other governmental agencies that must deal with this issue. The concern of each is to ensure that the public interest is served. If California designs a direct access program which meets the principles I discussed at the outset, it will achieve substantial consumer and public benefits, and many of the issues of jurisdiction will become simply matters of implementation.

The debate on direct access obviously will continue for many months and years. It is an important debate, a debate worth engaging in, because it deals with making fundamental changes in one of this Nation's most important industries. There will be some who say "it can't be done," and others who say, "do it quicker, do it now." Balancing those competing interests and views will take time, patience, and commitment.

But we believe it can be done. It would be naive to believe that any proposal will satisfy everyone. It won't. It can't. But if it is properly crafted, it will balance the interests of the parties sufficiently so that, over time, all the major objectives—which I believe most of us share—can be achieved. Ultimately, as we have seen in so many other major industries, competition will replace monopoly, and consumers will have choices.

Mr. SHARP. Mr. Divine? Very pleased to hear from you at this point.

STATEMENT OF DOUG DIVINE

Mr. DIVINE. Mr. Chairman, my name is Doug Divine. I am here today representing the National Independent Energy Producers as a member of its Board of Directors and Chair of its Marketing Competition Task Force. I am also the Director of Planning and Regulatory Affairs for Falcon Seaboard Resources, headquartered in Houston, Texas.

Mr. Chairman, as you prepare to retire from Congress later this year, NIEP wishes to applaud the extraordinary leadership and vision you have demonstrated on competitive electric power issues. Indeed, your legislative leadership over the last two decades has been critical to the growth and vitality of the competitive power industry as we know it today.

Since NIEP was founded by the pioneers of independent power in 1986 it has been on the leading edge of the push for competition in wholesale power markets as a best means of increasing benefits for consumers without sacrificing safety and reliability. At times, delivering this message has meant tackling head-on the difficult market power issues to be faced in making the transition from a vertically integrated monopoly industry to one characterized by competition.

NIEP thanks the committee for the opportunity to present its views on the future of the electric industry. We believe that competitive forces will continue to reshape the industry over the next decade. To reap the most from this restructuring, full and fair wholesale competition must be achieved. To do this three critical transition elements must be addressed.

First, regulators must see their role as stakeholders in and stewards of competitive markets with responsibility to guard against anticompetitive behaviors.

Second, performance-based regulation must be implemented to align the interest of utilities with their customers and the competitive forces of the marketplace.

Third, the divestiture of utility generation assets from transmission and distribution must be considered to reduce the regulatory burden of policing competitive markets. Furthermore, we believe there will continue to be a role for integrated resource planning in this competitive electric industry structure.

And finally, the importance of PURPA protections until the transition to this new energy structure is completed must be acknowledged.

Congress made the entrance of independents into competitive generation markets possible with the passage of PURPA and further embraced competition with the Energy Policy Act. The fact that independent power producers have built over 30 percent of the electric generating capacity added in the last 4 years shows how much the industry has changed. Yet the fact that only 6 percent of total generating assets nationwide have been procured through competitive processes demonstrates that we still have a long way to go.

We believe that full and fair wholesale competition has not yet been achieved. Many of the provisions of the Energy Policy Act are not yet fully implemented, including that address separating ownership from use in transmission. Utilities still wheel considerable market power through their control of transmission, distribution and the vast majority of generation assets.

There are impediments in the existing system that have yet to be overcome, particularly the relationship between ownership and use of the electric transmission infrastructure. The Federal Energy Regulatory Commission has just begun to implement the vision of open access transmission and pricing policy outlined in the Energy Policy Act through a series of recent promising orders, and we have yet to see how regional transmission groups and other institutional arrangements to foster a competitive electric market will develop.

For these reasons, NIEP believes that although the wholesale generation market is headed toward full competition it is not there yet.

What is the role of regulators during this transition period? Regulators, particularly at the State level, have a role to play now as stakeholders in and stewards of competitive markets with the responsibility to guard against anticompetitive behaviors. Typical examples of such behaviors include forced contract renegotiations and curtailments, manipulation of competitive bidding programs, and unfair pricing practices like cogeneration deferral rates.

NIEP believes that problems with wholesale competition should be addressed on a parallel track with industry restructuring. Changes in the industry caused by the forces of competition require comparable changes in the regulatory framework. Traditionally, utility cost-of-service regulation must give way to performance-based regulation where utilities are rewarded based on how well they run their plants, procure new resources and meet customer needs. Performance-based regulation is more compatible with an increasing competitive industry structure.

What will the future look like? The industry is clearly in a period of transition. Continued regulatory oversight remains critical to ensuring that competition generation is full and fair while the transmission side of the business continues to be treated like the monopoly it is.

NIEP believes that the task of regulatory oversight is made doubly hard if utilities are allowed to participate in the market as both buyers and sellers of generation services while they remain owners of transmission and distribution facilities. Therefore, NIEP has argued that the California commission consider separating ownership of generation from transmission and distribution by spinning generation off into independent new corporations as a better way to remove the incentives to thwart competition when utilities are under the current industry structure.

What is the role of integrated resource planning? NIEP has long supported utility adoption of integrated resource planning. Some commenters have expressed concerns that the expansion of competitive forces in electric markets would drive utilities to abandon IRP, environmental considerations and utilities' roles as providers of social benefits. NIEP strongly believes that this need not be the case.

The fundamental spirit of IRP is consistent with increased competition in the electric industry and the separation of functions once performed by vertically integrated utilities. NIEP believes that IRP can and should be the framework within which a robust and competitive electric generation market flourishes.

If utilities are still the appropriate entities to deliver social benefits, for example, programs to assist low income customers, then we need to work together to investigate new mechanisms that will enable them to do so without distorting the emerging competitive markets.

I understand that some members of the committee have asked whether the passage of the Energy Policy Act and the continuing changes in the industry obviate the need for PURPA. NIEP does not believe this to be so. The Energy Policy Act builds on PURPA rather than replaces it.

The Energy Policy Act starts with the assumption that a competitive wholesale market exists. However, the competitive pressures facing many utilities are driving them to attempt to close doors to independence rather than compete openly and fairly.

This demonstrates a continuing need for PURPA as a lever that holds open the door to electric generation markets so that those who have been on the outside can continue to enter. This is particularly important for renewables.

In conclusion, NIEP urges the House Subcommittee on Energy and Power to continue to support efforts that will lead the way to increased competition, reduce rates for consumers and improve the environmental performance in the U.S. electric industry. Competitive forces will continue to reshape the industry over the next decade.

To reap the most from this restructuring, full and fair wholesale competition must be pursued, performance-based regulation must be implemented, divestiture of utility generating assets must be considered, a role for IRP must be recognized, and fair transmission access and pricing policies must be adopted.

Thank you, Chairman.

[The prepared statement of Mr. Divine follows:]

STATEMENT OF THE NATIONAL INDEPENDENT ENERGY PRODUCERS

Mr. Chairman, my name is Doug Divine. I am here today representing the National Independent Energy Producers (NIEP)[1] as a member of its Board of Directors and as chair of the organization's Task Force on Markets and Competition. I am also the Director of Planning and Regulatory Affairs for Falcon Seaboard Resources, Inc., headquartered in Houston, Texas.

NIEP welcomes this opportunity to present its views on the future of the electric power industry to the Subcommittee on Energy and Power. Since its founding, NIEP has advocated increased competition and a greater reliance on market forces as the most effective means to deliver reliable, innovative, and cost-effective energy services in the United States. NIEP is pleased that the committee embraced competition in the Energy Policy Act of 1992 (EPAct). Support for this Act was based on the recognition that competition in wholesale power markets has increased benefits for consumers without sacrificing safety or system reliability.

However, NIEP urges the committee to recognize that the transition to competition in wholesale markets is far from complete. Many of the provisions of the EPAct are not yet fully implemented, including those that address separating ownership from use in transmission. Moreover, questions about the costs of the transition to a more competitive industry structure, such as the definition and allocation of stranded investment, are still unresolved.

Change has come to the electric power industry and it will accelerate in the next 10 years. The major forces driving change today continue to be fundamental economic ones—demands for consumer choice in the face of higher costs and a wider array of options. The evidence of change is clear. Independent power producers (IPP's) have built 53 percent of the capacity brought on line in the last 4 years. Over the last 15 years, IPP's have demonstrated that they can produce power as reliably and safely as utilities, and due to competitive pressures, as cost-effectively, if not more.

New business opportunities also are emerging, such as the development of power marketing and brokering services that match buyers and sellers. By contracting to purchase power from various sources, marketers can efficiently match supply and demand in the electricity marketplace. Power marketers and brokers include players new to the electricity industry, such as companies that have been successful at marketing and brokering in the natural gas industry, as well as IPP's. The rapid development of this segment of the industry illustrates the degree to which change has occurred.

[1] NIEP is an association of companies that generate electricity for sale to utilities and develop cogeneration projects for a variety of users. NIEP membership is comprised of publicly-traded and privately-held corporations which represent a broad spectrum of fossil fuel-fired and renewable technologies, including hydro, biomass, pumped storage, geothermal, wind, wood, waste coal and waste-to-energy plants, as well as oil, gas and coal-fired cogeneration and wholesale generation facilities.

Federal legislation has served as the catalyst for this change. Congress recognized the strength of the fundamental economic forces affecting the cost structure of the electric utility industry when it enacted the Public Utility Regulatory Policies Act of 1978 (PURPA). PURPA opened the wholesale electric generation market to competition by requiring utilities to look beyond construction of their own power plants to meet their customers' energy needs. With the enactment of EPAct, Congress took the next step toward a new electricity industry structure by acknowledging the benefits of competition in the provision of wholesale power and establishing the goal of harnessing market forces to expand these benefits.

Today, change is being driven by customer demands for greater energy service options and lower costs. High electricity bills are mobilizing political support for competition and consumer choice. The difference between average utility electricity prices and the costs of new generation are significant in many areas of the country. U.S. industry, competing in a global market, is demanding competitively-priced electricity. Many industries are trying to bypass their traditional utility suppliers through self-generation, direct purchase from cheaper suppliers through retail wheeling, or relocation.

These demands confront higher cost utilities with a dilemma. If they respond to their customers who have choices, e.g., industrials, and offer them lower rates, utilities must increase rates to residential and other captive customers in order to cover their costs. If utilities allow industrial customers to gain access to cheaper power from alternative suppliers, high cost utility generating assets lose value. Since under existing cost-of-service regulation, utility profits are based on returns on their aggregate investment in these assets, utilities currently have an incentive to try to stifle competition.

The wholesale competition goals envisioned by EPAct will not be achieved without restructuring and regulatory reform at the State level to give the regulated utilities a way out of this dilemma. Despite the dramatic changes in the last 10 years, the competitive sector of power generation is still small. Nationwide, only 6 percent of generation is unregulated. The great majority of generating assets remain in the hands of traditional utilities under a regulatory system which shields them from many of the market incentives for cost cutting. Tapping the potential cost savings in existing utility generating assets by shifting to performance-based regulation and removing disincentives to closing uneconomic plants are critical components in the solution to the problem of high electricity costs in the United States.

Utilities still wield considerable market power through their control of transmission, distribution and the vast majority of generation assets. In its recent restructuring proposal, the California Public Utilities Commission (California PUC) assumes that "[a] wholesale market for electric services, and the attendant institutional, financial and contractual arrangements necessary to facilitate a more vibrant market, currently exist and continue to develop." This assumption is premature. It is NIEP's view that, although the wholesale generation market is moving in the direction of full competition, it is not there yet.

There are impediments in the existing system that have yet to be overcome, particularly the relationship between ownership and use of the electric transmission infrastructure. Power pools as an institution are still dominated by transmission-owning utilities; many non-utility power providers cannot be members. When IPP's sell power outside the power pool structure, there may be significant barriers to access and efficient use of the transmission infrastructure.

EPAct may have laid the groundwork for severing the link between ownership and the use of transmission facilities, but this link has not yet been effectively severed. The Federal Energy Regulatory Commission (FERC) just issued the first final transmission access (section 211) order on May 11, 1994, and its policies with respect to the terms, conditions and pricing of transmission service are still evolving. NIEP and the numerous other interested parties who filed comments in the FERC's inquiry regarding transmission pricing are awaiting a FERC order that we expect will have significant impact on how transmission services will be provided in an increasingly competitive electricity industry. Further, on June 29, 1994, the FERC issued a Notice of Proposed Rulemaking on stranded investment, one of the most complex issues to be faced in the transition to a more competitive electricity industry. In addition, significant questions of jurisdictional authority between the States and Federal regulators have not yet been clarified.

Consequently, Federal and State legislators and regulators cannot take competition in wholesale power markets for granted. Regulators, particularly at the State level, have a role to play now as stewards of competitive markets with a responsibility to guard against anticompetitive behaviors, such as forced contract renegotiations and curtailments, manipulation of competitive resource procurement processes and unfair pricing practices, like cogeneration deferral rates. Some have argued that the problems with wholesale competition should be addressed before considering industry restructuring. NIEP believes that the two must be pursued on parallel tracks. Regulators must move today to reexamine the current regulatory structure to ensure that utilities that make good resource procurement decisions and perform well are rewarded commensurately.

The forces of competition have been unleashed in an electricity industry with a regulatory structure premised on the utility as a vertically integrated monopoly. While this structure has adapted considerably over recent years, the extent to which the industry has changed, particularly in the wake of EPAct, now requires a concomitant change in regulatory practices.

Performance-based regulation (PBR) is a more compatible regulatory framework in an increasingly competitive electricity industry. Traditional utility cost-of-service regulation continues to stand in the way of efforts to bring competitive discipline to the utility industry. Among other things, the continued linkage of utility profits to return on rate-based assets gives many utilities a "buy and die" attitude toward purchased power.

NIEP agrees with the California PUC that "Performance Based Regulation can provide considerably stronger incentives for efficient utility operations and investment" than traditional cost-of service regulation. However, the degree to which PBR will lead to the "lower rates and . . . more reasonable, competitive prices for consumers" that the California commission anticipates, will depend on how the PBR is structured and, at least during the transition and the early stages, require diligent regulatory scrutiny.

This presents regulators with the challenge of redefining their roles from being protectors of ratepayers against monopolies to architects of a framework that will harness competitive forces effectively. Replacing asset-based returns with performance-based returns serves everyone's interest.

Some have suggested that the passage of EPAct and the continuing changes in the industry make PURPA redundant. On the contrary, EPAct and the process of implementing full competition build on PURPA rather than replace it. Since 1978, PURPA has been the lever that pried open the wholesale generation market and let competition come in. Without PURPA, there is no guarantee that utilities will continue to open their markets to IPP competitors.

EPAct, by itself, does not require that utility investments in generation be tested by the market. EPAct starts with the assumption that a competitive wholesale market exists and then tries to perfect it by eliminating barriers to ownership and transmission access. However, competitive pressures facing many utilities are driving them to attempt to close doors to IPP's, rather than compete openly and fairly. This demonstrates the continuing need for the mandatory purchase provisions of PURPA. These provisions led many States and utilities to initiate competitive procurement processes as the most efficient way to allocate specific capacity needs to the non-utility suppliers brought forth by PURPA.

In addition, through the preferences granted to cogeneration and renewable resources, PURPA continues to serve the public policy objective of encouraging energy conservation through greater efficiency, and technology and fuel diversity in the generation of electric power. PURPA is particularly important for suppliers of renewable energy. Power generation from renewables is sustainable and relatively benign environmentally, but it must have a predictable market to survive and grow. The mandatory purchase provisions, among others, of PURPA ensure that renewables will have access to this market.

One of the fundamental preconditions for the long term success of a competitive wholesale generation market is the ability of investors to rely on the terms and conditions of power purchase agreements negotiated at arms length between buyers and sellers of power. The same principle applies to the ability of regulated utilities to rely on promises of cost recovery for prudently made investments. There is a simple reason why honoring these obligations is important. It costs a great deal of money to build power plants. The electricity industry is a voracious consumer of capital, second only to the automobile industry in the United States.

One of the benefits of competition has been the introduction of innovative financing methods that reduce the cost of capital for new generation without risking instability. However, if the electric generation industry is to be successful in competing for capital, investors must have confidence that regulators will respect the sanctity of contracts. In a number of States, IPP's with existing contracts have been pressured by utilities and regulators to renegotiate or terminate existing contracts. The competitive generation market of the future will be dominated by contract-based transactions by utilities, their unregulated affiliates and IPP's, and the sanctity of contracts will be essential for all.

Continued regulatory oversight remains critical to ensuring the successful development of a competitive generation market. This is particularly true if utilities are allowed to continue to participate in the market as both buyers and sellers of generation services.

There are several options to consider in moving to a new industry structure. They include:

Option I: Status quo for existing assets; development of new assets in separate wholly-owned subsidiary with no ability to ratebase investment for development, operation and maintenance. The subsidiary may sell power to its parent and to other markets.

Option II: Existing and new assets transferred to or developed by wholly-owned subsidiary with no ability to ratebase investment. The subsidiary may sell power to its parent and other markets.

Option III: Same as Option II except that the subsidiary may not sell power to its parent.

Option IV: Transfer of generation assets to new publicly-traded company with separate board of directors from the divesting utility. The new company may sell to the transmission and distribution ("transco/disco") utility which formerly owned assets and to other markets.

Option V: Sale of existing assets piecemeal to highest bidder.

In Options I through III, utilities are allowed to continue both to buy and sell generation services. In these scenarios, utility participation in generation will have to be tightly regulated to protect against the potential for self-dealing and cross-subsidization between the generation and transmission and distribution sides of the utility business. Utilities must not be able to compete unfairly with IPP's, power marketers and other non-utility entities by shifting costs between the two.

If allowed to keep one foot in the monopoly transmission and distribution arena and one in the competitive arena, utilities may be able to maximize profits to shareholders by shifting costs to captive customers and, at the same time, gain a competitive advantage over other generators wishing to sell to the utility. If the generating subsidiary is allowed to compete in the unregulated wholesale market with a portion of the costs of its facilities transferred to captive ratepayers of its regulated transmission and distribution divisions, the market will be distorted.

Some have proposed that ordering utilities to separate their generation from their transmission and distribution operations (functional unbundling) will be sufficient to protect against anticompetitive behaviors, when combined with enhanced regulatory oversight. However, accounting mechanisms that separate generation and transmission assets may not provide adequate safeguards against cross-subsidies. The regulatory review required to prevent abuses may result in a degree of micromanagement by public service commissions which the utilities will find intolerable. In addition, the regulatory scrutiny needed to prevent unfair competition by utilities is likely to prove unworkable and extremely costly.

NIEP believes that a better way to deal with these problems is to have generation spun off into independent new corporations without ties to the divesting utilities. The new corporations would have stock separate from the divesting utility and independent boards of directors, and they would be allowed to sell to the transco/disco utility and other markets without concern about conflict of interest. The transco/disco could seek the best deal from a broad range of competing suppliers. Divestiture of generation from transmission and distribution also offers an opportunity to address the stranded investment issue in a way that can minimize distortions to competitive markets.

Financial deregulation of all generating assets by a divestiture that separates sellers from buyers of power is likely to capture the full benefits of competition for both utilities and consumers.

Traditionally, electric utilities have been regarded as "public" institutions and as such have been used to deliver certain social goods to their customers and thereby society as a whole. These social goods included universal access through reduced rates for low-income, disabled and elderly customers; discounted economic development rates for troubled industrial customers; and activities that can be characterized under the heading of good citizenship, such as charitable donations and community investment.

In more recent history, electric utilities began to look beyond providing customers with electricity in the form of kilowatts and kilowatt-hours generated by their own power plants to the provision of electric energy services from a combination of utility-owned generation, independent power production (including qualifying facilities), and demand-side management investments. This broader approach to the development and acquisition of energy resources has come to be referred to as integrated resource planning (IRP). IRP has been widely acclaimed as a process, subject to public review and input, that enables utilities to evaluate a comprehensive array of resource options in a comparable manner and to select those that best meet the needs of their customers. NIEP has long supported utility adoption of IRP practices.

However, as the debate about the future structure of the electricity industry has moved forward, some have expressed concerns that the expansion of competitive forces in electricity markets will drive utilities to abandon IRP, environmental considerations and their role as providers of social goods. NIEP strongly believes that this need not be the case.

The objectives of IRP have helped to lay the foundation for an increasingly competitive electricity industry structure. They should be retained to guide its current transition. Nowhere is this more clear than in the area of environmental performance. Under existing IRP frameworks, IPP's have built new, clean and efficient generating facilities using state-of-the-art generation and environmental control technologies. While competition has driven IPP's to increase efficiency and reduce costs, it has also prompted IPP's to improve environmental performance steadily and pioneer the use of renewable technologies. IPP's have been rewarded by receiving credit in IRP processes that recognize environmental values and the insurance that these technologies provide against risk of future environmental control requirements.

NIEP strongly believes that the fundamental spirit of IRP is also consistent with the "unbundling" of functions once performed by vertically integrated utilities. While the locale of IRP may shift to the level of distribution companies and other resource aggregators, such as power marketers, NIEP believes that IRP can and should be the framework within which a robust and competitive electric generation market can flourish. However, as competition spreads in the electricity industry, it will be important to distinguish between IRP as a sound business practice and IRP as a regulatory process. The latter should diminish as the former is more widely accepted.

NIEP agrees with the California PUC when it says that "the spirit of IRP is, and ought to be, grounded in best business practices and sound public policy."[2] A more competitive electricity industry will likely reduce, but not eliminate, the regulatory role for oversight of resource procurement decisions. Today, small transmission and distribution utilities that do not own any generation operate according to IRP principles. While competitive forces will substitute for the heavy hand of regulation in many areas, utilities, particularly those who wish to succeed in a competitive industry structure, will need to continue to "carefully consider a wide range of investment options, balance short-term and long-term goals, quantify as many costs and benefits as possible, and not ignore those factors which resist quantification."[2]

NIEP distinguishes between the principles and practice of IRP, on the one hand, which we see continuing to be a fundamental part of a new electricity industry structure, and the traditional use of utilities to deliver social goods, on the other. NIEP agrees that competitive cost pressures will make it difficult for utilities to continue to deliver social goods not directly related to the provision of electricity services in the ways that they have in the past.

If legislators and regulators determine that utilities are still the appropriate entities to deliver these goods (e.g., programs to assist low-income customers), then it will be incumbent upon all of us to investigate new mechanisms that will enable them to do so without distorting emerging competitive markets. However, concerns about the provision of social goods should not be allowed to delay the transition to a more competitive electricity industry and the benefits of increased efficiency, technological innovation, reduced environmental impacts and resulting lower costs that competition can bring to customers, utilities, independent generators and society as a whole.

The benefits of full and fair competition cannot be achieved without reasonably priced transmission access, available with predictable and efficient terms and conditions, on both point-to point and network bases. The FERC's transmission pricing policy is a critically important component in removing barriers to a competitive wholesale power market as intended by Congress in the EPAct. NIEP has been an active participant in the FERC's formal inquiry into pricing policies for transmission services and has taken the following positions in a series of comments filed with the Commission.

---

[2] California Public Utilities Commission Order Instituting Rulemaking on the Commission's Proposed Policies Governing Restructuring California's Electric Services Industry and Reforming Regulation (R.94-04-031) and Order Instituting Investigation on the Commission's Proposed Policies Governing Restructuring California's Electric Services Industry and Reforming Regulation (I.94-04-032) ("Order"), p. 50.

[3] Order, p. 50.

Transmission pricing policy must work in concert with all elements of transmission access to support a competitive wholesale generation market. Pricing policy must be non-discriminatory, easy to use, practical and flexible and comprehensive.

Transmission pricing policy must be cost-based and provide fair and equal treatment to all similarly-situated transmission users. Users must be able to understand and rely upon the terms of any transmission pricing policy. Transmission pricing must recognize the exigencies of the rapidly changing electric power market. Finally, it must take into account the full range of ancillary services necessary to effect the transmission of power and encourage the efficient use of the full capacity of the transmission system, thereby maintaining the lowest possible unit cost of transmission.

An embedded cost methodology is appropriate for pricing transmission service under the current circumstances. A competitive wholesale generation market demands comparability in pricing among all wholesale transmission customers. The FERC has emphasized comparability in a series of recent orders.[4] Furthermore, a competitive wholesale market is a dynamic structure where the free movement of its participants renders distinctions among customers meaningless. All transactions affect one another, and no single transaction can be analyzed in isolation. The even application and systemwide orientation of embedded cost methodology provides the best fit with the current circumstances of this dynamic market.

Embedded cost-based transmission pricing will impose no economic harm on either native load customers or transmission owners. The benefit to be realized by electricity consumers of additional robustness created in the generation market by a simple and efficient transmission pricing policy will far outweigh any burden associated with the failure to allocate properly every cost under that pricing policy. The alternative, incremental cost-based pricing methodologies, are likely to be discriminatory, complicated and impractical in their application at the present time, thereby inhibiting competition. As generally proposed, they are potentially discriminatory because differential treatment is applied to customers, usually embedded cost pricing for the native load and incremental cost pricing for prospective transmission customers. Moreover, they are unduly complicated and difficult, relying on complex methods of cost calculation which are largely inaccessible to utilities' customers.

Ultimately, basic common sense principles of practicality must govern the formation of transmission pricing policy. Because the health of the power industry depends on a relatively stable regulatory environment, and because the competitive generation market is experiencing phenomenally rapid change, it is reasonable for the FERC to establish an initial pricing policy based on embedded cost methodology which can then be revised over time as the competitive generation market evolves. Future revisions must ensure fair treatment of customers and recognize that transmission, unlike generation, is a natural monopoly.

Above all, a transmission pricing policy must be administratively workable to be effective. A pricing methodology that attempts to account for every possible factor and compensate for every identifiable impact will inevitably fail. In trying to capture all efficiencies, a pricing regime will become so detailed that the costs of individual transactions will become impractical. Furthermore, a complex pricing policy could have the unintended effect of stifling competition.

NIEP urges the House Subcommittee on Energy and Power to continue to support efforts that will lead the way to increased competition, reduced rates for customers, and improved environmental performance in the U.S. electricity industry. We commend Congress for establishing market-oriented goals in EPAct, but additional work lies ahead, particularly at the State level, in order to obtain the benefits resulting from greater competition.

Competitive forces will continue to reshape the industry over the next decade. To reap the most from this restructuring:

—the implementation of full and fair wholesale competition must be pursued by:

—regulators acting as stewards of competitive markets and guarding against anti-competitive behaviors,

—the implementation of performance based regulation, and

—divestiture of utility generating assets;

—a role for IRP must be recognized; and

—fair transmission access and pricing policies must be adopted.

NIEP appreciates this opportunity to present its views before the subcommittee. We stand ready to respond to any requests for additional information and look forward to working with you in the future.

---

[4] American Electric Power Service Corporation, 67 FERC Para. 61,168 (May 11, 1994); Florida Municipal Power Agency, 67 FERC, para. 61,167 (May 11, 1994); Kansas City Power & Light Company, 67 FERC para 61,183 (May 13, 1994).

Mr. SHARP. Thank you very much, Mr. Divine.
T1Mr. Foster, very pleased to hear from you now.

STATEMENT OF ROBERT G. FOSTER

Mr. FOSTER. Thank you. Good afternoon, Mr. Chairman, members. My name is Bob Foster. I am Vice President for Public Affairs for the Southern California Edison Company. I want to thank you for extending the opportunity to present our views on electric industry restructure.

I would like to make those views known by making five points. The Southern California Edison Company approaches the issue associated with change in the utility industry from the perspective of a leader in promoting renewable energy technologies and energy efficiency.

We support consumer choice in electricity supply but we oppose forced retail access now because the means and institutions do not yet exist to ensure that, (1) all customers benefit from choice, and (2) utility programs to improve environmental quality, energy efficiency and technology can achieve other public policy objectives may be lost or damaged in this effort.

Wholesale competition, the third point. Wholesale competition is authorized by the Energy Policy Act in retail—in competition as proposed by the California Public Utilities Commission are fundamentally different.

The fourth point is a key to enabling all customers to obtain the maximum benefit from greater competition in the electricity industry is the creation of POOLCO, an independent entity that would be responsible for dispatching reasonable transmission and generation resources and providing nondiscriminatory access to all generation suppliers.

And finally, the complex and difficult questions of Federal and State authority and responsibilities should be resolved before retail access is mandated.

Let me address each in turn. From our perspective, we believe we are unique because we have a larger percentage of our power supply, 34 percent, coming from PURPA-qualified facilities, and that is larger than, we believe, any other utility in the country.

We have more renewable generation, 23 percent, than any utility, including almost all of the world's solar generation, along with 50 percent of the installed wind capacity and 30 percent of the geothermal energy nationwide.

As a result of the substantial investments Edison has made to increase reliance on clean and renewable energy sources and capture cost effective energy efficiency improvements, investments which in many cases were mandated by State and Federal law over the last 15 years, Edison has been able to reduce emissions by over 80 percent even though our service territory has grown by more than a million customers.

We support wholesale competition and consumer choice in electricity supply, but we cannot now support mandatory retail access. The proper institutions and means have not yet been developed to ensure that all customers receive the benefits from choice.

To compete in the short-term oriented electricity commodity market utilities will have no choice but to reduce research and development expenditures and pursue a least dollar rather than a least societal cost approach to the provision of electricity service. Unless alternative mechanisms are developed, programs to improve environmental quality, energy efficiency, resource diversity, outreach and procurement programs to minority communities, assistance to the disadvantaged, and other goals to which utilities have, in essence, been the agents of government would be jeopardized.

In this connection, it is important to understand that wholesale competition as authorized by the Energy Policy Act and retail competition as proposed by the California Public Utilities Commission are fundamentally different. We support vigorous wholesale competition and expanded consumer choice. In order to meet the challenge of this competition, we are cutting costs and offering our customers more options, including contracts customized to their needs.

A fully developed wholesale market featuring a regional pool mechanism such as we have proposed, would enable us to provide additional benefits to our customers without breaking apart the system of reciprocal obligations between and among end-use customers. Retail competition, by contrast, breaks apart those reciprocal obligations.

PURPA's mandatory purchase obligation, public power preferences and subsidies, and the traditional obligation to serve are not compatible with a retail access world in which retail franchises no longer exist. The creation of POOLCO would enable all customers to obtain the maximum benefit from robust wholesale competition.

Edison has proposed the creation of a new regional, privately owned company independent of all utility and generation suppliers that would be responsible for dispatching regional transmission and generation resources. POOLCO would ensure that electric power is provided at the lowest possible price and that all customers can become fully informed about the price of electricity.

The major advantage we see here is that this is a very transparent operation. All transactions are public and published and there are no secrets. The information is available to everyone.

With or without direct access, we and a growing number of utility experts that POOLCO or something like it is essential to the development of the most efficient competition in the electricity industry. Certainly direct access should not be ordered until all customers are able to make their own electricity supply choices in a transparent regional power market that is open to all buyers and sellers on equal terms.

And finally, the complex and difficult questions of Federal and State authority and responsibility should be resolved before retail access is undertaken. Great uncertainty exists as to what authority, if any, the States have to order retail wheeling. Similar uncertainty exists as to the authority and responsibility to recover costs that may be stranded by the transition to a new model of competition and regulation.

Millions of individuals around the country have invested substantial portions of their life savings for retirement or other needs in utilities like Edison. On the basis of the then current network of reciprocal obligations, these individuals reasonably believed that utility stock holdings were a stable, relatively low risk investment. I might add that many of them have been disabused of that notion.

Although both California and the Federal Energy Regulatory Commission have recognized the need to make provision for stranded costs, both proposals, in our judgment, fall well short of what is required. Other uncertainties and conflicts abound, including the issues of reciprocity and our continued obligation to acquire power under California State programs.

I might add that an example of how we get caught in the middle of this is at the same time the California Public Utilities Commission is moving to a more competitive market we were recently, 2 weeks ago, ordered to enter into contracts for additional capacity, which we presently do not need, for long-term contracts, the payment streams of which would total $16 billion over 17 to 30 years.

In sum, Congress and the Energy Policy Act set the course for expanded wholesale competition. Wholesale markets are developing quickly, but full development of these markets is not possible without new institutions such as POOLCO and without further guidance from FERC.

In the Energy Policy Act, Congress forbade FERC from ordering retail wheeling and did not confer upon the States any authority to require retail wheeling. If Congress now believes that retail transmission is desirable public policy, then we see no alternative to legislation that would address comprehensively and definitively the many complex questions raised by retail wheeling.

We do not underestimate the difficulty of such an effort, but the alternative, years of litigation and piecemeal efforts by individual States that may or may not conflict with what FERC is doing, threaten huge, unavoidable costs and jeopardize valuable institutions and programs, and clearly are an essential part of society's infrastructure.

I want to thank you for the opportunity to present our views, and I will be happy to answer any questions.

[The prepared statement of Mr. Foster follows. Attachments are retained in the subcommittee files.]

STATEMENT OF ROBERT G. FOSTER, VICE PRESIDENT, PUBLIC AFFAIRS, SOUTHERN CALIFORNIA EDISON COMPANY

Thank you Mr. Chairman. I very much appreciate the opportunity to testify before the subcommittee as it reviews the sweeping changes that are occurring and proposed in the provision of electricity service.

I would like to share today with the subcommittee: Edison's experience with competition and change in the industry; our concerns for a framework to support greater competition; the important differences between wholesale versus retail competition; our assessment of the California Public Utilities Commission Proposal on Retail Access; Edison's POOLCO proposal; and the issues that are vital for government to address first, so competition can occur efficiently rather than ill serving consumers, the environment, investors and the economy.

Southern California Edison is a fully integrated electric utility, providing generation, transmission and distribution services to more than four million customers located in central and southern California. We have a diverse resource mix, with over one-third of our energy provided by qualifying facilities and other independent power producers. Most of this energy is sold to us under long-term contracts mandated by the California Public Utilities Commission (CPUC) to implement the Public Utility Regulatory Policies Act of 1978 (PURPA)—Standard Offer 4 contracts—at prices that substantially exceed the cost at which other generation resources can be acquired.

Edison has been a leader in promoting clean, renewable energy technologies and in the pursuit of cost-effective energy efficiency improvements. We have virtually all (97 percent) of the solar generation in the country, 50 percent of the installed wind capacity and 30 percent of the geothermal energy capacity, nationwide. Fully 23 percent of our electricity in 1993 came from renewable resources. As of last year, Edison's energy efficiency programs had produced benefits for California's citizens which include:

—more than 6 billion kWh in annual energy savings—enough to serve the energy needs of more than 1 million homes each year, with savings to the State's economy, estimated by the California Energy Commission, of over $10 billion between 1987 and 1992;

—a 1.860 ton annual reduction in NO_x emissions from power plants, equivalent to the removal of more than 1.5 million cars from Southern California roads;

—a 3.6 million ton annual reduction Of C02 in 1993; and

—a $200 million savings in fuel costs—an average annual electric bill savings of $50 for each of our four million customers.

As a result of these efforts, over the last 15 years Edison has been able to reduce emissions by over 80 percent even though our service territory has grown by over 1,000,000 customers.

As you suggested in your letter of invitation, "the pace of change towards a 're-structured' electric industry operating in a far more competitive marketplace has been far more rapid than many had anticipated." We are increasingly concerned that the necessary institutional, regulatory and legal framework to support this rapid change is not in place. Developing the necessary framework requires an understanding of the fundamentally different nature and consequences of wholesale competition, as authorized by the Energy Policy Act of 1992, and retail competition, as proposed by the California Public Utilities Commission.

We support vigorous wholesale competition and expanded consumer choice. In order to meet the challenge of this competition, we are cutting costs and offering our customers more options, including contracts customized to their needs. A fully developed wholesale market, featuring a regional pool mechanism such as we have proposed, would enable us to provide additional benefits to all our customers without breaking apart the system of reciprocal obligations between and among end use customers, utilities and State regulators that has provided the public with the most reliable electricity service in the world.

Retail competition, by contrast, breaks those reciprocal obligations. The incremental efficiency improvements that would be gained by moving beyond a fully competitive wholesale market to retail access appear to be slight, but the adverse consequences could be far-reaching. New means and institutions will be required to accomplish the "quasi-public" responsibilities performed by utilities under the current retail service structure. Until those new institutions are created and tested, moving to direct retail competition poses substantial risks to our economy and our environment.

RESPONSES TO THE SUBCOMMITTEE'S QUESTIONS

1. Change is Irreversible and Pervasive

The Public Utility Regulatory Policies Act of 1978 and the comprehensive efforts of the California Public Utilities Commission to implement PURPA's mandatory purchase requirements have significantly changed the complexion of the electric utility industry in California and the West. In 1983, only 1 percent of Edison's energy mix came from non-utility power producers. By 1993, 34 percent of our energy mix was provided by qualifying facilities and other independent power producers. During the same time period, because the contracts under which this power is provided do not allow us to take advantage of our own, and other, lower-cost power, we have seen our purchases from other utilities fall from 40 percent to 5 percent, and generation from our oil and gas fired units fall from 33 percent to 23 percent.

With passage of the Energy Policy Act in 1992, competition among generators of electricity has become more widespread, affecting not only the structure of individual companies, but their customers, the services to be offered, the entities offering those services and the relationships among participants in the electricity industry.

Partly as a response to these changes, the California Public Utilities Commission on April 20 issued a "Proposed Policy Statement on Restructuring California's Electric Services Industry and Reforming Regulatory Policy." The CPUC proposal would restructure California's electric system by providing all consumers, eventually, with direct access to the electricity supplier of their choice. (In point of fact, "direct access" in the physical sense would not occur. Rather, it would be accomplished by means of accounting procedures).

Under the CPUC proposal, customer choice would be implemented in stages, with the first block of customers eligible in 1996, and all customers eligible by the year 2002. Customers wishing to remain with their existing utility provider would be able to do so: our obligation to serve them would continue. The CPUC's vision contemplates that the transmission and distribution functions would remain natural monopolies, subject to performance-based ratemaking oversight.

It is fair to say that the California proposal has sparked considerable debate and controversy within the State and around the country among utilities, regulators, lawmakers, consumers, environmental groups and the financial community. Edison, and many other stakeholders, have expressed support for certain aspects of the commission's vision, while voicing concern about others. We do not oppose retail choice. We do believe, however, that it should not be pursued until the necessary foundation has been put in place to ensure that: (1) all consumers benefit from meaningful supplier choice; and (2) the other responsibilities borne by utilities are not abandoned.

It is far too early to predict the outcome of the California effort, but one thing is certain: the provision of electricity services in California will be a different business, done in a different regulatory and competitive environment than exists today.

We would offer for the subcommittee's information our formal submissions to the CPUC.

2. Federal and State Regulators Must Manage the Transition To a Competitive Market By Taking the Time to Do It Right, and With Clear Understanding Of the Respective Roles, Responsibilities and Authorities of Each

The creation of a restructured electric services industry requires that numerous complex legislative, jurisdictional equity, and environmental issues be addressed in a comprehensive manner. There are no easy or clear answers to many of these issues, but if they are not resolved early in the transition, the benefits of competition for consumers could be more than overwhelmed by cost shifting, transition costs, and uncertainty about reliability.

A. Example 1: POOLCO

For the near-term, we at Edison have proposed a strategy for the transition to competition that benefits our customers by focusing on efficiency gains from existing assets—not on reassigning the costs of prior public policy choices or shifting costs from large to small customers. Our longer term strategy focuses on gaining efficiencies from optimal selection of new resources.

Key to this strategy is what we call POOLCO: POOLCO would be a new regional, privately-owned company—independent of all utilities and generation suppliers—responsible for dispatching, regional transmission and generation resources, and providing non-discriminatory access to all generation suppliers. POOLCO would ensure that needed electric power is provided at the lowest possible price, and that all customers can become fully informed about the price of electricity to aid them in decisionmaking. Utilities would be free to provide consumers with a broad array of competitive services, including generation, energy efficiency and new products using existing assets and utility resources.

With or without direct access, we, and a growing number of other industry experts believe that POOLCO, or something like it, is absolutely essential to the development of the most efficient competition in the electricity industry. Certainly, direct access should not be ordered until all customers are able to make their own electricity supply choices in a transparent, regional power market that is open to all buyers and sellers on equal terms. That is what POOLCO is designed to do.

As big as California is, however, a POOLCO established by California alone would not be most economically efficient. It should be done on a regional basis. It is far from clear, however, how such a regional entity could be established under the current State/Federal division of jurisdiction. Yet, without POOLCO, the benefits of a competitive electricity market are not likely to be realized by the vast majority of consumers.

B. Example 2: Stranded Costs

Today's electricity supply is driven by past decisions on investments and contractual commitments made by utilities to fulfill their service obligations to their customers. Those decisions were the products of public policy debate and extensive regulatory proceedings which resulted in the California Commission's issuance of Certificates of Public Convenience and Necessity in advance of the construction of each Edison power plant as well as extensive post-construction reasonableness reviews of our major utility investments. Our Qualifying Facility (QF) contracts were the product of additional regulatory oversight and mandates.

In our case, California has made important policy choices—high environmental standards, energy efficiency, renewable resources, energy resource diversity, and utility research and development support of environmentally beneficial electric technologies. Each of those choices has affected current rates. Each must now be addressed in the transition to competition.

Millions of individuals around the country have invested substantial portions of their life savings for retirement and other needs in utilities like Edison. On the basis of the then-current network of reciprocal obligations, these individuals reasonably believed that utility stockholdings were a stable, relatively low-risk investment. These small and large investors deserve consideration and fair treatment if the rules under which they invested are now to be changed fundamentally.

As messy as the treatment of transition costs was in the natural gas industry, dealing with transition costs in the electric industry promises to be far more difficult. This is because responsibility for addressing stranded costs is not clear, and because the potential amount of electricity industry stranded costs dwarfs the costs accompanying the natural gas industry's transition to a more competitive environment.

In the natural gas industry, virtually all the costs stranded by the transition to a competitive marketplace were in interstate (wholesale) rates, and therefore subject to disposition by FERC. In the electric industry, by contrast, the vast majority of costs likely to be stranded are in retail rates. Moreover, the intensive prudence inquiries and regulatory review applied to electric industry assets that may be stranded were largely absent with respect to take-or-pay contracts, the buydown and buyout of which were responsible for the bulk of the transition costs in the natural gas industry.

In its proposed rulemaking on stranded costs issued June 29 (Docket No. RM94-7-000), FERC indicated that it believes it has the authority to deal with stranded retail rate base costs, but it has proposed not to do so at this time, based upon legal questions and policy considerations. (Mimeo at p. 51.) With active wholesale competition a reality and with retail access now proposed by several States, we cannot go further without a clear understanding of how prudent investments, which may be rendered uneconomic by reason of statutory and regulatory changes, will be addressed.

C. Other Examples

Other examples of issues that must be addressed if the transition to competition is to be accomplished the right way, in a manner that is efficient, fair and beneficial to all consumers, include:

PURPA's Mandatory Purchase Obligation—Under a California resource procurement program, Edison recently has been ordered to acquire $16 billion of new generating resources, even though we have over 34 percent reserve margins. If utilities no longer have an exclusive retail franchise, what is the basis for obligating utilities to purchase power from certain providers?

Public Power Entitlements—If robust retail and wholesale competitive electric markets are to be created, there is no place for tax-exempt financing subsidies and preferences that give one power supplier guaranteed access to lower cost power than others. If this set of Issues is not addressed, how can competition among electricity suppliers in the West—where nearly 40 percent of the power is public—be truly fair or efficient?

Asymmetry Between the Obligation to Serve and the Obligation to Purchase—Utilities should not be expected in a competitive marketplace to remain obligated to serve customers who are not obligated to purchase from them. Nor is it fair to obligate utilities alone in a competitive marketplace to be the supplier of last resort. Who will be the provider of service to those unable to pay for, or otherwise unattractive to, other suppliers? Who will perform this and other historical utility roles, such as economic development, in a direct access world? Shouldn't this question be resolved before retail franchises are demolished?

3. Existing Mechanisms to Achieve Social and Environmental Objectives Cannot Be Expected to Work In a More Competitive Electricity Marketplace.

Edison is proud of its record in promoting sustainable energy and environmental policy. We do not believe these objectives should be abandoned. In our judgment, before direct access is implemented, alternative mechanisms should be developed to promote environmental and other goals. Otherwise, to compete in a short-term oriented, electricity commodity market, utilities will have no choice but to reduce research and development expenditures and pursue a least dollar, rather than least societal cost, approach to the provision of electricity service.

Unless adequate alternative market mechanisms can be developed, desired levels of environmental quality, energy efficiency, resource diversity, outreach in procurement programs to minority communities, assistance to the disadvantaged, and other goals for which utilities have, in essence, been the agents of government would be put at risk. Utility pricing regulation must be reformed to give utilities sufficient flexibility to compete in the marketplace. Transmission pricing must also be reformed to better reflect its value in the electricity service market.

Traditional cost of service regulation, premised on utility service to exclusive franchise areas, has come to involve significant micro-management by regulators of utility operational and planning decisions. As a consequence of both competitive generation markets and an overburdened regulatory regime, there is growing consensus that the old cost-of-service, command and control regulatory model is inadequate.

Edison believes that performance based ratemaking ("PBR") is the preferred regulatory approach. PBR would link earnings not to cost, but to performance, thereby providing powerful incentives for utilities to use existing assets more efficiently. Adoption of PBR for utility generation, transmission and distribution, as proposed by Edison, will:

—provide a transitional bridge to a more competitive long-run generation structure,

—substitute market-like incentives for regulation, and

—encourage the use of existing assets as productively as possible for the benefit of all customers.

We view the immediate adoption of PBR as a key companion to POOLCO in achieving lower costs for all consumers.

With respect to transmission pricing, as electricity systems around the world have been privatized or made more competitive, transmission has been found to be undervalued as a result of historic embedded cost-based regulation. If proper incentives for the use and construction of transmission necessary for a truly competitive marketplace are to be created, transmission must be priced at something closer to its value in the marketplace.

5. Critical Questions Involving-State and Federal Authority Over Electricity Transmission, Sales and Stranded Cost Recovery Must Be Resolved If the Transition to Competition is To Succeed From the Standpoint of Consumers.

A. May States Require Utilities to Perform Retail Wheeling?

In its proposed rulemaking on stranded costs, the FERC said that it has "exclusive jurisdiction" over the rates, terms and conditions of transmission in interstate commerce whether that transmission is wholesale or retail. FERC also said that, "Because of the highly integrated nature of the electric system, this results in most transmission of electric energy being in interstate commerce." (Mimeo at p. 49.) FERC declined to express an opinion on the precise question of whether a State could nonetheless order retail wheeling. Thus, large uncertainty exists as to whether States may order retail access, even though the rates, terms and conditions of that service would be regulated by FERC. To the extent such action is later determined to be without proper basis in law, tremendous disruption could occur in the electric service marketplace.

Other important retail wheeling-related issues involving State and Federal regulatory authority that need to be addressed include:

—Reciprocity. It seems unfair for one State to open its retail markets to firms in other States or nations, when those other States and nations do not offer equally open markets; however, it is questionable whether any one State could legally impose reciprocity requirements on neighboring States, let alone Mexico or Canada where the North American Free Trade Agreement further complicates matters. Should firms in Canada or Mexico be able to sell power into American markets when U.S. utilities are blocked from selling into those firms' markets? This is a very real issue in California.

—Jurisdiction Over Interstate Retail Sales. It is not clear who, if anyone, would regulate the retail sale of electricity from a supplier in one State to a consumer in another State.

B. Stranded Costs

FERC has proposed that the States deal with potential stranded costs in the retail rate base. It is unclear, however, what mechanisms the States could use to address stranded costs, particularly if, by virtue of FERC's exclusive jurisdiction over transmission, FERC could allow retail stranded cost recovery in rates for either wholesale or retail transmission services. If retail rate base stranded costs may not be recovered as part of FERC-regulated transmission rates, as some have argued, the existence and adequacy of State mechanisms to provide for such recovery are even more critical, but at this juncture, are completely untested. This illustrates the uncertainty associated with regulatory responsibility for managing the transition to a competitive electricity marketplace.

Many other State/Federal issues related to stranded costs will be addressed in the FERC rulemaking, but one of particular importance to Edison is:

—Stranded QF Contracts. To the extent FERC addresses wholesale stranded costs in its proposal, it focuses exclusively on "requirements" contracts. No provision is made for stranded costs related to contracts with non-utility suppliers, for example. In our judgment, a mechanism for "good faith renegotiation" of QF contracts should be provided.

Congress in the Energy Policy Act set the course for expanded wholesale competition. We, as an industry, are still in the process of digesting that. Wholesale markets are developing quickly, but full development of these markets is not possible without new institutions such as POOLCO and without further guidance from FERC. FERC has not yet determined how to deal with issues posed by municipalization, for example. Moreover, it is only beginning to develop what nondiscriminatory access to wholesale transmission requires, and has yet to establish rules of the road for transmission pricing or recovery of stranded costs.

In an analysis of electricity transmission done by the Department of Energy in 1992 in connection with the National Energy Strategy, the Department said:

With retail access, equity problems are far more difficult than with wholesale access. The reliability issues are much more difficult because of the huge number of potential retail buyers that could want access. Efficiency gains would be questionable, except perhaps for the largest of retail buyers willing to pay appropriate entry and exit fees. Given the crucial need to maintain the reliability of the transmission system and the unexplored cost accounting problems, retail access is not a recommended policy at this time. (Technical Annex 3, "Electricity Transmission Access," p.27.)

Congress was wise to heed this recommendation in crafting the Energy Policy Act. We were active participants in working with the Congress to develop the Energy Policy Act. We understood that when Congress forbade FERC from ordering retail wheeling and did not confer upon the States any authority to require retail wheeling, Congress did not expect FERC or anyone else to be in the business of ordering retail wheeling.

If Congress now believes that retail transmission access is desirable public policy, then legislation is necessary to address comprehensively and definitively the many complex issues and questions raised by retail wheeling. This legislation should address the appropriate roles of State and Federal regulation, the treatment and role of public power entities (including Bonneville Power Administration, Western Area Power Administration and municipal utilities), and who in this new retail access world will carry out the environmental and technology development programs formerly performed by utilities with retail franchises.

I do not underestimate the difficulty of such an effort, but the alternative—years of litigation and piecemeal efforts by individual States that may or may not conflict with what FERC is doing—threaten huge, avoidable costs and jeopardize important existing institutions and programs. We are, after all, dealing here with a service essential to every business and resident in the Nation.

Thank you for the opportunity to offer these views. We look forward to working with you on the challenges confronting all of us with the restructuring of the electric industry.

Mr. SHARP. Thank you very much, Mr. Foster.

Let me first return, as we inevitably will, to the question of the stranded investment or sunk cost problem. All of you, or at least three of you, as I recall, clearly touched on it.

And I asked the question, as you probably heard, of the previous panel where we had a couple of utilities represented for whom it would not entail as great a choice or pain as it probably does for the two representatives from California at the moment, and perhaps others that Mr. DeNicola is representing today.

I just want to let you outline what you think is a reasonable proposition. Perhaps you are already stating that at FERC or certainly at the California commission, as to how the techniques by which this might be handled, just sort of an indication of the policy avenues we might take.

You can go in any order. I don't care.

Mr. Glynn?

Mr. GLYNN. Well, we have a proposal on the table in California to do just that. The first thing that we have done is proposed to separate the costs that are a result of investments that we have made as a utility business for generation that supplies our customers in our service territories. And we separated that from any above market costs that are associated with contracts that we were caused to enter into by the actions of the CPUC, which are almost exclusively qualifying facilities contracts, not all of which, but some of which are priced above the market.

With respect to our own generation, we have proposed (a) to continue to sell a bundle of generation from the portfolio that we have. Contrary to popular opinion, not all of our supply comes from nuclear power plants. We have the largest privately owned hydro system in the United States. It produces very low cost power, much of it under a cent per kilowatt hour. I haven't heard anyone suggest that we divest that in a big hurry.

But the idea is that we would take the responsibility ourselves during a transition period into a direct access world to manage the costs of our own generation and not ask for an explicit transition charge, and that the consequence of costs that somehow wouldn't be borne by the market would be simply a matter between the management and the owners of PG&E.

With respect to costs that are——

Mr. SHARP. Excuse me. Did you indicate when you were saying that in your statement, though, that you were seeking compensation due to the factor of time? I thought you alluded to that. Is that asking the Commission to take more time to implement, or am I off base on that?

Mr. GLYNN. No. You are precisely on target. Whatever the ultimate time period is, and, of course, our proposal is premised on the time period that we have suggested, which is an orderly transition from today's State into a State when every customer has the opportunity for direct access.

Mr. SHARP. That is longer than the initial indication of the California commission. Is that correct?

Mr. GLYNN. It is longer than the initial indication of the CPUC. I think actually the concept of having the utility fully responsible for those costs and not be seeking transition charges is remarkably different from what was anticipated by the CPUC when they put their schedule together, however.

And the other half of it is for costs that are associated with contracts with third parties that we entered into where we are simply a collector between customers and those projects, that any costs that are not supported by the competitive marketplace simply be borne by the whole of the customer group without raising anybody's rates, but without providing any particular customer class or individual customer to escape those costs as long as they are in place. And they will be worked through in time.

Those contracts have finite terms. Some of them have some self-correcting mechanisms in their contracts and over time they will become—over some time, not a short amount of time, they will become a smaller component. That is the way we propose to deal with it.

Mr. SHARP. When you say over all of your customers, if you lose any of your large customers through their choosing down the table or somewhere else——

Mr. GLYNN. There have been plenty of folks offering to sell to them.

Mr. SHARP. Right. I assume that is not a part of the customer base any more you are alluding to. Or is there a leaving charge, or is there any way in which you are spreading that around?

Mr. GLYNN. The concept that we have for that is that to the extent that the customers are still connected to the PG&E system and are purchasing any services from PG&E, the CPUC would direct that the charges associated with formerly serving that customer that were above market would be separately billed to them.

I know that this is an area where there is a substantial amount of difference of opinion about whose jurisdiction in fact that would fall under, FERC's or the State's. And, as of this week, there is a new wrinkle on that as well.

The concept that we have proposed is that those costs would wind up being billed to customers. Now, customers can escape those costs. They can move their businesses out of our service territory. They can elect to self-generate so completely that they don't even purchase back-up from our system.

And I don't think it is likely that a lot of them would do that solely because of an energy bill. So fundamentally they would remain interconnected, receiving some package of services under both the jurisdictional entities of CPUC and FERC.

Mr. SHARP. I assume it is conceivable that another wire might be strung, but that is probably so difficult to get accomplished, to get outside your system unless they are on the edge of it. Is that correct?

Mr. GLYNN. That is the way we see it. Not that there is an absolute bar to that, but in terms of the broad economics of making a change—building new transmission lines is not an easy thing these days. You have not heard anyone clamor in here for the right to build new transmission lines in California or anywhere else. So, I think that would be the exception rather than the rule.

Mr. SHARP. We had a few people dreaming that in EPAct we might be able to solve that problem.

Mr. GLYNN. I recall some of those dreams.

Mr. SHARP. I think they were wonderful dreams but they were not politically realistic at the time.

But I guess what I am saying to you, in a sense then these customers in one sense, except to the extent that it might become economical for them to generate their own, they remain in a sense somewhat captive.

In other words, it is going to go over your wire even if they buy it from another generator. So, the opportunity remains there to put some element of that cost on. But obviously, if it is too great a cost, then it may shift them to self-generation, if they are in that position?

Mr. GLYNN. That is right. And our proposal attempts to split the baby by separating out those costs in which we have a direct financial interest and hopefully a return continuing from those where we are simply a middle person doing the bidding of legislative and regulatory mandates.

Mr. SHARP. Have you made estimates to the Commission as to how much you estimate this to be or what percentage of your costs? Or is there any measurement you could share with us?

Obviously, we are not going to be following your details. But the point is to give us some perspective on how big a problem this is viewed within your system.

Mr. GLYNN. The only number that I recall expressly providing the commission, the California commission, at their request, was our estimate of the amount by which our current QF contract payments might exceed a hypothetical current market price.

And I really want to emphasize for all, but particularly my colleagues in the independent power business, this is a hypothetical calculation that one could do. It doesn't suggest that there could be an overnight replacement of those sources.

And that number in 1994 is about $800 million per year, and its decline is towards zero. It gets very difficult to estimate the tail because what it depends on is what fossil energy prices are because that is the bogey that sets the market alternative principally. I mean the market alternative is a natural-gas-fired, combined cycle cogeneration plant in California, and I think in most other parts of the country.

Mr. SHARP. Let me let Mr. Foster respond similarly, if he would like, to the situation Southern California Edison finds themselves in and the argument. You outlined some of it in your opening statement, but I want to let you do that, and then I will recognize my colleague from California who undoubtedly has an intense interest in this issue.

Mr. FOSTER. Our situation is a little different from my colleague at PG&E. We basically have two components of stranded investment as well. One are the contracts with qualifying facilities which are at present substantially above market, and those are a fairly sizable number on our system.

I believe this year the overpayment on the QF contracts on our system will be $1.2 billion. And we have our indigenous generation which could be stranded in the event of competition as well. We believe both those on our system should be placed on some kind of charge to recover those stranded investments. And there is a difference between ourselves and PG&E in that regard.

Their major asset is not in rate base and ours is. So we believe that it is appropriate that we have a public needs and necessity, it went through a prudency review, it went through all the regulatory hurdles, and we have, we think, completed the regulatory bargain, if you will, and we think that is a legitimate cost to recover.

Mr. SHARP. I wasn't clear what asset you were referring to of theirs. That is not in the rate base.

Mr. FOSTER. I think it is clear that—the nuclear assets are the ones we are talking about.

Mr. SHARP. Yes, sir, Mr. DeNicola?

Mr. DENICOLA. If I can make one comment in addition to that. When you look at the industry as a whole, and it was just mentioned that all those facilities went through some type of regulatory process, and, in fact, over the last 12 years over $15 billion of that investment was written off and deemed to be imprudent or whatever.

So, when you look at the utility already paying some of that cost or sharing costs, I am not ready to concede that there is a piece there that needs to be shared. I think the sharing has gone on over the last 12 years.

Mr. SHARP. Let me let my colleague from California, if he wish, ask questions. I was going to come back and ask about some of the pollution aspects of what you are talking about.

Mr. MOORHEAD. Thank you.

In the area of nuclear energy, I was kind of interested in knowing, I know you have taken one of the generators off at San Onofre. Is that correct?

Mr. FOSTER. Unit 1. It is a fairly old unit. That is correct.

Mr. MOORHEAD. How are the rest of them doing?

Mr. FOSTER. Two and three are newer, larger and running very well. They are running at approximately 82 to 85 percent capacity factors.

Mr. MOORHEAD. Are you going to try to replace number 1?

Mr. FOSTER. No. Number 1 is in retirement and it won't come back.

Mr. MOORHEAD. What do you think the plan—are we giving up the nuclear as each generator dies or are we going back to it one day down the line?

Mr. FOSTER. To be very candid, I don't—in the foreseeable future, I can't imagine anyone constructing a nuclear facility, certainly not in California.

Mr. MOORHEAD. Well, of course, in the National Energy Act we made it potentially possible to do.

Mr. FOSTER. Oh, it is possible. It is just that it is cost prohibitive at this point.

Mr. MOORHEAD. Mr. Glynn, what is the situation with you?

Mr. GLYNN. I agree it is possible but not likely. We have two units operating at the Diablo Canyon nuclear plant. They are excellent running units. They are consistently on the Nuclear Regulatory Commission's "Best Plants" list and receive top ratings from the Institute of Nuclear Power Operations, and we don't have any plans to build new nuclear power plants either.

Mr. MOORHEAD. Probably will last another 20 years for you, at least?

Mr. GLYNN. On that order. I don't have the retirement dates in my mind, but on that order.

Mr. MOORHEAD. I go by San Onofre all the time as I drive to San Diego.

One thing we are interested in, I know Southern California Edison has bought an awful lot of alternative fuels. You have got one of the best records in the Nation as far as that is concerned. With some of these plants such as the thermal energy plants they have been restricted to 80 megawatts. We have gone on in the legislation to take that lid off because it would appear that it would help you get cheaper energy.

Obviously, if they don't have that lid on them they can produce thermal energy a lot cheaper than they are now. And unless the Commission increases the amount that you have to buy, which I doubt if they would, but unless they do, it is bound to help our consumers get cheaper energy. If you are going to buy the same amount, it can be produced cheaper because then it can be done in larger amounts for the same plant.

I just wondered if you support that legislation?

Mr. FOSTER. Well, we have not supported it. It is not a—the cap issue is not a very large one for us. I think the larger question for us, particularly utilities facing an often increasingly competitive environment, is the mandatory purchase requirements that are contained in PURPA.

We think if there is going to be any change to PURPA that a review of the Act and its purposes and whether it is relevant in the new world we are entering, I think that is the kind of review we would like to see.

Mr. MOORHEAD. Of course, the cap would go back on unless this bill is passed.

Mr. FOSTER. Again, the cap really doesn't—I don't think it makes energy purchase any more economical for us. I have never seen any indication of that. And very expensive in the past.

Mr. MOORHEAD. It won't make it any more expensive, though.

Mr. FOSTER. I really don't—I don't know that.

Mr. MOORHEAD. It will have to be passed on unless we pass that bill. That is the thing that we are concerned about.

I know one of your subsidiaries is kind of interested in seeing that that cap does not go back on.

Mr. FOSTER. Oh. Them again, huh.

Mr. GLYNN. Mr. Moorhead, one policy issue with respect to any thought of removing the cap is that any move that increases the mandatory purchase obligation of utilities in a time when every market force is going to be reducing the customer base on which the utility can then resell that power doesn't seem to make good policy sense.

Mr. MOORHEAD. Well, I don't think our legislation increases the mandate for you to purchase in anyway.

Mr. GLYNN. Well, if a facility right now is limited to a certain size, and the cap removal would allow it to go to a bigger size, and there would be no change in the mandatory purchase obligation, then it is expanding the base on which those mandatory purchase obligations would function.

Mr. MOORHEAD. That is only if the Commission orders you to buy more from them.

Mr. GLYNN. Over the years that has been a very substantial issue to deal with in California. I mean fundamentally the difficulty that you are placed in is by trying to rebalance the world for the renewables, and I certainly—PG&E has for many years championed renewables, and a California environmental climate is premised on a healthy dose of renewables.

But the market balancing mechanism that used to exist when utilities bought it all and sold it all, that is going away. So, the framework within which this needs to be done really needs to be modified, and it is the mandatory take that is really the big problem.

Mr. MOORHEAD. If you look at this thing, the real purpose of renewables—well, there are two purposes. One is it takes the load off of fossil fuels. We don't know the limit of what is available, but it does expand our capability.

The second thing is that we can get cutbacks or we can get raises in prices of natural gas or oil and other things, and if you have got some kind of a balance in the system, you can't get hit as hard at any one time, and that is another major purpose of it.

Mr. GLYNN. I agree with each one of those objectives. I think the principal difficulty is that those—the mechanism for accomplishing those objectives, the PURPA mechanism, made sense to achieve them in a world where you could look at the utilities and say they are the folks who either make or buy all the power. We will focus the effect of this legislation on them.

All of a sudden we have a very different world. Example—a dozen years ago in Northern California we had 100 generating plants that served our needs. We controlled and owned every one of them. Now, we have got over 500 and we still own and control 100, and 400 are under contract to us. But since we were essentially the only game in town in our service territory, that was the vehicle.

But there are going to be other games in our town, and so that is why there is an element of anachronism developing in the mandatory purchase idea which made good sense when it was initiated in today's utility environment.

Mr. MOORHEAD. As long as the structure of PURPA is in place, why should the renewable sources or technologies be disadvantaged under PURPA as compared to other sources of qualifying facilities such as cogeneration and so forth that we have? Should the cap go back on and have them disadvantaged?

Mr. GLYNN. What we support is the 2-year deferral of making the change so that there can be a more comprehensive evaluation of how renewables are going to continue to get into the generation mix. That is the alternative that PG&E proposes.

Mr. SHARP. Would the gentleman yield?

Mr. MOORHEAD. Yes.

Mr. SHARP. I just want to understand his answer.

Are you saying extend the cap for 2 years? Is that what you are saying? Just a new sunset on it?

Mr. GLYNN. And use that time to examine the structure of how we are going to get renewables to still be a component of the generation portfolio, but recognizing this new utility world that we have.

Mr. SHARP. Thank you.

Mr. MOORHEAD. Mr. Foster, what is your feeling about that?

Mr. FOSTER. I was hoping you were going to miss me.

Mr. MOORHEAD. I would never want to miss you.

Mr. FOSTER. I didn't think so.

Well, I don't want to make a real firm commitment on that. I think that we probably would acquiesce to that with the condition that there be a commitment for a real strong look at PURPA itself.

And I will echo Mr. Glynn's comments. It is really, it is questionable at best whether it has a place in a competitive environment. We think that whole structure needs to be re-examined.

Mr. MOORHEAD. You know the trouble with the 2-year span too is that, you know, there is always differences between the time the House and Senate does anything.

Mr. GLYNN. I have no advice on that whatsoever.

Mr. MOORHEAD. You could take the whole 2 years debating these things between the two houses.

Have either of your companies benefited from the 1990 legislation on the facilities that have been developed or do you think it has been a burden on you that you have had to use it?

Mr. GLYNN. I am sorry. From which? From 1992 from the Energy Policy Act?

Mr. MOORHEAD. From the 1990 legislation, yes. We are dealing with the cap.

Mr. FOSTER. I think for us it has had I don't think a major effect at all one way or the other.

Mr. GLYNN. I believe that the same is the case with us. I can't think of an event that has occurred since then where it has had a material impact on us.

Mr. MOORHEAD. You know, those of us that live in California want the cheapest prices for our people that we can have, but still with the stability that they need, for some kind of a guarantee that they are not totally dependent upon one source of energy.

You know, we hear that nuclear is with us for a while, who knows for how long? Hopefully, the 20 years at least. But we also want to make sure.

We have seen gasoline prices and natural gas prices go to the moon at times when there have been shortages and everything else. We would like a little bit of security back home, and I am sure you do too. And it would seem to me that having this cap off does give you a certain security.

Mr. FOSTER. If I may comment on that. We, as you know, are a strong believer in resource diversity.

Mr. MOORHEAD. I know you are.

Mr. FOSTER. As I mention in the early part of my testimony, 23 percent of our resources are in renewables, and they do provide that cushion for supply shocks that may occur in other fuels. I couldn't agree more.

However, the way—and I think this is an illustration of where PURPA may have some unintended consequences, it doesn't give you any price stability. It doesn't give you price protection because of the avoided cost formulas. Natural gas prices increase, so do the prices for renewables. It does not give you that protection. You don't get that protection because once in a while Bonneville wants—decide they are going to charge you about twice——

Mr. FOSTER. Yes. But the price of renewables are linked, are linked to any avoided cost formula, to gas basically.

Mr. MOORHEAD. One of the concerns regarding the lifting of the mandatory purchase obligations and PURPA is a loss of continued development of renewable energy sources.

Is your industry opposed to the continued development of renewables? I don't think so. Are they?

Mr. FOSTER. Not opposed.

Mr. GLYNN. No. Absolutely not.

Mr. MOORHEAD. How would you propose providing incentives for the continued development of renewables, particularly those that are not currently price competitive? And I understand that right now natural gas is several—a couple pen niece at least cheaper than some of the others.

Mr. GLYNN. One of the mechanisms for some continued development of those resources is the research that goes into them. And second is the concept of, for example, production credits or tax credits which provides the financial vehicle for those projects to move out of the laboratory and into field and actually attract and obtain financing.

It used to be the case that utilities were a good place to go get that certainty because we had a market that wasn't going to go anywhere so we could just blend it in and sell it. And those days are gone.

So the other supports, tax credits, production credits and R&D money from the very entities who will manufacture and sell and operate that equipment, those need to be the vehicles.

Mr. FOSTER. In a competitive market, I don't have an answer for that question. I honestly don't know how you protect renewables in a competitive market, because at least right now they would have a difficult time competing with natural gas.

Today, in addition to the things Mr. Glynn mentioned, in California you actually have a renewable set-aside in the law. As long as you have integrated resource planning, there will be a portion of new resources that are set aside for renewables.

Now, I think that that particular planning structure may evaporate with complete competition. I can't—I don't know. But I mean right now today they do have that provision, at least in California law. There is a set aside.

Mr. MOORHEAD. Well, I expect you are right. If you had total competition and were thinking only about today and not possibilities for the future, I think you are absolutely right. People would go to the very cheapest source. But that is not always the best policy in the long run. It can be a devastating policy if you get chopped off from a source that is subject to disturbances.

And surely both oil and natural gas are subject to that kind of policy. That is why discerning companies such as yours have planned to go into it.

I am very proud of the two major California energy suppliers because I think they have been the leaders in the country in all of these respects. And I think it is wise policy.

Mr. SHARP. Would the gentleman yield?

Mr. MOORHEAD. Yes.

Mr. SHARP. I would like to follow up on his questions on renewables that he was asking about, other techniques apart from either sticking with the less competitive markets, I mean uncompetitive retail market, or going to a tax credit so the government becomes the—even for longer use of the utility system as it is being used now to blend prices as a way to subsidize anything that is not market cost.

Are there voices before the Public Utility Commission—I realize this may not be a fair question to you and not to witnesses that probably sit through all those hearings all day long—maybe you do—before the Utility Commission in California on this rulemaking as to whether or not there are other ways in which to continue the California renewable set-aside?

I must say I want to join in with my colleague in praising both your utilities and the fact that California has been unquestionably the leader and done remarkable things for the rest of the country apart from its own interest on this area of renewable development, and many of us from a national perspective hate to see that lost and undermined even though we recognize that it has also been at a surcharge in many instances in order to carry it out.

Is it conceivably possible that the Commission might decide just simply as you were suggesting, Mr. Glynn, to allocate the above-market PURPA costs? That it simply would allocate those costs in your system, or in your system as well instead of calling that stranded investment, say, now for the next increment of power, everybody will pay what will be a spread out cost and do that?

Mr. GLYNN. That is very close to what we have proposed. The concept we have proposed is to the extent that California policymakers, or in this setting national policymakers, want to see certain end results occur, they simply select a vehicle that gets that to happen, and one way in California would be to have a line item on the bill of the connect charge that every electricity purchaser would have wherever they were buying their kilowatt hours from, that says if you are in California and you use electricity, then you are going to have a charge of *X* or *Y* amount and the purpose of that is going to be to see that California's renewable commitment continues. That is the way to do it.

What we have, as an individual utility company what we have objected to is being pushed into competition bearing those charges uniquely as well.

And I might add that the voices—you asked about the voices within California on this issue. The loudest ones, some voices that I think you are familiar with have said simply don't do it. Don't do direct access because it will make more difficult the process of continuing the very things that you have applauded on behalf of California's renewable commitment.

Mr. SHARP. Well, I am not in a position to judge what the State of California is attempting to do at this point, and obviously it is undergoing a great review of this process and it hasn't made final decisions very clearly one hopes. That is a little prejudging. But that is a time prejudgment, not necessarily the final product.

But it does strike me that it is not intellectually inconsistent that there are other alternative mechanisms, that change does not have to be the enemy of renewables and change does not have to be the enemy of demand side management. It may well be, but it depends on how it is managed.

Mr. GLYNN. Change does not have to be the enemy at all. And while I don't want to hit on a sore subject for anyone, but if you look at some of California's decisions with respect to vehicles, California has simply made some policy decisions over the years with respect to fuel performance levels, with respect to emissions, and now with respect to electrical or nonemitting or remote-emitting vehicles, or whatever you choose to call them. And they have simply said as a matter of policy we are going to push them into the marketplace.

They haven't had a utility to go to do that, but they have found mechanisms where there are multiple sellers and multiple buyers to make it simply a matter of law to have it happen. That can happen in renewables. It can happen in demand side management. It can happen in social programs, even in the construct of a direct access world.

Mr. MOORHEAD. I have no further questions. We didn't really mean to ignore Mr. DeNicola or Mr. Divine. It just happened to be a California——

Mr. DENICOLA. We understand that. But that is where the interest is right now.

If I could take just a minute and take my EEI hat off for just a second and put my Southern Company hat on for a minute, on one of the previous panels there was a comment made about a solicitation for general rates in the State of Georgia. That was changed. And that people had spent money bidding on that capacity.

And, in fact, what happened was we reevaluated our requirements for capacity, did not need them, and canceled the bidding. I would submit to you that is a part of the competitive process, and the people who participated in that knew that when they went into the process.

So, again, either you play in a competitive word or you don't.

Mr. SHARP. Most of us believe we should play, but we just don't like to have to play. Whether it is politics or the economy.

Mr. DENICOLA. Not being at all familiar with the situation in Georgia we have, our company, and certainly there are stories out there about shifting need dates on RFP's from 6 months to a year later when really we are talking about deferral of need, not elimination of need.

On behalf of NIEP, I would like to say on these issues we talked about today, obviously we have producers of renewable power who are members of NIEP. We are supportive, obviously, of the need to continue those types of projects going forward, and we think, as we have elaborated in our comments both here and in California, that wholesale competition combined with IRP processes will continue to forge a market for renewables that would be to the benefit of all the ratepayers.

Mr. SHARP. Well, thank you.

Mr. Foster, I had intended to just ask you to elaborate a little more on pool company but our room time here is running out—POOLCO, I guess it is called. Not so much because—again, I am not in a position to judge whether that is a direction the State of California ought to go or not but because we are trying to establish a record, and I think it is to some degree elaborated in your testimony, a record of what some of these alternative means of restructuring or of a regulatory approach to these issues are so that we have a fairly complete record.

If you are not comfortable with what is elaborated in your testimony, we would be happy to include in the record any additional statement that indicates. And I sort of mean that about all the different kinds of proposals we are getting, because I do think it is useful to try to get onto the table some of these ideas. They are not always adopted, but sometimes they force somebody else to think about a different way of doing it as well.

Mr. FOSTER. Well, thank you, Mr. Chairman. We have a more elaborate explanation of what POOLCO is, and I will make sure that gets submitted to the record. Thank you.

Mr. SHARP. Appreciate that.

Thank you very much for your time and attention.

Tomorrow we will have the Federal Energy Regulatory Commission before us.

[Whereupon, at 3:06 p.m., the subcommittee was adjourned, to reconvene at the call of the Chair.]

ELECTRICITY ISSUES

THURSDAY, JULY 14, 1994

House of Representatives,
Committee on Energy and Commerce,
Subcommittee on Energy and Power,
Washington, DC.

The subcommittee met, pursuant to notice, at 10:12 a.m., in room 2123, Rayburn House Office Building, Hon. Philip R. Sharp (chairman) presiding.

Mr. Sharp. The subcommittee will please come to order.

Today I am pleased to welcome all five FERC Commissioners to join the subcommittee in its discussion of changes in electricity markets and the challenges that regulators face as this industry evolves in an ever more competitive environment and the subcommittee has previously had the opportunity to hear from all five Commissioners, and I am very pleased to have them with us today. I particularly want to commend the President for indicating that he will renominate Ms. Moler to be the Chair of the Commission and to continue.

I must say, in my 20 years of sitting on this subcommittee, I have never heard so many comments from so many interested parties and opinion leaders of all stripes in this country about the quality of the FERC Commission, or the former FPC. While there have been many outstanding individuals over the years on the Commission and chairpersons as well, I find the comments are very refreshing in this day of cynicism and doubt about any of our institutions, because continually people comment on the high quality of the Commission individually and collectively and its timely and serious approach to the issues, and I simply wanted to indicate that I wish the Congress got as high marks across the country.

I must say that I have been consistently impressed with the serious, timely, and thoughtful attention this Commission has given to the implementation of the Energy Policy Act of 1992. I observed at yesterday's hearing that today it is hard to remember how very heated the debate was 2 years ago about whether to clarify FERC's transmission authority. Perhaps one reason for this is the masterful way in which FERC has tackled EPAct's implementation, leaving no doubt that greater competition is the rule of the day. As a result, the debate has shifted from whether the industry has changed to how best to accommodate and to facilitate that change.

One of the most difficult issues the Commission faces is the issue of stranded investment, or uneconomic assets, or sunk costs, as various commentators would have it, how should it be defined, how should it be treated, and which regulators should deal with it, whether at the retail or at the Federal or the State level. Recent State initiatives which we will focus on at next week's hearing have intensified interest in this issue as retail wheeling proposals are considered.

Once again, FERC is to be commended for approaching a central issue squarely and in a timely manner. In its recent proposed rule on stranded investment, the Commission outlined a thoughtful approach on handling such costs in this difficult transition period to an era characterized by more unbundling of services.

While there hardly has been time to review the decision issued earlier this week by the D.C. Circuit Court of Appeals in the Entergy proceeding, some already have predicted the end of stranded investment as a legitimate topic for regulators' consideration. I think such predictions are obviously, at best, premature, and I would like to point out a few key facts about the decision to help keep things in perspective.

First, the court did not issue a final order on the merits of the case. Instead, it remanded the case to FERC for further hearings. Second, the court did not question FERC's fundamental authority to deal with stranded investment. Rather, it said it did not find the Commission's explanation to be sufficient. Finally, it noted that things have changed since FERC issued the Entergy order early in 1992 because Congress thereafter enacted changes to the Federal Power Act as part of EPAct.

My point is that it would be unwise for any of us to overinterpret the meaning of this decision before FERC has had a chance to hold final hearings and issue a new decision. FERC is to be commended for issuing the recent proposed rule, and I look forward to hearing more about this and other issues today.

I particularly look forward to hearing from some of the wisest voices on electricity, those on our second panel, not to mention our first. Through practical experience as well as thoughtful observation, these three people should help us step back and look a bit more dispassionately at the issues facing the industry and consumers and States today. Certainly I welcome all of our panelists.

I would ask unanimous consent that an opening statement from Congressman Mike Synar be placed in the record. Mr. Synar had hoped to be with us but had to Chair other hearings on another subcommittee that he is Chair of, so he cannot.

[The statements of Hon. Mike Synar, Hon. Carlos J. Moorhead, and Hon. Michael Bilirakis follow:]

Opening Statement of Hon. Mike Synar

Mr. Chairman, thank you for holding this series of hearings on the future of the electricity industry after passage of the Energy Policy Act of 1992 (EPAct). Thanks largely to your outstanding leadership on this complicated issue, the sweeping reforms put in place 2 years ago have revolutionized the industry.

Electric utilities are beginning to operate in a vigorously competitive environment. But we need to finish the work we started in 1992 and tie up the loose ends that remain despite EPAct. It's time for the Congress to take another look at provisions of both the Public Utility Regulatory Policies Act (PURPA) and the Public Utility Holding Company Act (PUHCA) that no longer make sense in light of EPAct. And we should assure that all electricity producers and consumers receive the benefits of increased competition.

It's time to ask whether we still need the mandatory purchase requirements of PURPA after we have provided for increased competition in EPAct. How can we justify higher-priced electricity under PURPA at a time when at least half the States have already gone to a bidding system for electricity—a system which is the very essence of competition.

We should also look at the proper role for registered holding companies in this Brave New World of Electricity. Registered holding companies remain subject to severe limitations on this ability to diversify and grow, while exempt holding companies and operating electric and gas utilities which compete with them are free of such restrictions.

Because of their market power, we still need some safeguards where registered holding companies are concerned in order to protect consumers; but it's time to review their status, which was left basically untouched by EPAct.These changes would allow EPAct to reach its full potential and bring the benefits of competition to all Americans.

OPENING STATEMENT OF HON. CARLOS J. MOORHEAD

Mr. Chairman, I do appreciate the time that the chairman has taken to focus this committee's attention on the effects of competition on the electricity industry. Although yesterday's hearing was very enlightening, I feel that we have just begun to scratch the surface of some very complex issues.

I would like to thank all of the Federal Energy Regulatory Commissioners for their attendance today, as well as our distinguished panel of academics. I look forward to hearing our witnesses' view of how the PUHCA reform and transmission access provisions of the Energy Policy Act of 1992 are working. Although there have been relatively few wheeling orders requested so far, I commend FERC for its efforts to ensure that the transmission provisions of the act are meaningful and truly promote competition.

I look forward to working with the Commission on the evolution of Federal energy policy as competition continues to develop in electricity markets.

OPENING STATEMENT OF HON. MICHAEL BILIRAKIS

Mr. Chairman, I usually congratulate the chairman on the timeliness of his hearings, but in light of the D.C. Circuits' recent decision regarding the FERC's approach to stranded investment, this hearing could not be more timely. I thank the FERC Commissioners for being here today, and look forward to hearing their views on the impact of the court's decision.

I also look forward to continuing the dialogue that we started yesterday regarding changes that competition is making in the structure of the utility industry and utility regulation. Although we are having three hearings on these issues this month, I suspect that these hearings are just the beginning of an on-going review of Federal electricity policies in response to fundamental changes in the industry.

Mr. SHARP. With that, I certainly welcome our Commissioners, and, Chair Moler, we will leave it to you as to how you would like to proceed.

STATEMENTS OF HON. ELIZABETH ANNE MOLER, CHAIR, FEDERAL ENERGY REGULATORY COMMISSION, ACCOMPANIED BY VICKY A. BAILEY, JAMES J. HOECKER, WILLIAM L. MASSEY, AND DONALD F. SANTA, JR., COMMISSIONERS

Ms. MOLER. Thank you very much, Mr. Chairman. I am pleased to be here this morning.

First let me thank you for your very generous comments about the Commission. Regulators, like Members of Congress, don't get a lot of praise, so we are particularly grateful. If I had any sense at all, I would leave the Commission on this high note, but I will admit that I am addicted to this policy issue and am really looking forward to the opportunity to stay.

Mr. SHARP. I am reluctant to interrupt, but at least congressmen get to pay for their own praise in campaigns, and that is not available to those in other appointed positions.

Ms. MOLER. We have been working hard. The issues we face are extraordinarily complex and challenging. We are doing our best to define a solid, sensible regulatory reform program in a very tumultuous time.

The five of us today will describe the major activities and issues facing the Federal Energy Regulatory Commission as we try to accomplish the transition to the competitive bulk power market envisioned by the Energy Policy Act.

There is no doubt that the electricity industry is undergoing very fundamental change. Some of these changes are driven by regulators. More often, however, these changes are driven by the marketplace. The marketplace is becoming increasingly competitive. New generating capacity can be built and operated at costs that are less than many utilities' embedded costs. This simple fact is encouraging many users to seek access to the new, lower-cost sources of supply. We are seeing an increasingly open transmission system, and we are seeing States examine and experiment with retail wheeling.

The Commission is committed to developing a competitive, open transmission access, wholesale bulk power market. Our goal is to facilitate the development of competitively-priced generation supply options and to ensure that wholesale purchasers of electric energy can reach alternative sources of supply, and vice versa. This vision of the industry is one that we all share.

The transition to an increasingly competitive marketplace at the retail level is also under way, like it or not. Our Nation's regulatory structure for this industry is a complicated one to say the least. Both State and Federal regulators are endeavoring to meet our obligations to protect the public interest as this transition unfolds. We are all very much aware that both the financial stability and reliability of the industry are at stake.

There is no doubt that sellers and buyers of electric energy are responding to the competitive opportunities and pressures resulting from the Energy Policy Act. Exempt wholesale generators, or EWG's, are proliferating. Transmission access applications are on the increase. Traditional electric utilities are positioning themselves in the marketplace as they change their corporate structures and philosophies to survive in the very new world of competition.

There are five major areas of activity the Commission would like to focus on today. We have each agreed to focus on one particular area to avoid too much repetition. I will discuss pricing of transmission service, Commissioner Bailey will focus on State-Federal jurisdictional issues, Commissioner Hoecker will focus on our notice of proposed rulemaking on the recovery of stranded costs, Commissioner Massey will focus on regional transmission groups, and Commissioner Santa will focus on what we call the comparability principle for transmission access.

Four other issues of critical importance are covered in my written submission to the subcommittee. Taken as a whole, they show that significant changes are taking place in the electricity market. These developments and our response to them are a part of the larger picture. I want to emphasize that we are committed to making competition work within this industry.

Transmission access is a critical ingredient in that endeavor. Another critical ingredient is dealing with the costs transition. Our experience with restructuring the natural gas industry taught us that dealing with transition costs was critical to successful restructuring. Transition costs are critical to restructuring in this industry too.

Contrary to the impression some may have from the D.C. Circuit and from witnesses before this subcommittee yesterday, there will be stranded investments. Dealing with them properly is a critical factor in making competition work.

Let me now turn to transmission pricing. One of the most technically difficult issues facing the Commission is transmission pricing reform. Whether transmission is ordered under the new section 211 of the Federal Power Act, is voluntarily provided, or is a condition of approving a merger, the pricing of transmission must be consistent and coherent.

In light of the fundamental changes occurring in the industry, the Commission in June of 1993 began a comprehensive review of transmission pricing issues. We were asking whether fundamental changes should be made in the pricing methodology known as embedded cost postage stamp rates we have been using for 50 years should be made. We are asking about alternatives. A threshold question is whether the Commission's current pricing policy promotes or discourages efficiency and competition in the wholesale generation market.

In response, we received voluminous written comments. We also held 2 days of technical conferences which are our version of hearings, informal hearings, in April of this year. We learned a lot. We are still learning.

We heard two recurring themes at those conferences and in the written comments. First, the industry needs more pricing flexibility. Second, the industry and customers need more guidance as to what innovative pricing proposals might be acceptable to the Commission. In response to what we heard, the Commission intends to issue a policy statement on transmission pricing this summer. I believe there are a number of alternative pricing mechanisms that will be acceptable to the Commission. We certainly need to provide more flexibility than we have to date.

We are at the present time undertaking our discussions about what precisely will be included in this pricing policy statement, so the best I can promise is that we will hope to have it out this summer. I will not reveal its contents at the present time.

We are committed to dealing with this issue in a timely fashion in response to the voluminous record we have developed before the Commission and to the pleas from the industry to tell them what will be acceptable in this new environment. I am hopeful that after the policy statement issues we will see a number of innovative pricing proposals emerge and will be able to deal with them in a prompt fashion.

[The prepared statement of Ms. Moler and responses to subcommittee questions follows:]

Statement of Elizabeth Anne Moler, Chair, Federal Energy Regulatory Commission

Mr. Chairman and Members of the Subcommittee: I am pleased to be here this morning to discuss changes in the electric utility industry since the enactment of the Energy Policy Act of 1992. In particular, I would like to describe the major activities and issues facing the Federal Energy Regulatory Commission as we try to accomplish a smooth transition to the competitive bulk power market envisioned by the Energy Policy Act.

There is no doubt that the electricity industry is undergoing a fundamental change. The market for new generating capacity is a highly competitive one. New generating capacity can be built and operated at costs that are less than many utilities, current embedded costs. This simple fact is encouraging many users to seek access to the new, lower cost sources of supply. We are seeing an increasingly open transmission system. And we are seeing States examine, and experiment with, retail wheeling.

The Commission is committed to developing a competitive, open transmission access, wholesale bulk power market. Our goal is to facilitate the development of competitively priced generation supply options, and to ensure that wholesale purchasers of electric energy can reach alternative power suppliers and vice versa. This vision of the industry is one we all share.

The transition to an increasingly competitive marketplace at the retail level is also under way. Our Nation's regulatory structure for this industry is a complicated one, to say the least. Both State and Federal regulators are endeavoring to meet their obligations to protect the public interest as this transition unfolds. The issues we all face are both novel and complex. We are all very much aware that both the financial stability and reliability of the industry are at stake.

There is no doubt that sellers and buyers of electric energy are responding to the competitive opportunities and pressures resulting from the Energy Policy Act. Exempt wholesale generators, or EWG's, are proliferating. Transmission access applications are on the increase. Traditional electric utilities are repositioning themselves in the marketplace as they change their corporate structures and philosophies to survive in the new world of competition.

These changes are taking place at an ever-increasing pace. The nature, volume, and complexity of the Commission's electric workload has never been more challenging, nor the decisions of the Commission more important.

There are nine major areas of Commission activity we would like to discuss today. Each of the members of the Commission will focus on a particular area.

—I will discuss pricing of transmission services;

—Commissioner Bailey will focus on State-Federal jurisdictional issues;

—Commissioner Hoecker will focus on our Notice of Proposed Rulemaking on recovery of stranded costs;

—Commissioner Massey will focus on regional transmission groups; and

—Commissioner Santa will focus on what we call the "comparability" principle for transmission services.

My statement covers each of these issues in a summary fashion. It also discusses four other major areas. These are:

Non-traditional wholesale sellers of electric energy; Transmission information availability; Mandatory transmission access under section 211 of the Federal Power Act; and Voluntary transmission access post-Energy Policy Act.

All of these areas are critical to the successful implementation of the Energy Policy Act, and, not surprisingly, all but one of them relate to transmission access.

Transmission pricing. One of the most technically difficult issues facing the Commission is transmission pricing reform. Whether transmission is ordered under the new section 211, is voluntary under section 205, or is a condition of approving a merger under section 203, the pricing of transmission must be consistent and coherent.

In light of the fundamental changes occurring in the industry, the increasing number of voluntary open access tariffs being filed with the Commission, and the new section 212 pricing provisions, in June 1993 the Commission initiated a comprehensive review of transmission pricing issues. We are asking whether fundamental changes should be made in the historic pricing practices of the past half-century. This includes an inquiry into distance-sensitive rates, pricing of actual as opposed to contract power flows, spot pricing, and numerous other issues. A threshold question is whether the Commission's current pricing policy promotes or discourages efficiency and competition in the wholesale electric generation market.

The Commission received voluminous written comments in response to the inquiry. We also held 2 days of technical conferences in April of this year. We learned a lot. We are still learning. We heard two recurring themes. First, the industry needs more pricing flexibility. Second, the industry and customers need more guidance as to what innovative pricing mechanisms might or might not be acceptable to the Commission.

In response to what we heard, the Commission intends to issue a policy statement on transmission pricing this summer. I believe there are a number of alternative pricing mechanisms that would be acceptable to the Commission both legally and as a policy matter. No doubt, there are many complexities and implementation issues that will need to be addressed. We all have a lot more to learn, but I believe the industry has the expertise to resolve these complexities if given sufficient ability to experiment. I am hopeful that after the policy statement issues, we will see a number of innovative proposals emerge.

State/Federal issues. As the movement toward competition has accelerated, new jurisdictional issues have arisen. Most notable are two issues associated with unbundled retail wheeling. First, do States have the authority to order utilities to provide unbundled retail wheeling? Second, who has jurisdiction over the rates, terms and conditions of unbundled retail wheeling?

In the proposed rulemaking on stranded cost recovery, the Commission concluded that it has exclusive jurisdiction over the rates, terms and conditions of retail transmission in interstate commerce by public utilities. However, it did not address the issue of whether States have the authority to order such wheeling. Nor did it address the physical or jurisdictional distinctions between transmission and local distribution. These and related issues clearly are difficult ones, and the Commission currently is analyzing them.

Stranded cost recovery. The Commission's most recent generic initiative to address post-Energy Policy Act changes is a notice of proposed rulemaking to address recovery of stranded costs. These are costs that a utility has incurred to serve a wholesale requirements or retail franchise customer that are stranded when the customer stops buying power from the utility and simply pays for transmission services in order to reach a different supplier.

I would like to emphasize one particular issue. Stranded costs may be incurred when wholesale customers leave a utility's system or when retail customers leave. We call the latter retail stranded costs. As States begin to experiment with retail wheeling, these costs could be very significant. The proposed rulemaking concludes that while the Commission may have authority to address retail stranded cost recovery, the States also have the authority, as well as a variety of procedural vehicles, for addressing retail stranded cost recovery. We stated our strong policy preference that States deal with the issue. Their actions, after all, will largely be responsible for utilities incurring these costs. However, the proposed rule leaves open the question of whether there are circumstances under which the Commission should entertain requests for retail stranded costs recovery, and solicits comments on this issue.

Regional transmission groups (RTG's). RTG's offer a regional solution for addressing all of the transmission issues I've mentioned so far, including pricing, terms and conditions of service, transmission planning and reliability. On July 30, 1993, the Commission issued a policy statement encouraging RTG's and providing guidance on the minimum components of an RTG agreement.

We recently received our first two RTG proposals. A filing was made by the Western Regional Transmission Association in May and the Southwest Regional Transmission Association in June. I anticipate that the Commission can act on these in the fall.

Comparability. The most important change in the Commission's analysis of voluntary transmission tariffs involves undue discrimination. Since the Energy Policy Act, claims of undue discrimination have focused on whether it is unduly discriminatory for a transmission owner not to offer a third party the same or similar uses of its transmission system that it provides to itself.

In view of the changed circumstances in the industry, and the Commission's responsibility to analyze undue discrimination in light of these changes, in a May 11, 1994 order involving an open access tariff proposed by American Electric Power Service Corporation (AEP), the Commission stated its belief that a new open access tariff that is not unduly discriminatory or anticompetitive should offer third parties access on the same or comparable basis as the transmission provider uses its system. We call this the "Golden Rule" for utilities—do unto others as you do for you.

The Commission is in the early stages of analyzing what comparability is and how it will be implemented. There are a number of complex legal, technical and policy issues associated with comparability. These will be briefed by the parties in the ongoing hearings. We have a lot of hard work ahead of us as we further define the concept of comparability.

Non-traditional sellers of electric energy. Since the passage of the Energy Policy Act less than 2 years ago, the Commission has seen a steady growth in two types of nontraditional sellers: exempt wholesale generators and electric power marketers.

As of July 7, 1994, the Commission had received 159 applications for EWG status. We had granted 113 of them. The applicants include EWG's both in the United States and foreign countries. They represent a wide range of fuel types and sizes of electric power facilities. All EWG applications have been acted on within the 60-day statutory time limit. The Commission's final rule providing filing requirements and ministerial procedures for EWG applicants has been in effect almost 18 months. It appears to be working well.

In addition to EWG's, the Commission has seen a significant increase in the number of wholesale power marketers. Between the passage of the Energy Policy Act and July 7, 1994, the Commission had received 31 applications for market-based rate approval for power marketers. This includes the first three applications by marketers that are affiliated with traditional investor-owned utilities. These three currently are pending before the Commission.

Competition has permitted the Commission to lower barriers to market entry for new generators. In a Kansas City Power & Light Company case, the Commission recently concluded that competition in new generation is sufficient enough to allow the Commission to drop the requirement that sellers seeking marketbased rates show that they do not have market generation dominance in new capacity.

Transmission information availability. In September 1993, the Commission issued a final rule concerning transmission information. The rule, which is required by section 213(b) of the Energy Policy Act, requires that transmitting utilities submit to the Commission annually information to inform potential transmission customers, State regulatory authorities, and the public of potentially available transmission capacity and known constraints.

This rule will be a key component in developing competitive generation markets because it will open the "black box" of information which previously was available only to transmission owners. The information will assist both requestors of mandatory transmission services under 211 as well as requestors of voluntary services under section 205. It should help focus requests and minimize potential disputes. The first annual filings were made in April of this year. We received 124 filings, representing about 182 transmitting utilities.

Mandatory transmission access under FPA section 211. As of July 7, 1994, the Commission had received ten applications for mandatory transmission services under newly amended section 211 of the Federal Power Act. The Commission has issued two proposed orders and three final orders requiring transmission services. These orders involve services by traditional investor-owned utilities as well as by municipal utilities.

The Commission has attempted to process all section 211 applications in a timely fashion. We've endeavored to take initial action on these cases within 90 days of receiving a complete application, and to provide sufficient, but not prolonged, periods for voluntary negotiation prior to issuing final orders. We've also attempted, both in our July 14, 1993 Policy Statement on good faith transmission requests and responses and in case-by-case determinations, to provide clear guidance on the processing of section 211 applications.

As we gain further experience with section 211 applications, we will determine whether further generic actions are necessary. In the meantime, we are well aware that actions in individual cases may have precedential effects for other section 211 applications. For this reason, we have adopted a very liberal intervention policy in section 211 cases.

One of the most significant of the transmission applications was a request filed by a group of Florida municipals for so-called "network" service from Florida Power & Light Company. In effect, the municipals wanted transmission service which would allow them to use Florida Power's system the same way Florida Power uses the system to serve its own native load customers. In May of this year the Commission issued a final order requiring that the service be provided, and for the first time addressed how such service should be priced.

The Commission determined in the Florida Power case that if the customers wanted to use a share of Florida Power's system, they should pay a proportionate share of the costs. Accordingly, the network service was priced so that the municipals and the native load customers will pay a proportionate share of system costs, based on the load which each places on the system.

Voluntary transmission access tariffs. Prior to passage of the Energy Policy Act, a large number of utilities had filed open access transmission tariffs. Most of these were filed in order to mitigate market power and thereby obtain Commission approval of market-based rates or mergers or consolidations. Subsequent to the Energy

Policy Act, several major utilities began to file broad-based open access tariffs with the Commission voluntarily, without seeking market-rate or merger approvals. At the same time, customers began to raise new issues regarding these voluntary tariffs in light of the competitive changes taking place in the industry.

In response to these concerns, the Commission has begun to reevaluate its traditional analyses of undue discrimination and anticompetitive effects. The result is the new "comparability" requirement discussed earlier.

In closing, the Commission is committed to developing a competitive, open transmission access, wholesale bulk power market. We are working diligently to deal with an array of complex transition issues associated with that task.

In addition to my general testimony, I have included responses to specific questions raised in the subcommittee's July 1, 1994 letter inviting me to testify.

This concludes my remarks. I would be happy to answer any questions the subcommittee may have.

Answers to Chairman Sharp's Written Questions

Question No. 1: To what degree do you believe the electric utility industry has entered a period of irreversible change, and how widespread do you feel change will be?

Answer: The electric utility industry clearly has entered a period of irreversible change. The market for new generating capacity is a highly competitive one. That change began with enactment of the Public Utilities Regulatory Policies Act of 1978 and accelerated substantially with enactment of the Energy Policy Act of 1992. New generating capacity can be built and operated at costs that are less than many utilities, current embedded costs. This simple fact is encouraging many users to seek access to the new, lower cost sources of supply. we are seeing an increasingly open transmission system. Traditional investor-owned utilities will increasingly have to compete with nontraditional sellers to serve wholesale customers, and will have to open their transmission facilities to provide comparable transmission services to all wholesale purchasers and sellers of electricity.

It is unclear at this time whether the provision of comparable, unbundled wholesale transmission services will result in widespread changes in utilities, corporate structures. By that I mean the disaggregation of the functions of their systems into separate generation, transmission and distribution companies. That is certainly one logical outcome, and major utilities have begun or are contemplating restructuring or other innovations that would facilitate disaggregation of electricity services.

Another fundamental change will be that any wholesale obligation to serve will be defined solely and strictly by contract, and that most utilities increasingly will sell power at market-based rather than cost-based rates.

Finally, we are seeing States examine, and experiment with, retail wheeling.

Question No. 2: Do you have any recommendations as to how regulators should handle the transition period to a more competitive market?

Answer: Both State and Federal regulators must deal with the issue of stranded costs as early as possible, in order to protect the financial stability of the electric industry during the transition, and to provide as much regulatory certainty as possible regarding recovery of these significant costs.

Question No. 3: How can social and environmental objectives be adapted to a "restructured" industry, in an environment in which services once provided by a single vertically integrated utility are increasingly "unbundled"? To what degree should these objectives be retained, changed, or abandoned?

Answer: Social and environmental objectives in the electric area are predominantly State issues.

The increasingly competitive electric industry will require States to reinvent cost recovery mechanisms for their social and environmental agenda. It will be difficult, but not impossible. Utilities alone cannot be expected to shoulder these costs.

How these objectives should be adapted or addressed in the context of a utility selling unbundled retail services is a matter best left to the judgment of State commissions.

Question No. 4: How should pricing mechanisms be adapted to further carry out Congress, goal, embodied in EPAct, of a more competitive wholesale electricity market facilitated by open access to transmission services?

Answer: Unlike generation, transmission remains a natural monopoly. Therefore, pricing of transmission should be cost-based. However, there are a variety of cost-based pricing mechanisms other than those traditionally used which may be more appropriate for competitive markets. These include distance-sensitive as opposed to postage stamp pricing, pricing based on actual physical flows as opposed to contract path, and spot pricing, among others. However, there is an inherent tradeoff between some advanced concepts of transmission pricing which offer more accurate pricing but are substantially more complicated and difficult to administer. Given this inherent tradeoff, the Commission should allow as much flexibility as possible for experimentation with new pricing mechanisms.

Question No. 5: What are the most important questions involving State and Federal regulatory authority? Please feel free to offer any opinion you may have on the State of the law, or on the need for changes to, or clarifications of, the respective roles of State and Federal regulators.

Answer: The most important issues at this time are: (1) whether the States have authority to order unbundled retail transmission service; and (2) whether the rates, terms, and conditions of unbundled retail transmission in interstate commerce (whether voluntary or involuntary) is exclusively within the Commission's jurisdiction. Related issues include the physical and jurisdictional distinction between transmission in interstate commerce and local distribution.

As discussed in the preceding testimony, in our recent proposed rulemaking on stranded cost recovery the Commission concluded that it has exclusive jurisdiction over the rates, terms and conditions of retail transmission in interstate commerce by public utilities. However, the rulemaking did not address whether States have authority to order such transmission.

This latter legal question is in the most need of clarification. It goes to the heart of possible efforts by States to mandate retail transmission service. The Commission is clearly prohibited by the Energy Policy Act from ordering retail wheeling. However, the Congress did not clearly indicate whether it intended to occupy the field of retail transmission, or rather intended the States to have some authority in this area. In the absence of clarity under Federal law, this issue no doubt will be resolved only through lengthy litigation.

The Commission also noted in the proposed rulemaking that under section 201(a), the Commission does not have jurisdiction over local distribution. However, the rulemaking did not address, or decide, the physical or jurisdictional distinctions between transmission and local distribution.

Mr. SHARP. Let me now turn to Commissioner Bailey.

STATEMENT OF VICKY A. BAILEY

Ms. BAILEY. Good morning.

Thank you for inviting me along with my fellow Commissioners to testify today. I especially want to thank the chairman for continuing his leadership and foresight in holding these hearings and gathering a formal record on the changes occurring in the industry, and I have said it personally but not publicly, I will miss my fellow Hoosier at the helm of this subcommittee since you have announced your decision to not seek reelection. It has been my singular opportunity to work with you, Mr. Chairman, on issues impacting Indiana resulting from the Clean Air Act amendments and to now be a part of the team to implement a great deal of your national vision in the Energy Policy Act of 1992.

My transition from the Indiana Utility Regulatory Commission to FERC has been an interesting one. One of the most interesting aspects has been to study firsthand the different perspectives one gets from being at the State level and then being at the Federal level. We have all dealt with the inherent tensions in our form of government as a union of States.

With the passage of the Energy Policy Act of 1992 and the amendments to the Federal Power Act, FERC has expanded authority to order transmission service including ordering the expansion of transmission facilities. Yet the same act preserves with States the authority to permit such facilities to be built.

Tension does not have to be a bad thing, it can be a powerful, creative force, it can be a force for change. But the force of tension can also be counterproductive, and thus it should be properly channeled. In a regulatory context, tension between two entities with partially ambiguous and partially overlapping jurisdiction could produce gridlock, uncertainty, chaos. To avoid regulatory gridlock, it appears to me that FERC and the States need mechanisms for coordinating actions on issues such as electric transmission expansion, and I think we have as yet untried mechanisms for Federal-State coordination.

Transmission is probably the most glaring issue, but it is within a context of major change in the electric power industry. Federal-State regulatory relations will be tested on many levels. We have already seen this in cases concerning stranded costs.

Retail access is another arena where Federal and State interests will intersect. To me, the concept of comity seems most apt in this situation. The legal definition of comity is the principle by which the courts of one jurisdiction may give effect to the laws and decisions of another or may stay their own proceedings in deference to those in another jurisdiction. That definition would seem to apply to regulatory authorities as well as courts.

The Federal Power Act in its original formulation anticipated the need for Federal-State coordination. Section 209(a) refers to State boards that would be convened by FERC. Specific cases could be assigned to one or more States for initial review and data gathering.

Section 209(b) allows for joint FERC-State conferences on issues of mutual interest. Yet another mechanism is deference. FERC has deferred to State decisions in a limited number of cases where there were distinct local issues. More currently we have said that we would give deference to decisions by regional transmission groups that have the involvement of relevant State authorities. It is my view that these mechanisms can be further explored, applied, and tested as ways to coordinate Federal and State interest in the upcoming issues that will confront us.

If these mechanisms fail to perform, then the need to consider changes to our governing statutes will be evident. As we at FERC and our colleagues in the States address the significant changes in the electric industry, it will be important for us to communicate with each other about the goals we have for our regulatory proposals. It will be important to keep in front of us the goals of regulation.

My experiences as a former State regulator and now as a Federal regulator have proven to me that the phrase "in the public interest" is shared at both levels. This will guide our thinking about the changes that we are contemplating.

Part of what defines the public interest is the need for flexibility in this time of transition. We should try not to be overly prescriptive. Enormous dollar expenditures are contemplated by entities seeking to serve the Nation's electric consumers. We have to consider the derivative effects of our decisions on the environment, the reliability of the transmission grid, and the financial community, and we have to look beyond the boundaries of the United States and into our neighbors to the north and the south as the electric marketplace expands.

In the context of the changes to the electric industry that are creating more competitive markets, I think we should consider as part of our goals a different regulatory approach than we have in classical rate-based regulation.

Twelve years ago Judge Stephen Breyer wrote a book on regulatory reform in which he said that the problems accompanying classical regulation would seem sufficiently serious to warrant adopting a least restrictive alternative approach to regulation. Such an approach would view regulation through a procompetitive lens. These words are particularly on point as State and Federal regulators contemplate more competitive retail and wholesale markets.

One of the approaches that appears to embody the least restrictive approach is performance-based ratemaking. In addition to the much discussed direct access portion of their proposal, the California Public Utility Commission also proposed performance-based rates as the other key component for dealing with their tough economic problems in the State. Performance-based rates provide for flexibility to reward or punish a company much like a competitive market would do but in a regulated environment.

Again, the concept of comity comes to mind as California and apparently many other States contemplate retail competition. What should be the response from FERC? Well, first I think we continue to try to convene and talk about our respective goals. My colleagues and I participate in NARUC meetings, we have had involvement by State commissioners in some of our proceedings, such as our transmission pricing conferences.

Second, I think we have to exhibit this idea of comity. In other words, FERC respects what the States are trying to accomplish at the retail level. After all, the States are often in a better position to address issues associated with the planning, investment, and purchase activities of the utilities that they regulate, as we pointed out in our recent notice of proposed rulemaking on stranded cost recovery.

With the national perspective that one gains from being in Washington, D.C., it is not a giant leap for an Indiana Hoosier, a Wisconsin Badger, a Connecticut Yankee, an Arkansas Razorback and a beekeeper from Utah, the Beehive State, to see that the role of States is integral to the competitive vision. This is why I believe that some formal mechanism is needed for dialogue and coming to terms with this fundamental and integral State role.

One thing I hope we all keep in mind is that the electric power industry is one that affects the lives of all Americans. As such, it is important for us to keep in mind that the changing of the industry is not in and of itself the goal, the goal is to create a more efficient and economical system for the delivery of electric power to consumers.

Thank you.

[The prepared statement of Ms. Bailey follows:]

STATEMENT OF VICKY A. BAILEY, COMMISSIONER, FEDERAL ENERGY REGULATORY COMMISSION

Mr. Chairman and Members of the Subcommittee: I am pleased to appear before you this morning to discuss important issues concerning the changes in the electricity industry since the enactment of the Energy Policy Act of 1992. My remarks will focus on the relationship between the Federal Energy Regulatory Commission which is responsible for interstate wholesale sales of electricity and interstate transmission services and the various State commissions which are responsible for the regulation of electric delivery at the retail level. Regarding the questions posed by the Chairman in his June 30, 1994 invitation letter, I join with Chair Moler in her answers in her testimony.

The question of the proper relationship between the States and the Federal Government is as old as the Union itself. Our Constitution embodies this question with clauses that provide. this Congress with the power to regulate interstate commerce as well as amendments that reserve to the States all powers not delegated to the United States by the Constitution. To say that arguments have raged over the meaning of these words would be an understatement of huge proportions. The tension between the States and the Federal Government is part of the fabric of our Nation. It has defined who we are. We are from individual States, but more importantly, we are from States that form a Union. This fundamental dichotomy in our Nation pervades our daily life to this very day. Indeed, the questions posed by Chairman Sharp get right to the point on the relationship between the States and the Federal Government.

Having seen both sides, I know that State and Federal interests intersect, namely to serve the public interest. I would like to share my reflections on the Federal and State questions that will be so important as we continue upon the path of change for this industry. I am not the first FERC Commissioner to have served previously on a State Commission, nor will I be the last. This view from both sides has given me insights into the relative roles of regulators. I have spoken often in my speeches as a FERC Commissioner of the differences and the similarities in being at FERC and having been at the Indiana Utility Regulatory Commission.

I recognize that there are issues associated with generation and distribution, but today I would like to focus on the evolution of the transmission system. Federal and State issues become paramount as the transmission system attains a unprecedented level of openness as provided for by the Energy Policy Act and implementing FERC orders. The degree of openness will be defined to a great extent by the interaction of State and Federal regulators. Jurisdiction over the transmission system is, by design, shared between the States and the FERC. As such, it will be important that each level of jurisdiction be aware of the goals and objectives of the other jurisdictional elements.

I would like to take a moment and give my shortened version of a jurisdictional analysis of transmission facilities. The FERC has economic regulatory jurisdiction over all interstate transmission service provided by public utilities. This means that FERC approves the rates, terms, and conditions for interstate transmission service. Until recently, this transmission service jurisdiction was viewed primarily in a wholesale transmission context, retail transmission service was essentially non-existent, or enough so that jurisdictional questions were not raised. With proposed unbundling of transmission service at the retail level a new set of jurisdictional questions has been raised. As articulated in the Commission's recently issued Notice of Proposed Rulemaking on, Stranded Costs, the FERC has determined that it has jurisdiction over the rates, terms, and conditions of interstate retail transmission service.

But this is just part of the jurisdictional analysis. The other important piece concerns siting and certification of transmission lines. As affirmed in the Energy Policy Act, States have all property right approval jurisdiction for all transmission lines, both interstate and intrastate, and both retail and wholesale. In other words, FERC may now order transmission capacity to be expanded, but a State must give siting and certification approval for any new facilities that need to be built. This split in jurisdiction for new facilities is one of the fundamental reasons that States and FERC need to discuss transmission access issues more explicitly. This goes to the heart of a utility's ability to plan for new capacity when needed.

The final piece of the jurisdiction puzzle, is just that, a puzzle. In the Energy Policy Act, Congress removed FERC from the ranks of entities that can order retail access to occur. It also had a savings clause that kept States on the same jurisdictional grounds regarding retail access as before the passage of the Act. The question now becomes, did States have the ability to order retail transmission access before passage? This question is before many decision makers as we speak. I agree with those who have said that the courts will likely have the last word on this question. In the meantime, there are many policy decisions that must be made. Holding off on those decisions until the jurisdictional lines are crystal clear may be an exercise in futility.

Ultimately, each entity, States and FERC, must decide upon the correct policy approach to take for opening up the transmission system across State and regional boundaries. The policy approach adopted will have cross cutting effects on the other important issues in this industry. To name just a few that my colleagues will address, Federal approaches to stranded cost recovery, comparable levels of transmission service for wholesale and retail customers, the formation of regional transmission groups, and pricing of transmission services.

To determine the correct policy approach, one should first examine the statutory framework within which decisions must be made. For the FERC, that framework is the Federal Power Act. Individual States have their own statutes to examine. As I discussed earlier, the U.S. Constitution says that Congress has the power to regulate commerce between the States, but it also says that powers not delegated to the United States by the Constitution are reserved to the States. The Federal Power Act says that FERC has jurisdiction over interstate transmission and that States have jurisdiction over distribution. How much clearer do we need to be? A lot clearer, if the looming concerns of Michigan, California and other States contemplating retail transmission access are any indication.

Clearly articulating boundaries of statutes is important, but I think it is equally important to note that the Federal Power Act should not be used to "draw a line in the sand." There is room for discussion between FERC and the States about the best policy approach to the transmission questions. As articulated in the Commission's recent proposed rule on stranded cost, States are often in a better position to address issues associated with the planning, investment, and purchase activities of the utilities that they regulate. It will become even more important for the discussion to happen as the industry attempts to deal with the changes being wrought in the marketplace. FERC and the States will have to send clear policy signals so that industry entities can adapt structurally to the evolving market.

Transmission issues are taking on added significance as individual States contemplate access for retail consumers via the transmission system. Michigan has proposed a more limited type of retail access and many other States appear to be studying the retail access issue. These States are considering these important departures from past practices because of the problems or opportunities faced by electric consumers.

I should point out that retail access is not the only new idea under consideration, new approaches to regulating utilities (as opposed to substituting competition for regulation) are also on the table. The California proposal is an ideal case in point. While the retail access proposal has garnered all the attention, an equally significant departure from the status quo was made in the proposal to use performance based ratemaking in the place of rate-base regulation. To students of regulation, this is a proposal of large proportions. Rewarding a utility based on something other than the rate base that is studied in great detail by State Commission staff is reminiscent of the change from carburetors to fuel injection systems. (Remember that I am from the home of the Indy 500.) The consumer may not care one iota about the change, but will hopefully notice that the machine runs a bit more smoothly, especially on those cold winter mornings. Many of the concerns about the changes occurring are tied to traditional rate-based regulation and performance-based ratemaking requires a different mind-set. Performance-based rates provide some flexibility to reward or punish a company much like a competitive market would do, but in a regulated environment. Correct performance-based rates could be tied to an index that would provide the basis for the incentive to the company.

If regulators are going to adopt performance-based rates or allow for retail access or adopt some other basic change to the electric industry they should do so with conviction. Tentative signals will not help executives with billion dollar decisions to make. There is a growing need for State and Federal regulators to avoid conflicting signals, to provide more certainty, and to hopefully send unambiguous signals. With the National perspective that one gains from being in Washington, D.C., it is not a giant leap for an Indiana Hoosier, a Wisconsin Badger, a Connecticut Yankee, an Arkansas Razorback, and a beekeeper from Utah, the Beehive State, to see that the role of States is integral to the competitive vision. This is why I believe that some formal mechanism is needed for dialogue and coming to terms with this fundamental and integral State role. If a formal mechanism does not exist, we must design one. If the statutory framework does not allow for a formal mechanism then we may want to examine changes to that framework.

Without a means for State and Federal regulators to discuss transmission issues, I am afraid that what we will achieve is regulatory gridlock. If we are willing to follow the path of industry evolution we must ask whether the existing authorities split between Federal and State governments will allow the evolution to occur in a manner that captures the efficiencies available from more competitive markets. or, will there come a point in time where further evolution is hindered by these jurisdictional splits?

I am confident that we have not yet explored the full range of opportunities for formal State/Federal discussion. I am sure that there are innovative approaches for

State and FERC to coordinate that we can try within our existing statutory framework. The following are mechanisms that can be explored. (1) Section 209(a) of the Federal Power Act provides FERC with the option of assigning a case to a State board as if it were an Administrative Law Judge. FERC would then act on the initial decision of the State board just like it would on an ALJ's decision. (2) Section 209(b) sets out the option of a joint FERC-State conference on an issue of mutual interest. Although less formal, this option would allow FERC and the States to meet officially on a specific issue. (3) Another option is for FERC to provide deference to State decisions on issues that cross Federal State boundaries. FERC has provided deference in limited circumstances in the past. We have held out deference as an option in the Regional Transmission Group policy statement. There may be other options out there. We should do some exploring and see what works. If we try and fail, then I believe that changes to make our efforts successful will become evident.

Again, we should keep the goals of regulation in mind as we examine and test out new mechanisms. The public interest shared by both State and Federal regulators should guide our thinking about the consequences of our actions. We need to consider the need for flexibility in this time of transition and try not to be overly prescriptive. We must consider the need to plan for enormous dollar expenditures by entities seeking to serve the Nation's electric consumers. We have to consider the derivative effects of our decisions on the environment, the reliability of the transmission grid, and the financial community. And we have to look beyond the boundaries of the United States and into our neighbors to the North and the South as the electric marketplace expands.

Twelve years ago, Judge Stephen Breyer wrote a book on regulatory reform titled appropriately enough Regulation and its Reform. In the book he discussed several alternatives to classical regulation. I think those words are appropriate now as we at FERC and our colleagues at the State level address the significant changes in the electric industry. He said, The problems accompanying classical regulation would seem sufficiently serious to warrant adopting a "least restrictive alternative" approach to regulation. Such an approach would view regulation through a procompetitive lens. It would urge reliance upon an unregulated market in the absence of a significant market defect. Then, when the harm produced by the unregulated market is serious, it would suggest first examining incentive-based intervention, such as taxes or marketable rights, or disclosure regulation, bargaining, or other less restrictive forms of interventions before turning to classical regulation itself. It would urge the adoption of classical regulatory methods only where less restrictive methods will not work.

Judge Breyer's words are particularly on point and challenging as State and Federal regulators contemplate the more competitive retail and wholesale markets spawned by statutory actions like the Energy Policy Act. They should consider how their regulatory approaches should be modified in light of this competition. It would seem to me that Federal and State regulators should get together and ask ourselves some fundamental questions. What is the new regulatory system that should accompany these more competitive markets? Are all markets competitive or are some still natural monopolies? What other structural changes would dictate a different regulatory approach?

The structural changes within this industry are challenging both State and Federal regulators. A few years ago the debate was defined by IOU's, coops, and munis. Now we have to know about EWG's, QF's, and RTG's. Soon, we may have to think about the role of gencos, transcos, and discos as was predicted back in the ancient history of the 1980's. And, we must remember that these changes are not just insular, but that they effect the fabric of the Nation including the millions of people who have invested vast sums of money into this industry. A dialogue among the States and with the Federal Energy Regulatory Commission is becoming increasingly important as these changes become more prevalent.

Finally, we must always keep in mind that all of us require electricity to carry out our daily activities. we are talking about a fundamental element of our modern lives. When we change the very nature of the electric business, we have to remember that change in and of itself is not the goal. The goal is to create a more efficient and economical system for the delivery of electric service to consumers.

Thank you, Mr. Chairman, for the opportunity to testify before the subcommittee today. I am prepared to answer any questions that you or members of the subcommittee may have.

Mr. SHARP. Who is next, Madam Beekeeper?

Ms. MOLER. Commissioner Hoecker.

STATEMENT OF JAMES J. HOECKER

Mr. HOECKER. Mr. Chairman, members of the subcommittee, good morning. It is indeed an honor to be here today, and, Mr. Chairman, I join with my colleagues in wishing you well in life after Congress.

I am Jim Hoecker, a Commissioner at the FERC. I have served in that capacity now for nearly 14 months. This has, of course, been a stimulating time at the Commission as we have virtually concluded the restructuring of the service offerings and operations of the interstate natural gas pipelines after nearly 10 years and begun to focus on the emergence of competition in the electric utility industry.

In both cases industry developments and regulatory initiatives have followed the same general course plotted in other basic regulated industries. That is to say, the overall trend is for increased competition among resource options pursued by an increasingly diverse group of economic agents in the marketplace.

The Commission is participating in the metamorphosis of this industry, not necessarily driving the changes. Certainly the procompetitive Energy Policy Act has helped change the atmosphere surrounding the industry, but the electric power business had already begun to change when Congress acted in 1992. The economies of scale in generating units had vanished, new distributed generation and transmission technologies were appearing, environmental concerns were increasing, and the debate over U.S. industrial competitiveness had come to include the public utility sector.

I might also add that the developments called into question the adequacy of traditional rate-based regulation as a surrogate for the marketplace.

Now while it is probably fashionable to conclude that competition is in and regulation is out, I believe regulators at the State and Federal levels will play an increasingly important role in helping move the industry from the complexities and inefficiencies of monopoly power, service obligations, and rate of return regulation to more efficient market-oriented models. Hopefully we can provide to the subcommittee today a good sense of how the Commission is proceeding on several of these fronts.

My colleague, Commissioner Santa, has identified in his written testimony the factors and ideas associated with the gas pipeline area that are now so clearly relevant to our emerging experiences on the electric side. I would add to his comparisons some notable factors that I think distinguish the circumstances of electric utilities and how we regulate them from those of natural gas.

First, electric power is ubiquitous and it is the single most capital intensive area of the Nation's economy, so everyone will be affected by what happens to this industry.

Second, while utilities operating at wholesale are subject to regulation by the FERC, most of the industry must respond to the requirements of State legislators and regulators. In particular, utilities have historically been subject to a statutory obligation to stand ready to serve all citizens.

Third, the Commission's authority to foster competition, while important, is also limited. Because the Commission is, for example, prohibited from ordering retail wheeling, the future of competition is as much a State as a Federal matter.

Fourth, electric utility facilities may be employed to serve at wholesale and at retail contemporaneously. These arrangements may entail jurisdictional frictions and overlaps, as Commissioner Bailey has discussed.

More importantly, as it encourages competition and restructuring, FERC must be mindful of the retail native load customers for whom utilities are also responsible for providing an unfailing bundled service pursuant to local franchise arrangements. So from a Federal perspective, no native load problem of this magnitude complicated the restructuring of the interstate natural gas pipeline services. These factors suggest to me that the procompetitive transition in electricity may be much more complex than that endured by gas pipelines.

Now I devoted much of my written comments to the topic of transition costs, so-called stranded investment. I will therefore conclude today with some observations about the July 12 decision of the U.S. Court of Appeals for the District of Columbia Circuit in *Cajun Electric Power Co-op* vs. *FERC*.

The court remanded the Commission's 1992 order allowing Entergy Corporation's operating companies to sell wholesale power in negotiated market-based rates on the condition that the companies provide transmission services under open access transmission tariffs. But as part of the deal, Entergy was given an assurance by the Commission that if customers' ability to buy power elsewhere as a result of the open access tariffs left the utilities with excess capacity and related costs, Entergy could recover those stranded costs in the transmission rates charged those customers but only under certain specific circumstances.

At bottom, the court's decision to remand turns on the fact that there was an insufficient evidentiary record and that the Commission failed to adequately explain its approval of the stranded cost provisions in the Entergy tariffs and their potential effects on competition. The Commission must now evaluate how it might respond to the Cajun Electric decision.

The court certainly made clear its displeasure with the Commissions' failure to fulfill a basic procedural obligation to investigate genuine issues of material fact. We take that challenge to the Commission's orders very seriously. But it is not altogether clear, to me at least, what would be accomplished now by a hearing about the effects of the stranded cost portions of Entergy's tariffs on competition in the market.

The reason for my question about the efficacy of the court's instructions to hold further proceedings stems from the occurrence of events that have, in effect, overtaken the Entergy case.

First, pursuant to the Commission's recent proposed rule on stranded costs, we will lay the evidentiary foundation for the recovery of certain stranded costs in rates but this time in the context of the electric industry as a whole. The final rule will apply to all public utilities, including Entergy's.

The court in the Entergy case naturally did not have before it the full picture of what has been happening in the electric utility industry since 1992, including the effects of enactment of the Energy Policy Act.

In light of subsequent developments, most of which have been discussed in these hearings, the Commission's final rule will probably supersede what was decided in the Entergy decisions. It will address stranded costs in terms of changing operation and regulation of the industry in toto and not just the open access regime of one holding company.

In fact, the D.C. Circuit instructed the Commission to do precisely that when it remanded Order 436 to the Commission in 1987 to address the cost consequences to open access pipelines of long-term take-or-pay contracts. In this sense then, stranded costs is a matter that must be dealt with in the context of the transition to the more competitive environment.

Second, since the time Entergy was decided by the Commission, we have made major strides in addressing what is necessary to mitigate market power. The key to that decision is now the comparability requirement that will be discussed by Commissioner Santa. Such a requirement in future cases under which transmission owning utilities would be required to provide to others a quality of transmission service comparable to that which they enjoy on their own systems should allay many of the court's concerns regarding the possible vestiges of market power still left in the Entergy tariffs.

In sum, I do not view the Cajun Electric decision as entirely negative from the Commission's current perspective. The court appears to have no disagreement with the principle of mitigating market power through open access transmission tariffs, nor does it challenge the Commission's exercise of its conditioning authority to obtain open access as a prerequisite for market-based ratemaking.

Whatever the outcome of this case, I do not believe that the decision fundamentally undermines, and indeed in some respects it supports, the Commission's general policy goals to encourage competition in the electric industry.

Thank you for your kind attention.

[The prepared statement of Mr. Hoecker and responses to subcommittee questions follow:]

STATEMENT OF JAMES J. HOECKER, COMMISSIONER, FEDERAL ENERGY REGULATORY COMMISSION

Mr. Chairman and Members of the Subcommittee: It is my pleasure and privilege to share with you this morning my thoughts on the evolving state of the electric utility industry. I subscribe to the remarks already offered by Chair Moler, as well as her summary of the major activities undertaken by the Federal Energy Regulatory Commission since enactment of the Energy Policy Act of 1992 (EPAct). Moreover, I agree with her filed responses to Chairman Sharp's questions enunciated in his June 30, 1994 invitation to testify. However, I want to take this opportunity to add my own observations about them. Those answers are appended.

My primary task this morning is to address the regulatory treatment of certain costs associated with the transition to competition. These kinds of costs will in some measure arise from and affect the restructuring of the electric utility industry, just as they have accompanied fundamental change in other regulated industries.

As I see it, a number of factors have coalesced to promote change in the electric utility industry. There exist today competitive non-utility sources of electrical generation that compete with utility investments to satisfy customer demand. Utilities themselves were also given the opportunity by the Energy Policy Act to invest in wholesale generators without being subject to Public Utility Holding Company Act regulation. In 1992, the Congress nevertheless clearly recognized that market access was the key to successful competition among these sources of power. It therefore gave the Commission potent, if limited, authority to mandate access to the bulk power transmission system. The increasingly interdependent nature of the transmission system makes the use of the grid by the transmission-owning utility a matter of key economic interest for the entire industry. The pressure of competitive change exerted upon electric utilities was preceded by similar developments in other basic industries such as airlines, natural gas pipelines, trucking, and communications (not to mention the electric industry in the United Kingdom) and occurs in the context of the information and computer revolutions.

So, we appear before you today having just begun to address the depth and breadth of changes in the industry—many of which are occurring beyond our jurisdictional reach. The Commission faces numerous large and important issues and choices.

1. At what speed should the transition to a more competitive bulk wholesale power market occur? or, to put it in familiar terms, is an electric Order No. 63611 called for?

2. What are the quality of service implications of revised section 211 of the Federal Power Act? For example, how should regulators select from among standards of comparability, non-discrimination, network service, or flexible transmission arrangements?

3. Can a competitive bulk power market be achieved using only or mainly the specialized procedures of revised sections 211 and 212 of the Federal Power Act or must the Commission also reexamine the standards of sections 205 and 206 and how it remedies undue discrimination thereunder?

4. To what extent should the Commission permit use of advanced transmission pricing concepts, particularly in light of consumer interest in services that reflect actual costs?

5. Under what circumstances should the Commission favor pricing and access models that permit flexibility and customer-driven results as opposed to models that rely on traditional regulated utility central planning?

In less than a year, we have begun coming to grips with these issues and have collectively arrived at some critical decisions about the industry and our regulatory role. I would summarize them as follows:

1. The FERC may, under section 211 of the Federal Power Act, authorize high quality network service equivalent to the flexible use of the grid enjoyed by the transmitting utility; yet, anyone requesting such transmission service must bear system costs commensurate with those borne by the transmission service provider. Florida Power & Light Co., 67 FERC 61,167 (1994).

2. Although the FERC has sparingly used its new authority to order wheeling, section 211 has changed the dynamic of our electric transmission regulation and led the Commission to reexamine how it might also promote a competitive bulk power market pursuant to the traditional standards and procedures of sections 205 and 206. New England Power Pool, 67 FERC 61,042 (1994).

3. Third parties seeking to engage in purchases and sales of power from distant generation facilities are entitled to open access transmission services that are not unduly discriminatory or anticompetitive and that are provided on the same or comparable basis, and under the same or comparable terms and conditions, as the transmission provider's use of its system. American Electric Power Service, 67 FERC 61,168 (1994) (I discuss the comparability standard at length in a separate concurrence, 67 FERC at 61,-491-2).

4. While change in industry structure and operations primarily concerns competitive forces and the public good that competition—properly introduced and managed—entails, the transition to whatever will emerge as the electric industry of the future will be chaotic, unbearably confrontational, and slow unless regulators address up front whether and how utilities can recover the costs associated with the move from one mode of regulation, one set of contracts, or one kind of service obligation to another. These are the costs variously identified as "costs without a customer," "stranded costs," or—as I prefer to think of them—transition costs.

When competition comes to utility service areas, it will mean that customers once served exclusively by a local utility may choose their suppliers of electric power and obtain fair access to transmission service to bring that power to them. In other words, existing wholesale customers of a utility may suddenly be able to buy all or part of their wholesale requirements elsewhere. In addition, retail customers may be legally reconstituted municipalization) into wholesale power or transmission customers no longer subject to traditional obligations to purchase power at retail from the local franchise utility. Or, retail customers that take traditional service from the local utility may be permitted by the State to buy power from other sources and have it "wheeled" to them by the local utility.

In any of these cases, the utility that served the departing customer will arguably find it difficult to recover the costs incurred to serve that customer. Those costs will be borne either by a utility's shareholders or by the other customers who remain. Generally speaking, someone will invariably pay for legitimate, verifiable, and prudently incurred costs already in the market. Such costs might include the costs of capital investments, the costs associated with purchased power arrangements, the cost of mandated social programs, the cost of environmental externalities, or investments in demand-side management (DSM) programs.

The question presented is plain enough: Who will pay and why? How should the Commission and its State counterparts address this potential side-effect of the transition from an environment of bilateral transactions governed by a regulatory bargain-or a public utility obligation to serve, which generally ensured a utility of an opportunity to recover all its costs, to a competitive environment of multilateral and changing commitments, market risks, and new institutions? The answer is anything but simple. There is, for example, no universally accepted estimate even of the potential magnitude of stranded costs. Estimates range from zero to as high as $200 billion.

Those advocating recovery of stranded costs argue that utilities must be compensated for investments made when they operated as franchise monopolies subject to statutory service obligations or other regulatory bargains to serve customers that become, by virtue of regulatory or legislative actions, eligible to buy elsewhere.

Those who oppose recovery of stranded costs argue that utilities should have foreseen competition or have simply not competed effectively and do not deserve to be compensated for investments the market now determines to be uneconomic. Moreover, they contend utilities are compensated in their rates of return for any risk associated with the increased ability of customers, particularly wholesale customers, to access other suppliers. Even if the recoverability of stranded costs were not in dispute, there is no general agreement about how such costs ought to be recovered in rates or what regulatory body ought to provide for recovery of certain costs.

Under current Commission precedent, utilities may include reasonable cancellation provisions in power sales contracts to protect themselves. They charge "exit fees" if a customer departs before a contract expires. However, many contracts lack such provisions, having been negotiated in a less competitive era when there was an expectation that the utility would serve a customer indefinitely. The Commission therefore has said it would allow recovery of costs associated with service to a departing customer (1) if a utility could show it incurred costs with the reasonable expectation that it would continue to serve that customer, (2) if the costs do not exceed what the customer would have contributed to the utility as a wholesale requirements customer, and (3) if the utility has mitigated the effect of the stranded costs. This policy evolved in the context of the Commission's approval of an "open access" transmission tariff in Entergy Services, Inc., 58 FERC 61,234, order on reh'q, 60 FERC 61,168 (1992). See Cajun Electric Power Coop., Inc. v. FERC, Case No. 92-1461 (D.C. Cir., July 12, 1994), remanding the Entergy decisions for further proceedings.

On June 29, 1994, the Commission issued a Notice of Proposed Rulemaking (Proposed Rule) to address stranded costs as a transitional issue, which is of generic importance in light of growing demand for competitive access to bulk power transmission services. Recovery of Stranded Costs by Public Utilities and Transmitting Utilities, 67 FERC 61,394 (1994). The Proposed Rule addresses wholesale and retail stranded costs. With respect to costs incurred to serve a wholesale customer, the Proposed Rule treats differently the stranded costs associated with new contracts and those subject to existing contracts. The essence of the Commission's approach to recovery of wholesale stranded costs in either case, however, is to rely primarily on the protections afforded by private contract. Where contractual cost recovery provisions do not presently exist, the Commission proposes to establish a brief period within which such provisions can be negotiated.

1. If an existing contract explicitly addresses stranded costs through an exit fee or other stranded cost provision, the public utility or transmitting utility may recover the costs only as specified in the contract. Moreover, after the end of a 3-year transition period, recovery of stranded costs for any public utility will be allowed only pursuant to contract.

2. If an existing contract does not explicitly address stranded costs, the parties to the contract must make a good faith attempt to negotiate a stranded cost amendment within 3 years of the Proposed Rule. If negotiations are successful, and the selling utility under the contract is a public utility, the amendment must be filed with the Commission and, if approved, the amended contract will govern any future recovery of stranded costs. Otherwise, the selling public utility, may unilaterally file a stranded cost amendment.

3. Wholesale stranded investment may not be recovered in a transmission rate except when a departing customer gives proper notice that it wants to purchase unbundled transmission service from the selling utility and the service will begin prior to the end of the 3-year transition period.

Transition costs associated with retail service present a very difficult set of issues, due in part to the significance of the costs involved, the jurisdictional questions associated with transmission access to retail customers, and the differences in the evolution of competition among the States. The Proposed Rule, therefore, expresses a strong policy preference for State or local authorities to address recovery of costs left stranded due to the departure of a customer that now only wants wheeling service. The Commission could, on one hand, serve as a forum of last resort for stranded cost recovery, if the appropriate State or local authority does not explicitly address retail stranded costs or if there is a conflict among authorities within a State or among different States about stranded cost recovery. On the other hand, the Commission might choose not to entertain requests for recovery of stranded costs that were incurred to provide retail service not subject to the Commission's jurisdiction.

In either case, the Commission does not wish to discourage openly competitive retail markets. If, however, State action results in the stranding of utility investment, there is a strong argument that the States must step up to the problem, however difficult or unpopular the issue might be. There is a countervailing argument, of course. Because the Commission believes that the transmission services associated with retail access involve FERC-jurisdictional rates, it can be argued that the investment stranded by these wheeling arrangements (though State-authorized) must be recoverable in the transmission rate. In any event, competition should not be allowed to become a means to foist embedded costs onto other customers or other regions without a rational plan to spread the legitimate costs of the transition to competition among those likely to benefit from that competition.

The Commission will be dealing with this issue in its cases and generically in the coming year. I believe that this issue must also be fully ventilated at the State level as the industry is encouraged to be more competitive and market-oriented. We five Commissioners agree that the stranded costs issue must be taken on immediately and not after restructuring.

Mr. Chairman, thank you for the opportunity to testify before the subcommittee today. I am pleased to respond to questions.

RESPONSES TO CHAIRMAN SHARP'S QUESTIONS

Question 1: To what degree do you believe the electric utility industry has entered a period of irreversible change, and how widespread do you feel change will be?

Answer 1: The electric utility industry is changing rapidly, fundamentally, and irreversibly. The industry and its regulators are looking for new institutions, a vision of what the industry will become, and reasonable and economical means to make the transition. There is no received wisdom about how we get from here to there—or even where "there" is. Professor Richard Pierce of Columbia University has described the forthcoming transition as more "difficult, painful, complicated, and unpredictable than the same transition of gas pipelines." I find his further observation plausible:

First, the transition will be virtually complete within the next 10 years. Second, the pace of the transition will accelerate rapidly at some unpredictable point in the transition process when the consumer benefits of deintegration and competition become obvious to the public. Third, neither the path nor the pace of the transition will prove to be manageable by any single institution. Fourth, utilities with high costs when the transition sweeps over them will not survive the transition. They will be unable to reallocate to others enough of their massive transition costs to remain financially viable.

As significant as the process of change is, we are well advised to remember what has not changed. First, utilities that own transmission facilities are still regulated monopolies. Second, the Commission's overall regulatory mandate to protect consumers under the Federal Power Act—including sections 205 and 206—has not changed. The Congress gave the Commission neither the authority to certificate the construction of transmission facilities nor the explicit means to disaggregate the segments of the industry. Third, the overall operations and finances of the electric power business continue to be more directly affected by the States than by Federal regulators.

Question 2: Do you have any recommendations as to how regulators should handle the transition period to a more competitive market?

Answer 2: The FERC and our State colleagues are engaged in a quintessentially legislative (i.e., policymaking) function. Care, measured judgment, and a grounding in the record should mark our decisions. The industry will nevertheless continue to change—whether or not the Commission participates. For instance, new institutions, such as regional transmission groups (RTG's), new pooling arrangements (Poolco's), or hybrid transmission companies (Gridco's), are being considered or devised. The industry will also develop spot markets and electronic trading. Amidst these changes, Federal regulators must continue to encourage or order open access transmission at the wholesale level. They may seek to expand market-based pricing, especially for bulk power sales. They must evaluate the viability of new institutions and arrangements. Transition costs and transmission pricing also require attention.

There is an open question of whether it would be best to manage the transition all at once or slowly over time. The all-at-once school probably favors disaggregation of generation and transmission assets. This would create an opportunity to subject those assets to a market test and form the basis of stranded investment compensation. Yet, the Commission may lack the legal authority to approach this so neatly. Rather, we must gauge our regulatory responses to the evolving industry.

One thing is clear to me. Federal and State regulators must communicate and cooperate as never before. We must develop consulting mechanisms and, hopefully, some shared assumptions about how regulation of a newly competitive wholesale and retail market for electric power will function.

Question 3: How can social and environmental objectives be adapted to a "restructured" industry, in an environment in which services once provided by a single vertically integrated utility are increasingly "unbundled"? To what degree should these objectives be retained, changed, or abandoned?

Answer 3: The cost (some would say subsidization) of demand-side management, mandated social or conservation programs, and the idea of universal service are almost always State regulatory concerns and a matter of national interest for the Department of Energy. I would nevertheless venture to say that industry restructuring is not necessarily incompatible with the efficacy of social and environmental objectives. In a purely competitive, price-determined market, customers may not want to bear these costs, of course, and may exercise their choices to avoid them. Yet, the mandate of regulatory agencies to enhance the public welfare will persist, even in a regulatory environment more oriented to promoting efficient market behavior. In the past, the regulatory mandate resulted in State intervention to order or encourage certain behaviors or programs. Increasingly, legislators and regulators are seeking market-based solutions to social/environmental problems. Programs that reward internalization of environmental costs or which establish incentives to pollute less, such as $S0_2$ credits, are examples. Indeed, the basic approach used today is to give incentive to a power producer not to produce. An alternative way to achieve the environmental objectives is to create incentives for consumers to conserve, while allowing competition to create incentives for producers to do their jobs efficiently. Consumer incentives might include real-time pricing, real-time control of appliances, and so forth.

Question 4: How should pricing mechanisms be adapted to further carry out Congress' goal, embodied in EPAct, of a more competitive wholesale electricity market facilitated by open access to transmission services?

Answer 4: I completely agree with the Chair.

Question 5: What are the most important questions involving State and Federal regulatory authority? Please feel free to offer any opinion you may have on the state of the law, or on the need for changes to, or clarification of, the respective roles of State and Federal regulators.

Answer 5: As stated in the Chair's responses, State authority to order retail wheeling and the lack of effective distinctions between transmission and distribution functions are unresolved issues that have potential importance for the regulation of competition and transition costs. Although no conflicts have arisen, there is a potential conflict between any future order by the Commission to build facilities to carry out wheeling pursuant to section 211 of the Federal Power Act and States' exclusive control over siting.

The coming 2 to 3 years will also be revealing with respect to the effectiveness of RTG's in achieving consensual resolution regional planning and transmission pricing issues. Without legislation, it is impossible at this juncture to determine the degree to which the FERC and State commissions will choose to defer to the decisions of RTG's. Some degree of deference is, in my opinion, important to the success of these emerging voluntary institutions.

Mr. SHARP. Thank you very much.

STATEMENT OF WILLIAM L. MASSEY

Mr. MASSEY. Mr. Chairman, if you will allow me to begin on a somewhat lighter note, when I make these morning presentations I am reminded of the comments of former Green Bay Packers running back Paul Hornung who had scheduled a morning wedding, and when he was asked why he did that, he said, well, that way if things didn't work out, he wouldn't blow the whole day. That story always comes to mind.

Nevertheless, Mr. Chairman——

Mr. SHARP. Are we expecting the pricing policy this afternoon, or what is the deal?

Mr. MASSEY. Nevertheless, Mr. Chairman and members of the subcommittee, I welcome this opportunity to testify on the changes in the electric utility industry since the passage of the Energy Policy Act of 1992. In passing that landmark legislation, Congress set an unmistakeable course toward competition in this industry, and since then FERC, as you are all aware, has been very actively implementing its new authority. In a number of orders and policy pronouncements, the Commission has underscored its commitment to the goal of capturing for consumers the efficiencies that wholesale competition can bring.

It is clear to me that the industry has entered an era of irreversible change. The genie of wholesale competition is out of the bottle. This is not occurring in isolation. Similar changes have already taken place in other domestic industries such as air lines, trucking, telecommunications, and natural gas. Moreover, in other parts of the world as well, privatization and competition are sweeping the electric industry.

FERC's vision of the wholesale power industry's transition to competition contains several key elements, as others and I will discuss today. Ultimately there may be additional pieces, but for now these present a reasonably coherent vision or blueprint for the industry.

First, as Commissioner Santa will set out in more detail, we have adopted a new policy on comparability of transmission service intended to guarantee equal access to the transmission wires for all users. Buyers and sellers of power should be indifferent to who owns the wires and should be able to take advantage of the electron superhighway on the same terms as transmission owners provide to themselves. This is the essence of our comparability policy, which I believe will become the cornerstone defining this emerging era of competition.

Second, we are moving toward progressively lighter regulatory control over wholesale generation as competition spreads. The monopoly power of the industry is in the wires, not the generators, although we will remain vigilant to guard against monopoly abuse that may arise in generation.

The third element of the mosaic is our proposed policy on the recovery of stranded costs. As outlined eloquently by Commissioner Hoecker, reasonable recovery of wholesale stranded costs is absolutely critical in the transition to full competition.

Fourth, we will soon provide policy guidance with respect to transmission pricing. I believe and hope that the Commission will move toward a policy of somewhat more flexibilities with respect to transmission pricing. As explained more fully by Chair Moler, we hope to issue policy guidance this summer.

Fifth, as I will discuss in a minute, we have in place a policy strongly encouraging the formation of regional transmission groups, or RTG's.

Taken together, these five pieces form a fairly comprehensive vision of the direction in which this Commission is moving. Other pieces may fall into place as well and as appropriate, but the overall directions have been set. I am personally proud of the vision of this Commission.

Let me return to the subject of RTG's. The idea of a regional transmission group, or RTG, in its simplest form is to allow all of the operators and users of the transmission grid in a region jointly to decide the rules for using and expanding that grid.

Last summer FERC issued a policy statement encouraging the formation of RTG's and identifying the minimum components for an acceptable RTG. For example, membership should be open to everyone eligible to obtain or subject to a transmission order by the Commission. Members should provide transmission service to other members on essentially an open access basis. In this sense, an RTG is a proxy for open access. The RTG should include fair and nondiscriminatory governing procedures and voluntary dispute resolution procedures, and it should provide for adequate consultation and coordination with State regulators.

RTG's are important for a number of reasons. RTG's will be structured to reflect the interests of all the grids' users, not just some. They will allow for consensual solutions to regional issues instead of solutions imposed from Washington, D.C. They can function as regional laboratories for experimentation on transmission issues. They will provide a regional forum, which I think is a necessary predicate to regional cooperation. They can benefit all industry participants. Transmission have-nots will gain a greater say in the use and expansion of the transmission grid as well as quicker access to the grid. The transmission owners will gain deference from FERC on RTG decisions and better access to neighboring systems. State regulators will gain a forum for regional coordination and planning. Ultimate consumers will benefit from greater competition in the wholesale power market and the resulting downward pressure on prices.

For all of these reasons, I believe the Commission is completely committed to its RTG policy and will be constantly urging the industry in this direction.

Let me make just a few comments on retail competition. This subcommittee is aware of the proposals that are being considered in California, Michigan and elsewhere. At this point I am neither criticizing nor endorsing any State action in this area. I have not been called upon to weigh the competing policy choices that face State regulators.

A threshold obstacle to such proposals, however, is a legal one: Do the States have the authority to order retail wheeling? It seems to me that a persuasive argument can be made that retail wheeling is preempted by the Federal Power Act. Section 201 gives FERC jurisdiction over all transmission of electric energy in interstate commerce.

Since section 201 does not expressly limit our jurisdiction to transmission at wholesale, the argument is that we have jurisdiction over all transmission in interstate commerce, whether it is wholesale or retail. Several of your judicial decisions broadly construe FERC's authority in this area.

On the other hand, it is argued that States have always regulated transmission to end users as part of their jurisdiction over retail sales and that the mere unbundling of transmission service in the form of retail competition does not undo a State's jurisdiction. Proponents of this argument note that while the Energy Policy Act bars FERC from ordering retail wheeling, Congress also inserted a savings clause in the act to preserve whatever authority States may have, thus disproving the contention that Congress considered the field of retail wheeling to be preempted.

I think there are good arguments on both sides, frankly. I have reached no conclusions about the lawfulness of retail wheeling. Let me just say that absent congressional action to resolve this debate, the issue will ultimately be decided by the courts and perhaps even the U.S. Supreme Court. If so, the likely prelude to a decision may be years of regulatory uncertainty at a steep cost to all industry participants and ultimate consumers.

In closing, let me just make three points. First, as regulators it is important for us to keep focused on our ultimate goal. That goal is not competition for competition's sake but reliable electric energy at lower prices. Competition is a means to achieve that goal.

Second, as a Federal regulator, I am convinced that there is still much efficiency to be squeezed out of our wholesale system. In many ways our job has just begun. A lot of nitty-gritty regulatory work remains.

Third, the transition is unlikely to be a glide path. The policy choices may become even more challenging, but regulators must face the tough issues early in the process. I believe that this Commission is attempting to do just that.

Thank you very much. I would be happy to answer any questions.

[The prepared statement of Mr. Massey follows:]

STATEMENT OF WILLIAM L. MASSEY, COMMISSIONER, FEDERAL ENERGY REGULATORY COMMISSION

Mr. Chairman, and Members of the Subcommittee, I welcome this opportunity to testify on the changes in the electric utility industry since the enactment of the Energy Policy Act of 1992.

In passing this landmark legislation, Congress charted an unmistakable course toward competition in the wholesale bulk power market. The new law substantially lowered barriers to the entry of new wholesale generation competitors by allowing the creation of Exempt Wholesale Generators (EWG's) free from the regulatory legerdemain of the Public Utility Holding Company Act (PUHCA). In addition, Congress understood that buyers and sellers of bulk power would often require access to wholesale transmission wires, and enacted historic amendments to section 211 of the Federal Power Act to facilitate such access.

The Commission, in turn, has been actively implementing its new authority. As of July 7, 1994, the Commission had received 159 applications for EWG status, and had granted 112 of them., From this and other evidence, one can conclude that competition among new wholesale generation is flourishing. With respect to transmission access, the Commission has issued a number of orders and policy pronouncements which underscore its commitment to the goal of capturing for consumers the efficiencies that wholesale competition can bring.

It is clear to me that the industry has entered a period of irreversible change. The genie of wholesale competition is out of the bottle. This is not occurring in isolation. Similar changes have already taken place in other domestic industries, such as airlines, trucking, telecommunications and natural gas. Moreover, in other parts of the world as well, privatization and competition are sweeping the electric industries.

In order to avoid a duplication of testimony, each Commissioner has agreed to take the lead in explaining a particular area of Federal regulatory policy of importance to the subject of this hearing. my remarks will focus primarily on FERC's policy encouraging the formation of regional transmission groups (RTG's).

Before beginning my discussion of regional transmission groups, let me briefly outline where our RTG policy fits in a FERC vision for the future of this industry at wholesale. Five key pieces of a vision mosaic are either in place or quickly falling into place. There may ultimately be other pieces as well, but for now these five present a reasonably coherent vision or blueprint for the industry.

First, as Commissioner Santa sets out in more detail, we have in place a new policy on comparability of transmission service, which is intended to guarantee equal access to the transmission wires for all users. Buyers and sellers of power should be indifferent to who owns the wires, and should be able to take advantage of the electron superhighway on the same terms as the transmission owners provide to themselves. This is the essence of our comparability policy, which I believe will become the cornerstone defining this emerging era of wholesale competition.

Second, as our decision in the recent KCP&L case indicates, FERC is moving toward progressively lighter regulatory control over wholesale generation as competition spreads. The monopoly power of the industry is in the wires, not the generators, although we will remain vigilant to guard against any monopoly abuse that may arise in generation.

The third element of the mosaic is our proposed policy on the recovery of stranded costs. As outlined in more detail by Commissioner Hoecker, reasonable recovery of wholesale stranded costs is absolutely critical in the transition to full competition.

Fourth, we will shortly provide policy guidance with respect to transmission pricing. Several Commissioners have expressed interest in moving toward a policy of somewhat more flexibility with respect to transmission pricing. As explained more fully by Chair Moler, we hope to issue policy guidance this summer.

And fifth, we have in place a policy strongly encouraging the formation of RTG's, as I will explain more fully in my testimony.

Taken together, these five pieces form a fairly comprehensive vision of the direction in which this Commission is moving. Other pieces may fall into place as appropriate, but the overall directions have already been set. I am proud of this vision.

The idea of an RTG, in its simplest form, is to allow all of the operators and users of the transmission grid in a region—public utilities, municipals, rural electric cooperatives, IPP's, EWG's, and power marketers—jointly to decide the rules for using and expanding that grid. A consensual regional organization should have the flexibility to track the expected regional evolution of competitive bulk power markets. An RTG is a proxy for open access that gives the transmission have-nots a real say in formulating the transmission rules of the road, facilitates regional transmission and generation planning, and provides an important bridge between Federal and State regulators in the transition to a more competitive wholesale environment. The Commission intends to grant a measure of deference to RTG decisionmaking.

As the subcommittee is aware, legislation on RTG's was considered during negotiations on the Energy Policy Act. In fact, many of the interested parties eventually agreed on a "consensus" RTG proposal, but by then it was apparently too late for inclusion in the final legislation hammered out by the conferees.

Soon afterward, FERC solicited comments on the consensus RTG proposal. Then, in July of 1993, after reviewing all of the comments, FERC adopted a policy statement encouraging the formation of RTG's.

Our policy statement identified the minimum components for an acceptable RTG. For example, membership in an RTG should be open to everyone eligible to obtain, or subject to, a transmission order by FERC. An RTG should contain a geographic area large enough to allow for reliable, efficient and competitive transmission services. Members should provide transmission service to other members on essentially an "open access" basis. Members should develop and update a regional transmission plan. The RTG should include fair and nondiscriminatory governing procedures and voluntary dispute resolution procedures. And, it should provide for adequate consultation and coordination with State regulators.

Why are RTG's so important? First, in a competitive era, wholesale markets will evolve without regard to State lines, probably on a regional basis. Historically, however, the industry's regional institutions for managing the transmission grid, such as power pools and regional reliability councils, have been dominated by the grid's bigger owners, with little or no input from other users. In contrast, RTG's will be more balanced, reflecting the interests of all users, not just some.

In this sense, RTG's are compatible with FERC's recently adopted policy of comparability. Under this policy, as explained in detail by Commissioner Santa, whatever uses a utility makes of its transmission system, it must offer to other users on a comparable basis, including comparable prices.

FERC has not yet decided whether RTG's must achieve comparability. The issue may not arise, since the balanced governing structure of RTG's may produce access and pricing rules that obviously satisfy the comparability principle without a FERC requirement. If not, however, I think the starting point for FERC's analysis of an RTG's rules may well be comparability. In any event, our analysis will likely be informed, if not controlled, by the comparability policy.

A second reason why RTG's are so important is that consensual solutions to transmission problems, reached in consultation with State regulators, are likely to be better solutions for a particular region than those imposed from Washington, D.C. An RTG can bring the decisionmaking process a little closer to home.

Third, RTG's can allow for regional experimentation on a variety of challenging transmission issues. FERC certainly does not have all the answers. With appropriate signals from FERC and State regulators, RTG's can in fact become regional laboratories in which the best minds in the industry thrash through a multitude of problems and adopt bold and creative solutions.

And fourth, as industry participants working to form RTG's are confirming to me, a regional Process is a key element in a regional solution. It is invaluable for industry participants to sit down within a regional structure and at least attempt to reach agreement. A broadly inclusive, regional process is inherently valuable.

I believe RTG's can benefit all industry participants. Take the transmission have-nots, such as non-utility generators and municipals. RTG's will give them more say in the use and expansion of the transmission grid. Also, since an RTG is a proxy for open access, an RTG can reduce significantly the time and cost needed to obtain transmission service, compared to the 2 years of litigation it might take to obtain service through a FERC order under section 211 of the Federal Power Act. In short, RTG's should reduce the market power of transmission owners and, thus, allow the development of a more competitive market for wholesale power.

RTG's can also benefit transmission-owning utilities. RTG's will give those utilities a greater measure of control over their transmission destiny in this new era. Within the bounds allowed by the Federal Power Act, FERC intends to grant deference to RTG decisions, including decisions on reliability and transmission pricing.

Moreover, transmission owners can benefit from easier access to neighboring transmission systems, for the purpose of buying cheap but distant power or selling their own power to distant buyers. Taking advantage of these market opportunities will be increasingly important as competition spreads through the industry. More and more utilities, especially those with generation capacity to sell, are beginning to see RTG's as a good marketing tool.

And importantly, RTG's (alone or enhanced through the development of regional trading mechanisms or power pooling) may allow utilities to prove that there are still substantial efficiencies and consumer benefits to be captured through purely wholesale competition.

RTG's will prove good for State regulators. RTG's can provide a forum for States to cooperate regionally. They are not regional regulatory bodies, but they do provide a regional forum for transmission and even generation planning. States also may benefit from having access issues decided in the first instance by the RTG rather than by FERC. States would retain their traditional authority to grant or deny permission for construction of transmission facilities, and States may have more influence on issues within an RTG than they would as a party in litigation before FERC.

As Commissioner Bailey underscores in her testimony here today, there is a growing need for improved coordination between State and Federal regulators. For example, FERC can order wholesale transmission service at a reasonable price that promotes an efficient transmission and generation market, but the hoped-for consumer benefits will never materialize if State regulators refuse to authorize construction of the necessary transmission facilities. And even if the necessary facilities are built, if Federal and State regulators fail to coordinate their decisionmaking and do not allocate the costs of such facilities fully, and shareholders must absorb part of the utility's just and reasonable costs, we have guaranteed that the transmission owner will fight each future request for transmission services with every available tactic.

I see RTG's enhancing opportunities for Federal and State cooperation in a competitive era. In our policy statement on RTG's, we emphasized that consultation and coordination with the States are "critical to the successful implementation of RTG's." We also explained that "State involvement in RTG's can allow State agencies to improve communications with utilities and with each other in dealing with transmission concerns, and can facilitate coordinated treatment of siting issues among the States." And, the extent of coordination with States will be a key factor "influencing the degree of deference the Commission [will] give to consensual resolutions reached under an RTG."

As I see it, RTG's can be an important vehicle for State regulators in addressing transmission issues on a comprehensive, regional basis and in promoting and achieving regional transmission planning. And, the improved transmission access available through an RTG should allow local utilities to make more, and better, wholesale trades and thus reduce the cost of service to retail ratepayers.

I would note that the National Association of Regulatory Utility Commissioners (NARUC) passed a resolution in March this year strongly endorsing FERC's RTG policy. The support of State regulators will be enormously helpful in encouraging RTG formation.

RTG's are certainly good for FERC. We are not yet equipped or staffed to handle a large increase in transmission-related disputes. I hope and expect that the industry will resolve these disputes better and far more quickly than FERC would. And, we need experimentation on a wide variety of transmission-related issues.

Finally, and most importantly, RTG's are good for ultimate consumers. Protecting their interests is the driving force for regulation, and the essential reason for promoting RTG's. RTG's will promote competition in the wholesale power market and downward pressure on prices. As I said, RTG's are a way to squeeze as much efficiency out of the wholesale system as possible.

The relationship between RTG's and transmission pricing is also worthy of comment. The Commission intends to grant deference to RTG decisionmaking, including decisions with respect to transmission pricing. In our pending transmission pricing inquiry, many parties asked FERC to offer clear guidance on the allowed pricing methods and the proper goals and principles for RTG's to consider in setting a regional pricing policy. As Chair Moler has more fully explained in her testimony today, the Commission is moving in the direction of allowing somewhat greater pricing flexibility than has been allowed heretofore, especially within RTG's.

A number of interesting principles have been suggested for determining appropriate transmission pricing methods. Pricing methods that met certain goals would be included on a pricing menu. An RTG could choose from the menu or make innovative proposals that furthered the goals set by the Commission.

I agree that RTG's should be allowed to adopt a pricing method in this way. This kind of regional pricing flexibility, at least within the context of an RTG where the transmission have-nots have real and not just nominal voting power, is workable and need not be driven by the market power of transmission owners.

To summarize, the Commission's RTG policy is a key component in our vision of the future of this industry. Wholesale markets will evolve regionally, and RTG's will allow market participants to decide the rules of the road through local, consensual solutions, instead of through the slower and more costly process of FERC rulemaking and litigation. RTG's are also a proxy for open access and a bridge between Federal and State regulators, and provide perhaps the best opportunity for squeezing every possible efficiency out of our wholesale system. The potential is there. It is up to the industry, with the help of regulators, to fulfill that potential.

I am pleased to see that FERC's RTG policy is finally starting to bear fruit. In recent weeks, FERC has received its first two RTG filings. A number of other regions are actively pursuing the formation of an RTG. There is evidence that transmission owners have a steadily growing interest in RTG formation. I believe the Commission is committed to its RTG policy and will be constantly encouraging the industry to move in this direction.

My above testimony responds to Chairman Sharp's questions, except for the question concerning clarification of Federal and State regulatory authority. In my judgment, there is some uncertainty with respect to the issue of retail competition.

Each member of the subcommittee is aware of the raging debate over the concept of retail competition. Proposals for retail competition are being considered in California, Michigan and elsewhere.

At this point, I am neither criticizing nor endorsing any State action in this area. I have not been called upon to weigh the competing policy choices that face State regulators. A threshold obstacle to such proposals, however, is a legal one. Do the States have the authority to order retail wheeling? FERC has concluded that it has jurisdiction over the rates, terms and conditions of retail wheeling, but FERC has not been called on to decide whether States can order retail wheeling.

A persuasive argument can be made that retail wheeling is preempted by the Federal Power Act. FPA section 201 gives FERC jurisdiction over all "transmission of electric energy in interstate commerce." Since section 201 does not expressly limit

FERC's jurisdiction to transmission at wholesale, the argument is that FERC has jurisdiction over all transmission in interstate commerce, whether it is wholesale or retail. Several judicial decisions broadly construe FERC's authority in this area.

On the other hand, some argue that States have always regulated transmission to end-users as part of their jurisdiction over retail sales, and that the "unbundling" of transmission service in the form of retail competition does not undo a State's jurisdiction. Proponents of this argument note that, while the Energy Policy Act bars FERC from ordering retail wheeling, Congress also inserted a "savings" clause in that Act to preserve whatever authority States may have, thus disproving the contention that Congress considered the field of retail wheeling to be preempted. Even if these arguments fail, States may be able to use their regulatory authority in creative, as yet unforeseen, ways to achieve results identical or similar to retail wheeling.

I have reached no conclusions on the lawfulness of retail wheeling. Absent Congressional action to resolve this debate, this issue ultimately may be decided by the U.S. Supreme Court. If so, the likely prelude to a decision may be years of regulatory uncertainty, at a steep cost to all industry participants and ultimate consumers.

Thank you. This concludes my testimony. I would be pleased to respond to any questions the subcommittee may have.

Mr. SHARP. Thank you.

Mr. Santa.

STATEMENT OF DONALD F. SANTA, JR.

Mr. SANTA. Thank you, Mr. Chairman and Mr. Franks, for the opportunity to be here this morning to discuss the changes in the electric utility industry since enactment of the Energy Policy Act of 1992.

This is a terribly exciting time to have the privilege to serve as a member of the Commission. The combination of the procompetitive forces already at work in the electric utility industry and the Commission's new authority under title VII of EPAct have provided this Commission with a unique opportunity to shape the regulatory framework that will govern how electricity is produced, marketed, and transmitted in this country. What this Commission, what State regulators, and what the industry's leaders do over the next several years will set the tone for many years to come.

The Commission's electric agenda represents not only a tremendous opportunity but also a tremendous responsibility. The electric utility industry is a major sector of our Nation's economy. A healthy, competitive, and efficient electric industry is critical to our Nation's economic competitiveness. It is also critical to the welfare of individual consumers who expect reliable, reasonably priced electric utility service.

Finally, the equity of the electric utility industry represents the investment of millions of Americans who likely did not expect that the winds of competitive change would sweep the industry so quickly.

In summary, for me as a Commissioner and I am sure for my colleagues as well, this is a time that is both exhilarating in terms of the opportunity to shape the regulatory framework for a more competitive industry and sobering in terms of the responsibility that has been invested in us.

As Chair Moler explained, the aspect of the Commission's activities that I will discuss is the Commission's comparability standard for electric transmission. The place that I would like to start is with the Federal Power Act. The Commission's ratemaking authority is spelled out in sections 205 and 206 of FPA. The statutory standard is that the rates, terms, and conditions for jurisdictional services must be just and reasonable and not unduly discriminatory or preferential or otherwise unlawful.

Historically, the Commission's standard for enforcing that statutory prohibition against undue discrimination has been whether a utility was consistent in its treatment of similarly situated customers. In other words, similarly situated customers should be treated the same for purposes of the rates, terms, and conditions of the jurisdictional services that they receive.

Conversely, based on factual differences between customers, a utility could justify differences in the rates, terms, and conditions that would apply to those customers. This would not be undue discrimination.

Prior to EPAct, the similarly situated standard provided the Commission with a good yardstick for identifying cases where utilities were unduly discriminatory in their rates, terms, and conditions for jurisdictional sales and transmission service. This was because, prior to EPAct, claims of undue discrimination usually took the form of allegations that a utility had provided a third party with more preferential treatment than it had provided to the claimant.

However, in recent years market conditions in the electric utility industry have changed dramatically. These new market conditions include the emergence of nontraditional power suppliers and greater competition in bulk power markets. All of a sudden, transmission-owning utilities found themselves facing greatly intensified competition in the bulk power market. A transmission-owning utility might find itself having to compete with other sellers for an off-system sales opportunity or it might find itself having to compete with others to retain the load of one of its historic wholesale customers. In either case, the utility's control of transmission gave it a distinct competitive advantage in the bulk power market.

As a result of the new competitive realities of the bulk power market, the nature of the claims of undue discrimination filed with the Commission began to change too. The focal point of these claims shifted from claims of undue discrimination in the rates and services which the utility offered to other customers to claims of undue discrimination in the rates and services which the utility offered when compared to its own use of the transmission system. This came to the Commission's attention in connection with voluntary transmission tariffs filed by several utilities last year. The utilities characterized their tariffs as offering open access transmission service. This, however, did not stop intervenors from protesting the tariffs as being unduly discriminatory.

The basis for these claims was that the transmitting utility did not offer its customers the same quality of service as the utility enjoyed when it used its own facilities. The focus of this debate was the distinction between so-called point-to-point service which the transmitting utilities offered to others and so-called network service which generally describes the flexibility the utility enjoyed when using the transmission system on its own behalf.

In a New England Power Pool case this past April, the Commission announced that it was reexamining the nature of the statutory obligation to ensure that its transmission service is not unduly discriminatory. In explaining its decision, the Commission cited the changing competitive conditions in the electric utility industry and the change in the nature of the claims of undue discrimination that it was receiving.

In May the Commission built on the foundation it had laid in April's NEPOOL order. The Commission issued an order in an American Electric Power proceeding in which it enunciated the new comparability standard for analyzing claims of undue discrimination. The standard is this. All transmission owners must provide service that is the same or comparable to the service which the transmission owner provides itself.

In AEP, the Commission's preliminary analysis determined that the proposed open access tariff was unduly discriminatory because it offered transmission customers only point-to-point service and not the network service that the utility provided itself.

The Commission's adoption of comparability as the yardstick for applying the FPA's prohibition on undue discrimination stands, I think, as the most important post-EPAct change in the Commission's analysis of voluntary transmission tariffs.

The comparability standard has been dubbed the golden rule for electric utilities—do unto others as you do unto yourself—and when stated in these terms the standard is compelling in its logic and its simplicity.

The comparability standard sends an unequivocal message to electric utilities that the Commission believes that their monopoly in transmission should no longer confer any advantage in generation markets. Thus, while comparability for service as provided under generally applicable transmission tariffs was not part of EPAct's specific statutory mandate, the comparability standard is fully consistent with what the Commission perceives to be the congressional intent behind title VII of EPAct, that being an intent to foster greater competition in wholesale electricity markets by means of open access transmission service.

Now that comparability has been adopted as the standard for analyzing claims of undue discrimination, there remains the question of how comparability will be implemented in practice. In order to answer this question, the Commission needs to understand better how electric utilities operate their transmission systems. In AEP and in the subsequent cases where it has applied the comparability standard, the Commission has initiated investigations that are intended to build a record that will support and inform its ultimate decision.

The following questions were set for a trial type evidentiary hearing in AEP. First: What are the different uses that AEP makes of its transmission system? Particularly, what degree of flexibility does AEP accord itself in using the transmission system for different purposes? Second, are there any impediments or consequences to providing third parties with the same transmission service that AEP provides to itself? And, third, what are the costs that AEP incurs in providing transmission to itself, and would the costs be any different to provide the same service to third parties?

Implicit in these questions is the fundamental question of whether third party transmission customers should receive service that is the same or comparable to the transmission service embedded in a utility's delivered electricity service to its retail native load customers.

In summary, without a doubt the AEP order in which the Commission announced its comparability standard represents a landmark in terms of the Commission's analysis of how the statutory prohibition on undue discrimination applies in the post-EPAct market for electric generation and transmission.

My final point goes generally to the Commission's role in defining its part of the regulatory framework for the electric utility industry's competitive transition. The point is the need for regulatory certainty. Utility executives frequently have commented to me, "If you only tell us what the rules are, we will make the changes that are necessary to do business in the new environment."

Quite frankly, I am not so naive as to believe that regulatory policy can remove all or even most of the uncertainties created by the competitive forces that are transforming the electric industry. After all, the market has a mind of its own and it usually pays little heed to the designs of regulators.

Still, to the extent that clarity of regulatory intent can remove one element of uncertainty, it will promote a more orderly and less costly transition, and this, I submit, is in the public interest. Consequently, I think it is of critical importance that FERC articulate clearly what it is doing, why it is doing it, and the legal basis for doing so.

You have probably heard Nike's popular advertising slogan, "Just do it." Well, I think in light of the importance of regulatory certainty, FERC's corollary to "Just do it" should be, "And make sure you are clear about it."

As I hope has been demonstrated by our testimony today, this Commission has done a good job of articulating its evolving policies. Certainly one example of this is the resolution of FERC's standard for applying the statutory prohibition on undue discrimination that I described earlier. Because FERC enunciated clearly that it was modifying the traditional analysis in adopting the comparability standard, utilities now are on notice that any new transmission services will be governed by this standard.

I would like to conclude with one more comment about regulatory certainty, and that is the role of contracts in defining relationships in the bulk power market.

In its notice of proposed rulemaking on stranded costs, FERC recognized that contracts are the basis for purchase and sale obligations in the bulk power market. This recognition of contracts as the foundation of wholesale market relationships ought to apply as well in other contexts. In some cases purchasers may be parties to a contract that restricts their ability to purchase power from third parties. While FERC has the ability to modify such contracts where it finds them to be unjust, unreasonable, unduly discriminatory, or preferential, I think it ought to be circumspect in doing so.

My reason for this position is not to frustrate competition; rather, it is in recognition that contracts will continue to define the relationships between wholesale sellers and purchasers and that this foundation will be undermined if regulated entities perceive that regulators easily will be persuaded to abrogate contracts.

This concludes my remarks. Thank you again for the opportunity to testify, and I would be happy to answer any questions that the subcommittee may have.

[The prepared statement of Mr. Santa follows:]

STATEMENT OF DONALD F. SANTA, JR., COMMISSIONER, FEDERAL ENERGY REGULATORY COMMISSION

Mr. Chairman and Members of the Subcommittee: I am pleased to be here this morning to discuss changes in the electric utility industry since the enactment of the Energy Policy Act of 1992 (EPAct). As Chair Moler explained, each of the Commissioners will comment in detail on a different aspect of the Commission's activities in the area of wholesale electricity regulation. My testimony will be directed to the topic of the comparability standard for electric transmission that the Commission has adopted for purposes of enforcing the Federal Power Act's (FPA) prohibition on undue discrimination.

After describing the comparability standard, I will add some of my own perspectives on electric utility regulation in general. Finally, with respect to the five questions that were asked as part of Chairman Sharp's June 30, 1994, invitation letter, I join with Chair Moler in the answers that were appended to her testimony.

In discussing comparability, I will address: (1) the legal standard under the FPA; (2) pre-EPAct transmission access policy; (3) the evolution of the comparability standard; and (4) questions that flow from the comparability standard.

Under the FPA, the Commission has exclusive jurisdiction over the transmission of electric energy in interstate commerce and the sale of electric energy at wholesale in interstate commerce.[1] Pursuant to sections 205 and 206 of the FPA,[2] the Commission is charged with determining whether the rates, terms and conditions for such jurisdictional services are just and reasonable, and not unduly discriminatory or preferential, or otherwise unlawful.

Prior to the enactment of EPAct, the Commission acquired a greater appreciation for the importance of transmission access to an efficient bulk power market through its experience with mergers and market-based rates for electric generation. The seminal order in this area was issued in 1987, and involved the merger which created the public utility PacifiCorp (Utah Merger). In the Utah Merger, the Commission found that access to transmission was necessary to mitigate the market power associated with the merged company's increased control of transmission facilities.[3] The Commission also examined transmission market power in the context of permitting bulk power sellers to charge market-based rates, i.e., negotiated rates.[4] The Commission has found in these circumstances that a seller's lack of transmission market power ensures that the seller is unable to block potential competitors from reaching willing purchasers.

Based on its experience with mergers and market-based rates, and working within the context of Congress' goal in EPAct of fostering more competitive bulk power markets, the Commission recently set a new standard for transmission service providers. Pursuant to its authority under FPA sections 205 and 206 to ensure that rates, terms and conditions of transmission service are not unduly discriminatory or preferential, the Commission has determined that all transmission owners must provide service that is the same or comparable to the service which the owner provides to itself.

The Commission's adoption of comparability as the standard for applying the FPA's prohibition on undue discrimination stands as the most important post-EPAct change in the Commission's analysis of voluntary transmission tariffs. The comparability standard has been dubbed the "golden rule" for electric utilities—do unto others as you do unto yourself.

Traditionally, the standard for undue discrimination was whether factual differences justified differences in rates charged to, and terms and conditions applicable to, similarly situated customers. In a New England Power Pool case this past

---

[1] 16 U.S.C. section 824 (1988).

[2] Id. at sections 824d, 824e.

[3] See Utah Power & Light Company, PacifiCorp, and PC/UP&L Merging Corp., Opinion No. 318, 45 FERC, para. 61,095 at 61,283-61,295 (1988), order on reh'q, Opinion No. 318-A, 47 FERC para. 61,209, order on reh'q, Opinion No. 318-B, 48 FERC para. 61,035 (1989), granting review in part sub nom., Environmental Action, Inc. v. 57 FERC para. 61,363 (1991).

[4] See, e.g., Public Service Company of Indiana, Opinion No. 349, 51 FERC para. 61,367 at 62,198-99, order on reh'q, Opinion No. 349-A, 57 FERC para 61,260 (1990), appeal dismissed, 954 F. 2d 736 (D.C. Cir. 1992).

April,[5] the Commission announced that, in order "to respond to changing conditions in the electric utility industry,"[6] it was re-examining the nature of its obligation under the FPA to ensure that transmission service is not unduly discriminatory. In explaining its decision, the Commission observed that, due to changing conditions such as the emergence of non-traditional suppliers and greater competition in bulk power markets, it was being presented with new types of undue discrimination claims. The Commission stated that, in many of these cases, the focal point had shifted from claims of undue discrimination in rates and services which the utility offers different customers, to claims of undue discrimination in the rates and services which the utility offers when compared to its own use of the transmission system.[7]

The Commission's recent order in American Electric Power Service Corporation (AEP)[8] built on the foundation laid in NEPOOL. The Commission in AU enunciated its new standard of transmission comparability. Under this standard, transmission owners are required to provide service to third parties that is the same or comparable to the service which the transmission owner provides to itself. In AEP, the utility had filed what it termed an "open access" transmission tariff for third-party use of its transmission system. Although this service was voluntarily offered, the Commission's preliminary analysis determined that the proposed tariff was unduly discriminatory because the tariff offered only so-called point-to-point transmission service and not so-called "network" transmission service. (Point-to-point transmission service is a lower quality service than network service; network service allows a transmission customer some flexibility with respect to the receipt and delivery points of scheduled power and energy.)

Now that it has enunciated the comparability standard, the Commission is investigating what it means in practice. The Commission in AEP opined that transmission owners may use their systems for a variety of purposes, such as serving native load customers, participating in the bulk power market, serving wholesale requirements customers, and other purposes. Recognizing that such differences in transmission system usage may exist, and that certain uses of the system may create operational or reliability constraints, the Commission set for evidentiary hearing the following key questions:

(a) What are the different uses that AEP makes of its transmission system, particularly, what degree of flexibility does AEP accord itself in using the transmission system for different purposes;

(b) Are there any impediments or consequences to providing third parties with the same transmission service that AEP provides to itself; and

(c) What are the costs that AEP incurs in providing transmission to itself, and would the costs be any different to provide the same service to third parties.

Implicit in these questions is the fundamental question of whether third-party transmission customers (which traditionally have been wholesale electricity purchasers and sellers) should receive the same or comparable transmission service to the transmission service "embedded" in a public utility's delivered electricity service to its retail native load customers. The AEP case—and subsequent cases in which the Commission has applied the comparability standard[9]—should inform the Commission as to how public utilities use their systems and, accordingly, how the Commission should regulate the marketplace based on these realities.

In addition to these questions concerning how comparability will apply in practice, the new standard raises other questions which the Commission may ultimately face. For instance:

(a) Will the comparability standard prove workable? In other words, based on its investigation of how individual utilities use their systems, will there emerge an objective measure of comparability that can be applied to all utilities across the board? In the alternative, what happens if the Commission finds that there are significant differences in the ways that utilities use their systems?

(b) Does comparability of service terms and conditions mean price comparability too? If it turns out that all transmission service is the same (wholesale and retail), won't transmission customers be entitled to the lowest price? what if it turns out

---

[5] New England Power Pool, 67 FERC para. 61,042, reh'q. denied, 67 FERC para 61,314 (1994) (NEPOOL).

[6] 67 FERC at 61,132.

[7] Id. at 61,132.

[8] 67 FERC para 61,168 (1994).

[9] Florida Power & Light Company, 67 FERC 61,326 (1994); Commonwealth Edison Company, 67 FERC 61,325 (1994) ; Northern States Power Company, 67 FERC 61,240 (1994); Kansas City Power & Light Company, 67 FERC 61,183 (1994).

that this is the "price" set by State regulators for the transmission that is embedded in the rates for retail sales?

(c) Will the comparability standard ultimately be effective in "opening up" the transmission system in order to achieve the goal of a competitive bulk power market? Or will more be necessary? For instance, will the Commission ultimately have to order (or provide a strong incentive for) the functional unbundling of transmission from generation?

I now would like to share with you some of my perspectives on electric utility regulation based on the past year's experience:

1. EPAct's Effect on the Electric Utility Industry and Electric Power Regulation.

My initial view of title VII of EPAct was from a lawyer's perspective and, as it turns out, was considerably narrower than what is now recognized as title VII's effect. On its face, title VII amended discrete sections of the FPA and the Public Utility Holding Company Act concerning case-specific authority to order transmission of wholesale power and the corporate organization of wholesale generators. But this narrow technical view does not do EPAct justice. The message that was received by the market, and later by both Federal and State regulators, was that EPAct was a broad endorsement of the use of access to the transmission grid as the means to promote competition in the generation sector of the electric industry.

While the root causes of the demand for access to transmission and competitive generation have built up over time, the catalyst for the current restructuring of the electric utility industry clearly has been EPAct. Whether Congress fully intended that EPAct produce this result, that is the market's perception, and the perception has become the reality. This perception affects not only the Commission, but also the States. For example, the California Public Utility Commission's (CPUC) retail wheeling "blue book" is full of references to EPAct as a motivating force behind the action proposed by the CPUC.

2. The Importance of Regulatory Certainty.

During this restructuring period, electric utilities and other stakeholders in the electric debate need to know what rules apply to them so that they can plan accordingly. Consequently, I think it is of critical importance that FERC articulate clearly what it is doing, why it is doing it, and the legal basis for doing so. An example of this is the evolution of FERC's standard for applying the statutory prohibition on undue discrimination that I described earlier in my testimony. Because FERC enunciated clearly that it was modifying the traditional analysis and adopting the comparability standard, utilities now are on notice that any new transmission services will be governed by this standard.

3. Importance of Contracts.

In its notice of proposed rulemaking (NOPR) on stranded costs,[10] FERC recognized that contracts are the basis for purchase and sale obligations in the bulk power market. The message in the NOPR is that FERC intends to hold utilities to the benefits and detriments of their contracts when it comes to wholesale stranded cost recovery. The NOPR proposes that extra contractual recovery of stranded costs be allowed only in limited situations; among other things, when it can be demonstrated that there was a "reasonable expectation" that the relationship with the purchaser would extend beyond the term of the contract.

This adherence to contracts as the foundation of wholesale market relationships ought to apply as well to other contexts. In some cases, purchasers may be parties to contracts that restrict their ability to purchase power from third parties. While FERC has the ability to modify such contracts where it finds them to be unjust, unreasonable, unduly discriminatory or preferential, it ought to be circumspect in doing so. My reason for this position is not to frustrate competition. Rather, it is in recognition that contracts will continue to define the relationships between wholesale sellers and purchasers, and that this foundation will be undermined if regulated entities perceive that regulators easily will be persuaded to abrogate contracts.

4. The Relevance of the Commission's Natural Gas Experience.

In my opinion, the Commission's experience with the restructuring of the natural gas industry offers a useful analytical framework for addressing issues arising in the context of the competitive evolution of the electric utility industry. Still, differences in the statutory framework, the allocation of jurisdiction between Federal and State regulators, industry structure, and the physical characteristics of electricity and natural gas mean that the competitive evolution of the electric utility industry will not exactly parallel the natural gas experience. Clearly, the natural gas

[10] Recovery of Stranded Costs by Public Utilities and Transmitting Utilities, 67 FERC para. 61,394 (1994).

restructuring model should not be viewed as a template that can be thoughtlessly applied to the electric utility industry.

There are four parallels that I think make the Commission's natural gas experience relevant to its current electric agenda:

(a) The emergence of competitive sectors (the market for gas as a commodity and the market for bulk power);

(b) A public policy decision that such competition is in the public interest and ought to be promoted (the NGPA and the Wellhead Decontrol Act for natural gas, and PURPA and EPAct for electric power);

(c) Recognition that in both cases the transmission function remains a monopoly and needs to be regulated in order to prevent monopolists from frustrating the development of the competitive sectors; and

(d) The need for regulatory policy to manage the transition from comprehensive regulation of all aspects of the business to regulation focused solely on those functions that remain monopolies. This includes assigning responsibility for the costs incurred as part of the transition to a more competitive industry structure (natural gas take-or-pay costs and electric utility stranded costs).

I believe that the Commission's electric policies have been influenced by the natural gas experience. For example, the comparability standard enunciated in the AEP order parallels the principles for non-discriminatory, open access natural gas transportation in Order No. 636. Also, the natural gas experience with take-or-pay costs and other transition costs has impressed on the Commission the importance of providing up-front guidance as to the standards for determining whether such costs will be recoverable as part of Commission-approved rates. The result is the Commission's stranded cost NOPR.

In conclusion, the natural gas experience provides the Commission with a useful analytical framework for evaluating issues arising in connection with the competitive evolution of the electric utility industry. In the end, however, electric policy issues must be addressed on their own merits.

This concludes my remarks. I would be happy to answer any questions that the subcommittee may have.

Mr. SHARP. Thank you very much, Mr. Santa.

You struck a chord with the notion of those urging you and the Commission to provide regulatory certainty. The same is often urged on us in the Congress to provide certainty of the rules.

I have noticed a distinct development, however, among people who advocate that certainty, that the minute they have it they often don't like it and therefore they are lobbying us to change it, and so while it is a wonderful goal and certainly to be commended and sought after, peoples' goals seem to change in this political world at a very rapid pace. You don't need this little 101 lecture.

One of the things over the years I have come to feel about asking Commissioners questions is that it is often difficult to engage in a dialogue because you folks are naturally restrained on any specific cases from making any comments and often prudently restrained from making decisions and announcing them in a public forum such as this, but let me ask some questions and see if I can elicit some kind of discussion from you.

One of the—clearly our intent in EPAct, many of us were fighting for and have been generally pleased with the outcome, was to try to bring more competition to the bulk power market and to head in a different direction from what we historically had focused on.

But we also were, in EPAct, in other sections, engaged in other goals that we were seeking in energy policy, and not unlike the California Commission which has announced, I think, its five principles. Many people have commented that they are wonderful principles but having them all is a little difficult to accomplish, and there may be some conflict among those goals. This is not a criticism of California; it is natural to our desires in individual and collective life to have multiple goals.

But one of our other goals has been over the years in energy policy to promote the development of alternative sources of energy—solar, wind, geothermal, renewables, others—which we recognized were higher priced than the marketplace. There have been endless fights and discussions about what techniques would help advance them until they became price competitive, and of course one of the goals of PURPA had been to do that within the electric utility system, but with it being heavily dependent upon decisions by the States. California is one of the most dramatic and prominent examples of a State which actively has promoted the development of these resources; and many of us have applauded that from afar. We haven't had to live necessarily with all the consequences of it, but we thought it was wonderful for the Nation, although some consumers are not as keen on it in California, I recognize. But it has been a very significant impact on the development of the renewables in this country, and many of us have fond hopes and believe it is important to us environmentally and economically in the future that we continue to make progress in this area.

Obviously, that goal can come in direct conflict with simple price competition if we had a highly priced competitive market. One of the things that happened in the seventies and eighties was, people who had been highly suspicious of the monopolies in the utility system in this country found they could be warm and friendly and cuddly for specific purposes because you could use them as instruments to carry out social and environmental policy more effectively than you can in some other ways, and so that is somewhat at risk under the proposal of California, and this is intensely debated in California.

But you folks may get drawn into this role as well, because I guess what I want to partly lecture you and partly put to you is the question of whether those goals are in conflict, in fact, Ms. Moler—and you may want to respond—what you have in your written comments to our question of social and environmental conflict of goals with competitive goals.

But one of our concerns is, can we press forward with competition and still have a State or the Federal Government bringing about the imposition of these alternative goals? Many of us believe that is possible, but it really requires cooperation of State regulators and possibly Federal regulators to accomplish it.

In yesterday's hearing one suggested way to do it was, of course, to put the cost on the wires on transmission. And this gets to the question Mr. Massey was raising, could the State of California do that?

Well, having sort of outlined this obvious potential conflict, let me ask you if you have any comments on whether or not or how we might work it out that a State, in particular, California, if it decides to proceed in some version of its retail wheeling, will run afoul of the FERC, for example, if it should decide to use an approach—there are many approaches, possibly many, I don't know if there are—but of spreading those costs out on the wires.

Ms. MOLER. Well, it is very difficult to reconcile promoting noncompetitive fuels in a competitive marketplace. That is, at bottom, what you are discussing. It is difficult to have competition across the board and subsidize certain sectors of the marketplace.

States are actively promoting some kinds of fuels and some kinds of projects in preference to others, and they have done this through a device that they have developed called environmental externalities in competitive bids where they weight their bidding criteria to favor certain desired fuels or desired results or desired environmental goals.

As I observed in my written comments for the subcommittee's hearing record, we have not been directly involved in the externality debate or in the questions of subsidies for some types of fuels versus a competitive marketplace.

If States impose retail wheeling requirements—without getting to the question of whether they have authority to do that—if they do it, there clearly will be costs in old contracts, whether they are PURPA contracts or whether they are demand side management programs or the like, that they have engaged in in the past that are not viable in a competitive marketplace.

I think what we have said in our stranded cost rulemaking proposal is that we hope and pray they will deal with this issue. We believe they have the authority to do that.

I personally would not be adverse to a surcharge of some sort on transmission rates to deal with these types of transition costs if the State were to ask us to do so. We have a very lively debate going amongst ourselves and in the industry on how far we should go in accommodating what are really the results of retail decisions.

The D.C. Circuit case this week calls into question the idea of doing transmission surcharges. I don't believe that is the last word on the subject, but I think the record that we get in our stranded costs rulemaking will give us some guidance on how the world at large wants us to address this issue.

But as I said at the outset, it is very difficult to reconcile promoting noncompetitive fuels in a competitive marketplace.

Mr. SHARP. Do others wish to comment on that?

Mr. Massey.

Mr. MASSEY. The question you raise is whether cheap electrons are the highest value in society or whether there are other values that are just as important, and I think we would all say that there are other values that are just as important.

The question you raise is the second most difficult question raised by the California proposal. The first is the law and whether retail competition is preempted. The second question is: How do you merge these two goals of cheap electrons versus additional programs that increase costs? I don't think it is impossible. It will require a lot of Federal-State cooperation to do it, and, as Commissioner Bailey has pointed out, I believe that we have to explore other creative mechanisms for ensuring that Federal regulators and State regulators are working together because if the State seeks to impose a charge, the question is, will we accept it in rates? They are very, very difficult questions, but I think it can be done.

If, however, the highest value is, in fact, cheap electrons, it can't be done.

Mr. SHARP. Anyone else?

Mr. SANTA. Yes, Mr. Chairman. I would add that another avenue for the States in dealing with this is the fact that the Federal Power Act makes a distinction between transmission and distribution, with transmission being subject to Federal jurisdiction and distribution remaining subject to State jurisdiction.

Where the line between transmission and distribution is drawn is unclear at this point, but I would tend to say that what is going on in California and the debate it creates will probably result in some focus on that question and it will get resolved.

To the extent that there is something there that is distribution that is State jurisdictional, I would tend to think that to the extent that a State, either its legislature or its regulators if they have the authority to do so, there is the possibility that some of those charges, whether they be the charges associated with stranded costs or the charges associated with various social programs or other values that were being reflected previously in the utilities rates, could be collected as some type of an adder or a surcharge to a distribution charge.

Of course, for the States there also is the possibility of trying to accommodate those values in ways that maybe are not part of the rates charged for utilities service. For example, there could be a direct subsidy. It could be something that was paid out of tax revenues, which I realize is not an easy answer because the States have got significant problems there, just as the Federal Government does. I merely mean to point out that there are a variety of options.

Mr. SHARP. Do you think that it is possible that the distinction does seem to begin to blur between transmission and distribution, between wholesale and retail, it seems to me, in the system, but there is at least a legal line in the Federal Power Act. Do you think it is possible that at some point the Commission might come to sort of a legal fiction as to what is in and what is out which would help resolve that question?

I mean I guess in part I am asking, is it really necessary to have to revisit the fundamental law, since we are not always swift or secure in doing that, or is this a possibility where simply, as in so many other things, it can be worked out and see if the Circuit Court will accept it?

Mr. SANTA. I don't think it is necessary for Congress to revisit the law. I think that the Commission could draw that line, and I think that if it did so in a basis which was reasoned, and which was consistent with the intent of Congress, then hopefully the courts would pay the Commission deference and that would be upheld.

Mr. SHARP. Anyone else?

Mr. HOECKER. Mr. Chairman, as my colleague Commissioner Massey said, certainly demand side management, conservation, social programs are important. They represent important values, but they are generally cognizable in retail rates and are not the kinds of issues that generally come before the Commission.

These programs, I think, are clearly at risk in the environment of retail wheeling, and I think that preservation of those kinds of programs may, in fact, require legislation to establish incentives to conserve, and to provide support for programs that are now supported through utility rates, and so forth. That is probably not something that the FERC would participate in.

The debate, as the Chair appropriately said, in our notice of proposed rulemaking on stranded costs, is, to what extent stranded investments—such as investments in demand side management—should be cognizable in a wholesale transmission rate. It is a live issue before the Commission, and I am sure we are going to get plenty of comment on that.

Ms. BAILEY. Mr. Chairman, I would join in the comments of my colleagues. The assumption behind the question in light of the more competitive electric power industry is that the utilities' automatic reaction will be to shed load, and part of that load that they will shed are the DSM, conservation, and other social-environmental programs. It will require, as Chair Moler States in her written testimony in answer to the question, that States reinvent cost recovery mechanisms for social and environmental agenda, and, as she says, it will be difficult but it is not impossible. But then, as my other colleague has stated, we may not be able to look to utilities to bear all the costs and we will have to look at other creative mechanisms—incentive regulation. Building in incentives for them to pursue social, environmental agendas is one possibility.

Mr. SHARP. Thank you.

In a moment I will recognize my colleague from Connecticut.

It just strikes me that FERC can obviously be drawn into it if this distinction over transmission and distribution is such that you have all kinds of large customers that write off the transmission and not in theory off the distribution network. It depends on both the technical mechanics, it seems to me, as well as how one thinks about it. FERC could make policies that limit what States can do, and this may require a great deal of work on your part in terms of what you, Ms. Bailey, said in your opening statement about one option of the Federal Power Commission in the Federal-State relationship is deference, deference to the States, and some of the legal questions. It might be interesting to see if you were able to come to conclusions, policy conclusions, how they would hold up in court if the Commission were to affirmatively indicate that it is deferring.

I mean I am sure in some instances, but I am thinking in more of a broad policy way what might be deferred to a State, and then I assume it would be somebody who would challenge that in the courts, but it might be an interesting legal way to proceed.

Let me recognize my colleague from Connecticut.

Mr. FRANKS. Thank you, Mr. Chairman, and as the chairman mentioned earlier, too, thank you for your enlightening testimony today, and, as the chairman would know, I do have some concerns about PURPA.

Yes, the environment is very important, and yes, energy efficiency is very important, but cost is also very important because the ultimate costs would be passed on to our consumers.

Yesterday it was suggested that PURPA contracts that are above the market should be treated as stranded investments. What is your view about this?

Ms. MOLER. In our recently issued notice of proposed rulemaking seeking comments on how we should deal with stranded costs, we asked specifically to have commenters tell us what kinds of costs ought to be covered, whether they are above-market nuclear power plants—which is not a hypothetical, I can assure you—or whether they are above-market PURPA contracts, or whether they are unamortized demand side management plans. There is a plethora of possibilities. We are seeking comment on that.

I really don't have a judgment at this point what will be the best way for us to proceed.

Mr. FRANKS. I am directing my question about PURPA contracts.

Ms. MOLER. Clearly PURPA contracts are among the many kinds of things that may be stranded costs. In California in particular—California, Maine, and New York in particular—have had a lot of PURPA contracts that are by any measure above market now. These contracts form the basis for project financing in many instances. For these qualifying facilities, if you don't have the ability to recover those costs, your project financing is in big trouble. On the other hand, they are not competitive in a competitive marketplace.

It is a very difficult problem, and I don't have a solution to suggest to you today other than to say we are looking at the issue.

Mr. FRANKS. Mr. Massey.

Mr. MASSEY. Yes, Congressman, I think you can make very good arguments that utilities who purchased power from QF's at above-market prices were doing so often because the law required them to do that, and I think you could make very good arguments that it would be unfair if they ultimately cannot recover those costs.

Mr. FRANKS. Because once again the ultimate costs would go to the consumer, obviously.

Mr. MASSEY. Well, it is a real challenge. These are the costs of expensive generation that could not sell at a market price. The costs of high-priced QF contracts are, in fact, sunk costs, and the question is who in our society is going to bear the burden of those costs that are incurred by utilities with an obligation to serve at retail, an obligation to stand ready to serve customers? Is it now fair to change those rules and disallow the recovery of those sunk costs?

Mr. FRANKS. Mr. Santa.

Mr. SANTA. Yes, Mr. Franks, I would agree with the comments of my colleagues and also point out the fact that as Commissioner Massey has alluded to, those utilities that are purchasing power from QF's are doing so in some ways due to a double legal obligation.

On the one hand, there is the mandatory purchase obligation imposed by Federal law in PURPA, and then also they are purchasing at avoided cost rates, which is the standard set in the Federal law, and those rates were approved by the State Commissions.

So there is both a legal and a regulatory blessing for those purchases, and I think that would be an argument that would make the case that they should be recognizable as stranded costs.

Mr. FRANKS. Thank you.

Mr. Hoecker.

Mr. HOECKER. Thank you, Mr. Franks.

I believe these costs may ultimately be eligible as stranded costs. Certainly it will be a matter that will be debated in our proposed rulemaking proceeding. I would simply point out that our proposed rule on stranded costs posits, as one of the basic solutions for stranded cost recovery, the importance of contracts and provides that contracts should provide for exit fees, for notice provisions, and that parties should arrive at ways of dealing with costs that may become uneconomic for one reason or another over time.

The PURPA contracts are contracts that are entered into in part because of a legislative scheme, but they have become in some cases uneconomic over time. The question is whether or not now regarding those as stranded costs is a fair way of addressing them, because a contract is a contract.

Mr. FRANKS. Is there an inference there, in your opinion, that we should look at PURPA or rereview PURPA?

Mr. HOECKER. Oh, I am not making any recommendation about reviewing PURPA.

Mr. FRANKS. I thought the inference was there. I just want to follow up on that.

Mr. HOECKER. Well, I don't want to make that big a splash.

Mr. FRANKS. OK. There aren't that many people here.

Mr. HOECKER. But I do think that you have some conflicting objectives here about the importance of adherence to private contracts over time, and certainly the fact that some PURPA contracts have become uneconomic may not necessarily make them stranded.

Mr. FRANKS. Ms. Bailey.

Ms. BAILEY. Thank you.

Congressman, I am always timid to make conclusions, especially on issues that we are trying to seek comments on, as we are in our stranded costs recovery NOPR, but I would agree that there is a legal and regulatory bargain in which those contracts were made, and I would probably view them as stranded costs, not as a failure on their part to compete, as has been used in the Cajun decision.

So I think, along with agreeing with the comments that have been made, that I think they would be eligible for stranded cost recovery, and as far as PURPA goes and how QF's view themselves, I think there probably is a need in this new competitive environment to at least ask questions, reevaluate positions. Other than that, I won't go any further.

Mr. FRANKS. OK. One last question. How competitive are wholesale electricity markets?

Ms. MOLER. Very.

Mr. FRANKS. Very. OK would you like to—any other comments on that, or would you like to elaborate on that?

Ms. MOLER. I would elaborate at length.

In every solicitation that I am aware of——

Mr. FRANKS. Because my question was going to be why hasn't FERC received more requests for orders for transmission services.

Ms. MOLER. I hope it is because we have made clear that we would grant such orders and so the parties are working them out themselves without coming to 825 North Capitol Street. For every solicitation I know of where a utility has gone out for bids for power, they have received an abundance of bids. It is a very competitive environment.

We have made a determination in a recent case that for next entrants, that the generation market for new capacity is competitive and thus we don't need to regulate the rates for it.

Mr. FRANKS. Mr. Massey.

Mr. MASSEY. Yes, I think there is a lot of evidence that deals are being cut without having to come to FERC to force the utility to wheel. My friends in public power tell me that the new law and FERC's implementation of it has completely changed the psychology of negotiating with their suppliers of power and that all of a sudden they are able to negotiate new contracts at a substantial reduction. They think it is because the supplier knows that they can come to us and seek a wheeling order to secure supplies from another supplier. Every order we issue that tells the world we are serious about implementing our new authority within the bounds set out by Congress moves the ball forward.

It is not necessarily true that the 211 filings that we have represent the totality of the changes that are occurring out in the industry because only 10 or 11 cases have been filed. But our bold orders with respect to those filings, I think, are changing the psychology of this industry.

Mr. FRANKS. OK.

Mr. SANTA. As Commissioner Massey noted, I think it is somewhat remarkable that to date the Commission has only received 11 applications for 211 orders. I think there are probably three factors I would point to as an explanation for it. The first is, as Commissioner Massey and I think Chair Moler noted, given the fact that the Commission has vigorously enforced section 211, there probably are a lot of instances out there where the parties are arriving at negotiated deals and need not come to the Commission for a section 211 order.

The second thing I would point out is that, by law under section 211, the Commission cannot order access if there is an existing obligation on the part of the utility that would be the transmission provider to sell power to the requesting utility, so therefore we may have cases where perhaps a couple of years from now when the contract expires it would be ripe for a 211 application.

The third thing I would cite is that a growing number of utilities have offered voluntarily open access transmission tariffs, and as I referenced in my comments on the comparability standard, as that becomes the standard, as that is enforced on those utility systems, the likelihood that someone would ever have to seek a 211 order is greatly diminished.

Mr. FRANKS. Thank you.

Mr. HOECKER. Well, as usual, my colleagues have it about right. When I mentioned in my oral testimony that the Energy Policy Act has sort of changed the atmospherics in the industry, that is exactly one of the areas I was referring to.

I am assuming, although it would be nice to know in more concrete terms, that a lot of open access transmission arrangements are being negotiated and are in progress. I think 211 implies a lot of leverage for transmission customers in the marketplace that didn't exist before.

Obviously the jury is out, but our response has been to be firm, to make very clear as we did in the Florida Power and Light case, exactly what the implications of a request for network service are, how we would act on those requests, and what the pricing implications are. We are being clear, as Commissioners Santa and Massey have both said, and I think that will delegate to the marketplace the ultimate decision about the workability of open access.

Ms. BAILEY. I'm certain that negotiations abound. I think that to allow FERC, allow us to keep our 211 hammer in the closet, is the idea. Going forth with litigation is very time and resource intensive, and I think voluntary agreements have a lot of appeal in light of that. So I have nothing to add.

Mr. FRANKS. OK, and I yield back, Mr. Chairman.

Mr. SHARP. Thank you very much.

Well, I must say that this is one place where the Chair feels compelled to say we may have gotten it about right in Congress for once, because oftentimes what we anticipated to happen didn't happen, but many of us were convinced that if we gave the authority to FERC on 211 and if they exercised it, as they have, with the clear indication that it would be used, that we would have more voluntary activity and that the recalcitrants would simply begin to negotiate instead of fight, and the fact is that the critics—at the point we were trying to get what at that point seemed to be so unbelievably controversial into the law, the critics claimed that this was a nightmare bureaucracy that we were about to establish, that the Commission would be overwhelmed with requests, it would not be able to deal with them, and it would not be able to complete its actions on natural gas or on anything else because of orders on 211.

So I can't resist indicating this was one claim of bureaucracy that did not so far come to pass. I do have to admit there are some people who claim there was never a problem anyway, but there was just too much anecdotal evidence that there were recalcitrant utilities in this country who would not cooperate at least on what seemed like a fair basis with some of the folks who were trying to buy and sell.

With that, let me turn to a couple of questions if I might. Let's turn back to this stranded investment just for a moment. Mr. Franks has raised a very important question of the PURPA contracts which are, as I think one of the Commissioners testified, in California and New York particularly, and especially California obviously with the Commission's relevance, but in fact as Ms. Moler just indicated, there are many kinds of high costs within these systems. In some instances it is nuclear power plants; in some places it might be a coal plant.

There are some States that actually try to promote their own local resource, and that may have high cost consequences. I don't know that it does, but it might because of environmental costs that are incurred. So what will become a stranded investment really will depend on the system.

Did you get in your comments much indication as to categories of these costs and how much the estimates are? I am sure it is like in natural gas, they are wildly divergent as to what might constitute a stranded investment.

Ms. MOLER. We have just issued the notice. The comment deadline has not occurred yet, so we don't have any record yet. I am sure we will have wide variations in the estimates. The biggest difference in the guesstimates is whether you assume full implementation of retail wheeling and whether you assume that all of those who are capable of self generating do so.

Mr. SHARP. Those are huge assumptions too, obviously.

Ms. MOLER. That is right, and that is why the numbers range from tens of billions to hundreds of billions.

Mr. SHARP. Right.

Also, you have the situation where, I mean it really isn't a stranded investment unless it cannot be recovered, the cost. It is not until a State balks, or a sale cannot be made, that it is a problem. I mean I can understand people sitting there holding these assets who worry, but in fact until you cannot get your money out of the system, it may keep you up at night but it is not the practical problem. Isn't that correct? I mean at some point there has to be an action that prevents us from getting our recovery.

Ms. MOLER. We also recognize that there are utility plants, for example, whose current cost of power is in excess of the market. If they seek stranded cost recovery for that plant, and still use the plant and sell power out of the plant at a lesser price then could they recover the difference between their above-market costs and what they are able to sell the power for? That is discussed in the notice of proposed rulemaking.

We have some experience on this in the gas side, their take-or-pay contracts. Say the cost is $7 or $8, the market price, and they are actually selling the gas at $2 while they attempt to renegotiate the contract or while they wait until the contract expires. Where their producer has been unwilling to renegotiate the contract, we allow them to recover the difference between the $7 and the $2. It is called a pricing differential cost as a transition cost under Order 636 and is analogous here.

Mr. SHARP. Is there likely to be much dispute over what is the actual cost in the system of a particular power plant, a nuclear power plant, or is that reasonably easily settled?

Ms. MOLER. I believe that the actual cost basis of the power plant is reasonably easily settled, as you describe it. What will be the big debate is whether it was foreseeable that the plant would be stranded, whether competition was foreseeable and who should pay the costs. And there would also be a prudence issue regarding plants, the prudence of which has not been adjudicated at the State level. And probably even if there has been an adjudication, this will be a very lively debate simply because the dollars are so huge.

Mr. SHARP. Sure.

Well, I sort of have a prejudice, coming from the natural gas debate, that it is very, very important for regulators to keep the pressure on and not just buy into the notion that whatever is put on the table is a stranded investment because the truth is that all kinds of investments, both in the utility and the nonutility part of the economy, become uneconomical either initially when a mistake is made or over time, and change is inevitable, and I don't think the sole goal ought to be just to protect.

I realize in the utility industry because of the obligation to serve and whatnot that is different, but the obligation to serve does not lead to each and every decision. That gets to the prudence question that is made, and in some instances we had enormous turmoil in the natural gas sector of course, and it took FERC several iterations in order to figure out how to get a policy on that. We had a terrible time in Congress deciding what approach ought to be taken on it.

But the fact is that just the intense pressure led all kinds of people to figure out ways to handle it, and if FERC had jumped in and said 100 percent recovery, why, we would have seen less negotiation and less difficulty.

I realize the two are somewhat different. The two are not analogous exactly because the systems work differently, but I just hope we won't—that competition is designed to bring the pressure—is changed in bringing the pressure for efficiency, and if we protect everything, if the transition is too long or the protection is too great, why, we just go on protecting uneconomic things, I think, too intensively. But that is just my advice.

I don't know if any of you wish to comment on what we have been discussing.

Mr. MASSEY. Well, Mr. Chairman, regardless of where we end up at the Commission on this question of stranded costs, we have decided that it is an issue that we should deal with early, and I appreciate your comments about the issues that we should consider.

There has been a raging debate about the concept of stranded costs. Some believe that regulators should just stand out of the way and let the chips fall where they may and that utilities will, in fact, figure out a way to deal with these costs. Others believe that if we really want a smooth transition to competition and if we want a fairer transition to competition, that regulators need to decide these questions early in the debate, and that is the policy that we have chosen.

Mr. SHARP. I don't disagree with your general proposition of trying to decide a framework early in the debate. I think that probably is useful, but I don't think anybody should assume that for everybody—and I am sure you don't—it could possibly be smooth or meet everyone's individual test of fairness.

I have noticed that the level playing field issue always comes up by those who think they don't have it, and they never seem to get it.

Mr. MASSEY. Yes.

Mr. SHARP. We never seem to achieve it, and while fairness is to be desired, it masks sometimes peoples' unwillingness to change.

Mr. MASSEY. Yes, I appreciate that, and I noticed in yesterday's testimony Mr. Rogers from PSI testified that his company—his shareholders had already eaten $2.7 billion in costs. That was their way of dealing with the stranded cost question, and so utilities have different views about the way in which we ought to proceed.

Mr. SHARP. It is certainly much easier for those that are not likely to have above-market costs. I mean for them it is an easy proposition, for others it is not.

But it is true it is also not a brand new issue. I mean the fact is that in many States and many utility systems there have been investments that had to be eaten or something done about them over the years. But the expectation here is, of course, that the problem of problematic assets is going to increase with competition.

Mr. Santa, did you want to comment?

Mr. SANTA. Yes, one comment that I would like to add, Mr. Chairman, is an answer to your question as to the magnitude of the costs. While the estimates of the magnitude of stranded costs vary widely, one thing that I think there is probably near universal agreement on, is that the majority of those costs are at the retail level and not at the wholesale level.

Therefore, getting to your point as to what extent do these estimates and when do they become real stranded costs, I think the extent to which they manifest themselves will have a lot to do with what the States do about competition at the retail level and how quickly they do it.

While the forces of competition are somewhat irresistible and can't be thwarted, another point that has been made is that these are, for the most part, depreciating assets. To the extent that more time goes by, the amount of underpreciated costs that the utility has goes down.

A final point I would like to make is that in the Commission's notice of proposed rulemaking, the approach that we have put out there with respect to wholesale costs is very much a contract-based approach. It looks first to the contract between the wholesale purchaser and seller, looks to see whether or not they have dealt with the issue either through an exit or a notice provision, provides a 3-year window for some type of contract renegotiation, and I think sent a signal that we would look pretty strictly to the contract and what were the reasonable expectations there in terms of evaluating claims for wholesale stranded cost recovery.

Mr. HOECKER. Mr. Chairman, you are absolutely right that this is an issue where the opinions about what is stranded and what isn't vary widely. The equities are going to be hotly debated. But I certainly agree that not all uneconomic costs are stranded within the meaning of our proposed rule.

Just as not all risks are the result of fundamental changes in service obligations, costs that become uneconomic may simply be a result of the failure to compete. It seems to me that when we talk about stranded investments in the context of this Agency's willingness to allow some form of compensation, we must keep in mind that those costs must bear some relationship to the transition to competition, and that is especially true where that transition is ordered by or mandated by regulators.

Ms. BAILEY. I think there are probably four things that I would want to point out as important to keep in mind. I think the fact that we are trying to deal with it up front and early is important, not just because of the dollar amount and substantial impact of stranded costs but as somewhat of a guide to the States and showing some leadership in that area, because stranded cost recovery has been such a hot potato. It is a very critical issue at the State level. It is the issue of who will and how much the remaining customers will have to share, be it shareholders and/or customers who will have to share in these costs. I think it is a very critical issue for them.

The financial integrity of the industry is at stake here also. I think that is a very critical consideration to keep in front of us. The idea of potential bankruptcies is not pleasant, and there are several other reverberating effects of possibly not getting recovery for stranded costs or stranded investments, stranded assets, whatever you wish to call them, and keep in mind that that wholesale entity supplies retail customers.

So the issue is very much tied and interrelated as to Federal and State regulators and how we deal with that problem.

Thank you.

Mr. SHARP. Let me turn to one last issue, which, again, I am sure you are not prepared to make any final disposition of, you could not, and that is the question of California retail wheeling, and ask you if you can envision—this is really speculative, I guess, since we don't know exactly what they are going to do. But let's assume for the moment that they decide on some policy that will be identified as coming with a retail wheeling proposal. How do you envision that coming before your Commission where you would have to make determinations about—you have raised the question, several of you have alluded to, as to whether it is legal or not legal. I am just trying to envision if there are alternative scenarios by which it might come before you.

Ms. MOLER. In our recent notice of proposed rulemaking on stranded costs, we did lay out the legal theory under the Federal Power Act that rates for service, transmission in interstate commerce, must be filed with FERC and cited all of the relevant cases.

If California were to require third party wheeling, to the extent that it involves interstate commerce transmission, those rates would have to be filed with the Commission. So the three public utilities that would be affected by the CPUC order, PG&E, SOCAL Edison, and San Diego Gas and Electric—and those are the only three, of course, who are subject to the order, the rest of the utilities in the State are not—would have to file wheeling rates with the Commission. It would come before us in that context.

At some point we may—and we are obviously evaluating right now the issue of where to draw the line on local distribution. But where you have a third party supplier going through the utility to an ultimate user, it is very difficult under the existing Supreme Court case law to say that those rates do not have to be on file under the Federal Power Act.

So we would have rates filed by the California utilities with the Commission. Whether the CPUC would order them to file those rates, for example, whether they would voluntarily do it, whether they end up doing a POOLCO, which some of the utilities have suggested, obviously a POOLCO is subject to our jurisdiction.

Mr. SHARP. Then you would be making the determination as to whether the rate is a correct, a fair one, a just and reasonable rate. Is that correct?

Ms. MOLER. Yes, sir.

Mr. SHARP. That the charge is in your view, and at that point is where the issue might rise as to what they were putting into that charge, if they were putting in the PURPA contracts, the above-market piece of the PURPA contract, or of a nuclear power plant, or of any other asset. I mean that is where you might have to confront the issues that I have been raising as to where you might be able to shut off if you chose one avenue over another.

Ms. MOLER. That is true. The CPUC's notice, of course, said that they will issue guidance on how they will spread these transition costs very broadly to all consumers, and they have not done so yet. But we would get those to review rates.

Mr. SHARP. What you are articulating is that their decision to spread those rates once you receive those rates, in a sense, will have to be judged by you in order to determine the just and reasonableness of those rates. Is that correct?

Ms. MOLER. Yes, unless they come up with a theory that they are rates for local distribution, for example.

Mr. SHARP. If they are able to find a way to draw a distinction that you accept between transmission and local distribution, you are saying.

Ms. MOLER. Right. That is one possibility. They could also do it in terms of some sort of exit fee. There are a variety of ways.

Mr. SHARP. The customer believing that seeking the new supplier, a big manufacturing plant that is seeking a better electric rate off of another competitor, the exit fee they would have to pay to the previous seller, is that what you are saying?

Ms. MOLER. Yes.

Mr. SHARP. That would remain——

Ms. MOLER. The exit fee they would have to pay to the utility they are leaving.

Mr. SHARP. Right.

Ms. MOLER. Yes, and those are also fairly common provisions in wholesale contracts as well.

Mr. SHARP. But those exit fees would then be strictly a local question.

Ms. MOLER. I wish the answer was that simple. It depends on how they are structured.

Mr. SHARP. OK. I don't know if any of the rest of you would comment. It is not necessary to make any comment on these questions.

Well, ladies and gentlemen, I really appreciate your time and attention and applaud your diligence in dealing with these very complex issues and wish you well. Thank you very much.

Ms. MOLER. Thank you.

Mr. SHARP. Why don't we take a 10-minute break, and then we will begin our next panel.

[Brief Recess.]

Mr. SHARP. The subcommittee will please come to order.

The Chair would ask unanimous consent that the opening statements of the distinguished ranking member of the full committee, Mr. Carlos Moorhead, and the distinguished ranking member of the subcommittee, Mr. Bilirakis, be placed in the record at the appropriate point.

I am very pleased now to welcome our second panel for today. I am very happy to have back with us Professor William Hogan with the Kennedy School of Government; Professor Charles Stalon; and Professor Ashley, three well known individuals in circles discussing these matters these days, and, as I referred to earlier, some of the real bright lights. There are some other bright lights, some of whom were invited and some of whom will be here next Thursday, I guess, when we continue. We are very pleased to have these three gentlemen with us.

I think you are all familiar with our processes. Dr. Hogan, we would be very pleased to start with you.

STATEMENTS OF WILLIAM W. HOGAN, PROFESSOR, KENNEDY SCHOOL OF GOVERNMENT, HARVARD UNIVERSITY; CHARLES G. STALON, PROFESSOR, OKEMOS, MICH.; AND ASHLEY C. BROWN, PROFESSOR, AND EXECUTIVE DIRECTOR, HARVARD ELECTRICITY POWER GROUP, HARVARD UNIVERSITY

Mr. HOGAN. Thank you very much, Mr. Chairman. It is a great pleasure and a privilege for me to have an opportunity to join you and this committee which was so important in the process of passing the Energy Policy Act, starting and advancing the subject which we are trying to deal with today.

I, as you know, have worked on this problem over the years and particularly through the Hardware and Electricity Policy Group of which you have participated in the past as well, and involved with many others in the industry. I emphasize the usual disclaimer, I am just here speaking for myself today, not on behalf of anybody else.

I prepared for you a written statement which I would propose to summarize rather than to read in, and you can use the full written statement at your pleasure, and I would like to take my allotted time to try to basically address some of the questions that you were talking about earlier this morning with the Commissioners here, which was a very informative and, I think, helpful discussion for me to hear, and to talk about actually three points that I would like to make.

There is obviously much more that we could talk about, but I want to emphasize three things at least in my opening remarks, and those three things are to issue a call essentially to remember that the world is going to be very different as it evolves in this competitive market, and we have to go back to basics in many ways to try to understand what is going on. The second point is that the wholesale market in designing—to have a fully competitive wholesale market will be the biggest challenge that we are going to have, and I will explain further why that is important. I think it is obvious, but I would like to expand on that. And then third is to come to the direct access, or California retail wheeling, as you called it earlier, and there to suggest that this problem is actually not as big a difficulty as many people have suggested and that there is a way to deal with it, and many of the problems that we think it might raise that are actually quite simple, and that will be important because it feeds back on how we deal with the wholesale market and so on.

So those are the three points: The back to basics, the wholesale market, and direct access.

The back to basics point I think is pretty straightforward. Things are changing in a radical way, and they are changing in producing new environments that we haven't fully anticipated. You had remarked yourself earlier that many of the things that happened after the Act passed surprised you but you are pleased with the way it is going, and so am I.

But much of what we have taken for granted in the past is not going to be the case in the future. I have quoted in my remarks my friend Charles Stalon who says we will be enjoying the pleasure of working on a problem where, "Most things that everyone knows to be true aren't," and I think that is going to be the world we are involved in, and some of the legal fictions of the past that served us well don't serve us well in the future, and much is going to have to change. The contract path for transmission is a prominent example that is now recognized as a fiction, and something else is going to have to be there.

In my written testimony, I listed off a variety of things: The distinction between energy and capacity may vanish; separate pricing of generation and transmission may be impossible; we are going to have to change the way we do cost-effective conservation or portfolio planning, environmental protection; we are going to have to redefine the obligation to serve; much as you have heard many people say here already, how we address and deal with concentration and market power is going to be different; everything is going to be different in a competitive market when we change all of the rules for access and all of the rules for pricing; and we have to look very closely and go back to examples and lay out in some detail what is going to happen when we are thinking about new institutions so that we are not deflected by implicit assumptions that are we are making which turn out not to be true; and I won't elaborate on it too much, but I will come back to it later, particularly when I get to the direct access issue.

But I think this is a very fundamental and important point. It means that what FERC is doing and State regulators are doing in developing principles to how to think about the problem is especially important.

The comparability standard as announced by FERC is a very good example of the kind of thing that should be done, and I think it is good for two reasons. One is, it is a good principle, it is a compelling argument, and it seems quite justifiable and persuasive, their comparability standard.

The second is that it is not trivial, it makes a big difference, it has an impact on what actually happens, so it is not an abstract principle that doesn't bite, it bites a lot, and that is also a good thing, and we need more of that thinking to be extended, and that is the process that is under way, and these hearings contribute to that, and so on. So back to basics and development of the fundamental principles.

The second is the wholesale market competition and the question Mr. Franks asked earlier about do we have a competitive wholesale market, and the answer that Chairman Moler gave was that it is very competitive, but if you listen closely to her answer, she was talking about competition for new construction in competitive bidding processes. That is a very competitive process, but that is not the wholesale market in total. As a matter of fact, it is a very small fraction of the wholesale market, and I think it is an open question yet as to how truly competitive and open access all the rest of it is, and that is an issue we need to explore in the future, and we plan to explore through the research and the Harvard Electricity

Policy Group, and there are important innovations going on around the country.

The discussions in Michigan and California are an example, but especially in California where we have surfaced and raised the level of attention to this idea of a POOLCO that was suggested by San Diego Gas and Electric and Southern California Edison and is being discussed vigorously in that environment. Whether that turns out to be the right way to go I think is an open question.

The alternative, bilateral trading models that people are talking about, these should be tested and compared against each other so that we can understand them, but they definitely should be pursued because there are, in fact, many complicated technical problems in the electric sector that have to be addressed, and these are mechanisms for dealing with those technical problems without thwarting the move to competition and open access, and that is something that we need to explore. It is going to be a different world, but it is something which we can do and should do, and I would encourage you and everyone else to support the process that is under way in California to explore those ideas.

One of the things that will happen in that new world in the wholesale market and electricity, just as we have seen in the wholesale market for natural gas, we will have developed arm's length spot prices that vary hour by hour at various locations around the country, just as we see at Henry Hub in Louisiana and so on.

Whether that evolves through a POOLCO or a bilateral market is not important at the moment. What is important is that we will get it somehow to one degree of efficiency or another, and the creation of that transparent arm's length spot price in the wholesale market, it will happen, it is necessary, and it will provide an enormously powerful tool for people to use who are worried about commercial arrangements and signing, developing contracts and providing all kinds of services and so forth, and thinking about how the world is different when you have that wholesale market operating and you have these transparent prices, and as an example of something where when we understand it and come to understand it better, it will change the way we think about other problems.

That brings me to my third point which is California retail wheeling, or direct access. I don't probe into the Commission's mind, but I think that they chose the term "direct access" to escape from some of the baggage that goes with "retail wheeling," which really is a misnomer and "direct access" actually is a better term, and a lot of the problems that people are worried about to the edge of paranoia about retail wheeling really come from a misconception about how it will operate in this competitive market, and the fundamental misconception is that power from one power plant goes to a particular customer, and then they might switch to another, and if we have an efficient market, as many people have pointed out, there is not going to be any wheeling of electricity, it is just wheeling of dollars and so on, and that criticism is actually true, but it doesn't mean we shouldn't do it, it just means we should think about it differently, and in a world of direct access of the type that California has proposed, the existence of an arm's length spot price in the wholesale market provides a mechanism, a mechanism available to the Commission under its State authority for implementing what I have called their efficient direct access, which is to allow for direct access which gives you every economic incentive for all kinds of commercial arrangements and investments that we should be doing that is the functional equivalent of what they do in England and provide direct access, or what we think of as retail wheeling, for their customers, but doing so without creating the fiction that I am buying from one power plant versus another, and the device for doing that is a very modest and simple change in the rate design tariff under State authority where all customers stay with their local distribution company but their rate design includes a tariff which is a time-of-use rate where the time-of-use price is the arm's length spot price.

That simple tariff change, which is clearly under the State jurisdiction, there is no problem doing it, requires no new legislation, is all that is needed in order to create efficient direct access for customers. I can answer questions later, and my paper goes through the details, but now with that device and that change, people can sign all kinds of contracts for long-term delivery of power and hedging contracts and create new instruments, all the tools are now available with that very simple change, and that form of efficient direct access requires no reclassification of customers, raises no jurisdictional conflicts with FERC or the States, requires no new legislation and strands no assets. It is a way to get around many of the problems that people have been concerned about with retail wheeling without compromising the objective that we all have described and that the Commission emphasized in achieving customer choice and access to the market.

That brings us back then to this question of the wholesale market. What we really want to do is focus on how to get the wholesale market to operate efficiently, so we have this arm's length spot price and the ability to do this form of efficient direct access.

If we think about direct access in this way, I think eventually it will relieve some of the paranoia, because when you recognize that that form of customer choice and access to the market is a very modest change, not a dramatic change, and does not produce this tremendous problem of incremental stranded assets and the inability to deal with demand side management, it doesn't threaten environmental protection, all of these problems suddenly are not on the table any more, then it becomes possible to have a rational discussion about how to deal with the wholesale market as opposed to having everyone take strategic positions because they are worried about what is going to happen with later retail wheeling decisions.

So I think these things are related to each other. I think the new world of the competitive market, if we can get an efficient wholesale market operating, we will completely change the way we think about it. It will actually make it easier to implement some of these customer choice and direct access methods, and the real challenge is to get the States and FERC cooperating on how to develop this wholesale market.

The initiative in California is a very important part of that process, and the receptiveness of the Commission here to State proposals and initiative in this matter will be, of course, very important.

With that, I will stop, and I will be happy to answer any questions.

[The prepared statement of Mr. Hogan follows:]

STATEMENT OF WILLIAM W. HOGAN

The Energy Policy Act of 1992 (EPAct) defined a critical milestone in the march towards greater reliance on competition and changed regulation in the electricity industry. Implementation of the law to provide transmission access and support competition in the wholesale market has already had far reaching effects in a process that is still in its early stages. Initiatives in many States, especially the recent proposals in Michigan and California, have expanded the vision for competition to include the possibility of extending to all classes of customers and transactions. The forces in motion are powerful and will sweep over those who ignore them, hence the value of this hearing and related investigations underway elsewhere. The shape of the new world can be perceived, but there are major challenges in capturing the promise of the vision. The key ingredients are the creation of an efficient wholesale market and the development of better performance based incentive regulation for the remaining monopoly elements. Once the wholesale market has come into operation, expansion to include competition for retail customers could be achieved without major jurisdictional changes, legislative initiatives or disruptions of other programs.

The California Public Utilities Commission (CPUC) has gone the furthest in defining a vision for the future of the electricity market. The "cornerstone" of these proposals is provision of "direct access," a term introduced to replace the misnomer of "retail wheeling" but covering the principle of customer choice in the market for electricity products and services. My summary of the principal benefits promised and that can be achieved through such expansions of competition includes:

—Reduction of the role of economic regulation and replacement with market forces;

—Reduction of long-run cost for all consumers while achieving environmental benefits, maintaining universal access, and promoting economic growth;

—Expansion of products and services through consumer choice and direct access to a reliable, open and efficient electricity market; and

—Prevention of cost shifting and uneconomic bypass while preserving the franchise utility's financial integrity.

At the same time, the CPUC's proposal recognizes what cannot be accomplished without sacrificing some or all of the objectives above:

—Provision of direct access with no more than unbundled rates for the embedded costs of transmission and distribution;

—Immediate and large rate decreases for customers who can bypass the franchise electric utility; or

—Movement to a more open and competitive market without provision for all the necessary elements of the transition and efficient operation of the new market, which is not yet in place.

The promises made can be kept. The promises not made need not concern us, although it is useful to mention what cannot be accomplished to avoid any confusion about what is intended. Realization of the benefits of an extended vision for the new electricity market built on the cornerstone of consumer choice through direct access is possible with a careful and comprehensive effort that should determine the nature and timing of the transition.

The broad strategy to expand the boundaries of competition and refocus the attention of regulation surely marks "the end of one era and the beginning of another." To make the new era better, everyone will need to recognize that it will be different. Old questions may need new answers. We are likely today to underestimate just how different some things may be, and in the transition to the new era there will be many opportunities to revisit and revise the conventional wisdom. The delight will be in working on a problem where "most things that everyone knows to be true, aren't." For example, by now everyone knows that the old truth of the contract path for transmission was only a workable fiction with no relation to reality and which is now collapsing under the pressures of the competitive market. This is only the tip of the iceberg. Many other old assumptions and convenient fictions will need to be revisited. A partial list includes:

Energy and Capacity: The need for separate energy and capacity planning and pricing arises in part from the incentives of cost-of-service regulation. The new incentives of the market may eliminate the practical need for maintaining any distinction.

Generation and Transmission: In the short run, with economic dispatch in an integrated network, generation and transmission are parts of the same function. Separate pricing of each may be as unnecessary as it would be difficult.

Cost-Effective Conservation: Criteria for evaluating cost-effective conservation investments must change with a move to marginal cost pricing that should be inherent in a more competitive electricity market. The important market failure created by average-cost pricing will be eliminated and incentives recast.

Portfolio Planning: Under the closed cost-of-service system, the franchise utility was the only source of diversification and portfolio planning for the long-run. With competition and open access to the essential facilities, the portfolio management function for commodity electricity can move to private markets.

Environmental Protection: Programs to improve environmental performance can be compatible with customer choice. Regulation would shift from restricting customer options to expanding environmental opportunities. Explicit subsidies to environmentally preferred technologies can replace command-and-control procurement directives.

Obligation to Serve: Cost-based regulation of a monopoly creates the need for an "obligation to serve." Redrawing the boundaries of the monopoly creates both the opportunity and the necessity to recast the obligation to serve to apply only to the monopoly elements. With the residual monopoly of transmission and distribution, the new world calls only for an obligation to deliver the commodity energy that can be purchased in the competitive market.

Market Power: Market power may remain in regional generation markets. However, the new institutions of the competitive market will create alternatives for regulating generation that can prevent monopoly pricing while preserving competitive pricing, both of which will differ from cost-of-service pricing.

We could add to the list, but the point would be the same. Everyone will have to go back to basics in reconsidering many of the assumptions about the most obvious facts of the electricity industry which may not be facts at all.

The process of the refining and moving towards the new vision would be enhanced by expanded public policy direction on the guiding principles. The Federal Energy Regulatory Commission (FERC) has been developing such principles for transmission, and the CPUC outlined further criteria for expanded competition. A report from the POWER group of the University of California discussed principles under the categories of "economic efficiency, equity, technical efficiency of the transmission and distribution network, quality of service, and externalities/public goods." There is much in this discussion that goes beyond pointing to issues, by recommending actual criteria for comparing the proposals, with which I would agree, such as "Price transmission to reflect costs . . . The cost of transmission includes two components: line losses and congestion costs." But other parts of the power discussion, e.g. under "Obligation to Serve," are still in the mode of pointing to problems rather than defining the criteria for solutions. Elsewhere I have discussed such principles for the critical transmission function.

Examples of basic principles that have emerged with special importance in the case of electricity include:

Separate Ownership from Use of Essential Facilities: Everyone should have equal access to and use of essential facilities, particularly transmission, with the rights of ownership limited to compensation consistent with opportunity costs in a competitive market;

Separate Physical Delivery from Financial Transactions: Market institutions and operations should allow for financial transactions that implement bilateral commercial arrangements while separately coordinating the spot market interactions of physical delivery of electricity in the transmission network;

Coordinate Federal and State Regulation: Pricing and access rules for the regulated monopoly essential facilities should be consistent between Federal and State jurisdictions.

The respective legislatures and regulatory Commissions can continue the process by distilling the many comments that have been or will be provided, looking for principles and criteria that can separate good from bad ideas. For example, the CPUC expressed interest in the development of a wholesale electricity market based on a pool. Many of the respondents supported the idea. The "Blue Book" comments from San Diego Gas & Electric and Southern California Edison go further to lay out some of the details of recommended criteria for operation of a competitive wholesale market through a pool. It would be useful to spend the time and effort needed to integrate and expand on such recommendations across all elements of the new vision, resolving conflicts and looking for gaps in the edifice. The need is not for the government alone to work out the details of the market institutions and operations.

Rather the focus should be on expanding the criteria to a point where different designs can be evaluated.

The degree of unbundling and competition, whether wholesale or retail, is a policy choice. In designing a competitive market, a focus should remain on providing nondiscriminatory, equal access to the essential facilities. The thrust of the EPAct of 1992 is that transmission is this essential facility and open access to the transmission system is a necessary requirement for development and operation of a competitive market.

The essential facility of electricity transmission involves much more than simple connection to the wires. With an eye towards identifying these analogous components in the case of electricity, the Figure 1 goes beyond the usual Genco, Transco, and Disco trilogy. The least obvious feature in Figure 1 is the separation of "Transco" into Poolco and Gridco. This choice is Competitive Wholesale Electricity Market Structure made in part with an eye towards the institutional structure of regulation in the United States, where the Gridco(s) construction and maintenance of transmission wires might be subject to State regulation and Poolco control of dispatch to regional or national supervision. However, the principal reason for the distinction is to accommodate the most important implication of the technical characteristics of electricity supply. In particular, the free-flowing grid requires coordination of short-term operations to maintain system stability and achieve least-cost dispatch.

Figure 1

Competitive Wholesale Electricity Market Structure

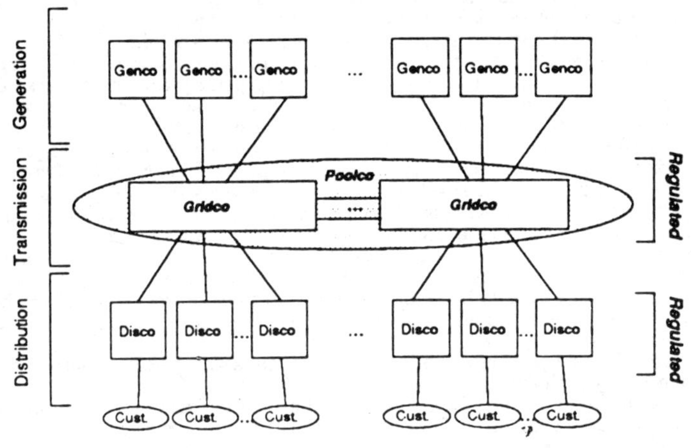

There is a strong case for building the institutions of the wholesale market through nondiscriminatory participation in a pool-based system." Economic dispatch defines the limiting case of the efficient market in which there are no unexploited opportunities for profitable short-term trading of energy. The Poolco model provides everyone, large and small, an equal opportunity to enjoy the benefits of efficient trading and the other services that are essential in the complex interactions of an integrated power system. Such a Poolco for the wholesale market which could support direct access for all market participants.

These advantages of the Poolco and its critical role in providing reliability and facilitating the operation of long-term contracts and investments are only beginning to be explored and understood. If the CPUC's remarkable success in leapfrogging what might otherwise have been a gradual and difficult evolution leads to an earlier implementation of this form of an efficient market, this alone will justify the decision to break with the past. However, even without further implementation of a pool, a vigorous pursuit of the pool idea, examined again from first principles, will do much to clear away the underbrush in the debate about what is possible and what is necessary for successful operation of the electricity market. Ultimately,

FERC will have jurisdiction over the rules and pricing of the wholesale market. However, State governments and the market participants have a major role to play in developing a workable approach that is consistent with the long-run vision.

Usual practice applies the label "retail wheeling" to an expansion of competition to include sale of electricity to retail customers. This label appealed to a comfortable fiction that suggested power could be directed from one source to another destination by "wheeling" through the wires of intervening utilities. For a variety of reasons, this is an exceptionally bad and misleading model of the actual operation of an electricity market. Either to avoid these errors, or to escape being burdened with all the accumulated baggage of retail wheeling, the CPUC introduced the term "direct access" to cover the principle of customer choice in much of the provision of electricity products and services. Efficient direct access to the wholesale price is a better and simpler concept that can support customer choice and a competitive market.

Direct access offers benefits that are hard to achieve in any other way. It will be difficult for the regulator to anticipate all the preferences of customers, or to avoid the clamors for redress when the inevitable surprises occur. Customer choice through direct access could shift these responsibilities to where the decisions can best be made. However, despite the attempt to refocus the debate, the CPUC did not insulate itself from the heat associated with retail wheeling. The fear of customer choice flows from a misunderstanding of the changes needed to implement and enjoy the benefits of efficient direct access. In part, this is a problem caused by another workable fiction from the past that serves us poorly in thinking about the future with a competitive market. The fiction is in the pervasive view that a particular customer buys power from a particular supplier or set of generating plants which are part of the franchise utility. And by changing the source of the supply from the franchise utility to another generator, the customer would leave the system. Hence, with this model, opting for direct access requires an essential change of status that moves the customer partly away from the State regulator's jurisdiction, and in the extreme view gives FERC complete control over all rates, right up to the final connection to the household.

Of course, as the engineers will tell us, the electric power comes from everywhere, and its impossible to identify which generating plant served which customer. The customer who remains connected to the wires never leaves the system in any physical sense. Hence, in the competitive market, direct access will be much simpler and much less disruptive. Not much more is required than is already in sight with the mandate to create a competitive wholesale market. Direct access is not hard to do; it may be harder to stop.

There are only two essential actions needed to make Efficient Direct Access operational, and one more to exploit its potential. All the steps are underway or appear obviously to be within the purview of the State regulatory commissions. There is no jurisdictional problem here, and no immediate radical change. The first step is to implement a competitive wholesale market with a readily available spot price against which generators can sell power at a profit or a loss. The details are important, but for the present all that is required is the existence of a transparent, arm's length spot price that is available to all suppliers and not subject to manipulation. The regulator can help by supporting the development of an efficient wholesale market, which the regulator should do no matter what the view on direct access.

The second step to Efficient Direct Access would be a change in rate design. The State regulator could modify traditional cost-of-service ratemaking to accommodate time-of-use pricing where the energy component of the bill would be based on the arm's length spot price. Other costs of service would be included as now in customer demand and volumetric distribution charges under familiar ratemaking principles. Customers who choose to ignore the spot price would receive a bill that in effect averages the spot price over the month, and they would see little that is different from today. Customers who want to exploit the pricing information, after the necessary metering, could pursue any of a variety of old and new demand management options.

This relatively modest rate innovation should be well within the regulator's authority. No customers would leave the system or change their status. Yet, the customers would become de facto direct access customers in every way that can be justified consistent with an efficient and competitive generation market. Faced with the uncertainty and volatility of the spot price, customers may look for longer-term arrangements that would reduce or fully eliminate the risk. Generators, selling into the same market for the spot price, would have the complementary interest and would be the source of long-term contracts for those customers who felt the need. This would happen automatically and in the private market. These contracts to protect against price changes would be the functional equivalent of the long-run contracts for price differences relative to the pool that are at the heart of the long-run market in the U.K. system. There would be no need for the State regulators to take any further action. The regulators at FERC would not be directly involved. With the competitive wholesale market, arm's length spot prices, and time-of-use rates, Efficient Direct Access would be in place. No new legislation or authorities would be required.

The third step, a natural corollary, would make Efficient Direct Access effective. With customers facing the spot price through the new rate design, and seeking to make their own private arrangements to provide the long-term certainty that they are willing to pay for at market prices, there would be no need for the franchise utility to make further long-term generation or power purchase contracts to be included in the ratebase. Gradually, old commitments to longterm generation included in the ratebase and cost-of-service would be replaced by new long-term generation commitments occurring outside the realm of regulation. The transition would avoid creating new stranded assets under the incentives of cost-of-service regulation.

Worthy programs and costs other than for the commodity energy could continue much as before. Investments in energy efficiency would be made when justified, and included in the cost of service applied to all customers separate from the time-of-use energy charges. Investments in subsidies for renewable or other environmentally preferred alternatives could be made and included as part of the cost of service without complicating the operation of the wholesale market for short-term sale and trading.

Given the operation of a competitive wholesale market, with a transparent and arm's length spot price, Efficient Direct Access is not a radical step. There are real benefits from direct access, but not necessarily big risks. Most of the major fears seem more directed at an old model of the industry that was built on a fiction that no longer serves us well. With the new model of an efficient wholesale market, we can see direct access as it should be seen, as an evolutionary change that opens up long-run benefits without requiring major short-run costs.

Mr. SHARP. Thank you very much.

Professor Stalon, we are very pleased to have you back with us, and in my positive comments about the current Commission I indicated there have been many bright lights over the years at FERC, and you certainly were one of them, one of the real minds that came to FERC and brought a perspective and an intensity of examination that not everybody appreciated, but it should have been appreciated for its quality.

Dr. Stalon, please.

STATEMENT OF CHARLES G. STALON

Mr. STALON. Thank you for those kind remarks, Mr. Chairman.

Again, I speak only for myself, and I would like to start by endorsing the arguments that Professor Hogan made. I also am convinced that once we reconceptualize this industry and realize what it is rather than what it used to be before it was all an interconnected industry, many of our problems turn out to be arguments over words and not over reality.

Having said that, I want to focus my attentions on certain dimensions of the regulatory problem today, but I would like to start by noticing that because the natural monopoly circumstances that justified and made possible regulation of electric generation have faded into near irrelevance as transmission networks became robust and network control technologies increased in sophistication, it is certainly desirable and it is probably necessary that the U.S. electric industry be deintegrated vertically and that competitive markets for generating services be created.

Second, because price projections for natural gas support a larger role for gas and electric generation and technological improvements in combined cycling generation of lowered gas-fired combined cycle unit power costs below the embedded costs of many utilities, a strong sense of urgency has developed in the industry and its regulators for defining and confronting the essential issues in industry restructuring.

Third, because some States will almost certainly permit some large power users, one way or the other, to buy power directly from generators, especially in the context that Professor Hogan has just mentioned, there is some urgency in ensuring that both wholesale and retail markets function efficiently and reliably. The objective of this transformation is not to create merely a competitive market, it is to create an efficient market with the understanding that competition is a useful tool for doing that.

Fourth, because the current system of piecemeal regulation makes it impossible for any regulatory agency to define a vision of a coherent industry and regulatory structure which it, acting alone, can implement, current changes in the industry appear at first glance to be much more chaotic than they really are.

I believe that FERC, for example, has made several aggressive and constructive decisions which point in the correct direction, and many other States, notably California, Michigan, and New York, have initiated debates and/or proceedings and/or discussions heading in the same direction.

Fifth, because current laws do not appear to give Federal regulators powers and responsibilities appropriate for the task of restructuring the industry and the industry's regulatory system so as to exploit efficiently potentials for new technologies and new relative fuel prices, some Federal legislation will soon be desirable. In effect then, the structure of current industry is not as appropriate for exploiting existing technologies as are alternative attainable structures, and the structure of the industry's regulatory system is not as appropriate for regulating either the existing industry or the likely new industry as alternative attainable regulatory structures.

The first point was recognized by the Energy Policy Act, and we have started a restructuring of the industry. The second is widely recognized by students of the industry, but we have not yet properly recognized in law the need to restructure the regulatory system.

Now a backdrop generalization. Economic efficiency requires that the structure of the industry be consistent with the underlying economics of the industry, and regulatory responsibilities need to be consistent with underlying economic and political theories of regulation, and objectives of regulation need to be consistent with widespread public values. The current structure of the industry and its regulatory system offends all three of these principles.

In brief, the current industry and the current regulatory system is surviving largely on institutional inertia. It is time we now at least address the regulatory deficiencies.

Let me talk about some of the problems that I think are involved in trying to address those deficiencies, first noting that while I think the Congress did a fine job, correct legislation appropriately timed in the Energy Policy Act in initiating reforms of the industry, it really overlooked, probably deliberately, the problems of restructuring the regulatory system.

Let's keep in mind what I consider to be the appropriate problem we have before us; that is, the essential technological characteristics of the industry. First, we have been looking for an industry. We really have it, but we are looking to bring it out in the open, and we are generating companies, and distribution companies in three nations under all forms of ownership are required to interact across interconnected transmission networks.

Consequently, some regulatory body, whether it be regional or Federal or international or either Government or private or a mix of government and private, must ensure that the pieces of the system fit into a coherent whole that operates without significant inefficiencies.

The principal regulatory bodies that ensure that today are essentially the industry in its self-regulatory regional reliability councils operating with some kind of quasi-approval or at least recognition of the Congress and the FERC.

Four generalizations are useful for framing the discussion of regulatory restructuring. The first one is the obvious one; it has been mentioned before. In my view, the most important and pressing energy issue in the Nation today is the transformation of the electric industry.

Second, this industry transition is not being driven by Government actions, and therefore the timing of it is probably not controllable by the Government, it is being powered by the demands of the municipals for access, of the nonutility generators for access, of the large industrial customers for obvious reasons I need not detail, by those investor-owned utilities that want to get out from underneath the responsibility to build generation under regulation, and by those investor-owned facilities who see a great deal of potential and making money in the nonutility generating sector.

The third point: As we contemplate this restructuring, let's keep in mind that we are not talking about creating competition in everything. There are three natural monopolies to be preserved. There is the natural monopoly of distribution, the natural monopoly of transmission, and the natural monopoly of system control. Two of those natural monopolies, transmission and control, are not only interstate, they are international, and therefore it is the Federal responsibility to deal with this.

My fourth point here is that if one assumes the Congress will not soon impose a new structure on the system—and I don't see any need why it should—the needed changes must be guided into existence by the only two Government agencies currently capable of taking a national and international view, and that is the FERC and the Department of Energy, because our decisions affect both the Canadians and Mexicans and we must deal with the State regulatory communities with the NERC and with the regional reliability councils if we are to bring this transition off.

Four topics briefly dealing with what I consider to be the problem of transmission, looking at it from the regulator's viewpoint. The first issue is an amusing one. For the first time in the history of the U.S. power industry, we are now going to see builders of generating plants who will be influenced in their plant location decisions by prices charged by transmission services. The security blanket that has allowed regulators for a century to price transmission services however they choose with the firm knowledge that no utility will pay any attention whatsoever to those prices when it locates its generating plants but, instead, will locate those plants to minimize the total cost of production, that security blanket is being removed. Prices have assumed much greater importance than they have in the past.

The second point, the Federal Power Act denies to the FERC the necessary authority to assure the construction of efficient transmission networks.

Third, in the United States a natural monopoly called transmission networks are not owned by a single legal entity. If that form of ownership is permitted to continue, any just and reasonable earning requirements will have to apply to separately owned segments of the network, which is an unnecessary obstacle.

Fourth, current technology and methods of operation create much more value added in the generating sector than they do in the transmission sector. One of the ways that the industry has operated in the past is to substitute transmission capacity for generation capacity in minimizing the total cost of service. With generators now owned by one set, it becomes vitally important that the transmission capacity be built in order to minimize total cost and satisfy many of the objectives in the past. That will be a major Federal responsibility.

Let me now look at a couple of reform objectives, focusing again on the transmission networks. There are several outcomes that I think we first need to avoid. It is nice to talk about what we want to accomplish, but a regulator always has to think about a fail-safe system—what do we want to avoid?—and the first thing is, since transmission is a natural monopoly, we don't want bilateral competition in transmission services. There is no reason whatsoever to believe that the regulatory fiction of a contract path, if built upon, would be anything other than subversive.

We obviously need to expand the role of the FERC over the pricing of transmission services. One tactic that would be very convenient for doing that should be discussed at length with the States because I have a feeling it would be controversial with some of them, would be simply to have the FERC say that we have an obligation, as one of the Commissioners said very persuasively this morning, to ensure not unduly discriminatory rates and services for customers using the transmission system.

If the FERC were to say in order to accomplish that objective we must have all transmission services unbundled, we would immediately put into play movements toward several purposes I think are very important. It would lay the groundwork for integration by contract, it would further weaken the idea that corporate vertical integration is a natural or essential form of organization of the industry.

Second, it would highlight FERC's responsibilities for defining and pricing all transmission services, and it would make the FERC fully responsible for adequate revenue for all transmission assets.

Third, it would force a debate, a very important debate that we have seen here already today, what is a transmission service? Where do we draw the line between a distribution company and a transmission company?

Fourth, it would make it easier for the FERC to argue that it must assume responsibility for planning transmission networks, and clearly it must do so since there is no one else other than the industry itself that can do so.

Fifth, it must encourage those who prefer or think they prefer alternative proposals to formulate those proposals with great care; and, sixth, it would make it easier to do something that I think is vitally important, and that is persuade transmission owning utilities to transfer the ownership of their transmission assets to incorporated transmission companies.

One of the problems of the U.S. system of regulation is, it is designed to regulate selected transactions of selected legal entities. We have a task here of regulating a natural monopoly which is a network with pieces of that network owned by many, many legal entities all demanding and deserving under law something called just and reasonable rates of return.

We may be creating a regulatory problem that is larger than the regulators can possibly handle. The simple way out of it is to create transmission corporations and lodge in them, in the same legal entity, the control network, the control system.

It would also introduce a debate of minor importance, although it turned out to be important in the United Kingdom, and that is who should own stock in a transco, should it be limited distribution companies, or who should do it.

Now there is one piece of legislation I think I want to mention, and then I will stop. I have listed a few others, but I will talk about one. We need to redefine the concept of electric utility. I noticed in the discussion this morning, as I have noticed in many others, that we still have not redefined our language. We are still not reconceptualizing the industry to recognize the changes which are being pushed on it.

But clearly today a generating company selling power into competitive markets to unaffiliated buyers should be thought of the same way we think of an independent coal company. That is, it may provide services which are essential to the efficient functioning of the industry, but their obligations are only those provided by contracts and those which are necessary for the efficient and reliable functioning of the networks.

Similarly, a genco who is a self-generator or a cogenerator selling power, which would include industrials who sell surplus power, should be viewed the same way. "Utility" as a term should be limited to the distribution company and perhaps the transmission company and the control monopoly; I would also include all brokers and resellers in that.

Rather than continue down my list, let me recapitulate very quickly. The economic and technological changes forcing restructuring of this industry are very powerful. They do not appear to be reversible at any time in the foreseeable future. Furthermore, with the potential benefits so large and so obvious, it is hard to imagine a reason why we would want to reverse this trend, but the magnitude and complexity of the developing transformation exceed those by far initiated in the gas and the railroad industries and approximate those initiated in the telecommunications industry in the eighties, as we know are still under way.

Electric industry regulators today must simultaneously facilitate the restructuring of the industry as well as the restructuring of their own jurisdictions among themselves. Furthermore, as we replace that set of regulators that have never been clearly and explicitly recognized in law—that is, the industry's self-regulating aspects, power pools and regional reliability councils—it will be necessary to create some regional regulatory bodies in order to satisfy all the potent interests involved. Realistically, electric industry restructuring is a Federal initiative, and Federal leadership is essential.

Thank you.

[The prepared statement of Mr. Stalon follows:]

STATEMENT OF CHARLES G. STALON

Because natural monopoly circumstances that justified and made possible regulation of electric generation faded to near irrelevance as transmission networks became robust and network control technologies increased in sophistication, it is certainly desirable and probably necessary that the U.S. electric industry be de-integrated vertically and that competitive markets for generation services be created.

Because price projections for natural gas support a larger role for gas in electric generation, and technological improvements in combined-cycle generation have lowered gas-fired, combined-cycle unit power cost below the embedded unit cost of many utilities, a strong sense of urgency has developed in the industry and its regulators for defining and confronting essential issues in industry restructuring.

Because some States will almost certainly permit some large power users to buy power directly from generators soon, there is some urgency in ensuring that both wholesale and retail power markets function efficiently and reliably.

Because the current system of piecemeal regulation makes it impossible for any regulatory agency to define a vision of a coherent industry and regulatory structure which it, acting alone, can implement, current changes appear at first glance to be more chaotic than they really are. The Federal Energy Regulatory Commission (FERC), for example, has made several aggressive and constructive decisions which point in the correct direction, and many States, notably California, Michigan and New York have initiated debates and/or proceedings and/or actions.

Because current laws do not appear to give Federal regulators powers and responsibilities appropriate for the task of restructuring the industry and the industry's regulatory system so as to exploit efficiently potentials of new technologies and new relative fuel prices, some Federal legislation will soon be desirable.

In recapitulation, the structure of the current industry is not as appropriate for exploiting existing technologies as alternative, attainable structures, and the structure of the industry's regulatory system is not as appropriate for regulating either the existing industry or the likely new industry as alternative, attainable regulatory structures. The first point was recognized in the Energy Policy Act of 1992 (EPAct). The second is widely recognized by students of the industry but not yet properly recognized in law.

Economic efficiency requires that the structure of the industry be consistent with the underlying economics of the industry, and regulatory responsibilities need to be consistent with underlying economic and political theories of regulation, and objectives of regulation need to be consistent with widespread public values. The current structure of the industry and its regulatory system offends an three of these principles:

—The generating sector is not a natural monopoly, and with proper regulation of the natural monopoly functions of transmission and distribution and proper applications of antitrust laws generation services can be a competitive, efficient industry.

—Public policy, with many exceptions, has supported reliance on competitive markets when doing so promises efficient outcomes, and competitive markets for power, with well-designed and implemented regulation for the natural monopolies of transmission and control promises efficient production, transmission and distribution of power.

—There are three natural monopolies in the industry that need to be regulated: They are the distribution function, the transmission function and grid control functions. The current system of regulating distribution functions is adequate, but the regulatory systems for regulating the other two functions are obsolete.

In brief, the current industry and regulatory system is surviving largely on institutional inertia. I think it is tolerably accurate to observe that the Congress recognized these facts when enacting the Energy Policy Act of 1992, but it did not recognize the speed with which economic forces were changing the industry. It is now time for the regulatory deficiencies to be addressed.

The Congress, in the Energy Policy Act, did not address the necessary restructuring of the industry's regulatory system. Clearly, the forces released by competition in power markets will coerce substantial changes in the responsibilities of State regulators and will almost certainly increase those of Federal regulators.

It is useful to keep in mind the essential technological characteristic of this industry, namely, generating and distribution companies in three nations, under all forms of ownership, must interact across interconnected transmission networks. Consequently, some regulatory body—regional, Federal or international, either government or private or a mix of both—must ensure that the pieces of the system fit into a coherent whole that operates without significant inefficiencies.

Four generalizations are useful for framing the discussion of regulatory restructuring:

1. The most important and most pressing energy issue in the Nation today is the transformation of the electricity industry initiated by Energy Policy Act. The opportunities for improving the efficiency of the industry are substantial, and the dangers of gross inefficiencies from improper policies are also substantial.

2. The industry transition is not likely to be driven by government actions; instead it win be powered by the demands of munis, coops, NUG's, large industrial customers, those IOU's who do not want to build any additional generating assets under regulation and those IOU's who see potential profits in the non utility generating sector.

3. When restructuring the industry's regulatory system it is essential to keep in mind that two natural monopolies, transmission and control, are not only interstate, they are international.

4. If one assumes that Congress will not soon impose a new structure on the regulatory system, the needed changes must be guided into existence by the only two government agencies capable of taking a national and international view, the FERC and the Department of Energy (DOE), although these agencies will have to work with Canadian and Mexican regulators and through the National Association of Regulatory Utility Commissioners (NARUC), the NERC and the RRC's.

Four transmission sector problems are spotlighted by the restructuring movement: They call for important policy decisions by regulators:

1. For the first time in the history of the U.S. power industry, some builders of generating plants will likely be influenced in their plant location decisions by prices charged for transmission services. The security blanket that allowed regulators to price transmission services with the knowledge that utilities would pay no attention to such prices when locating generating plants, but would, instead, locate such plants so as to minimize the total cost of delivered power has been removed by the EPAct.

2. The Federal Power Act denies to the FERC the necessary authority to assure the construction of efficient transmission networks.

3. In the United States, the natural monopolies called transmission networks are not owned by a single legal entity. If that form of ownership is permitted to continue, any "just and reasonable" earnings requirement must apply to separately owned segments of the network.

4. Current technology and methods of operation create much more value added in the generation sector than in the transmission sector. In a regulatory system in which the use of prices to ration capacity is not widely accepted, an important element of economic efficiency is positive reserve margins for both generating capacity and transmission capacity. Both, however, are expensive, and to some extent they are substitutable for one another. In such a circumstance, it is logical to substitute the cheaper input (transmission lines) for the more expensive input (generating plants) extensively. Making this trade-off will be an important responsibility of Federal regulators, acting perhaps through regional transmission groups.

Although PUC's, in most cases, have substantial authority to regulate vertically-integrated utilities, they have never had any meaningful control over the operation of transmission networks. For almost half a century, these network have been planned, constructed and operated by utilities acting cooperatively, primarily in reliability councils and power pools. In section 205 of PURPA, the Congress reaffirmed the importance of denying to States the right to disrupt this system of voluntary cooperation. Consequently, it seems implausible to expect State regulators, even if they could act collectively, to provide the leadership needed to redefine the role of transmission and control systems.

There are, however, certain outcomes that all ought to seek to avoid. Five such outcomes are worth noting:

The transmission system is a natural monopoly. It should be planned and operated as one. There is no meaningful sense in which competition between "owners" of imprecisely-defined rights to use the system are likely to contribute to efficiency of the industry by offering competing services to buyers and sellers of power. Further attempts to convert the regulatory fiction of a transmission contract path into a vehicle for competition in transmission services will produce not only inefficient use of the system but also defensive actions, e.g., phase shifters, from other TOU's that will lower further the efficiency and reliability of the system.

The managers and regulators of control system(s) need to be able to assert credibly that the combination of dispatching within control areas and trading between control areas is as good an approximation to maximal efficiency as current communications and control technologies will allow. There is no reason to settle for a second best arrangement wherein control centers overlap. Obviously, if overlapping control areas are a necessary step on the route to larger control areas, overlapping area may be have to be tolerated for a transition period of uncertain length.

The system of bilateral trading among utilities and between utilities and NUG's is a system grafted onto a basic system in which most power moves in response to efficient dispatch requirement. The objective of the restructuring should be, among other things, to improve the efficiency of use of the generating sector. In order to do that the Nation needs to reaffirm the primacy of the dispatching system and to support reforms than protect and enhance that objective.

It would be a disservice to the Nation to impose a major restructuring on the electricity industry with the announced objective of improving the efficiency of the system and simultaneously impose an inefficient system for pricing transmission services.

Since retail wheeling is very likely to increase in importance, policies should seek to ensure that physical bypass of discos, including municipalization, is done only when it contributes to efficiency. State regulators, however, have a responsibility to avoid price differences that offer strong encouragement of uneconomic physical bypass.

For those who see the creation of efficiently functioning transmission and control systems as the principal problem calling for regulatory restructuring, the overall strategy for the DOE, the FERC, the Rural Electrification Administration (REA), the TVA, the BPA, other PMA's, and other transmission owning utilities (TOU's) and State regulators should be to transfer to one or more regional agencies or to some Federal agency, most likely the FERC, plenary authority to regulate the planning of transmission systems. This does not require that the regional agencies or the FERC have sole authority to site transmission lines; that power might be left, in part at least, to States to exercise under constraints designed by the FERC, or possibly by the regional agencies.

One tactic that appears to be useful to encourage the necessary jurisdictional changes is the following: Since the FERC has jurisdiction to price unbundled transmission services, it could impose non discriminatory pricing standards on TOU's for unbundled and currently bundled services and argue that only by unbundling what is now bundled can the TOU demonstrate that it satisfies such standards. Such a tactic would serve several purposes:

1. It would lay the groundwork for integration-by-contract and further weaken the idea that corporate vertical integration is a natural or essential form of organization in this industry.
2. It would highlight the FERC's responsibilities for defining and pricing all transmission services, and it would make the FERC fully responsible for just and reasonable earnings on all transmission assets.
3. It would force a debate on definitions of "transmission services" and a debate on distinguishing between disco and transco assets and functions.
4. It would make it easier to argue that the FERC must assume responsibility for planning transmission networks in an open access system.
5. It would encourage those who prefer, or think they might prefer, regional regulation of transcos to FERC regulation to develop and publicize their proposals.
6. It would make it easier to argue that TOU's should sell their transmission assets to one or more corporate transcos.
7. It would introduce the debate as to who can own stock in a transco, anyone or only discos with a retail obligation to serve.

Since the powers of the FERC and the DOE to mold the desired structure of the industry have not been tested, assertions about needed Federal legislation must be made with caution. Still, there are three areas where legislation could facilitate the transition.

The concept of an electric utility needs to be redefined.

A genco selling power into competitive markets to unaffiliated buyers should be thought of as an independent coal company is thought of today, that is, as providing services essential to the efficient functioning of the electric industry but whose obligations are determined by his contract and those needed for reliable functioning of interconnected networks. Similarly, a genco who is a self-generator or co-generator selling power should be viewed similarly.

Similarly brokers and resellers of power unaffiliated with a disco should not be thought of as utilities with obligations other than those assumed by contract.

A responsible Federal regulator, presumably the FERC, must gain the ability to order the construction of transmission and control assets and to approve operating rules and pricing policies for control areas that are interconnected with other control areas.

Regional transmission networks and control centers must evolve with great powers to plan and some discretion in pricing, but such centers will be interdependent and some agency must be equipped to settle disputes among them.

Customer-owned and government-owned generators must be incorporated into the evolving competitive power markets. They will have much to gain by buying and selling for the benefit of their stakeholders and they will add depth to power markets. Customer-owned and government-owned transmission assets must be sold to or otherwise integrated into transmission networks to add strength to networks and reliability to their operation.

The economic and technological changes forcing restructuring of the electric industry are powerful, and do not appear reversible any time in the foreseeable future. The Congress accelerated the transformation to competitive power markets in the EPAct, and current natural gas prices and turbine technology provide powerful incentives for large users and States seeking to attract large users to extend free choice of gencos to users within franchised territories of distribution utilities. With potential benefits so large and obvious, it is hard to imagine a reason to reverse the trend.

The magnitude and complexity of the developing transformation exceed those initiated in the gas and railroad industries and approximate those initiated in the telecommunications industry in the 1980's, which are still underway. Electric industry regulators must, simultaneously, facilitate the restructuring of the industry and the reallocation of jurisdictions among themselves. In fact, it will be necessary to create new regional regulatory bodies to satisfy all the potent interests involved.

Realistically, electricity industry restructuring is a Federal initiative, and Federal leadership is essential. Legislation redefining the concept of an electric utility would be helpful. Federal regulators and the DOE need to work closely with the NARUC, especially the NARUC's Committee on Electricity, but, more important, the needs to publicize tentative operational objectives so the debate has substance.

Mr. SHARP. Thank you very much, Dr. Stalon.

Professor Brown, we are very pleased to hear from you now.

Mr. BROWN. Thank you.

Mr. SHARP. A former State utility commissioner and, as I recall, head of NARUC and, again, engaged before this subcommittee on many occasions.

I don't know if you will be bringing the State perspective today or not.

STATEMENT OF ASHLEY C. BROWN

Mr. BROWN. Well, actually, as usual, the first thing I have to do is convince you I am speaking only for myself, or, paraphrasing Groucho Marx, I wouldn't trust anyone who would trust me to speak for them.

But actually, the subject of this hearing reminds me a lot of a story my grandfather told me about a matchmaker in a small Jewish community in the Catskills right after the war. The matchmaker came to the rabbi, and he said, "Rabbi, I have got a bride for your son. He is of marriageable age, and I've got a marriage for him," and he said, "Well, who?" And he said, "Princess Margaret." The rabbi goes, "Princess Margaret? She is not Jewish. My son can't marry Princess Margaret." He said, "No problem." He said, "It will be wonderful for the Jewish people. Think of Queen Esther when she married the emperor of Persia. It turned out she saved our people. Rabbi, you should think about it." "All right, I I'll think about. You come back in 1 week, and I'll give you an answer," and 1 week later the matchmaker woke up, and he ran to the rabbi's house and said, "Rabbi, rabbi, a week has gone by. Have you made a decision?" And he said, "I have. I have thought about it. I have spent a lot of time thinking about it, and I considered all sides of it, and finally I have concluded that you are right, my son should marry Princess Margaret" and the matchmaker got all excited, and he goes, "Great, I've half a deal."

And that is what we have after the Energy Policy Act, and I haven't been in academia long enough to be too theoretical, so I wanted to be somewhat practical in terms of observing what I think are some of the critical regulatory problems in trying to deal with the half a deal.

The first is to look very quickly at the jurisdictional matrix that we have currently for transmission. Usually I think about this in terms of four issues—pricing, access, siting, and planning—but actually there is a fifth one, and that is who actually is regulating and the operation of the grid, and what is interesting is, it is completely incoherent as to who does what and it is happenstance. On pricing, it is clear wholesale. Pricing is Federal jurisdictional, as is wholesale access.

Of course, we are debating who has pricing jurisdiction over retail transmission services. I have taken a scientific survey and determined that 92.3 percent of all lawyers who practice energy law in Washington have concluded it is Federal jurisdictional and 92.3 percent of all lawyers who practice in State capitals have concluded it is jurisdictional to the States, and those non-self-interested folks have spoken with scholarly precision on the subject, and retail access is obviously another question. I think traditionally we have thought that was a State issue. My sense is, Congress thought it did when it put the savings clause into the Energy Policy Act, but others have a very different view. So we can say that is somewhat controversial.

The question of siting and regulatory jurisdiction over the siting, the physical siting and the planning of transmission facilities, is fairly simple. With a few minor exceptions those are exclusively subject to State jurisdiction, and the question of who regulates controlling and dispatch, the answer is for the most part, with a few exceptions, it is the States. They are the ones who, in their fuel clause proceedings, generally outreview the dispatch and operations of the grid by the utility.

The exceptions are in the case of the registered holding companies who have entered system agreements that are reviewed by the FERC, and there are also obviously some exceptions in the operation of a couple of power pools, but for the most part that tends to be State jurisdictional, but it is somewhat split.

So how do each of those jurisdictions approach the rather confused set of jurisdictions, and the approaches couldn't be more different. In fact, I suspect what you should have done was got an anthropologist to discuss the cultural aspects of the difference in approaches.

States approach transmission issues, if they approach them at all, to be quite frank about it, on a bundled basis. They provide bundled retail prices to customers who until recently have been generally captive, and to the extent to which they have some options now, they still tend to be, for transmission purposes, captive customers.

The second is, once they decide on siting or approving a transmission facility, or they—the way they treat it is not through pricing that as a service but, rather, by throwing that into retail rate base along with generating assets, distribution assets, and the like, so you end up with this bundled, blended rate at the end, but you are not sending—no one ever intended in State regulation, there is no State that has ever intended, to send some discrete pricing about transmission, and by doing it the way they do it, they have imposed essentially an obligation somewhat akin to that which the Federal Government has assumed on the savings and loan industry, and that is, they are the ultimate guarantor of the revenue requirement of the grid.

I think most States did that on the understanding that they would then have priority of access and wholesale use of the grid would become—was essentially incremental in the times when there was capacity available but obviously as a time warp notion in view of the emergence of competition in the bulk power market.

The last two: The States generally approach—make the critical transmission decisions not in rate cases or in contract filings as the FERC but do it in planning and siting processes, that is where the critical decisions are made, and so they are approaching this in an entirely different context than the Federal regulators.

FERC, on the other hand, tends to view—FERC, first off, is presented not with a grid that it has to meet the revenue requirements for and get serious about pricing. In fact, it is handed a grid and says, "Here, all incremental use of this grid you will price," and it is not surprising, as was described by the Commissioners earlier, that for 50 years FERC has used kind of an average embedded cost pricing arrangement. It kept transmission rates reasonably low, and the utilities weren't unhappy because their revenue requirements were met by somebody else, and we haven't gotten serious about transmission pricing.

The FERC, in their NOPR, may well do that now, but their approach has been, here's the grid, all we need is somebody to price incremental use of it, and that is precisely what they have done. They have approached transmission, unlike the States, on a discrete unbundled basis for the most part and view it in very discrete terms, and, unlike the States, FERC has tended to formulate certain transmission policies and approached it in that discrete way and in that unbundled sort of way.

Where it exercises that jurisdiction, as I said, is quite different than the States, it is not early on in the planning and the siting proceeding, but it has generally done it in the case of rate filings or contract filings or through mergers and acquisition cases where it has some ancillary—where it is concerned about the effects of the use of the grid. So it approaches it in entirely different kinds of proceedings, and FERC—the final area where FERC and the States differ is, FERC's objective has tended to be over the years almost exclusively economic efficiency, and certainly, as Chair Moler pointed out this morning, she and, in fact, the FERC have not been involved in the debate over externalities, they have not been involved in IRP questions, they have not been involved in, if you will, social programs like low-income programs where States have non-economic objectives, but they have tried to use their jurisdiction over the electric utility industry to accomplish, and so that is another very important difference. The States' objectives have tended to be a little more complicated than the Federal objectives because of the nature of the jurisdiction.

So where does that leave us? What I want to do is focus on four key problems here that I think have really emerged not just from the Energy Policy Act but also from simply the transition in the industry and the changes in the industry. One is what I think most commentators would call the black hole of the Energy Policy Act, and that is what happens when FERC issues a 211 order and the utility convinces everyone that they don't have the capacity to meet that obligation, and they go to the State to get the line sited.

The dilemma for the State regulator in that circumstance is rather intense because traditionally State regulators, who may, by the way, may or may not be the PUC—in many States it is some other agency or some combination of agencies—but State regulators have historically made two decisions in the siting process. Decision one is, should this line be certified and built? That is, is the need such that it outweighs any sort of environmental aesthetic or health effects that may be adverse or whatever other issues they want to consider. And the second, is who pays for the line?

The implication is once the State certifies that line for the utility, it will go in the retail rate base and the residual revenue responsibility will be made by retail customers. Well, presumably a 211 access order is going to be for the benefit of a nonjurisdictional customer, and so the question for the State is, number one, what impact does that essential need determination by the FERC through the issuance of a 211 order—how does that impact the need determination by the State that the line is needed, and then the second question, which is a totally different question, is, why should these State regulators—and I am being a State regulator—why should my ratepayers have to pay for a line that provides them no benefit?

So on the one hand I shouldn't be, as a matter of public policy, parochial on the question of whether the line is needed and siting it. On the other hand, I am not doing my job and protecting ratepayers if I am imposing costs on them that are caused by somebody else who, in effect, by my decision, could avoid paying the freight for the costs he or she imposes on the system. So that is the dilemma number one I wanted to point to.

Dilemma number two is the question of what happens in the pricing arrangements. As I said, FERC can do whatever it wants with pricing, and when it does it, it is incrementally pricing the grid, pricing incremental service on the grid.

The problem here is, think about the utility's incentives. If a utility is woefully inefficient or takes a, "Hell, no, I'll never provide service," transmission service, which you have precluded them from doing now but which some of them may have done in the past—but if a company took that position, there is no financial penalty associated with it because the full revenue requirements for their grid investment is recovered from retail ratepayers.

Let's take a utility that uses the grid to the optimum, whatever that means. They use it about as efficiently as one can reasonably expect. Do they get rewarded for that? No. What will happen is, the State regulators will say, "Look, our ratepayers, the retail ratepayers, basically guaranteed you the revenue requirement, they absorbed all the risks for this line, and so therefore the ratepayers ought to get the benefits of the incremental revenues that came to you by the use of that line," and so generally those benefits get recaptured, and so the utility is indifferent to whether it is efficient or whether it is inefficient, it really doesn't make any difference to its bottom line.

So the problem here is not only that the utility doesn't have the correct incentives, the other question is that if we upset this system in some way, then the utility may well find, or any investor, even if we allow NUT's into the business—nonutility transmitters—if we allow those folks into the business, then the question, why would anyone invest in a transmission circumstance with a State regulator, and I should point out, NARUC is on record as raising this question, why should retail ratepayers continue to bear the residual revenue responsibility? So if they are not, if the States are saying, "Maybe we shouldn't bear this responsibility in the face of a competitive bulk power market and, on the other hand, we don't know what FERC's pricing arrangements are, maybe we will know more 6 months from now, but we don't know now," why in the world would I ever want to invest in transmission? And the answer is because you are a NUT, because the uncertainty is such that it isn't clear that it would be a prudent thing to do from your investors' point of view.

The fourth and what I think are the two biggest ironies of the situation we are in for the third and fourth items which are really ironic, one is this question about what happens if, in fact, it is true—if it is true, and I say it because I tend not to believe this myself, but if it is true that States lack jurisdiction to order retail transmission access or, if you will, to unbundle transmission service and then make that available as a discrete service to retail customers, what is the effect of that?

Well, the first question is, who would raise that question? Well, apart from perhaps some bureaucratic imperialist in Washington, somebody else who might feel offended by retail wheeling may raise that question.

Let's take a hypothetical high-cost utility for example, a purely hypothetical example, in a State that may have contemplated retail wheeling and issued an order suggesting that it might do that in the future or was thinking about doing it, and hypothetically, of course, on an experimental basis so we limit the hypothesis. We have a high-cost utility which says, "I'm going to challenge your jurisdiction," and they win, and then the CEO discovers he forgot to ask the lawyers what they win when they win, because if you are sitting on a State commission and this question comes to you, the instant thing you do is, "My God, my predecessors and I have been wrong for 100 years; we have been putting transmission in rate base, and now we find out we have no jurisdiction." So what we do is, we take it out of rate base, and some intervenors would undoubtedly come in and make the irresponsible argument, "We are entitled to 100 years of refunds because you never had jurisdiction to impose those costs on us."

So what happens is, instantly the company finds transmission is not in rate base, and the answer is, you go to FERC, which presumably under this sort of scenario has jurisdiction to set transmission rates, we have a highly contested hearing which lasts 2 or 3 years, and by the time we get the rates, the confusion in sending price signals into the grid and the evolution of a bulk power market in that State is going to be set back considerably, and the utility investors are the ones that are going to be the ones that have been put at risk.

"And, by the way," say the State commissioners while they are contemplating this issue, "you know that high-cost nuclear plant that you have got? Now that you have opted yourself for disaggregating your assets, and disaggregating yourself for ratemaking purposes and you want to avail yourself of the FERC transmission regime, we think now that you have made that decision and you have beaten us in court and we don't have the authority to prevent you from doing that, we are going to give you the benefits of that, so therefore we will take your high-cost nuclear plant, and you can sell it to us, but you have got to compete. So what you are going to sell, it is at market value, not book value, and you have chosen the disaggregation route, and that is the route that we are going to follow."

That is one of the big ironies. For utilities, for example, they may take the position that States don't have this authority. It could be the approach that is the quickest way to disaggregating themselves as a vertically integrated utility.

The final irony, and it is one that you, Mr. Chairman, raised this morning, and almost funny, but it is true. If you have an intrastate transaction where an end user is selling directly to the customer, and you asked Chair Moler this question in regard to where the voter's level of an industrial customer takes may be transmission level, and where is there any State jurisdiction in that transaction, and the answer is, FERC and the States have switched roles, because in that example, if FERC has the ability to price retail transmission, then FERC prices it right down to the customer's premises.

On the other hand, the generator now selling to an end user not making a sale for resale has now removed himself from Federal jurisdiction and become State jurisdictional because as long as he is in the same State, because it is not a sale for resale, so in essence what you have, rather than unifying the jurisdictional regime, we have sort of reversed roles, and so the State regulates the generation and the Feds regulate the transmission, and then the question about that also becomes: What is the utility's obligation to serve in that circumstance? What if the utility says, "This guy doesn't want to buy from me, and so therefore I don't have an obligation to serve," and I think that is a legitimate question for the utility to raise, because there is kind of an asymmetrical obligation here.

Does the utility have an open-ended obligation to serve when the customer chooses not to buy? It wouldn't be reasonable to expect a utility to continue to build as if that customer were there when, in fact, that customer is not there.

So where does all this—where do these dilemmas lead us and these ironies lead us? And the answer, I think—I am not convinced (a) that legislation is needed or (b) that legislation would necessarily be a good thing to do on this subject.

The experience of trying to deal with the issues of regulatory Federalism in the context of legislation is not one that leads me to think it has great promise for resolution, definitive resolution, in the future, but I do think—and I want to reiterate what Charles Stalon said, that there are agencies, particularly DOE and the FERC, that can take a lead on this issue, and I think there are within the regulatory authorities that are held either by the FERC or by the States or by there is the ability to pool this together in some sort of coherent process.

As Commissioner Bailey noted this morning, there is a provision for joint boards in the Federal Power Act as it exists. It has never been used. FERC's staff lawyers for years have had problems with it. Maybe there aren't problems with it, but there are other things they could do apart from the informal dialogues that have gone on and off over the years, a formal task force, joint hearings, raising the question of regional regulation.

FERC is promoting, as Commissioner Massey noted, the formation of regional transmission groups. It is not a very far step then to say let's go one step beyond that and let's bring the Governmental regulators, State and Federal, together into some kind of regional forum to address these kinds of issues. Joint protocols and mutual deferences, for example. Why couldn't FERC ask the States in exchange for—why couldn't there be an exchange between FERC and the States where the FERC says, "We will defer to the States who have comprehensive planning proceedings on transmission, such as Wisconsin. As long as they are held on a certain periodic basis, we will not issue 211 orders, and we will require potential applicants to participate in the State planning process because they will get access that way and it is consistent with sound planning over which we, the FERC, have no jurisdiction."

The flip side is, FERC could ask in exchange, and it would be reasonable to ask in exchange, that State siting agencies when there is a 211 order not go back and relitigate the need issue and say that is determinative of need, and they simply recognize that, and then they evaluate the rest of what they have to evaluate in siting decisions.

I use that as an example. There are a whole series of protocols that could be developed between FERC and the States that I think would allow the development of coherent policy, but what is necessary is, there has to be a collaborative approach. Academic debates or legal academic debates—worse than academic debates, legal academic debates over who has jurisdiction over what transaction and asking the courts to resolve it, I think, is a very dangerous step because I think what gets lost in that is, we make narrow legal arguments as opposed to trying to develop sound policy options, and I think the regulators need to collaborate, and it would be very useful, Mr. Chair, for you and your colleagues on the Hill to strongly encourage, if not implore, the FERC and the DOE, which is in the bully pulpit, and the States to get together and cooperate.

Thank you.

[The prepared statement of Mr. Brown follows:]

STATEMENT OF ASHLEY BROWN, EXECUTIVE DIRECTOR, HARVARD ELECTRICITY POLICY GROUP, JOHN F. KENNEDY SCHOOL OF GOVERNMENT, HARVARD UNIVERSITY

Mr. Chairman and Members of the Committee, thank you for the opportunity to testify on changes in the electricity industry since the passage of the Energy Policy Act (EPAct) in 1992. I would like to congratulate the committee on choosing to hold a hearing on such a timely subject. The effects of the EPAct on, and the enormous changes in, the electricity industry merit a great deal of attention. We are certainly devoting a great deal of effort to studying the subject at the Harvard Electricity Policy Group at the Kennedy School of Government at Harvard University, where I serve as Executive Director. Prior to joining Harvard I served as commissioner of the Public Utility Commission of Ohio and Chair of the Electricity Committee of National Association of Regulatory Utility Commissioners (NARUC) . My comments today, of course, reflect my own opinions, and do not necessarily represent the views of the NARUC or the Harvard Electricity Policy Group.

The passage of the EPAct facilitated and expedited the preexisting-trend toward a new competitive era in the electricity industry. The development of a competitive wholesale market is well under way. Competition is also a reality in many retail markets, and some State Public Utility Commissions (PUC's), like those in Michigan and California, are proposing to implement plans for retail open access. Even without retail wheeling, large industrial customers, local governments, and others are already seeking to use other competitive means to reduce the price of electricity through procurement of power through alternative sources to the local franchise utility by such means as moving production lines, self-generation, cogeneration, or municipalizing. Competition in electricity is already present and is certain to intensify in the future.

The task ahead for both the Federal Energy Regulatory Commission (FERC) and the PUC's is to develop coherent policies which will smooth the transition from the traditional vertically integrated monopolies to the emerging competitive industry. While the trend is clear there are major impediments in the way of the transition to and perhaps even to the evolution of a fully competitive electricity industry.

It can be stated unequivocally that transmission policy can never achieve coherence or predictability without coordinating Federal and State policies and practices. The current jurisdictional allocation and practice is simply incapable of providing coherence. The FERC has jurisdiction over wholesale pricing and access to transmission facilities. On the other hand, the FERC has no authority over the planning and siting of, and retail access to, transmission facilities. These are domains over which the PUC or, in some cases, other agencies in State government, have exclusive jurisdiction. While the pricing of retail transmission services has been done by the State commissions for over a century, proposed unbundling of those services have led to more controversy over whether States have had the jurisdiction they have been exercising for so long. Furthermore, in exercising their respective jurisdictions, the FERC and the PUC's take remarkably different approaches. The FERC approaches transmission as a discrete service providing the incremental use of existing facilities for which the residual revenue responsibility has been assured by native load, primarily retail ratepayers. PUC's, on the other hand, approach transmission issues as part of unbundled electric service for which retail customers bear all prudently invested costs. Any revenue garnered from non-retail use of facilities (e.g. wholesale wheeling) serves to offset the retail customers. This current structure, quite obviously, leads to a fragmented, highly contentious, and often unpredictable approach to regulatory decisions affecting transmission services.

A clear example of the consequences of the fragmented approach occurs when the FERC mandates access to transmission facilities and its order falls into the "black hole, in the EPAct, namely the interface between the FERC's access powers and State siting authority. The EPAct says that, on request, FERC can mandate that a utility provide access to an eligible party under section 211 who seeks it, but if that party from whom access is sought can show to FERC's satisfaction that they cannot provide that access with existing facilities, they then have a good faith obligation to go to the State siting authority and get that line sited. The State, even without any parochialism or NIMBY perspective, is then faced with a real dilemma.

The State's siting decision makers have historically made two decisions when they sited and certified a line. First, they have decided there is a need for that facility and that the facility ought to be built. Even if you consider this decision to be preempted by the FERC access order, there still remains a whole other set of issues that States have examine including: environmental issues, aesthetic issues, and public health concerns relating to the siting of transmission lines. These issues should be balanced against the need for the facility.

The second decision is whether the transmission facility should be included in the retail rate base, thereby placing retail customers in the position of guaranteeing the residual revenue requirements to the utility for that facility. These issues—how to site and how to pay for transmission—present two very different decisions. On the one hand, there is a very compelling argument for the FERC to be able to order access once it has established there is a need for it. But, on the other hand, does that mean that the State then ought to make retail customers bear the residual revenue responsibility if FERC happens to use a pricing scheme that doesn't fully compensate the utility? Can the utility assume it can come to the retail customers to pay for it whether or not the State finds that the retail customers need this facility?

The traditional duality in State siting decisions may need to be decoupled by separating access orders from the question of who pays. Decoupling could well mean that a utility might be ordered by the FERC to build a line, without knowing whether it would be compensated for the facility. If I were managing a utility I'd be very concerned about being subject to an access order when I have no idea whether I am going to be adequately compensated for it. But from the State regulator's perspective, even if you believe the line is needed, or are willing to accept that proposition as inherent in FERC's access order,if you do not believe there is a retail need for the facility, why in the world would you ever want to impose the residual revenue responsibility for it on retail customers? Even without an access order, States can force this issue, although none have yet done so by doing load flow studies and disallowing from rate base that proportion of transmission investment which is used for non-jurisdictional transactions. As long as the scope of Federal and State jurisdiction remains uncertain in this area, and recovery of investment is made less certain, underinvestment in transmission facilities, which could eventually reduce the vitality of a competitive bulk power market, seems a probable result.

In terms of transmission siting, cooperation between the States is needed as well as cooperation between State and Federal regulators. Some form of regional regulation should be encouraged since transmission planning and siting is a regional rather than simply a local issue. Regional Transmission Groups (RTG's) can provide some of the benefits of regional planning, but it must be remembered that RTG's are self-regulating institutions and they cannot either replace or determine the decisions made by government regulators. Persuasive evidence of this is the fact that no RTG has emerged which professes to decide important transmission pricing issues. There is not enough unity of interest on difficult jurisdictional issues for the RTG's to be an effective decisionmaking process in those areas where governmental authority is required.

A second example of jurisdictional conflicts occurs in the context of transmission pricing. On the one hand, the FERC has jurisdiction over wholesale sales of electricity in interstate commerce under section 201(b) of the Federal Power Act (FPA). The FERC examines transmission pricing in discrete terms, namely as one of many unbundled services a utility may provide. On the other hand, PUC's deal almost exclusively with vertically integrated utility services. PUC's typically use bundled pricing by throwing transmission costs into retail rates and treating it like any other utility asset in the rate base. Without ever thinking about it, what States have done is to impose on the retail ratepayer the residual revenue responsibility by essentially guaranteeing the utility recovery of the cost of its transmission investment. If the present system continues, States may well start to ask questions such as: if Ontario Hydro sells a thousand megawatts to New York State and an average 40 percent of that transaction crosses the transmission system in the rate base paid for by Ohio ratepayers and drops not a single penny of revenue to offset Ohio's revenue responsibility, why should Ohio's ratepayers continue to pay for 100 percent of the cost of that facility?

The introduction of retail wheeling complicates the jurisdictional equation even further. It makes little sense to price unbundled retail transmissions services differently than unbundled wholesale services. That is one of the arguments advanced by proponents of Federal preemption of all transmission services. on the other hand, the centerpiece of transmission pricing for the entire history of the industry has been the assumption of the residual revenue responsibility by native-load, predominantly retail, customers. To alter that arrangement, and the case for change is compelling, without careful coordination between Federal and State regulators would necessarily complicate and discourage investment in transmission pricing, particularly given FERC's historic commitment to average, embedded cost pricing. Unbundling by preemption is almost certain to yield chaotic economics in transmission for years and seems almost certain to make the siting of new facilities, already a difficult task, a virtual impossibility. That is the case when preemption is easily done; it is by no means clear that it is easily done in the area of transmission pricing given the interrelationships of pricing and access itself, and the terms and conditions thereof.

Another area in which there is a need for collaboration between Federal and State regulators with regard to jurisdiction is in the context of registered holding companies. First, as exemplified by the Ohio Power [1] case, there is an overlapping jurisdiction between the SEC and the FERC with regards to registered holding companies. The effect of the Ohio Power decision is that the FERC, and by extension the PUC's as well, are preempted from disallowing purchases approved by the SEC. However, leaving ratemaking decisions in the exclusive hands of the SEC is not wise considering that the SEC has little experience in these types of decisions and, as noted in footnote 1, rarely holds administrative hearings to inquire into the propriety of such transactions.[2]

Another example of the jurisdictional confusion over registered holding companies by the courts is the Mississippi Power [3] case. In this case, the Middle South Utilities (MSU) holding company formed Middle South Energy in order to construct the Grand Gulf Nuclear Plant and to sell the electricity generated to MSU subsidiaries. The FERC approved the plan but the Mississippi PUC would not allow the costs to be passed through to the retail rates. The Supreme Court ruled that the costs approved by the FERC cannot be "trapped" by the Mississippi PUC and that States are prohibited from substituting their own determination of what is "just and reasonable" for the FERC's determination.

By denying the PUC's the authority to determine if parent-subsidiary transactions should be passed on to retail consumers, both the Mississippi Power and the Ohio Power cases are anomalous because they reinforce monopoly power in an era of increasing competition. Indeed, the effect of those decisions has also served to shield very high cost investments by utilities from prudence reviews. These jurisdictional decisions by the courts can only retard or confuse the development of a competitive industry.[4] In order to properly supervise transactions involving registered holding companies which are not at arm's length, both the FERC and the PUC's should work together. For utilities that are not registered holding companies this can be accomplished by codifying or otherwise formalizing the Pike County [5] doctrine. This doctrine allows State regulation to determine if wholesale purchases between a parent and a subsidiary should be passed through to retail customers. This protection is needed since parent companies may purchase power generated from their subsidiaries when cheaper sources of power are available, or utilities are simply being imprudent in their purchasing activities. For registered holding companies with their multistate integration requirement, Pike County ought to be replicated on a regional basis.

The failure to codify the Pike County doctrine has produced three critical results: (1) it increases the attractiveness of retail wheeling proposals as a way of disciplining retail rates; (2) it makes IRP less meaningful since the State's power to enforce IRP is unclear for most utilities and absent in the context of a registered holding

[1] Ohio Power Company v. FERC, 954 F.2d 779 (D.C. Cir. 1992). In order to prevent abuses involving transactions between a utility holding company and its subsidiary, the SEC has authority, under section 13 of the Public Utility Holding Company Act (PUHCA), to set prices for the sale of coal to from Southern Ohio Coal Company (SOCCO) to Ohio Power. The SEC set the price for the coal at cost, but the FERC used a market test and found that comparable coal could have been purchased at a cheaper price. Ohio Power claimed the FERC was precluded from disallowing the purchases because they had already been approved by the SEC. The Court of Appeals for the D.C. Circuit, on remand from the Supreme Court, held in favor of Ohio Power by declaring that the SEC had rightfully exercised its authority and that the FERC must respect this grant of authority to the SEC by not interfering with approved purchases, even though the SEC has failed to review either its test or the prudency of the costs incurred in its original 1970 decision.

[2] Ironically, in this case, the SEC claims it set the price at cost on a recommendation from the Ohio PUC but the Ohio PUC has no record of ever giving such advice.

[3] Mississippi Power & Light Co. v. Mississippi ex rel. Moore, 487 U.S. 354 (1987).

[4] Indeed, the case make it clear that if jurisdiction over retail transmission were left to the courts to decide, clarity may never emerge.

[5] Pike County Light & Power Co. v. Pennsylvania Public Utility Commission, 465 A.2d 735 (Pa.Cmwlth. 1983).

company;[6] and (3) it diminishes diversity of regulation on the purchasing side of the market which in turn retards the growth of a viable competitive wholesale market.[7]

It is readily apparent that on these issues of transmission access and siting, transmission pricing, and registered holding companies there is a compelling need for collaboration and cooperation between the FERC and the PUC's. There is, however, no need to legislate at present if regulators themselves take the initiative to find ways to collaborate. Absent collaboration the problems inherent in the transition to competition will multiply quickly and legislation may become the only solution. As the current jurisdictional system now stands, neither Federal nor State regulators have sufficient jurisdiction of their own to formulate and carry out coherent policies which ease the transition toward a competitive marketplace. Historically, since 1981, the FERC has consistently refused to meaningfully collaborate with the States on matters of substance. The FERC has treated State regulators as parties rather than peers in comity and hid behind ex parte rules to avoid serious discussions. It has never utilized the Joint Board provision in the FPA, nor ever conducted any type of alternative joint proceeding other than formalized discussion sessions. The new membership of the Commission currently in place, however, offers the real possibility of putting substantive cooperation and collaboration ahead of bureaucratic turf protection. The Commissioners should be encouraged to take the statesmanlike approach.

In this respect, I would like to propose ways in which the Federal and State agencies might be encouraged to work together. First, there is built in to the FPA a heretofore unused joint mechanism to handle difficult jurisdictional problems. This mechanism, or, if it is deficient, some other cooperative mechanism, should be explored and taken advantage of. Second, protocols or mutual deferrals can be used whereby the FERC maintains the jurisdictional authority to review State decisions but defers to the judgment of the PUC's in the absence of bad faith or abuse of power. This was the compromise reached in the FERC's Consolidated Edison decision.[8] Other types of deferrals could be established where States agree, for example, for purposes of siting a transmission line to allow a FERC section 211 access order to be determinative of the "need" in exchange to FERC agreeing to defer to comprehensive State transmission planning in lieu of separate and redundant section 211 access proceedings. Similarly, RTG's could be greatly enhanced if FERC and the States in a given region convened Joint Boards, or other collaborative mechanisms, to contemplate RTG regional pools or other regional actions in order to remake needed governmental decisions that are, at least in part, subject to multiple jurisdiction (e.g., an IPB selling directly to an end user, siting a new transmission facility made necessary by a section 211 Access order by the FERC, utility mergers). Joint activity, in such areas as transmission pricing terms and conditions, as well as determinations on the recovery of stranded assets, would be far preferable to protected legislation over bureaucratic turf. Third, a measure of comity should be developed between the FERC and the PUC's. State PUC's should be treated not as mere parties before the FERC but should be given deference similar to what the Federal courts give to prior State court decisions. States should also work with each other to a greater extent on issues like transmission siting and planning.

Finally, the decision-making process itself should be reformed. Most of the decisions regulatory agencies make are legislative in nature—prospective decisions like ratemaking and tariff filing—and therefore administrative decision-making should be less judicialized.[9] To accomplish this objective, ex parte rules should be relaxed to allow the FERC and the PUC's to consult informally on important technical matters of mutual decisionmaking interest and develop a coherent approach for the regulation of a rapidly changing industry. Furthermore, sunshine laws should be read in ways that maximize the opportunity for Federal and State regulators to meet privately in order to coordinate or reach compromises on important policy issues. This would also avoid sending premature and misleading signals to the investors, utilities, and consumer groups. Of course, these observations are equally true with regard to intra-PUC communications and inter-PUC communications as they are with regard to FERC-PUC communications.

---

[6] It is one of the great ironies of EPAct that it requires PUC's to consider the implementation of IRP but fails to vest commissions with the authority to make IRP determinations binding.

[7] An irony worth noting is that retail wheeling deprives FERC of jurisdiction over a generator which might otherwise be engaged in a sale for resale. Hence, while some have argued that States lose transmission jurisdiction when retail rates are unbundled, it appears that States may gain more jurisdiction over generation in a retail access regime.

[8] Consolidated Edison of New York, 15 F.E.R.C. 61,174 (1981).

[9] See Appendix A, "The Overjudicialization of Regulatory Decisionmaking" by Ashley Brown.

A related Federal/State issue in the electricity industry which needs to be addressed is whether the utilities should be allowed to recover stranded assets and who should decide that question. Good arguments can be made on both sides of the issue. Those who advocate that no recovery should be allowed for stranded assets point out that structural change in the electricity industry was foreseeable and investors knowingly took risks and that these risks are reflected in the allowed rates of return. On the other hand, those who advocate partial or complete recovery of stranded assets declare that the investments were undertaken on the behalf of ratepayers, were declared to be prudent, and that the swiftness and the magnitude of the regulatory changes was unforeseeable.

Whatever you think of the arguments on both sides of the issue, the ultimate question in this debate is who is going to bear the risks of competition—the investors or the consumers. Clearly, who bears the risk of competition is fundamentally a policy decision and the question should be treated as such. Sound public policy, however, dictates that to whatever extent government action caused actions to be stranded, that some level of government be responsible for bearing the consequences produced by the decision. In short, if State policy causes investments to be stranded than State governments should decide the question of who pays for stranded assets. The same principle holds true for Federal actions. A regulatory regime should not be created where one level. of government makes the decisions and another is forced to deal with the consequences. Accountability requires jurisdictional symmetry.

In conclusion, the jurisdictional mechanisms that regulate and provide policy guidance for the industry are neither reflective of nor very useful for the emerging structure of the industry. The debate over the EPAct in 1992 made it clear that regulatory federalism is not an easy area for Congress to legislate. It is clear, however, that Congressional action may not even be necessary or desirable if the regulatory community gets its own house in order for dialogue, collaboration, and coordination. It could be desirable for the Congress to encourage, indeed, charge the FERC and PUC's with the responsibility of producing durable policy guidelines that investors, consumers, and other can rely on without a lengthy and protracted dispute over jurisdiction and turf. This Nation deserves statesmanship rather than bureaucratic turf battles. Congress would do well to encourage such an approach.

Mr. SHARP. Well, thank you very much.

Obviously you have brought us plenty to think about. A hundred years of imprudence makes some of these problems look small. But I appreciate your ironies, and as several of you recognize, there might well have been other issues. Indeed, at least one of you on the panel urged us to deal with some of the other issues when EPAct was before us, and rationally I think that was a wise recommendation; politically, it didn't seem as easy at the time. Our ability to master either intellectually or politically these broad things is very, very difficult.

But I was struck, and let me just quickly return to that question. I am not sure that Dr. Hogan said anything about legislation, but I know both of you indicated there were possibilities, but I got the feeling that it is not imperative, that you see, working out lots of these issues, it is not as if we have to do something. Let me just express my own concern about these kinds of things. Maybe having been here for 20 years, it has given many a sense of reluctance that many people say, "Well, gee, why don't you just legislate and solve this problem," and it is usually because, given the nature of our pluralistic system, multiple interests always have an interest in what you are doing, and therefore you rarely do what is asked exactly, and you usually do many more things as well, so you usually don't get the result that one starts out with. Therefore, my own theory is, when you don't have to rely on Congress, you are better off.

But my real question is whether you see with us a legal structure in PUHCA or FPC—I mean the Federal Power Act—where there is some fundamental need that, say, within the next 3 or 4 years that really must be addressed if we are to unravel this, some major hurdle that you feel strongly about.

Mr. BROWN. Just two quick points. One, in my written testimony I do address one area that I do think needs legislation, which I know you are working on, which is the pike—which is not the Pike County—Pike County would be helpful, but actually it is the Ohio Power fix. That is a problem. We have got the SEC and the FERC, and presumably the States are in the same position as the FERC in that. That is a real problem.

Mr. SHARP. As you know, Pike County was a classic example where—the Mississippi case where we just couldn't get enough people on the same side. You might have a different perspective on that. It seemed to us at the time that we couldn't get enough people on the same side to deal with it in the context of 100 other issues.

Mr. BROWN. The one I really can't resist mentioning about that is another provision in the EPAct. You are essentially requiring States—Congress is requiring States to consider doing IRP, and then in another provision it is making sure it doesn't have the authority to make sure that IRP decisions are then binding on the utilities.

Mr. SHARP. You know, we ran into obviously the Federal limitation we had from PURPA to begin with, which was, we sought over the years to be sensitive to State power, that we could not demand them to do certain things.

Mr. BROWN. But let me answer on the subject of what you said, and that is, as you correctly pointed out and as you anticipated during the EPAct debate, once FERC had the ability to order access, they wouldn't have to use it very much, they could keep that power largely sheathed, and we know there have been only 11 requests yet transmission access is obviously a lot more—it is greatly facilitated today than it was before the act.

But what is interesting here is, in many ways it might be the same thing here. If Congress were to look at this issue of the problem of regulatory federalism, my guess is that the regulators, both FERC and the States, not knowing what you were going to do, not each understanding fully what the political forces that are going to be unleashed by that debate are, may well find ways of cooperating that heretofore have been objectionable, and the fact that you are looking into that issue and raising the specter of legislation that may gore somebody's ox or may gore everybody's ox—presumably that is what it will do, everybody will be upset over some provision—it seems to me would be very useful.

It would also be useful, I think—and then DOE has begun to do this by creating this electricity forum in November, and presumably there will be follow-ups. But DOE is clearly in the bully pulpit. They have no real power here, but in terms of providing some leadership they could certainly do that, and I think it would be useful, especially if they worked in connection with some proceedings that were going on on the Hill, looking at these issues and trying to decide whether to act and, if so, how so.

Mr. SHARP. I didn't mean to rule out, because I fully agree with you, the threat of congressional action, rational or irrational, sometimes does amazing things.

Mr. STALON. I would mention three areas that I think we need to think about legislation, and while I think it is correct that we haven't truly tested the powers of the FERC to bring about all of these changes, I also think that if we reconceptualize this industry as Professor Hogan has suggested, what we see is an industry where power doesn't flow from a company's generation through its transmission lines through its distribution companies, power flows from a generator into a grid and out of a grid into distribution companies, and there is essentially where reliability for all of us depends upon the robustness of the transmission grid and the law of large numbers. We have a lot of generators attached to that.

If we think of that model for a moment, we recognize that the principal regulators giving us reliability for the last almost 60 years and clearly for the last 30 have been what I call QUANGO's stealing an anachronism from the British, quasi-autonomous non-governmental agencies. They have truly planned this transmission system, they have planned the rule of the road, they have planned how it is going to function.

The QUANGO's are being subverted by competition and rivalry among the generator owners. We need to make sure that we preserve those organizations until we get something in place.

Clearly the FERC has many, many chips based on the fact that the regional transmission groups will be able to continue to take over that function, because you must coordinate not merely the investor-owned utilities but also the munis the co-ops, the TVA, the Bonneville Power Administration, the Canadian utilities, both privately and publicly owned, and the northern Mexico utilities. There is a sense in which this is an international regulatory organization that the industry has evolved.

If that organization is, as I believe, being subverted by competition and the RTG's, as I believe, are not likely to have the clout if we think we are going to get all parties involved, we are going to have to make sure that we impose on somebody—and my candidate would be the FERC—full responsibility for preserving and maintaining the ability to have a transmission grid that is robust and it can provide us the service that we have come to expect of the utility industry. That I think is important.

The other two I would mention very briefly. I still think one of our serious problems is that we have laws at many, many State levels defining electric utilities that make it very difficult to create a competitive, efficient generating industry as long as a generator becomes a utility as soon as it sells power to a particular person in that State. That means that the only generators who gain great freedom are those who always sell to somebody in another State, and I think that is an inefficiency which we need to try to overcome, and it is maybe possible that DOE and FERC could provide leadership to persuade the States to change it.

The third one I would very quickly say that only the Federal Government can do, and that is, we have to integrate all of those customer-owned utilities and publicly-owned utilities like the Tennessee Valley Authority and the Bonneville Power Administration, and the other power marketing agencies, the REA's, the munis, all of those have to be integrated into the system. Their roles have to change.

I think when Congress drafted the Energy Policy Act, they were focusing on the investor-owned utility sector, properly so. They own about 70 percent of all the assets and produce about the same amount of power. But as a matter of fact, they changed the economic environment in which all of those Government-owned utilities and customer-owned utilities must operate as well. Their futures have been changed, and they do not fall neatly under any of the existing regulatory bodies. They have been players in the QUANGO's, and they will continue to need to be players, and I think that one of the reasons that FERC has placed so many chips on the regional transmission groups is, that is the only way they know, given their limited authority, to create a new regulatory body that can continue the functions of the old QUANGO's. They want a new QUANGO.

I will stop there.

Mr. SHARP. Dr. Hogan, I don't know if you had a comment on the legislation question.

Mr. HOGAN. I would say I haven't thought very much about new legislation and I have been concentrating—I think there is a great deal to do within the authority that has already been granted, and that is where I have been focusing. I haven't confronted anything yet that I am sure requires legislation.

I can imagine scenarios, some of the ones you have heard about, where we would go down with the current authority and decide that it didn't work and you needed to do something else, but I am open to the possibility that we can still do it within the current legislation for some of these things. So I am focusing on that problem.

Mr. SHARP. I would like to return to you on what you just briefly indicated, your proposal on the tariff schedules that the State of California might use as one technique to get at the question of direct access. Could you elaborate on how that works and how that is distinguished from the POOLCO proposal? Obviously they are two quite different proposals, but I would like to hear.

Mr. HOGAN. We can set aside the pool operation. That is: How does the wholesale market operate? Whatever we do in the wholesale market will get an arm's length spot price, that will happen. So now let's assume that that has happened, which we don't quite have yet, but I read yesterday or today before that Citizens Power up in Massachusetts was in the process of trying to put together an index which would be the spot price index of just this type for just this purpose so that people could start writing contracts around it and all that kind of thing.

Perhaps the simplest way to explain it is to compare two things. One is traditional regulation, and the argument made by people who are opposed to direct access is that we should keep the current regulation so everybody should stay as a customer of the distribution company, the utility, and the distribution company should purchase and then sell on to them, and if we did that, then we would be able to preserve demand side management, environmental protection, recovery of historical investments, no new stranded, and all of those things would more or less—problems would be greatly reduced. It wouldn't solve every problem that the industry has, but the argument they make is that we could continue that.

Let's accept that. Suppose that they are correct in that argument, and if everybody stayed as customers of their local distribution company, all of these other problems would be greatly reduced or vanish completely.

Second, let's look at the British system and how do they provide direct access, or retail wheeling, to customers who want it. The details are a little different but in essence what happens is, the local distribution company charges a fee for everybody for managing the wires and delivering things to them, and then they monitor how much electricity the customer consumes every half hour, and they bill them the spot price from the British spot market for that half hour price, and that is what the local distribution company does. Then the customer independently and separately goes and signs a long-term contract, if they choose, with some generator, and the structure of those contracts is what they call contracts for differences where, if you want to get a fixed price for the electricity that is coming in, you sign a contract for, say, four cents with the generator, and then if the spot price turns out to be three cents, you buy from the spot market for three cents, the generator sells to the spot market for three cents, you owe them a penny, and the reverse happens when the price goes above, and this is a way of getting stability in these commercial arrangements. It also means that on days when the generator isn't running, it doesn't affect the customer, they still buy from the spot market, and they still do the financial transaction with the generator. If the customer takes more than the contract amount, they pay the spot. All the other services that come in the system are handled and provided by the pool, and it is consistent with what Charles described before of everybody selling into the grid and then everybody is buying out of the grid and you are not actually trying to keep track of matching individual generators against individual customers. It is a great system they invented there, and they use these contracts for differences. It is a very fundamental idea, it is a different idea. It was mentioned in the California order that there was an interesting idea here.

Now you go back to our system in the United States, and you say, well, why do we need to get contracts for differences? Well, suppose we took our current system and we have customers that are customers of the local distribution utility, and they have a program, and they have cost of service and so forth, and one of the things they do now is, they have time-of-use rates for customers. The only difference is that now they kind of estimate what they think the marginal cost—there's a lot of complicated things that they do to figure out what to charge them every half hour.

Suppose we took that form of regulation but, instead of estimating what these charges were every half hour, we charged them the arm's length spot price, that is all, nothing else changes, they still pay demand charges, and customer service charges, they are part of the demand side management programs and environmental—whatever you choose to do can be as part of the distribution company's service, you don't reclassify the customers, they don't leave the system in any sense, and we know they are not really leaving, they are really there and the electricity is still flowing in the same way it did before. So there is no reclassification of the customer, there is no separation of a transmission charge, there is no appeal to FERC, it is just using rate design, which we have been doing forever, with a very modest change where the marginal cost that is used at the half hour is the arm's length spot price.

Now if you think about that, you say, well, that makes this customer exactly like that customer in England relative to its market. He is paying a fee for using, you know, the rest of this cost of service, he sees this spot price every half hour, there's generators on the other side of the deal that are selling into that same wholesale market. If they want to, they can seek long-term contracts for differences completely in regulated contracts. All of the backup services and everything are going to be provided through that wholesale market in effect and through these spot sales.

So you can create the functional equivalent, exactly the same thing, and everybody gets the right incentives for investments and all these kinds of things that are consistent with the British model, but you can do it within the American context of State regulation by using it as this tariff mechanism, and you don't have to then separate the customers and create these politically vulnerable problems of competition, transition charges as access charges to the wires and it has the label uneconomic, we are going to collect the uneconomic costs this way, which would last about—I don't know. I mean the fear is, it won't last very long and very politically vulnerable.

But if you believe the previous argument that if everybody stays as a customer of the local distribution company and we continue cost-of-service ratemaking with this one little tariff change, then all of these problems become much less severe and maybe even vanish, and so when you look at it that way, you realize that direct access really is a much more modest institutional change than we often think. It gives you great advantages in terms of commercial operations and customer choice, but it isn't that hard to implement and it doesn't raise all of these problems that we think about when we go through the fiction of saying that the power came from that power plant and not from one over here and I am now separately wheeling through the wire. None of this happens, that is all fiction, and we create this elaborate fiction called retail wheeling which then runs into our fictions about how we distinguish between Federal and State, and real money is going to be shifted around based on those fictions. Why don't we just describe what it actually is, which is direct access to the spot market price and the wholesale market.

It turns out you don't have to create the fictions, that is what really is there, and it doesn't raise all of these jurisdictional questions and you can continue whatever other programs you choose for demand side management, environmental issues, and so on.

Mr. SHARP. Well, that seems to have a lot of appeal. The question that I don't think I understand or am not clear about is why it eliminates all these problems from the high-cost generator. I mean it seems to me the high-cost generator is in trouble if there is available to the buyer, whether it is only in the paper transaction as I hear you describe it or in the more physical effort to wheel—isn't that generator going to run into some kind of problem that he no longer—that spot price is probably going to be less than all those average nuclear and PURPA contracts, let's say?

Mr. HOGAN. Let's separate the case of those situations where you are talking about high-cost generators who don't have retail customers. That is part of the wholesale market now, and that is a different problem. We are talking about the retail——

Mr. SHARP. I am thinking about one particular utility in California, but there may be a couple in California that worry about this sort of thing.

Mr. HOGAN. Truly. They should worry about it. Those high-cost plants are part of their rate base. That gets included in their cost of service.

Mr. SHARP. To every customer.

Mr. HOGAN. To every customer.

Mr. SHARP. All right. So the key here is that that industrial customer out there today in California who is in there testifying and hoping to be free of that rate base is, in fact, simply this would reduce its opportunity to substantially cut it costs.

Mr. HOGAN. It certainly would have all of the opportunities it has today in terms of negotiating and so forth.

Mr. SHARP. Right. But their goal is a bigger opportunity. My impression is, political and economically, what they see is a bigger differential that they can seize. But of course what California Public Service Commission, to make its plan work, may have to do is reduce that anyway. Obviously, I am creating the fiction of what they think, but—and yours is just a different technique as to how to spread those costs.

Mr. HOGAN. It is functionally equivalent to what was called the competition—if you did what the Commission said it was going to do and if you believed in the political will to preserve it forever, it is the same thing; it is the functional equivalent of that. But it is very different structurally in the form that it takes, and it provides a way of doing it within our institutional structure that doesn't raise all of these jurisdictional and legislative problems and is naturally phased and so on.

Now it does not—and I said in my written testimony, one of these things that this does not do is, it does not produce an immediate reduction in rates, and the Commission in California, I think, recognizes that, and that is one of the characteristics of their proposal, it doesn't produce an immediate reduction in rates, but it does provide the opportunity for all of the choices that get made by customers for investments and new investments to be made in a way that they have direct access to the market and then they can phase in over time.

We can deal separately with the question, do you spin off the existing generation or not. Those are all important matters, but they are really separate from this question of how do you provide direct access. So I think there is actually an easy way to do it.

Mr. SHARP. It strikes me that what we are dealing with here—I mean what I have obviously gotten focused on is the expectations game. Some of the most intense advocates for direct access have this hope that they can get over to a low-cost generator as opposed to a high-cost.

But the State of California, in hoping to provide this new opportunity, is recognizing—or one presumes it will certainly be bombarded with the argument that, wait a minute, you just can't let them all run away from these costs, we will have to spread those costs somehow, how do you think we ought to do it, and what you are saying is this would—I don't think any solution is going to reduce the window of opportunity that some of the staunchest advocates are hoping for.

Mr. HOGAN. That is correct. The California commission called it—their way of dealing with this, the competition transition charge and the uneconomic assets, and if you could do it exactly, you know, and put it on and everyone could figure out the wires, estimating that number is hard, and so on. There are a lot of technical complications, but conceptually it would work.

But the problem is that in our system the act of unbundling and separating that and then saying the customer has now left the utility, no, they didn't actually leave, they are right where they were before, and the power is flowing the same way, but if we claim that they left the utility, then FERC is put into the position of having to exert jurisdiction perhaps and so on, and we are into all that debate.

Why don't we just recognize that they didn't leave, they are still there, and so just treat them as customers but give them this tariff, and that tariff provides the economic equivalent of what we were talking about the other way.

Mr. SHARP. Let me recognize the distinguished ranking member of the full committee, Mr. Moorhead.

Mr. MOORHEAD. Thank you, Mr. Chairman.

Since you are talking about my State, you have perked my interest up.

Many people in California believe that because of the problems we have with the environment and needing to cut smog and so forth, we need to build up alternative sources of energy such as thermal and solar and wind machines and all kinds of things, and Southern California Edison has been buying quite a bit of that electric power.

Now if you give people, especially big business, the direct access to the spot market, you are going to leave them out from this type of energy. Many people think that if you do away with alternative sources of energy, one, you may have more smog, unless you buy the product outside of the area, and if there are shortages of fossil fuels you may have shortages, so if there is a social purpose or maybe an economic purpose of having those alternative fuels, and we all know that, there are some atomic energy plants that have cost more money than they are ever going to get out of them probably, and, so some way or other, they have to be paid for, how do you get around these problems?

Mr. HOGAN. Well, they are good questions, and they are hard problems.

Let me say first, when you go to a more competitive market, some of these things will be taken care of in part by the fact that people will pay real prices that represent the real costs for these activities that are internalized in the marketplace. Average cost prices that we have used for 100 years is a distortion in itself, so removing that will be helpful, but it will leave the possibility of substantial environmental and other effects of the type that you are talking about that would not be incorporated naturally in the market, and it is widely recognized that that is a problem, a market failure problem and conceptually the way you deal with that problem is pretty straightforward. You cannot rely on the market to solve problems which, by definition, are not addressed by the market, or you have to do it some other way. The some other way is typically through some kind of governmental action, and fundamentally you can package it however you wish, but fundamentally it comes down to, you have to raise the money some place, and the place where you are going to raise the money, by definition, has to be on whatever is the remaining monopoly.

Now you can't collect additional funds from the market, you can only collect it from some monopoly or from taxation power. The remaining monopoly is the local distribution utility. That local distribution utility could collect in its bundled rates funds to support demand side management. It could collect in its bundled rates funds to support renewable plants which could then be subsidized and compete in the marketplace.

Mr. MOORHEAD. You are going to drive every large business in the country out of that monopoly one way or the other, because they have ways of getting around it now.

Mr. HOGAN. Then you would choose not to do it, but you can't have it both ways. What you cannot have is above-market activities that are paid for but nobody pays. You are going to have to have somebody pay if you want to have it done, and that is a decision the State regulators can make, which is how much demand side management, how much environmental renewables, do they want, and how much can they pay for?

But there is no problem in doing it in a way that is compatible with allowing competition in the commodity market for electricity. They are not incompatible with each other, but you do have to collect the additional cost on the residual monopoly, and that is the distribution utility.

Mr. BROWN. Can I just add something to that because that is where we run smack into this jurisdictional problem. It isn't at all clear who can put those additional costs there. I think we heard the FERC this morning assert essentially the jurisdiction to do that or at least to pass ultimately on it.

I think there are certainly a number of State commissions that would take a different view of that and exactly which social programs are important, whether it is benign—environmentally benign energy sources that you are giving examples of or whether it is the feeling that the utility ought to recover for stranded investment for nuclear plants or for contracts they entered into with nonutility generators or, for that matter, for simple programs to make sure that low-income people don't get their electricity turned off in very cold or very hot weather. Those costs need to come from somewhere, and the normal argument most economists make is, well, the easy thing to do is let the Government subsidize it, let's tax the service and subsidize it.

Well, as you and the chairman know, imposing taxes is not the world's most popular thing to do or the easiest thing to do, and obviously that coffers of the Federal Government and most State Governments, particularly in California, are not exactly bottomless, and so that is very difficult to do. So the pressure on regulators to impose costs on their residual monopoly is growing at the very time that their authority to do it is most open to question.

Mr. MOORHEAD. They have actually reached the bottom in many instances.

I got in a little late on the discussion of Professor Stalon, but I was interested in the comment you made about Government-owned utilities such as one category—you didn't put them in categories, but certainly TVA and Bonneville are wholesalers for the most part and in a different connection than munis.

I happen to come from a district that is virtually all muni. The Department of Water and Power now, I guess, is the biggest muni in the country. You said they were going to have to come under some entirely different set of regulations or changes than they do now. They are not regulated as far as price by the State, but actually there is little difference between any other form of taxes paid and what they pay for their water and electricity, it is a form of taxation, but they get a product in exchange, and they run the budget in such a way, I am sure, that it all has to come out even in the end. What would you do to change that?

Mr. STALON. I was calling attention primarily to the fact that electric utilities, whether they be municipal or whether they be BPA or TVA or they be investor owned, cannot operate without operating interdependently. Therefore, the industry many years ago evolved into self-regulatory systems laying down the rules as to how transmission lines will be built, how they will be integrated, how generators will be operated, certain obligations that every generator has to provide power upon call of the dispatchers.

This whole regulatory system, while it has been approved by States and probably falls within the category of State jurisdiction, has really never been regulated by States because it is a multiple-State activity. The FERC has never regulated, although it has explicitly approved it.

The problem that we approach today is that the generators in this game are being forced into competition with one another and the forms of control that have existed in the past as self-industry regulatory control is falling victim to the rivalry among the generators who have power to sell and cannot always reconcile their disputes easily.

So my argument was, we must evolve new systems for regulating what might be called the traffic cop function and that air traffic control function, for a good analogy, and in that new regulatory scheme we have to make sure that we don't use the reliability and we improve the efficiency of the old system.

My guess is, as I said earlier, the FERC is trying to accomplish that through the regional transition groups, create new kinds of self-regulatory schemes. I am skeptical that will work, because there are too many players in the game. That is a true natural monopoly, a true utility in the traditional sense of the word. I would like to see its regulation simplified and brought under a corporation where it can be handled.

Mr. MOORHEAD. To some extent these agencies regulate themselves. I know that California is a different type of a situation than you find most places in the country because of our environmental problems, but most of the munis use—they have the capability of producing electric power, and they do produce some, but that has almost become a supplemental source of energy for them. They buy from other agencies, from Bonneville, or they even buy from Southern California Edison, and there is a good relationship there between the privates and publics. I don't know whether you are going to change very much actually as far as production is concerned.

Mr. STALON. We are not talking about changing production, we are talking about viewing the transactions that occur. In reality, what they do is to back off their own generators and take more power out of the grid and make a side payment to a generator some place else to put more power in the grid. That is not going to change. That reality remains as it has been.

The question now is, what are their obligations for operating their generating assets and what are their obligations for building transmission lines. In the past, they have worked that out with the industry self-regulatory sides, and we need to make sure that, whatever we remove there, we add something of comparable importance in the new system.

The generators, recall, in the past were treated as a utility. We didn't pay too much attention to antitrust laws when the owners of the generators sat down together. Now they always drew a sharp distinction, they were only dealing with reliability, another regulatory fiction which we all found convenient.

It is going to be more and more difficult to have owners of generation sitting around planning the system because they are now going to be a competitive operation. They are already intensely rivalrous, and we need to think of a new system for coordinating the control system, the building and operation of transmission lines.

Mr. MOORHEAD. Well, we have got just barely enough time now to go over and vote, but it is a very interesting discussion.

Mr. SHARP. Thank you, Mr. Moorhead.

Gentlemen, I would love to continue. You might not feel the same way since you did not get lunch. I cheated and had a quick snack in that short break we took, and I realize the discourtesy to you and some of the people in our audience that these things are easier to deal with when one is well fed. But I appreciate your time and attention and think you have provided us, and I know you have done it in a number of other fora as well, with some very thought-provoking views, and it is very clear to me by the intense interest, both here and yesterday and the comments we receive in our audience, that people in this industry and associated with it are intensely interested in what you and others have to say, because they are trying to figure out where this world is going to go and whether they should panic or whether they should seize the opportunity.

But I certainly appreciate your time and attention to this, and the subcommittee will meet next week, further continuing our saga. Thank you very much.

[Whereupon, at 1:45 p.m., the subcommittee was adjourned, to reconvene at the call of the Chair.]

ELECTRICITY ISSUES

THURSDAY, JULY 21, 1994

HOUSE OF REPRESENTATIVES,
COMMITTEE ON ENERGY AND COMMERCE,
SUBCOMMITTEE ON ENERGY AND POWER,
Washington, D.C.

The subcommittee met, pursuant to notice, at 10:15 a.m., in room 2123, Rayburn House Office Building, Hon. Philip R. Sharp (chairman) presiding.

Mr. SHARP. The subcommittee will please come to order.

I would like to welcome our witnesses to the third in the subcommittee's series of hearings on changes in the electric power industry. The quality of our testimony has been uniformly high and exciting. It is exciting to hear a wide variety of perspectives on changing electricity markets and approaches to managing the transition period and the restructured industry of the future.

At our last hearing, we heard from all five of our Federal Energy Regulatory Commissioners, as well as prominent consultants and academics. Much of that testimony dealt with the difficult issue of potential stranded investment. How should that term be defined? When is it a legitimate cost of business? And under what circumstances should it be included in rates? Last, but not least, when regulatory bodies have jurisdiction—which of the legislative bodies have jurisdiction over the parts of the wholesale/retail puzzle?

We also, in the previous hearing, dealt extensively, of course, with the other two issues: the potential conflict between competitive and near-term efficiency goals and longer-term environmental and other social goals, and in addition to that, the complex jurisdictional questions between Federal and State regulation.

Today we are very fortunate to have several witnesses who can help us tackle these questions from a State perspective, Commissioners from California and Michigan, which have already undertaken major initiatives in retail wheeling or direct access, and Wisconsin, which is contemplating doing so in the near future.

As even a casual observer of the electricity industry knows, the subject of retail wheeling or "direct access" involves many of the most difficult issues facing the industry today, including the question of stranded investment. And there is debate as to whether direct access is the correct response to market changes or whether it would, in fact, cause more problems than it cures. And I look forward to our witnesses' perspectives on these complex questions.

In addition, much of our prior testimony has been focused on the question of how to maintain social and economic values, such as those embodied in the Public Utility Policies Act, or PURPA, of 1978, in a more competitive electricity market. For some, this amounts to no more than an accounting problem. For others, it raises very fundamental questions about equity and the fundamental character of the more competitive electricity market which everyone or most people have embraced.

We also are fortunate today to have a panel of witnesses with a Wall Street perspective. As we pursue the interesting question of how a restructured industry should function and how to best protect consumers, we should obviously not lose sight of the fact that investors' and ratepayers' interests are not always in conflict.

Without market confidence, utilities, their cost of capital, of course, rises and this translates into higher rates for the average customer. Regulators are trying to adapt a market in transition in a manner that protects ratepayers, but which also takes into consideration the investor's understandable desire for the predictable.

Finally, the hearing will step back a bit from the larger picture issues we have been focused on and look at a more near-term concern; the question of the Solar, Wind, Waste and Geothermal Power Production Act of 1994 which makes permanent the removal of the size cap on small power producers under PURPA.

In 1978, PURPA imposed a 50 megawatt size limitation on renewable generators for qualification as a small power producer under the act. The size cap was removed by Congress in 1990 with a sunset provision of 4 years. The sunset provision expires this year, and our bill, the one that has been under consideration in the subcommittee, would simply remove the sunset provision making the size cap removal permanent.

The category of generators we are discussing here are among the cleanest. It is the policy of the administration to encourage these resources as a part of the Climate Change Action Plan, and it is important that we be sure that these clean generators have a role in our energy future. The kind of encouragement that PURPA incentives provide has been essential to the success of this kind of generation.

Let me now recognize our distinguished colleague from Idaho, Mr. Crapo.

Mr. CRAPO. Thank you, Mr. Chairman.

I don't have an opening statement, but I do have a statement by the Idaho Power Company that they would like to have just inserted in the record of the hearing, if that would be possible.

Mr. SHARP. We will certainly make that possible.

Mr. CRAPO. Ask unanimous consent for that.

Mr. SHARP. Without objection.

Mr. CRAPO. Thank you.

[The prepared statement of the Idaho Power Company follows:]

STATEMENT OF IDAHO POWER COMPANY

The Idaho Power Company (Company) appreciates the opportunity to comment on The Public Utilities Resource Policy Act (PURPA) of 1978, The Company applauds the leadership shown by Chairman Sharp and the other members of the Energy and Power Subcommittee in pursuing and implementing the Energy Policy Act of 1992 (EPAct) and for holding these oversight hearings to ensure its goals are being met.

By way of information, Idaho Power serves approximately 320,000 customers in Southern Idaho, Northern Nevada and Eastern Oregon. Its service territory encompasses approximately 20,000 square miles. Although recent years of drought have changed the Company's fuel mix, the long-term average is 60 percent hydropower, 40 percent thermal and purchased power. According to a study by Dillon Read Equity Research in 1993, Idaho Power's total retail rates are the lowest of any investor-owned utility in the country.

Even prior to the passage of the EPAct, the Company had been preparing itself to compete in a less regulated environment. The Company believes it is in a favorable position to meet the new challenges placed in motion by EPAct. However, if it really is in the public interest for Congress to encourage utilities to move away from a regulated environment, it is necessary for Congress to also recognize existing regulatory policies that will act as an impediment to operational efficiency. One of those regulatory impediments is PURPA.

PURPA had two fundamental goals: (1) to encourage the development of new sources of power other than the traditional utility generation plants and (2) to use all resources effectively without harming the consumer. The vehicle for implementing these goals was a legal mandate that utilities purchase all of the electricity generated by PURPA qualifying facilities (QF's). The setting of the appropriate purchase prices for each utility is at the discretion of the individual State commissions, but PURPA requires that the purchase price be equivalent to each utility's "avoided cost." Avoided costs are the costs an electric utility would otherwise incur to acquire power if it did not purchase electricity generated by QF's.

Since the inception of PURPA, Idaho Power has purchased a substantial amount of QF power. Idaho Power began receiving QF energy in 1982 with purchases of approximately $242,000. By the beginning of 1993, the Company was purchasing energy from 57 QF projects amounting to 64 average megawatts (MWa) of non-dispatchable energy. QF purchases through 1992 have totalled approximately $200 million at an average cost of 5.9cts per kilowatt-hour. The almost $200 million in QF purchases through 1992 pales in comparison to the potential purchase amounts on the horizon as both the number and the size of potential QF projects continue to grow. Besides the 57 operating projects mentioned above, there are eight additional projects scheduled to come on-line in the very near future, These projects will result in additional annual Company expenses of approximately $28 million by the end of 1996, which is an average cost of 5.4cts per kilowatt-hour. The total annual QF expenses would increase to $60 million. PURPA resources have and will continue to represent a major part of Idaho Power's resource portfolio.

Electric utilities are facing a period of great uncertainty as the Nation addresses the issues of deregulation and competition in the electric industry. While energy competition in electric generation has existed in various forms for many years, the potential for direct competition between suppliers for sales to retail customers poses many difficult questions. Electric retail competition would simply alter the "regulatory compact" if not eliminate it entirely. In a competitive retail market with no service area boundaries, the market price for energy will dictate the utility's generation-related revenues, regardless of the cost of its resources. The regulatory authority would no longer take the place of the market and the wholesale and retail prices of energy would rise and fall to meet market conditions. Obviously, utilities that purchase or build resources whose costs exceed expected future market prices will lose in this competitive environment. Utilities who commit to buy at long-term fixed prices that are subsequently found to be above the current market prices, will not be able to recover their costs regardless of how attractive these price levels were when the resources were acquired. Utilities cannot make long-term commitments to acquire generation resources at above-market fixed prices without reciprocal obligations from their customers to take service and pay rates that assure recovery of costs incurred to serve them. If retail competition becomes a reality, those customers that have the ability to acquire their electricity from non-utility suppliers at lower costs will do so, leaving the remaining customers to share the embedded costs of long-term, above-market contract commitments. In Idaho Power's case, QF purchases represent the single greatest source of long-term fixed rate, above-market purchases.

The independent power generation industry has evolved beyond the point where PURPA's mandatory purchase obligation is necessary to create an artificial market for non-utility generation projects. With the evolution of exempt wholesale generator (EWG's), the necessity for power project developers to avoid regulation by achieving QF status has been eliminated. In addition, State regulatory authorities and utilities have become much more sophisticated in integrated resource planning and resource acquisition. The vast majority of utilities now acquire resources in a competitive arena utilizing competitive bidding, competitive negotiation or a combination of both. This allows EWG's to compete with potential utility-owned resources on a level playing field. Idaho Power attempts to acquire resources that will meet its future service obligations at minimum cost. In its integrated resource planning process, the

Company, with substantial regulatory Commission oversight and public participation, identifies least-cost resources. Experience in the Pacific Northwest has demonstrated that utilities mill contract with EWG's and other independent power producers to supply large portions of future resource needs. Idaho Power is no exception. When the time comes for Idaho Power to acquire new supply-side resources, the Company will include EWG's, IPP's and all other potential resources in its portfolio of potential resources. To utilize competitive resource generation to its fullest, utilities should not be burdened with a mandatory purchase obligation if that purchase obligation results in the acquisition of uncompetitive resources. The benefits and drawbacks to PURPA can and will be debated. However, it is clearly unfair and economically inefficient to expect utilities to compete in an environment of reduced regulation while at the same time, maintaining a regulatory mandate that requires the purchase of resources at uncompetitive prices.

Mr. SHARP. The distinguished gentleman from California, Mr. Lehman.

Mr. LEHMAN. Thank you, Mr. Chairman.

And likewise, I do not have an opening statement. I look forward to the testimony.

I would like to especially welcome Chairman Fessler from California here today. I appreciate the opportunity to meet him, look forward to talking to him.

Mr. SHARP. I would like to ask unanimous consent that we put in the opening statement by the gentleman from California, Mr. Moorhead, who I am sure will try to come by. He is intensely interested in these questions, but unfortunately has major responsibilities on the Judiciary Committee as well.

[The opening statements of Hon. Carlos J. Moorhead and Hon. Michael Bilirakis were received for the record:]

OPENING STATEMENT OF HON. CARLOS J. MOORHEAD

Mr. Chairman, I would like to thank the chairman again for holding this timely series of hearings on the future of the electricity industry. During the last two hearings, the discussion has concentrated on the effects of increased competition, with a special focus on competition at the retail level.

As a result, I am glad to have before us today representatives of State commissions that are taking steps to increase retail competition. I am particularly gratified that Chairman Fessler, of the State Public Utilities Commission in my home State of California, could join us. Over the last several months, the industry restructuring proposal of the CPUC has raised the debate over the future of electricity industry to a new level of intensity.

When we passed the Energy Policy Act, our objective was to create an atmosphere in which wholesale competition could flourish. Similarly, the CPUC proposal attempts to allow the discipline of market forces to work at the retail level.

I also would like to thank the chairman for assembling the third panel, which will address issues raised by a bipartisan bill cosponsored by the chairman and myself that would permanently remove the PURPA cap on the production of solar, wind, and geothermal energy.

When PURPA was enacted, renewable technologies were in their infancy—essentially untried and unproved. In PURPA, Congress imposed a cap on the size of these facilities that would qualify for mandatory purchase requirements. In 1990, by unanimous consent, Congress eliminated the cap for a 4-year period.

In the absence of legislation, an 80-megawatt cap will be imposed on "qualifying" renewable energy sources under PURPA at the end of the year.

Removing the cap would allow new renewable resources to qualify under PURPA while taking advantage of economies of scale, thus lowering their cost of generation. It is important to note that there is no size restriction on cogeneration facilities that qualify under PURPA.

I know that many people would like to raise other issues regarding the structure of PURPA. This bill would simply make sure that renewable facilities are not handicapped by a cap that does not apply to cogeneration facilities under PURPA.

This does not mean that, in the future, we will not consider and address concerns regarding the structure of PURPA and its role in increasingly competitive electricity markets. It may be that PURPA is not the best vehicle for encouraging the development of renewable power. However, we cannot address those issues with this bill.

We must act to preserve the existing cap exemption for renewable resources before it sunsets in December. The removal of the 80 megawatt cap would allow Americans to continue to reap the full benefits of clean domestic energy from these renewable resources. I have committed to work with the member of this subcommittee to address any concerns that you may have regarding the permanent removal of the cap.

Again, I would like to thank the chairman for providing the forum for the full and fair discussion of these issues.

STATEMENT OF HON. MICHAEL BILIRAKIS

Mr. Chairman, I would like to thank the chairman for holding this, the third in a series of hearings on the rapid changes taking place in the utility industry in the wake of the Energy Policy Act. In the last 2 hearings, we have heard that these changes are in part due to market forces, and in part due to regulatory changes at both the State and Federal level.

The commissioners before us today represent States that are at the cutting edge of the movement toward retail competition. Although I am in favor of increased competition, I am concerned about the transition to this new world. These States have many difficult issues to face with regard to issues such as "stranded investment."

In that regard, I look forward to hearing from our witnesses that represent the investment community. In a very real sense, they are the representatives of those whose capital has been, and will continue to be, used to build the majority of our country's electrical system, whether that system is constructed by investor-owned utilities or independent power producers.

For investors, as with the companies themselves, competition will preset the opportunity to fail as well as the opportunity to succeed. It is incumbent upon us, as lawmakers and regulators, to ensure that those opportunities are equal for all. We must carefully evaluate all of our policies to make sure that some competitors are not given an advantage or disadvantage by programs that were designed in a regulated environment.

In this vein, I am pleased that we have before us witnesses on the issue of whether or not we should eliminate the 80 megawatt cap on renewable facilities that qualify for mandatory purchase under PURPA. In many cases, States have set prices for power subject to the mandatory purchase obligation at above-market rates. In a competitive environment, especially where retail competition is present, the requirement to purchase this power will hamper the ability of utilities to compete with those that have no such obligation.

We will not be able to address the general problems with the structure of PURPA before the exemption from the 80 megawatt cap on renewable power projects expires in December. As such, I will be interested in hearing the views of our panelists regarding effect of the cap on both the utilities that must buy the power and the renewable power industry.

I thank the chairman for holding this hearing so that we may fully investigate the impact of the 80 megawatt bill in the context of increasingly competitive electricity markets.

Mr. SHARP. We are very pleased to welcome our first panel of witnesses. We are delighted to have the Honorable Daniel William Fessler, the President of the California Public Utilities Commission with us; the Honorable Ronald E. Russell, the Commissioner of the Michigan Public Service Commission; and the Honorable Scott A. Neitzel, the Commissioner, Wisconsin Public Service Commission.

Gentlemen, we are very pleased you could take time from what we know are very intense responsibilities that you have in your respective States. But what you are doing not only has significant consequences, as you are well aware, for your ratepayers in your industries in your States, but has certainly captured the attention of people all across the country as to both national and other State policies in other States. So we are very pleased to have you—your willingness to come here and share with us your views and concerns.

Mr. Fessler, we would be very pleased to start with you.

STATEMENTS OF DANIEL WILLIAM FESSLER, PRESIDENT, CALIFORNIA PUBLIC UTILITIES COMMISSION; RONALD E. RUSSELL, COMMISSIONER, MICHIGAN PUBLIC SERVICE COMMISSION; AND SCOTT A. NEITZEL, COMMISSIONER, WISCONSIN PUBLIC SERVICE COMMISSION

Mr. FESSLER. Thank you very much.

Mr. Chairman and members of the subcommittee, I am delighted to accept your invitation to participate in this important discussion of change in the electric services industry. I do so fully recognizing that much of the reform has its origin in steps which the Congress took in 1978 with the passage of the Public Utilities Regulatory Policies Act.

The introduction of competition in the realm of generation unhinged the foundation of the vertically integrated monopoly which had dominated both policy and practice for nearly a century. More recently, the members of this subcommittee played a vital part in framing the terms of the Energy Policy Act which sought to extend the presence of competitive providers of generation and secure their access to a nondiscriminatory transmission grid.

Mr. Chairman, I have read the prepared remarks of several of the distinguished witnesses who appeared before this body last week. Much of what they said was framed with a greater precision and clarity of expression than would have been the case were I to seek to make the same points.

In general, I find myself in agreement with most, if not all of the sentiments expressed before you. There is one exception. In the course of remarks which I otherwise found most instructive, former Maine Commissioner, David Moskovitz made the following statement: "The movement to a competitive market is being and should be led by regulators."

Mr. Chairman, I respectfully disagree. I believe that if the movement was ever centered in government aspirations or leadership, it has long since moved to the marketplace and is in the custody of those men and women who are the multifaceted industry. Given this perspective, I find greater accuracy in the remark of Professor William Hogan before the committee that, "the forces in motion are powerful and will sweep over those who ignore them." But in embracing the Hogan view that we may influence that which we cannot stop, I would like to speak to the disturbing possibility that the several States and the Federal Government are possessed of an ability to frustrate and distort the development of market institutions, and in the course of a misguided effort, betray the public trust each of us so diligently seeks to advance.

Such a disagreeable outcome is ensured if the private sector genius for innovation and reform is paralleled by a public sector penchant for jurisdictional debates and authority disputation. Given this concern, I would ask your permission to concentrate my remarks upon the second and fifth of the questions propounded in your letter of invitation.

As a preliminary, may I briefly emphasize the obvious, that I am 1 of but 5 members of the California Public Utilities Commission and my thoughts are my own and are not put forward as embracing the sentiments of my colleagues. Further, my Commission has presently pending both a rulemaking and an investigation which touch upon virtually all aspects of this discussion and has been clearly acknowledged by virtually every witness, that constitute one of the precipitating factors which occasion these important hearings.

I am also conscious that my Commission is a party to multiple proceedings before the Federal Energy Regulatory Commission, and I am certain that you and your colleagues will sympathize with my ambition, that in my multiple capacities, I say nothing today that foreshadows my function as a decisionmaker nor alters formal submissions of my Agency before Federal tribunals.

Mr. Chairman, you have asked if I have any recommendations as to how regulators should handle the transition period to a more competitive market? And you have further propounded the question: What are the important questions involving State and Federal regulatory authority?

I find in the juxtaposition of your two questions, the invitation to frame what I hope will be a more productive chapter in the ongoing attempt to define the hallmarks of modern federalism. As each Member of Congress is aware, that debate has had an origin that dates to the very formulation of the constitutional framework which we erected to replace the dysfunctional Articles of Confederation.

The Constitution of 1787 contains two phrases upon which ground the authority of the national government in the economic sphere. Article I, section 8, Clause 3, grants to Congress the power "to regulate the commerce with foreign nations and among the several States." While Article VI, Clause 2 contains a declaration that the Constitution and laws of the United States made in conformity shall be the supreme law of the land and take precedence over the constitutional laws of any State.

Standing as a perceived limitation on this sweeping proviso is the subsequent act of our people in 1791 amending the Constitution in 10 particulars. Of interest to me is the last of those amendments which declares that, "The powers not delegated to the United States by the Constitution, nor prohibited by it to the States, are reserved to the States respectively, or to the people."

Now, over the course of 200 years, the provisions of the original text and terms of the 10th Amendment have jousted for ascendancy in judicial forum. Without doubt, Chief Justice Marshall was the leading proponent of an emphasis upon national power which virtually ignored the 10th amendment. The literal if not figurative high watermark was achieved in *Gibbons* vs. *Ogden*. But 13 years later, the stock of the 10th amendment was clearly in the ascendant. In *New York* vs. *Miln,* the court announced judgment in a case heard, but decided after Chief Justice Marshall's death. Justice Barbour wrote for the majority rejecting the contention that it sought to distinguish *Gibbons* on the premise that the New York legislation then under review did not directly affront an act of the Congress.

The court responded: "We do not place our opinion on this ground. We choose rather to plant ourselves on what we consider impregnable positions and. They are these: That a State has the same undeniable and unlimited jurisdiction over all persons and things within its territorial limits, as any foreign nation, where that jurisdiction is not surrendered or restrained by the Constitution of the United States. That, by virtue of this, it is not only the right, but the burden and solemn duty of a State to advance the safety, happiness and prosperity of its people and to provide for its general welfare by any and every act of legislation, which it may deem to be conducive to these ends where the power over the particular subject or the manner of its exercise is not explicitly surrendered or restricted in the manner just stated. That all those powers which relate to merely municipal legislation or what may be called 'internal police,' are not thus surrendered or restrained; and that consequently in relation to these, the authority of the State is complete, it is unqualified, and it is exclusive."

Now, less dramatic examples could be cited in the jurisprudence of this century. I recognize clearly the tides and ebbs and flows, and in recent years has turned again in favor of a recognition of national authority as the Court has struggled to sustain economic policies and programs beginning with those of the New Deal.

And yet, Mr. Chairman, there are those who would go further with their aggressive claims of exclusive Federal jurisdiction respecting the objectives which my Commission has advanced for discussion. Indeed, taken at their extreme, there are those residing within Washington who would find an assertion of Federal supremacy in the mere silence of the Congress which would nullify the authority of the several States to promote the prosperity of citizens, to say nothing of the efficiency of the electric service industry by implementing some form of retail competition.

Gentlemen, I would like to respond to those who find in our rulemaking a challenge to accustomed concepts of Federal authority. The concern is misplaced. Neither Justice Barbour nor Chief Justice Taney has a modern descendant upon the California Public Utilities Commission. Yet, I do find myself in communion with Justice McKenna who, writing for the court in *Hoke* vs. *United States* made the following common sense observation: The year was 1913. "Our dual form of government has its perplexities, States and Nation having differing spheres of jurisdiction—but it must be kept in mind that we are one people and the powers reserved to the States and those conferred on the Nation are adapted to be exercised, whether independently or concurrently, to promote the general welfare, material and moral."

May I respectfully suggest that McKenna's call for what I would term "cooperative federalism" is the only appropriate resolution of this jurisdictional inquiry. I thus endorse the call for cooperation asserted by Chair Moler and by Commissioner Bailey in testifying before this distinguished body.

I also embrace the specific proposals advanced by former Commissioner Ashley Brown. And yet I would seek to go beyond the description of the State versus Federal issues depicted in their excellent testimony to assert a view of the electric industry as having assumed physical reality which defies the traditional jurisdictional claims of any existing institution of government in North America.

Our common purpose is to harness efficiency gains in an electric grid which, as we speak, encompasses 11 Western States, two Canadian Provinces and the Mexican States of Sonora and Baja California Norte, and most likely the Mexican State of Chihuahua. Seen in this perspective, it is evident that the authority of my Commission, or any instrument of the State of California, is inadequate, if not irrelevant to electrons which show no respect for those who would assert authority under the 10th amendment.

And yet it is equally plain that the jurisdiction of the FERC and, indeed, the Congress, is at once too broad and too limited for the governance task which confronts our people. Put bluntly, Mr. Chairman, we face a regional market in the western part of North America which is institutional, but it is international in its character. In such circumstances, both common sense and common decency urge the surrender of the coercive writ in favor of cooperative solutions.

In closing, I would like to recur to one aspect of Justice Barbour's statement in *Miln,* where he addressed what he termed the "solemn duty of the State to advance the safety, happiness and prosperity of its people." On the day my colleagues and I issued our rulemaking, we declared that while our minds remain open to modifications of our proposal, or even the substitution of what we might come to be convinced are superior ideas, we are single-minded in our objective: to lower the cost of electric service to California's residential and business customers without sacrificing the financial integrity of our investor-owned utilities.

The debate over Federal/State relations cannot be allowed to obscure your appreciation of facts which must be accommodated if we are to promote the general welfare. Electric rates in California are 143 percent of the national average. While it is true that a combination of our Mediterranean climate and aggressive conservation efforts have blessed our householders with electric bills which are below the national average, those who cannot avoid significant consumption, the commercial, industrial, and agricultural sectors of my State's economy point to a crushing burden.

In yesterdecade, a vibrant defense industry may have prompted a level of prosperity which allowed us in California to bear this burden without vocal complaint; 800,000 highly skilled jobs later, we recognize that we must rebuild our economy in a competitive climate which has zero tolerance for those who are indifferent to inefficiency.

Mr. Chairman, whatever you may have heard here, in the nearly 25 hours of public hearings which I have conducted to date, and in the written submissions of more than 90 parties, I have yet to encounter the phrase "it ain't broke, so don't fix it."

I asked the members of this committee to inform their colleagues that in pursuing our study of the industry, and in our effort to reinvent regulation at the State level, California seeks nothing at the expense of our sister States, nor of jurisdictions in Canada or in Mexico. What we do seek is self-improvement in facilitating the emergence of market mechanisms and regional efficiencies which can only improve the competitiveness of a region vital to North America's posture in the global economy. In the words of Mr. Justice McKenna: Our goal is to promote the general welfare, material and moral, and I look to this distinguished committee and to the Congress for encouragement in this endeavor.

And I would be delighted to respond to any questions to the best of my limited ability.

Mr. SHARP. Thank you very much, Mr. Fessler, we will hear our other two panelists and then we will have comments from our panel.

Mr. Russell, we are very pleased to hear from you now.

STATEMENT OF RONALD E. RUSSELL

Mr. RUSSELL. Thank you, Mr. Chairman, members of the subcommittee. My name is Ronald E. Russell. I am a commissioner with the Michigan Public Service Commission. I appreciate the opportunity to speak to you today on changes occurring in the electric industry, specifically the retail wheeling experiment recently ordered by the Michigan Public Service Commission.

While retail wheeling has been defined in a variety of ways, we in Michigan use the term when referring to a local utility's transmission of power to an end-user located within its territory. This type of transaction differs from full electric service in that the power is not provided by the utility out of its own system resources. Instead, the retail wheeling customer assumes the responsibility of arranging for the purchase of power from a third-party supplier and the transmission of that power to a local utility system.

The ability to retail wheel could allow those customers using large amounts of electricity to obtain power at lower prices and/or better terms than might be available from the host utility, providing that an efficient and economic transaction exists.

In August 1992, the Association of Business Advocating Tariff Equity, ABATE, an organization of large utility customers in Michigan, sought ex parte approval from the Commission for a retail wheeling tariff for a consumer's power company, one of Michigan's largest electric companies. The Commission denied ex parte approval and instead commenced the contested case proceeding for consideration of an experimental retail wheeling program for both consumers and Detroit Edison Company.

After considerable deliberation, the Commission determined that a persuasive case had been made for conducting a limited retail wheeling experiment. An interim order was issued in April 1994, establishing parameters for the design of an experimental program: the size and duration of the experiment must minimize the risk of unforeseen consequences, sufficient safeguards must be established to protect nonparticipating ratepayers, and participating retail wheeling customers must be willing to assume the risks associated with participation in a nonregulated generation market.

The Commission found, however, that the record and the proceeding did not provide adequate information to establish a price for retail delivery service. Therefore, the record was reopened for the limited purpose of determining appropriate rates and charges.

Initial testimony is scheduled to be filed by the two electric companies on August 25th, followed by the Commission's staff and other interested parties on October 13th. Only at the end of this proceeding will the Commission consider giving final approval for the experiment. In the interim order—I won't go into all of the details of the order, but that will be included in the basis of my testimony.

I would like to emphasize that the Michigan Commission's interim order finds that a retail wheeling experiment is in the public interest, but does not attempt to find retail wheeling is in the public interest. From this experiment, the Commission will gather information which would be used in any future deliberations to determine if retail wheeling should ultimately be included as an element of retail competition.

The experiment should provide information and data on the extent to which retail wheeling impacts the utility and its ability to provide safe, reliable service at reasonable rates to its retail ratepayers, how retail wheeling affects the utility's financial status, and whether retail wheeling actually attracts participants who are willing to assume the risks associated with participation in the competitive generation market.

While it is entirely appropriate to examine retail wheeling closely to further understand how it impacts all relevant parties, as we are proposing in Michigan, it is also necessary to stand back and view it in conjunction with other opportunities becoming available to utilities, non-utility generators, and customers as the industry undergoes significant change.

In this context, retail wheeling is considered by many in the industry to be a logical outgrowth of the availability of competitive-priced generation, and open access to transmission, both results of the Energy Policy Act of 1992.

Many of the utility customers, "chomping at the bit," for the opportunity to retail wheel, are the same customers with other options for supplying their generation needs: on-site generation, municipalization, shifts in production, cogeneration, options that, in some cases, could remove them completely from utility systems.

Given this broader perspective, State commissions find themselves weighing the consequences to the utility and its ratepayers of losing the revenue from the entire sale if a customer converts to a nonhost utility generation versus losing the revenue from generation portion of the sale, but continuing to collect transmission revenues, as would occur in a retail wheeling transaction.

In this light, retail wheeling may be a reasonable option; however, and this is a significant "however," if retail wheeling is to be a realistic option, the transmission price must include all cost associated with the provision of service, and the retail wheeling customer must assume all of the risks associated with participation in the market. Retail wheeling must not be subsidized by the utility's full service ratepayers.

State utility commissions are facing the challenge of effectively balancing the need to protect the interests of ratepayers of our State's public utilities, while adapting to the competitive forces facing regulated utilities. The competition which is beginning to evolve in the electricity generation may result in net gains and efficiency nationally, but perhaps at an increased cost to some retail customers.

Those without the option, at least a currently available option, to leave the utility's full service system may be left with higher costs. The only way these ratepayers can be adequately protected is through careful State oversight.

In order to effectively provide this oversight, State commissions need to maintain their jurisdictions over transactions occurring within their States. It appears that the debate is intensifying regarding a State's ability to regulate the transactions occurring within its boundaries.

The Federal Power Act grants the Federal Energy Regulatory Commission exclusive nondelegatable jurisdiction over, "the transmission of electric energy interstate commerce," "the sale of electric energy at wholesale and interstate commerce".

It additionally states that the "FERC shall not have jurisdiction over facilities used in local distribution." Transmission in interstate commerce has been interpreted by the courts and the FERC as any instance in which "energy commingles in a bus," i.e., the point of interconnection between two utilities. That is part of a transmission grid interconnected with and capable of transferring energy to and from other utilities in other States, thus, the bright line separating Federal and State jurisdiction places wholesale transmission within the FERC's jurisdiction and retail transactions within each State's jurisdiction.

Using this framework, the regulation of rates, terms and conditions of retail electric service by a local utility to end-users located within its service territory is indisputably a function of local distribution and thus within State jurisdiction.

Similarly, States have historically recognized this responsibility by including in retail rates charges for those transmission facilities used for retail transactions. The fact that the retail customer has the opportunity to wheel power from a third-party supplier and purchase only "unbundled" transmission service from the local utility does not change the regulatory status of that arrangement. It remains a matter of State jurisdiction as delineated by the Federal Power Act.

I should note that this opinion of jurisdiction is not shared by all. In a June 13, 1994 letter to Secretary of Energy Hazel O'Leary, the Honorable Edward Markey suggested that State commissions ordering retail wheeling, "may well find themselves reaching beyond their authority. Once their actions result in a movement of electrons across State lines, they may find themselves in a realm of interstate commerce clause and FERC jurisdiction."

Let me reiterate, this position ignores the historical duality of Federal/State regulation which recognizes FERC jurisdiction over wholesale transactions and State jurisdiction over the delivery of power to end-users, or retail customers. Properly understood, retail wheeling is the unbundling of the full retail service that local franchise utilities have historically provided and State commissions have historically regulated. It is difficult to understand why, as a matter of either law or policy, an "unbundled" component of retail service should be treated differently than the "bundled" service for jurisdictional purposes.

The Energy Policy Act amending the Federal Power Act has further clarified the jurisdictional framework for regulating retail wheeling transactions. Section 212(g) states that no FERC order can be inconsistent with any State law that governs the retail marketing areas of electric utilities, insuring that States would make the final decision on whether retail wheeling would be allowed within the State. Section 212(h) expressly prohibits the FERC from issuing any order conditioned upon or requiring the transmission of electric energy directly to an ultimate consumer. Thus, the FERC cannot order retail wheeling.

Furthermore, this provision carries with it a savings clause stating that, "nothing in the subsection shall affect any authority of any State or local government under State law concerning the transmission of electric energy directly to an ultimate consumer".

The savings clause clearly expresses congressional intent to leave the provisions of transmission service to end-users and retail wheeling transactions within the State's jurisdiction. It is interesting to note that the Honorable Richard Lehman in a May 26th letter to Daniel W. Fessler, President of the California Public Utilities Commission, provided another interpretation of Congress' intent in section 212(h). Congressman Lehman stated that, "Congress did not grant to or confer upon the States, authority to order retail access to ultimate consumers." Congressman Lehman may be technically correct as section 212(h) does not contain an explicit grant of jurisdiction to the States.

However, it is significant that section 212(h) expressly disavows any intent to preempt the States from exercising their jurisdiction over the delivery of power to a retail wheeling customer. Because the provision in the Federal Power Act reserving to States jurisdiction over facilities used in local distribution were left unchanged, the most fair and reasonable understanding of the Energy Policy Act is that States continue to have jurisdiction over retail wheeling.

I also note that in the recent FERC Notice of Proposed rulemaking on the Recovery of Stranded Costs by Public Utilities and Transmitting Utilities, dated June 29, 1994, the FERC claims it has, "exclusive jurisdiction over the rates, terms and conditions for the retail interstate transmission services." I submit this statement is in direct opposition with the framework set out in the Federal Power Act and Energy Policy Act.

Unless there is express language to the contrary, a pervasive scheme, or a conflict between implementation of State and Federal law, the Energy Policy Act limits the jurisdiction of FERC to wholesale transactions.

This potential jurisdictional conflict is representative of issues resulting from the change in the electric industry. The evolving competitive electricity industry is not merely "business as usual" with a few changes, but the foundation of a complete restructuring of how the electric industry will function, State and Federal commissions will regulate, and perhaps more importantly, under what rates, terms and conditions customers will be served.

In my opinion, the change is gaining momentum and there is no turning back. Competition in generation may not prove to be beneficial to all utility ratepayers, especially if it results in higher capital cost, stranded investments and the inability of utilities to plan efficiently. State commissions must have the ability to ensure the interest of ratepayers are protected as the industry matures. Consistent with this, State abilities to act as incubators of new regulatory approaches should not be inhibited.

In closing, I would like to again refer to Congressman Markey's recent letter to Secretary O'Leary. Congressman Markey suggests that section 722(h) of the Energy Policy Act states that, "no order issued under this act, be conditioned upon or require the transmission of electric energy, one, directly to an ultimate consumer," should be interpreted as discouraging State implementation of retail wheeling. Congressman Markey cites a variety of concerns, encouraging the departments and States, "to consider the myriad of concerns and problems with retail wheeling before taking premature actions."

I would like to reassure all of the subcommittee members that given the full authority reserved to States in the Federal Power Act and reinforced in the Energy Policy Act, the Michigan Commission has carefully considered the myriad of concerns and problems that could arise through implementation of a limited wheeling experiment. If record evidence is persuasive, the Michigan Commission is fully capable of testing the thesis that large customers should be provided with an option to traditional utility service while at the same time protecting ratepayers and providing the Nation with data on the impacts of retail wheeling. Whether customers choose that option is another matter.

Thank you.

Mr. SHARP. Thank you very much.

[The prepared statement of Mr. Russell follows:]

STATEMENT OF RONALD E. RUSSELL, COMMISSIONER, MICHIGAN PUBLIC SERVICE COMMISSION

Mr. Chairman and Members of the Subcommittee: My name is Ronald E. Russell. I am a Commissioner with the Michigan Public Service Commission. I appreciate the opportunity to speak to you today on the changes occurring in the electricity industry, specifically the retail wheeling experiment recently ordered by the Michigan Public Service Commission.

While retail wheeling has been defined in a variety of ways, we in Michigan use the term when referring to a local utility's transmission of power to an end-user located within its service territory. This type of transaction differs from full electric service in that the power is not provided by the utility out of its own system resources; instead the retail wheeling customer assumes the responsibility of arranging for the purchase of power from a third party supplier and the transmission of that power to the local utility's system. (Opinion and Interim Order Remanding to the Administrative Law Judge for Further Proceedings, Case Nos. U-10143 and U-10176, April 11, 1994.) The ability to retail wheel could allow those customers using large amounts of electricity to obtain power at lower prices and/or better terms than might be available from the host utility, providing that an efficient and economic transaction exists.

In August 1992, the Association of Businesses Advocating Tariff Equity (ABATE), an organization of large utility customers in Michigan, sought ex parte approval from the Commission for a retail wheeling tariff for Consumers Power Company (Consumers), one of Michigan's largest electric companies. The Commission denied ex parte approval and instead commenced a contested case proceeding for consideration of an experimental retail wheeling program for both Consumers and The Detroit Edison Company (Detroit Edison).

After considerable deliberation, the Commission determined that a persuasive case had been made for conducting a limited retail wheeling experiment. An interim order was issued in April 1994 establishing parameters for the design of the experimental program: the size and duration of the experiment must minimize the risk of unforeseen consequences, sufficient safeguards must be established to protect non-participating ratepayers, and participating retail wheeling customers must be willing to assume the risks associated with participation in the non-regulated generation market. The Commission found, however, that the record in the proceeding did not provide adequate information to establish a price for retail delivery service. Therefore, the record was reopened for the limited purpose of determining appropriate rates and charges. Initial testimony is scheduled to be filed by the two electric companies on August 25th, followed by the Commission staff and other interested parties on October 13. Only at the end of this proceeding will the Commission consider giving final approval for the experiment.

In the Interim Order, the Commission established terms and conditions for the experiment, including:

—Consumers and Detroit Edison must offer firm retail delivery service with capacity limits of 60 Megawatts (MW) and 90 MW respectively, or 1 percent of each utility's peak demand.

—The experiment will commence with each utility's need for new capacity and will continue for 5 years. At the end of the 5 year experiment for each utility, retail wheeling customers will be able to return to full utility service without penalty. Retail wheeling customers withdrawing from the program prior to the end of the experiment may only return to full utility service by paying the incremental power supply rates, the highest priced power.

—Retail wheeling customers would be limited to 2-10 MW of transmission or subtransmission service at each location and no single customer can take more than half of the available capacity for each utility's program.

—Stand-by service will be provided by the utility at existing tariff rates unless a more appropriate rate is approved by the Commission.

—Full service and retail wheeling loads must be separately metered.

—Third party providers will be required to obtain a certificate of public convenience and necessity as required by Michigan law. (MCL 460.501 et seq.)

—Retail wheeling customers will be required to obtain Commission approval of all agreements.

I would like to emphasize that the Michigan Commission's interim order finds that a retail wheeling experiment is in the public interest, but does not attempt to find retail wheeling is in the public interest. From this experiment, the Commission will gather information which would be used in any future deliberations to determine if retail wheeling should ultimately be included as an element of retail competition. The experiment should provide information and data on the extent to which retail wheeling impacts the utility and its ability to provide safe, reliable service at reasonable rates to its retail ratepayers, how retail wheeling affects the utility's financial status, and whether retail wheeling actually attracts participants who are willing to assume the risks associated with participation in the competitive generation market.

While it is entirely appropriate to examine retail wheeling closely to further understand how it impacts all relevant parties, as we are proposing in Michigan, it is also necessary to stand back and view it in conjunction with the other opportunities becoming available to utilities, non-utility generators, and customers as the industry undergoes significant change. In this context, retail wheeling is considered by many in the industry to be a logical outgrowth of the availability of competitively priced generation and open access to transmission, both results of the Energy Policy Act of 1992 (Energy Policy Act). Many of the utility customers "chomping at the bit" for the opportunity to retail wheel are the same customers with other options for supplying their generation needs—on-site generation, municipalization, shifts in production, cogeneration—options that in some cases could remove them completely from the utility's system. Given this broader perspective, State commissions may find themselves weighing the consequences to the utility and its ratepayers of losing the revenue from the entire sale if a customer converts to non-host utility generation versus losing the revenue from the generation portion of the sale, but continuing to collect transmission revenues, as would occur in a retail wheeling transaction. In this light, retail wheeling may be a reasonable option; however, and this is a significant "however," if retail wheeling is to be a realistic option, the transmission price must include all costs associated with the provision of the service, and the retail wheeling customer must assume all of the risks associated with participation in the. market. Retail wheeling must not be subsidized by the utility's full service ratepayers.

State utility commissions are facing the challenge of effectively balancing the need to protect the interests of the ratepayers of our State's public utilities while adapting to the competitive forces facing regulated utilities. The competition which is beginning to evolve in electricity generation may result in net gains in efficiency nationally, but perhaps at an increased cost to some retail customers. Those without the option, at least a currently available option, to leave the utility's full service system may be left with higher costs. The only way these ratepayers can be adequately protected is through careful State oversight.

In order to effectively provide this oversight, State commissions need to maintain their jurisdiction over the transactions occurring within their States. It appears that the debate is intensifying regarding a State's ability to regulate the transactions occurring within its boundaries. The Federal Power Act grants the Federal Energy

Regulatory Commission (FERC) exclusive, non-delegable jurisdiction over ". . . the transmission of electric energy in interstate commerce," and ". . . the sale of electric energy at wholesale in interstate commerce." [16 USC 824(b)(1).] It additionally states that the FERC ". . . shall not have jurisdiction . . . over facilities used in local distribution." [16 USC 824(b)(1).] Transmission in interstate commerce has been interpreted by the courts and the FERC as any instance in which "energy commingles in a bus" (i.e. the point of interconnection between two utilities) that is part of a transmission grid interconnected with, and capable of transferring energy to and from, other utilities in other States. [*Federal Power Comm.* vs. *Florida Power and Light Co.,* 404 US 453, 463; 92 S Ct 637; 30 L Ed 2nd 600, 609 (1972). Thus, the "bright line" separating Federal and State jurisdiction places wholesale transmission within the FERC's jurisdiction and retail transactions within each State's jurisdiction. [See *Federal Power Comm.* vs. *Southern California Edison Co,* 376 US 205; 84 S Ct 644; 11 L Ed 2d 638 (1964).] Using this framework, the regulation of rates, terms, and conditions of retail electric service by a local utility to end-users located within its service territory is indisputably a function of local distribution, and thus within State jurisdiction. Similarly, States have historically recognized this responsibility by including in retail rates, charges for those transmission facilities used for retail transactions. The fact that the retail customer has the opportunity to wheel power from a third party supplier and purchase only "unbundled" transmission service from the local utility does not change the regulatory status of that arrangement. It remains a matter of State jurisdiction as delineated by the Federal Power Act.

I should note that this opinion of jurisdiction is not shared by all. In a June 13, 1994 letter to Secretary of Energy Hazel O'Leary, the Honorable Edward J. Markey suggested that State commissions ordering retail wheeling ". . . may well find themselves reaching beyond their authority—once their actions result in a movement of electrons across State lines they may find themselves in the realm of the interstate commerce clause and FERC's jurisdiction." Let me reiterate, this position ignores the historical duality of Federal/State regulation, which recognizes FERC jurisdiction over wholesale transactions and State jurisdiction over the delivery of power to end-users, or retail customers. Properly understood, retail wheeling is the unbundling of the full retail service that local franchise utilities have historically provided and State commissions have historically regulated. It is difficult to understand why, as a matter of either law or policy, an unbundled component of retail service should be treated differently than bundled service for jurisdictional purposes.

The Energy Policy Act, amending the Federal Power Act, has further clarified the jurisdictional framework for regulating retail wheeling transactions. Section 212(g) states that no FERC order can be inconsistent with any State law that governs the retail marketing areas of electric utilities, ensuring that States would make the final decision on whether retail wheeling would be allowed within the State. [16 USC 824k(g).]

Section 212(h) expressly prohibits the FERC from issuing any order conditioned upon or requiring the transmission of electric energy directly to an ultimate consumer. [16 USC 824k(h).] Thus, the FERC cannot order retail wheeling. Furthermore, this provision carries with it a savings clause stating that "[n]othing in this subsection shall affect any authority of any State or local government under State law concerning the transmission of electric energy directly to an ultimate consumer." [16 USC 824k(h).] The savings clause clearly expresses Congressional intent to leave the provision of transmission service to end-users in a retail wheeling transaction within the State's jurisdiction. It is interesting to note that the Honorable Richard Lehman, in a May 26, 1994 letter to Daniel W. Fessler, President of the California Public Utilities Commission, provided another interpretation of Congress's intent in section 212(h). Congressman Lehman stated that "Congress did not grant to or confer upon the States . . ." authority to order retail access to ultimate consumers. Congressman Lehman may be technically correct that section 212(h) does not contain an explicit grant of jurisdiction to the States. However, it is significant that section 212(h) expressly disavows any intent to preempt the States from exercising their jurisdiction over the delivery of power to a retail wheeling customer. Because the provisions in the Federal Power Act reserving to States jurisdiction over facilities used in local distribution were left unchanged, the most fair and reasonable understanding of the Energy Policy Act is that the States continue to have jurisdiction over retail wheeling.

I also note that in the recent FERC Notice of Proposed Rulemaking on the Recovery of Stranded Costs by Public Utilities and Transmitting Utilities dated June 29, 1994, the FERC claims it has ". . . exclusive jurisdiction over the rates, terms and conditions for the retail interstate transmission services." (Docket No. RM94-7-000, pg.47) I respectfully submit that this statement is in direct opposition with the framework set out in the Federal Power Act and the Energy Policy Act. Unless there is express language to the contrary, a pervasive scheme, or a conflict between implementation of State and Federal law, the. Energy Policy Act limits the jurisdiction of the FERC to wholesale transactions.

This potential jurisdictional conflict is representative of issues resulting from the change in the electric industry. The evolving competitive electricity industry is not merely "business as usual" with a few changes, but the foundation of a complete restructuring of how the electricity industry will function, State and Federal commissions will regulate, and perhaps most importantly, under what rates, terms, and conditions customers will be served. In my opinion, the change is gaining momentum and there is no turning back. Competition in generation may not prove to be beneficial to all utility ratepayers, especially if it results in higher capital costs, stranded investments, and an inability of the utility to plan efficiently. State commissions must have the ability to ensure the interests of ratepayers are protected as the restructured industry matures. Consistent with this, State abilities to act as incubators of new regulatory approaches should not be inhibited.

In closing I would like to again refer to Congressman Markey's recent letter to Secretary O'Leary. Congressman Markey suggested that section 722(h) of the Energy Policy Act states that "(N)o order issued under this Act shall be conditioned upon or require the transmission of electric energy (1) directly to an ultimate consumer . . . ." should be interpreted as discouraging State implementation of retail wheeling. Congressman Markey cites a variety of realistic concerns, encouraging the Department and States ". . . to consider the myriad of concerns and problems with retail wheeling before taking premature actions." I would like to reassure all of the subcommittee members that given the full authority reserved to States in the Federal Power Act and reinforced in the Energy Policy Act, the Michigan Commission has carefully considered the "myriad of concerns and problems" that could arise through implementation of a limited retail wheeling experiment. If record evidence is persuasive, the Michigan Commission is fully capable of testing the thesis that large customers should be provided with an option to traditional utility service, while at the same time protecting ratepayers, and providing the Nation with data on the impacts of retail wheeling. Whether customers choose that option is another matter.

Mr. SHARP. Mr. Neitzel, very pleased to hear from you now.

STATEMENT OF SCOTT A. NEITZEL

Mr. NEITZEL. Good morning and thank you, Mr. Chairman, and Mr. Moorhead, for the invitation to address you this morning. As the son of a grocery store clerk and a retired factory worker, I cannot tell you how honored I am to be here and I only hope that my comments will make a positive contribution to your efforts. Let me first briefly summarize my testimony for you and then try to address the five questions that you gave me to respond to.

First of all, I believe that the competition genie is out of the bottle for the electric utility industry. So we as policymakers have probably a couple or three options: (1) we can choose to ignore it; (2) we can attempt to put it back in the bottle; or, (3) we can decide to use this competition genie to achieve our objectives.

I would submit that we as policymakers should choose the latter and try to request from that genie what three public policy wishes we might want to see taken forward as we implement change in this industry.

My three wishes are: One, that we—if we are going to move to competition, that we create a truly competitive marketplace. Absence of regulation—in the absence of regulation, only a full competitive market can really protect the interest of consumers.

Two, I think we have to build retail competition on the firm foundation of wholesale competition so that as we move forward, we don't get ahead of ourselves and we make sure that the competition that we have is sustainable.

Three, I think it would be a serious mistake to overpromise and underdeliver the benefits of moving from a regulated to a competitive marketplace. Regulation and competition are just two different ways of delivering better service to people.

We are very experienced with the regulatory model, therefore we understand all its faults. What we are looking to is a new model, the competitive model, and what we have to understand is that that will have positives and negatives, just as regulation does.

So I think we have to make it clear to the public that what we are doing here is trying to find a more efficient, better way to meet their needs, but that there will still be problems that have to be addressed. And that gets into the questions that you asked. I think I have pretty much stated that I feel competition is here and we have to deal with it.

Number two, how should regulators handle the transition? I think it is very important that not only regulators, but all policymakers, not get caught up in the rhetoric of what is going on. We have to look to reason.

Social programs. The social programs which have been implemented through a regulated utility are all things that we really like. They are demand-side management, conservation, more environmental awareness, programs for low-income customers.

In Wisconsin, you cannot get cut off in the middle of winter from your electric or gas whether you have paid your bill or not, and, in fact, even if you haven't paid it by the next fall, you get reconnected. Now, those are things that we could do in a regulated world because there was a captive market to push those costs onto.

We have to realize if we are going to go to a competitive market, that there will no longer be those captive customers to push those costs onto. So if these things are important to us as a society, we have to figure out a new way to deliver those programs and pay for them out of some different societal pot of money.

As far as the pricing mechanisms to further advance the competitive wholesale market on access, I think at least the Wisconsin Commission is very much in step with the Federal Energy Regulatory Commission. We have found ways to use our jurisdictions complementary when it comes to wholesale access of the transmission system: One, we have used our authority to order our utilities to file transmission tariffs, wholesale transmission tariffs at the FERC. We don't set out the price or the conditions. We just say we want it filed.

Two, we have recently made a decision to order the utilities also to file comparability tariffs. There again, file the tariff. We will participate in the FERC, along with the utilities, to try to figure out what comparability is. But I think that is a way that we can mesh jurisdictions, at least in that area.

Going forward, however, how do we address the jurisdictional problems that we foresee? I think it would be a mistake to think that if we are going to restructure the industry, we can leave the regulatory jurisdictions untouched. If we are going to have a fundamental restructuring of the industry, then I think we also need to have a fundamental restructuring potentially of jurisdictions.

From my perspective, I think there is enough jurisdiction to go around. I think it is just a matter of making sure that it makes sense in the new world. So I think what that means is that hopefully States will have a chance, as Commissioner Russell said, to incubate some ideas. I would like States to have the chance to have some, by State agreements, some regional-type regulation. That is new to us also, and I would hope that the Federal Energy Regulatory Commission would give those types of things a chance to work before necessarily imposing a Federal solution on all these problems.

Finally, one might ask, why is Wisconsin even looking at retail wheeling? We have some of the lowest rates in the country. We have some of the strongest-rated financially utilities in the country. And I do get the question quite a bit; why do you want to change?

Our industrial customers are not asking for retail access, but I think the time to make the change is when good public policy can be the driver of the change and not economic or political expediency. I would compare it to when you are on top is probably the best way—or the best time to figure out what you are going to do to stay on top, rather than waiting until you reach the bottom and try to figure out how you are going to get out of the hole.

So I would be happy to answer any questions and I hope my testimony was useful to the committee.

[The prepared statement of Mr. Neitzel follows:]

STATEMENT OF SCOTT A. NEITZEL, COMMISSIONER, PUBLIC SERVICE COMMISSION OF WISCONSIN

Good morning. Thank you Mr. Chairman and members of the subcommittee for giving me this opportunity to testify before you today on the changing nature of the electric utility business. The views expressed in this testimony are my own and should not be interpreted as the policy of the Public Service Commission of Wisconsin (PSCW).

As a State utility regulator in Wisconsin, I have inherited a proud tradition of regulatory excellence. Wisconsin has some of the lowest-cost providers of electricity in the country. This has been achieved largely through intelligent planning by Wisconsin utilities within the context of our State's Integrated Resource Planning process, one of the most thorough and mature processes of that type in the country. Furthermore, our utilities' rate structures are, by and large, based upon marginal cost principles, and recognize the necessity for rates to be competitive in rate classes where competition exists.

This puts Wisconsin in a very different position than other States, where costs, resource availability and past utility or regulatory decisions have resulted in rates that are high in comparison. Looking to the future, our immediate concerns are also very different. We are not worried about losing industries to lower cost areas—we are concerned about maintaining the price advantage our State has achieved by dint of a lot of careful work. In a nutshell, we want to maintain and improve the competitiveness of our prices, while providing a structure that is open to new choices for our customers. We also want to continue to satisfy as many of the concerns of our customers as we can.

It is my belief that the competition genie is out of the bottle for the electricity industry. While the wires will remain a monopoly; electron production and sale will probably not. State and Federal regulators and Congress have made fundamental changes in the generation sector and the wholesale sector. We cannot expect that the retail sector will be left behind. Therefore, policymakers have two choices: (1) attempt to put the genie back in the bottle wasting valuable time and resources or, (2) request those policy outcomes we think will most benefit society.

I suggest we choose the latter option and embrace competition in the electric utility industry.

Now the difficult part. We have to decide on our three wishes. While we might like more than three wishes it must be understood that in the competitive world, just like the regulated world, we can't have everything we desire. We have known the regulated world for about 60 years and as they say, "familiarity breeds contempt." Now we are looking in earnest at a new regime. Our goal should be to make sure it is better than our current system.

I submit to the committee three wishes we should request from the genie that State and Federal policymakers pursue as we shape a competitive model for this industry.

Wish No. 1: That we create a truly competitive market, not just allow firms to compete.

Absence of regulation does not equal competition. Further, public policy often confuses competing with real competition. My definition of competing is a high school football team playing against the Dallas Cowboys. There may be a level playing field, and both teams would play by the same rules, I would even bet those kids would compete their hearts out and maybe, just maybe, by chance they might win a game. But probably only one out of 100. This, however, I would not classify as robust competition. This would only be competing. We must avoid a similar structure in the electric industry. In the absence of pervasive regulation, robust competition is the only other option to protect consumers. (N.B. Competition, unless it is illegal, is always regulated to some extent).

Also, in my view of a competitive world, all customers would have choices of electron and service providers. As a matter of policy we should not leave any customer captive to a single provider. This would not be consistent with a truly competitive market. If customer choice is good for municipal utilities, then it should be good for large industrial customers, and if it's good for industrial customers it should be good for commercial and residential customers. We should also avoid the massive dissatisfaction of a majority of customers who may have more choice than they want, by preserving the option of plain old bundled service from current providers until we are sure that the bundled option is a serious impediment to the development of a competitive market or new technologies.

I have a choice of tire suppliers whether I am buying one tire or two hundred. While my choices for buying one tire may be fewer and more expensive, I still have a choice in a competitive market. If in the electric utility industry we cannot give everyone choice, then I would question whether or not it is truly a competitive industry or just competition living a cross-subsidized parasitic existence on a large captive market.

My vision of customer choice also includes the ability of customers to shop for supplies from renewable providers, and alternative demand side management suppliers. In a regulated world State regulators have to make a judgment about these supply mixes and we may be selling demand for clean energy resources short. Allowing customers the ability to make independent decisions may actually be a big boost to the market penetration of renewables and demand side management. If customers want these alternative supplies, we have to ensure that a market will provide them. If customers choose not to purchase these services in the market place, then we may have been wrong in our assumption that customers were demanding them, or we will find customers were willing to demand them, just unwilling to pay for them.

In a free market, we have to have the courage of our own convictions and realize that people have the right to make their own decisions.

In creating a truly competitive industry, the sharply defined lines of each State's boundaries must begin to disappear. States should be given some time to establish regional solutions, or bi-state reciprocity agreements. Premature Federal intervention in forcing such solutions may unduly restrict the flexibility that is needed to meet each region's unique existing circumstances. Without a multistate perspective, the possibility of true competition will be artificially restricted by State lines and difficult to realize.

Wish No. 2: That a firm foundation of a competitive wholesale market be in place before we begin retail competition.

There is always a rush to get ahead of ourselves. While time may be of the essence we should also remember the quote: "there never seems to be time to do it right, but there is always time to do it over." We must ensure that, if a competitive market is created, it will be sustainable. Anyone who has built a house knows the feeling. The fun part is picking out carpeting, appliances and paint colors. The drudgery is watching the foundation being poured and the 2x4's being put in place. However, if the foundation and infrastructure is not done well, all the beautiful decorating won't amount to much in a couple of years when the house collapses.

The initial hard work has to be done at the wholesale level before we can move on to retail competition. By the same token, we must realize that competition in the wholesale market will inevitably force competition in the retail market. The most important part of this is having an open access transmission system that allows all users to obtain comparable services and prices as the transmission owners. The transmission system was developed by the means of eminent domain and should therefore benefit the public at large and not enhance the bottom line or competitive position of its private owner(s). I believe the transmission system and the distribution system are the only natural monopoly components of the electric utility industry. Wisconsin has aggressively implemented wholesale power market competition while vigorously protecting the ratepayer and the environment.

(1) In 1986, before EPAct gave the FERC the authority to order the filing of transmission tariffs, the Public Service Commission of Wisconsin (PSCW) ordered the Wisconsin utilities to file wheeling tariffs at the FERC. The PSCW actively participated in these proceedings to support the development of tariffs that would facilitate wholesale competition. All Wisconsin utilities now have wheeling tariffs on file with the FERC.

(2) In 1990, in an interim planning proceeding, the PSCW required that power purchases be considered as an alternative in all generation construction cases.

(3) In 1993, the PSCW initiated a competitive bidding process for selection of generation additions that has resulted in lower costs for Wisconsin consumers.

(4) In a recent transmission construction case the PSCW has ordered the three major utilities involved to upgrade their tariff to the new FERC standards of network service and comparable service. The PSCW will participate in these proceedings at the FERC to contribute to the development of tariffs that are conducive to improved wholesale power competition.

(5) The PSCW is participating in the Commonwealth Edison and Northern States Power tariff cases to again push forward the development of the new FERC standards of network service and comparable service so that these standards are defined in a way that is conducive to improved wholesale power competition.

As a result of these activities Wisconsin already has created a foundation of competition in the wholesale power market. This is not the situation in many areas of the country. Wisconsin consumers are already receiving the benefits of lower prices because of increased competition.

But the PSCW is not resting on its laurels; we plan to initiate a proceeding to investigate the potential benefit and possible approaches to further implementation of competition in the electric industry. This will include an investigation of the potential benefits of implementing retail wheeling.

My point in presenting this list is to show that much foundational work has been done in Wisconsin. Therefore, while Wisconsin may not be forced by economic circumstances to implement retail competition we may be in the best position to design and sustain a retail wheeling scenario because of our past efforts at the wholesale level.

This list also illustrates how State and Federal jurisdictions can work hand in hand to implement a competitive paradigm for the industry.

Wish No. 3: That we not over-promise and under-deliver the benefits of competition. Too often in our attempt to gain public approval for our actions we make claims that nirvana is just one policy decision away. We cannot, I believe, promise only the good from competition and hide the possible negative aspects from view. If the policy decision is to have electricity suppliers act in a competitive manner then it should be understood the utilities may no longer be "the off budget United Way" to fund any number of social objectives. This is a decision which needs to be clearly laid out for the public. Some of those programs we like the best (DSM, conservation, environmental programs, etc.) may no longer be able to be dictated through the regulated utility. The decision must be made as to whether or not these programs have societal benefit. If they do, then programs funded by all providers of electricity, all producers of unwanted pollutants or externalities will have to be implemented in the new world. I would suggest these decisions may best be made by elected representatives acting on behalf of their constituents.

As we make these policy decisions it should also be our objective to have processes such as this that are as open as possible. Energy policy affects everyone in every State. The public policy test has to be something more than "it's new and different." We must work to create something better than the current situation. The public deserves a full debate.

We should also realize that, as in the movie Aladdin, we cannot ask the genie to make someone love us. In fact, no matter what path we choose there will be, as always, those interests upset by our actions. However, I would submit, that good public policy should be defined as being something greater than just the sum of the individual interests.

Mr. SHARP. Well, thank you very much, gentlemen. You have provided us some very thought-provoking and important perspectives on some essential questions of our day. And all of us have been not shocked, but pleasantly surprised at the widespread attention to these hearings, and other fora going on around the country which is an indication of how intensely these issues are being viewed and how significant they are.

And indeed, it is interesting that just the three perspectives of those of you who are on the cutting edge of these issues among the States, that you come at them from such widely different bases, which I think confirms in part what Mr. Fessler and all three of you were talking about, is the quite different nature of our States and the continuing value of the Federal system.

Mr. Neitzel, you were just indicating that contrary to what Mr. Fessler faces among his ratepayers, that you don't have this major cost differential that would drive customers in appropriate fury to come to you and others and say: Please, bring about change.

Mr. Russell, I don't know in your State if that is different. I trust that the large users are seriously concerned about cost and that is partly what is driving this; is that a fair assessment?

Mr. RUSSELL. That would be correct. We are not as high-priced as California, but we are not as also low-priced as Wisconsin.

Mr. SHARP. Well, let me follow up on Mr. Fessler's call for cooperative federalism, and I know all three of you addressed that with some degree of specificity, but I would like to see if you might lay out for us any more specific suggestions about how we might do that.

If you saw Ms. Bailey's testimony, she was outlining sort of general propositions, which have been under discussion in the country but seldom used in the past. And I wondered if you could be a little more specific, whether you think there is a point for joint boards, whether there is—there are techniques or just joint conferences and discussions, or whether there is a philosophy of deference by the Federal Energy Regulatory Commission, except where they see that there is a true problem for other States or for the interstate system.

Mr. Fessler.

Mr. FESSLER. Mr. Chairman, I could not spend a greater effort in praising Commissioner Bailey for not only the content, but also the tone of her statement. I think it is exactly what in California we are looking for and what we find to be most encouraging.

I believe that what we are attempting to do in California is to provide a forum, a forum for the discussion of issues that have been debated behind closed doors, have been parsed in boardrooms, but the public has rarely been able to hear what is at stake and what is potentially there in terms of change.

We have also tried to explicate what a world of retail access might look like, and to invite parties literally from across the country, and indeed our invitation has been taken up beyond our wildest dreams as we have had witnesses come from around the world to participate.

We recognize that we do not have a self-sufficiency in my Commission, in my Commission augmented by an agreeable legislature or augmented by an agreeable executive department of California's government. We are but one component, albeit the largest consumer component, of a regional market.

And so the first thing I think that we need to do is have a period of time to explore with some degree of specificity, so that we are not discussing cliches, but we are discussing proposals, and to begin to see if we can build a regional consensus, that this would be desirable. Because it cannot be desirable only if it works for California, if it is disagreeable to Idaho, or is it disagreeable to Oregon or Arizona, the other States.

But as I have also indicated, we find that we are pushed well beyond the traditional realm of discussion of State versus Federal jurisdiction, where our jurisdiction would end and those of our colleagues in Nevada would begin, because we obviously share a border and a transmission grid with Baja, California Norte.

We are the major counterpart trader with British Columbia's hydro-energy. We are part of an interdependent system. The FERC clearly has the dominant regulatory authority over what I will call "transjurisdictional issues" that take place there. But what I have been asking in California, what I ask you and your colleagues to pursue here, is to see that it will be insufficient to the needs of western Americans if we come to some understanding with the FERC or the Congress or the Supreme Court of the United States, if we do not also recognize that the market that we are talking about accommodating includes vital participants from Canada and from Mexico.

And I think that at once adds an elaboration to our problem, but it adds a dimension to our opportunities because it forces us to begin with the premise that no one has the coercive capacity to say: So let it be written, so let it be done. That era, if it ever existed, has passed. And now government, in cooperation with the private sector, governments across jurisdictional lines that have previously been thought to be steady and secure, must now begin to work towards solutions.

We have opened a forum. You have opened a complementary forum. Other forums will be opened in other jurisdictions.

If I may be permitted to indulge the idea that when you have left the distinguished period of service in this body, I would hope that you would continue to be interested in these issues and would continue to exhibit leadership. This is a matter that will not be resolved in California and it will not be resolved in Michigan or Wisconsin. Each of us is trying different things.

I envy my distinguished colleague from Wisconsin. I don't think I will ever be in public life on the point that ratepayers come to me and complain that their bills are too low or that things are just perfectly agreeable. Such is not my life.

I am looking to help a dialogue to formulate institutions that will treat California's ratepayers better than we have all collectively through good-faith efforts been able to accomplish to date. That involves reaching out to the jurisdictions that are around us.

It emphatically involves reaching out to the Federal Energy Regulatory Commission, to the Department of Energy, to the Clinton administration, and most certainly to this distinguished Congress.

Mr. SHARP. Thank you.

Mr. Russell.

Mr. RUSSELL. Thank you, Mr. Chairman.

In Michigan we are fortunate in that we have an exemption to the Open Meetings Act. My colleagues and I can go into a room and deliberate on what we will decide on a particular order.

We can bring in staffs to explain the technical aspects of that, and then the only way we can act is through our order. There are certain benefits from that. You get a better record. You get those who are appointed to decide, in fact, deciding. You get less speculation from Wall Street from open debate.

The complexity of the issues, the speed of deliberation, the reduction in staffs and resources all indicate a need to have discussions with each other as opposed to double hearsay, where "he said, she said" types of debates go on as you try to affect these types of policies.

A concern that I would raise is the inability, one, of the FERC to talk to itself. You have to have a quorum of members. If you have a quorum of members, then you have to have public notice and you have to go through the whole process of the Open Meetings Act, and not to say that the deliberations in closed meetings are more secretive and therefore something bad is going to come out of the back end. It just means that the collective wisdom of the appointees is increased by their ability to converse with each other.

That Open Meetings Act is an impediment, I believe, in my opinion, to the discussions of the types of reforms and the types of changes that are currently going on in this country. And to have to go through a third person in order to dialogue, I would pose to you, if you had—you and your colleagues had to deliberate on the same types of issues that we deliberate today through a third person, without the ability to talk to your colleagues, I would virtually assure you that you would come to a standstill in trying to deal with the complexity of these types of issues.

Mr. SHARP. We are close now.

Mr. RUSSELL. And so there is an impediment there, with the FERC and its ability, State commissions and their ability. That needs to be addressed, the ability to talk.

Joint boards should be explored. We currently have on the books, legislation that authorizes joint boards. It has not been explored by the FERC. It has been explored at the FCC. But there may need to be some types of adjustments in those joint boards arrangements, smaller subunits in order to deal with the regionalism as opposed to dealing with a macro-transmission on a national basis in that type of organization.

The third piece that needs to be done is that there needs to be some type of deference given to State decisions by the FERC, and as long as States have a range and provide a forum and opportunity for all the parties to participate, to have their day in court, to be second-guessed by a Federal agency who is, at most, distant from the action with respect to facilities, with respect to facilities, ratepayers, parties and transactions, where States are more close to that action, that deference should be given to States. And I would agree with my colleagues that there is some responsibility in NARUC, through which I am involved, is also trying to work with—trying to assure that parochialism by the individual States do not surface to subvert the types of changes that were taking place.

And finally, I would look to the development of the RTG's as an example of the need for the type of cooperation and the type of coordination that is going to be required in order to move this system of a duality, of facilities that have dual purposes, dual regulators, and that deference be given to States if, in fact, they can provide those forums and those opportunities.

Thank you.

Mr. SHARP. Thank you, Mr. Russell.

Mr. Neitzel.

Mr. NEITZEL. Well, I am always caught in between on these things. In an earlier life, I worked at the FERC and now I am at the State Commission, so I have kind of seen it from both sides.

I think—and let me assure my colleague from California that even if you get rates low, ratepayers will still find many reasons not to like you. They have me. I don't take them personally.

The joint board, I think, is something that needs to be explored or at least some way that we can discuss these things. I am not sure that a conference is the way to go. But then the question becomes who serves on the joint board; and I know that the FERC is sometimes more interested in that than less interested in it. But I think in this era, it would probably be a real positive thing to explore because it would give at least some State regulators or our leadership the chance to converse with FERC just to know kind of where things are going and have maybe a common strategy.

The other thing is maybe in the interim, maybe we want to leave FERC at some point in some appellate jurisdiction to say, we are going to set these things aside and leave them to the States until such time as the States are hurting each other or one State files a claim against the others, and then they take it on that basis. Because there are certain things that I am not sure that the FERC would want to get into.

I know in their prognostications on stranded costs, they said, well, we have—we probably have jurisdiction over retail wheeling and the rates charged for retail distribution. And even if you agree with that and accept their rationale, I am not sure that they want that. I am not sure that they wouldn't want to let the States try to work that out first before they got it.

It seems to me if we have learned anything on the telephone side is that, you know, we have massive competition on the telephone side, and one of the things is that you can now own your own pay telephone. So now I stop at what is called the COCOT and it says if I have a problem, I am supposed to call the FCC.

Well, I had a problem a couple weeks ago. I am a State Commissioner, I didn't even call the FCC. I doubt if a regular consumer would. But I know that regular consumers call the commissions in the State very, very often. And so I think we need to look at these things, like I said, in a rational way.

Sure, we can say, yes, we have jurisdiction here and, gosh darn it, we are going to take it, or, no, you don't. I think the joint board might give us at least a forum to start to explore these things, say what in the real world makes rational sense.

Mr. SHARP. Well, thank you gentlemen.

I am not sure—you had indicated, Mr. Neitzel, the need to restructure the regulatory system at some point. Obviously, the issues come up but nobody has a plan, and I am not sure we are ready or it would be wise to have a plan at this point.

I do think what you raise—before I recognize my distinguished colleague from California, I want to indicate that I really think we are in an era which is going to take a great deal of skill and leadership at the State and Federal level to work these things out, and it behooves the public which doesn't normally wish to focus very closely on this kind of thing. That is why they delegate the responsibility to us and to you, to make sure that we, in fact, are providing the resources and that Governors and presidents, to the extent these are appointee positions, are recognizing the compelling need to make sure we have folks like yourselves who are intensely interested, knowledgeable and committed to examining these questions.

I personally believe we are very fortunate at the moment in the decisions that have been made in the many appointments at FERC, because we have an unusually high-caliber team. We have always had. We are into serious policymaking, not just whether or not you marginally improve or hurt a class of consumers in a very marginal way on a rate case. We are in big-time stuff.

It does strike me kind of interesting, do all three of your agencies have jurisdiction over telephones in your area and how about natural gas as well? It is interesting, when you speak with the—you often get the analogy of telephones when you speak with the State Commissioner. Because we divide that authority at the Federal level, the tendency is to be with gas and electric and therefore the analogy is made to that. Maybe there is need for more cross fertilization here on that.

I find my own disability is that I am focused on natural gas and electricity and struggle like the average consumer to figure out what on earth is happening in telephones, not to mention in natural gas and electricity.

But let me recognize and share time with the distinguished gentleman from California, the ranking member of the full committee, Mr. Moorhead. Mr. Moorhead, we put your opening statement in the record, recognizing that you had other duties this morning.

Mr. MOORHEAD. Thank you.

I did want to say especially that I was glad to have before us today representatives of the State commissions that are taking steps to increase retail competition. I am particularly gratified that Chairman Fessler is here from California, my home State.

Over the last several months, the industry restructuring proposal of the CPUC has raised the debate over the future of the electricity industry to a new level of intensity. I think we can tell that by the conversations here this morning.

When we passed the National Energy Policy Act, our objective was to create an atmosphere in which wholesale competition could flourish. Similarly, the CPUC proposal attempts to allow the discipline of market forces to work on the retail level.

Chairman Fessler, the overriding theme of the Energy Policy Act transmission access and PUHCA provisions was to promote competition as the way of reducing electricity costs. I am painfully aware of California's economic problems, as you are. These have resulted in 11 percent unemployment in our State.

I think it is down slightly from that level now, more like 9, but it is still very severe. The CPUC restructuring proposal states that California's electricity rates are 143 percent of the national average and identified retail competition as a way to get these costs down to keep businesses in California. Could you please explain what effect high electricity rates will have on California's economy if we can't get these rates in line?

Mr. FESSLER. I would be delighted, Congressman, although what I will be reviewing known to your good self and Congressman Lehman, are very disagreeable consequences that we are experiencing in the agricultural, commercial and industrial sectors of the economy.

There are many problems cited by businessmen and women doing business in the State of California. The one that is of great concern to me and is the focus of your inquiry today is the fact that electric and gas rates are higher in our State than the national average, but it is even worse in the comparative sense of looking at the States that border California. Arizona and Nevada and Oregon are below the national average, so that the situation is exacerbated by the proximity of States to us which furnish alternative sites for the siting of plants.

I have had businessmen and women come to my office and inform me of closures that were taking place, or of expansions that were going to occur, but the closure would be of a facility in California, the expansion would be of a facility that would be headquartered in New Mexico, or in Arizona, or in Utah, or perhaps Nevada, and invariably what they pointed to were the savings that they would accomplish in industry.

We have had a debate in California for more than a decade that we have high rates and low bills, but the bills that are below the national average are householder bills, and if you are condemned by virtue of the nature of your economic enterprise to be a substantial user of electricity, then the fact that we have high rates becomes a significant impediment toward finding California to be a competitive climate to retain your business.

Now, what we are discussing—I hold out no illusions of an instant cure—if a retail access component shows up in a final proposal from our Commission and if it has a sufficient jurisdictional predicate or cooperation from those who do that, it becomes effective, it is the best we can offer, that it would set in motion a series of events that would promote greater efficiency in generation, and through that greater efficiency, would lower those rates.

I also believe that it would move our utilities, which have already made substantial progress, in the direction of seeing themselves as businesses, businesses that are very interested and conscious of costs. So that the efficiency gains will not only be on the generation side where there is already competition, but along with the performance-based ratemaking aspects of our proposal, which are a package of reforms, we would begin to see a greater efficiency, and in that greater efficiency, lower costs.

But the short answer to your very pointed question, what are the consequences of having 143 percent of a national average. When you sit on the edge of States that are below the national average, you lose jobs. You lose infrastructure, and that infrastructure is very, very difficult to rebuild, and our economy is rocked by many other factors, but it is exacerbated by our energy rates.

Mr. MOORHEAD. At the present time, are industrial customers currently subsidizing small customers?

Mr. FESSLER. In the historic pattern of ratemaking in the State of California, it would be true to say that the rates that have been set for industrial and large commercial users can—the point is a contentious one, but I think that on balance, when all of the arguments have cleared, it could be said that we have set rates over years which have tended to favor householders. That has been true in telephone, it has been true in gas, and it is true in electricity.

The Commission has been moving toward long-term marginal cost as the basis for ratemaking, and we have substantially realigned rates in California. And so today it would be to a much lesser extent than was historically the case that there would be any cross-subsidization going on.

Mr. MOORHEAD. My own district is not directly affected by CPUC to any great extent, but whatever happens in the other districts of the State has to be reflected very shortly by the city councils in my district who run their own municipal electric power company. How will CPUC's restructuring proposal work to reduce cost?

Mr. FESSLER. Well, you raised two issues, Congressman. One is an important one to note for the record, and that is that in California, the California Public Utilities Commission does not have ratemaking authority over public utilities, such as municipal utilities in Pasadena, and therefore the direct consequence of any orders that we might make in the ratemaking area, as opposed to safety, would have no immediate consequences because our authority simply does not run to those entities.

But those entities are very active and vital participants in the discussions that are going on in front of the Commission at the present time, because they recognize that they also have a vital stake in containing cost. And so when we speak of jurisdictional issues, one jurisdictional issue we have to address, Mr. Chairman, is the jurisdictional issue within the State of California, before we begin to worry about it beyond. How will they participate in lowering cost as a consequence of this?

Again, if it should eventuate that generators from outside of California or generators located other than in historic service territories that your city council would have been dependent on, can offer electricity at lower cost, and if we can move to a merit generation dispatch so that the need of a society at any given period in the day is met with the lowest cost sources of generation willing to go on line at that point, there will be cheaper electricity flowing into the system. And I am certain that the constituents that you have would insist that lower bills come out of the other end of the system. Whether that is out of an investor-owned utility or a utility that is responsive to the direct wishes of voters.

Mr. MOORHEAD. I have a question that can be commented on by all of the commissioners.

The Energy Policy Act of 1992 is a major step in furthering the development of competitive power markets. The acquisition of bulk power resources in most regulatory jurisdictions is now subject to competitive bidding, and States such as many of yours are promoting retail competition. In this evolving, competitive world, do you feel the playing field is level? Are there further changes to Federal laws that need to be made to make the competition fair?

Mr. RUSSELL. At this point, I don't know that we need further change in legislation that will increase competition.

What I would suggest is to let the dust settle but let the States have an opportunity to adapt and to adjust to what the playing field is at this point in time. I don't believe that unless you need further clarification—now, there are some other pieces. We are looking at other pieces of legislation that fix gaps, Ohio Power, fix where we are trying to deal with registered holding companies.

The Congress was faced at the 12th hour on the Energy Policy Act by regional transmission groups. We are now in the process of trying to develop and bring together those types of organizations.

I think what we need is—there are some problems, some gaps that could be fixed with regulation, but level the playing field only means that we can fill those particular types of gaps.

I would suggest that you allow the States to have the flexibility to adjust and to adapt and see what happens there. I think you will be pleasantly surprised as we take responsibility for our own destiny on these issues.

Mr. NEITZEL. I think I would agree and say that we need to see how things shake out, but I think we also need to realize that, you know, the utilities—utility industry has been a real workhorse and now we kind of want to change it into a thoroughbred, and we want to be fast and sleek and mean, but yet we still loaded it up with a lot of old rock in the wagons, and so we got this Clydesdale there and now we can't understand why they are not beating the thoroughbred to the finish line.

Well, if we are going to move towards competition, I think we have to have the courage of our own convictions and say, OK, this monopoly used to be a real advantage. Now, it actually may be a monopoly over the customers that nobody else wants to serve or the high-cost, low-revenue customers.

So there is your monopoly, have fun with it. That doesn't really create a level playing field. But by the same token, I think we as regulators have to make sure that there is more to this than just leveling the playing field. The analogy I have always heard is that a high school football team can play the Dallas Cowboys and that will be on a level playing field, but it won't necessarily be real competition.

So I think that is what we need to be concerned about. You know, to get to real competition, do we want to keep Emmett Smith out of the game, or do we want to give Troy Aikman to the high school team, something like that. Those are the questions that go beyond just leveling the playing field.

Mr. MOORHEAD. Mr. Fessler, do you want to comment on that?

Mr. FESSLER. I would say, Congressman, as you heard, when one comes from Wisconsin and rates are so low, one has the luxury of developing an elegant means of expression.

I think I would align myself with Commissioner Russell in suggesting to you that I don't think that there is a proven case at this moment for further Federal legislation.

86-340 95 - 10

I think that the conversation and colloquy that you had last week with the FERC commissioners in which you inquired as to how your act had done with regard to having a sort of tonic or physic effect on the transmission grid, and you were being told that you had struck a very useful balance, because even the most recalcitrant utilities were now coming along and opening the grid, suggests that you have put some very useful tools in place, and that the industry—and I think this is the important point because we continue to talk about organs of government—that the industry is rising to the challenge. And I would echo the call for a period of—as we try to figure out some of these factors, at least don't move the polar stars. That would—that I think would just confuse everyone at this point.

So I would hold current—Congressman, with the current Federal legislation. I agree with the Chair's characterization of both the talent and the temperament of the current members of the Federal Energy Regulatory Commission. And I would also say that you will find that the States are not approaching this from the vantage point of believing themselves to be a California, Michigan, or Wisconsin uber alles approach.

We, too, recognize that things have changed and that the industry that we deal with, once we got rid of absolutely sacrosanct, vertically integrated monopolies with service territories, all traditional thinking was under assault by simply the forces of reality.

Mr. SHARP. Thank the gentleman.

The gentleman from California, Mr. Lehman is recognized.

Mr. LEHMAN. Thank you very much, Mr. Chairman.

And let me follow up on Mr. Moorhead's line of questioning to some degree and I certainly don't wish to enter into any legal debate, but only to try to get the facts on the record and shed some light on this issue for the public on what is happening, and what I believe is a very important policy decision in my State and nationwide as well. And, obviously, I have some personal concerns as the author of the amendment on retail wheeling when the Energy Policy Act was before this committee, and I have written Professor Fessler on it as he knows, and I appreciate your prompt response to me in that regard. I certainly think we share a desire to provide consumers with effective and economic service and I am very pleased that you are interested in cooperating with the Federal Government.

The order proposed by the CPUC was a very strong statement and an assertion of authority over issues that really haven't been fully debated here in this body, and to many of us, that fired shots across our bow, and I guess the bow of the utility industry as well. I felt compelled to respond.

The proposal that the California Commission has put forward makes passing references to the need for changes in State law and then claims that Congress in its savings clause cleared the field over retail competition with an implicit request that the States occupy it.

I am concerned, first and foremost, obviously, for my California constituents who fear that overaggressive proposals and the ensuing uncertainty might undermine their electricity service.

Already the stock of the three major utilities in California have experienced a $3 billion decline in shareholder value following the CPUC proposal, and obviously you represent many investors as well who believe they were buying into conservative utility stocks and weren't expecting that kind of volatility. Looking at that, to what would you attribute the apparent free-fall in the value of those stocks?

Mr. FESSLER. Congressman, I don't know exactly what the predicate is for the computation which leads to the figure of $3 billion. I think that utility stocks in general in the United States hit their high watermark for value in the fall of last year, and had begun to substantially decline.

It is through—you will be hearing from investor advisors who are far more expert at these prognostications than I, but the movement in interest rates was said to be a considerable factor, an uncertainty about interest rates.

It is also undeniable that the uncertainty that followed in the wake of our April 20th proposal has contributed in California to further declines in the value of the securities of the three investor-owned utilities. And I absolutely want you to know that not only I, but all members of the Commission, share your concern with those share declines.

Those investors are an important part of the equation in maintaining the service and institutions that we have in the State of California, and it is also worthy of note, as I am sure you have taken into account, that one of the statements and goals articulated in the proposal was the preservation of the integrity of California's investor-owned utilities.

Those who advise investors, in my limited experience, Congressman, don't like uncertainty. They very much prefer that things remain on an even keel and remain quite familiar. The introduction of any form of uncertainty is an anathema to the markets.

Yesterday, the Chairman of the Federal Reserve came to the Hill and made a statement and the markets reacted instantly and nervously. That type of uncertainty is regarded by me as a very unfortunate consequence of pursuing any discussion of change. And I think that it is a consequence, albeit a disagreeable one, of the fact that we have elected to discuss these changes and the potential changes going on within the industry and their consequences on the industry in public, rather than attempting to suppress these things behind closed doors.

I also would say to you, Congressman, what I said in New York about 10 days after our proposal was made in April. When I went back at the invitation of the investor community and I complained in what I hope was a well-received message, I indicated that I was shocked at the degree to which certain investing houses within 24 hours after our proposal was made were telling their clients that the California public utilities had ordered the implementation of retail wheeling.

A distinguished investment house said that this would be a day that would be remembered in history as the day that retail wheeling was ordered in California. Of course, not even the most venial acquaintance with the document that we put out could be characterized as an order.

And in my business as a lawyer, I tend to carry a rather high profile of malpractice insurance. I do not know whether there is malpractice insurance in the financial community, but if there is, there may be claims that will be generating.

What I am saying to you is that the order was not well-interpreted. Now, much of that must be our fault because when one speaks, one has the obligation to attempt to be heard and understood, and I regret and apologize for any deficiencies on our part.

But also I think when one speaks, listeners have an obligation to try to listen and understand, and it eludes me how people can walk out of a meeting with a proposal of this magnitude and proclaim dispositively on the plot lines of plays they have not seen or hum the tunes of music they have never heard nor read, and that I think has contributed to part of the problem that we are having.

Mr. LEHMAN. Well, I think, you know, investment advisors usually push, if you will, utility stocks as relatively stable income-generating investments, and I think many of them were probably afraid that they were going to be accused by people, to whom they have recommended those stocks on that basis, of misleading them in the wake of the publicity around CPUC's proposal.

Mr. FESSLER. The Florida Power & Lighting dividend cut, which occurred weeks before we acted, is also cited as a major factor and exciting concern within the investment community.

Mr. LEHMAN. I recognize the other factors involved here. But you as well have recognized the possibility that this had an impact. Certainly it did. In the future then, if we embark down the path that the CPUC is headed or appears to be headed, or again, without prejudging every detail of what is going to happen, we could expect utility investments then to be more volatile than they have been in the past, correct?

Mr. FESSLER. No, I don't know that I would necessarily agree with that. They will be volatile in the sense that California's investor-owned utilities, even as you and I are speaking, are in the process of reinventing themselves. They are in the process of seeing themselves not only as the supplier of electrons or natural gas molecules, but also as major movers in the suppliers of what are called energy services. They are involved in attempting to help restructure our economy.

Two of our investor-owned utilities have recently gone into candidly the banking business in helping to front to industrial consumers, commercial consumers funds and assistance in upgrading and modernizing their plants and equipment which will be repaid by the maintenance of historic bills, but will, over time, return to the utility not only its investment but a return on that investment.

So I think that it would be something that you would not wish and it would be something emphatically that I would not wish to do, is to contribute to the idea that the only thing that can happen to California investor-owned utilities in this process is negative. What is happening to California investor-owned utilities as they respond to this and other challenges is, I think, full of creativity and full of positive implications.

Mr. LEHMAN. Now, and just following up on that, then, what about the general public consumer of electricity who is not a commercial enterprise, who is not going to be able to go out and negotiate with some independent power producer, who is still going to be dependent on PG&E or Edison or whoever provides their power? Can you tell me that their rates are not going to increase as a result of this kind of policy that allows for, in effect, cherry-picking of major consumers out of the rate base?

Mr. FESSLER. I would wish to be understood as speaking exclusively for myself, and under the obligation to respond to you in good faith with a recognition that I am still in the process of acting as a decisionmaker in this matter insofar as it is under the authority of my Commission.

The staff proposal contained in this blue book said that there should be no cross subsidization and that there should be no cost shifting that would occur in the process of this industry reform legislation. So the staff proposal is that the answer to your question should be that we could assure you that there would not be a consequence in which the—as my colleague from Wisconsin put it, that there would emerge a class of consumers that were high cost and no one served and it was on their backs that we loaded all of the cost of this transition. That would be wrong. Indeed, this proposal envisions that all customers would have the direct access opportunity.

Please remember that the preface to this document says that we were open to other ideas. We have received some very interesting ideas that may be of interest to you. We have received a proposal that says that we could accomplish virtually everything that is set forth here and provoke no jurisdictional issues at all if we simply cooperate with our sister States and with the Canadian provinces and the Mexican states, with the Federal Government, to absolutely maximize the existence of a wholesale regional market for electricity in our area, use the investor and municipally owned utilities within the State of California to shop that market, would dispatch electricity on a merit system that would be weighted in favor of low-cost providers, and would then allow customers in California to have time-of-use rates across the board so that they would see what the cost of electricity was, and for the first time be able to influence their rates by changing their patterns of behavior. That has been an idea advanced by Professor Hogan. It was immediately dubbed virtual retail-wheeling by Amory Lovins, who was the succeeding witness before us.

That idea, if it were to be the one that we attempted to implement, would say that there would be no phasing of this order, all customers would be able to go to this type of an arrangement and partake of the advantages of that type of a wholesale market and that, I think, would be a very interesting thing, because as Professor Hogan reminded you all last week, there are many fundamental misconceptions of how this industry works, and one of them is that under a retail-wheeling scheme as might be envisioned or would be sensible in my mind, that people would actually physically leave the system. They don't. They remain connected to the system, and the issue is the terms under which they pay for the service they take off that system.

So I think that it is within our possibility working together, to answer your question, that there will be no residual class of ratepayers upon whose backs this reform would be accomplished.

Mr. LEHMAN. Thank you for that statement.

Mr. SHARP. The Chair just wanted to inquire if the gentleman has many more questions.

Mr. LEHMAN. I have a couple more but I would be happy to take them up after other members have had a chance.

Mr. SHARP. I wanted to particularly give Mr. Moorhead and Mr. Lehman time because of the intense importance this has, obviously, in their parts of the country and Mr. Klug may feel the same way.

The gentleman from Wisconsin, Mr. Klug is recognized, and I will come back to Mr. Lehman after.

Mr. KLUG. Actually I want to follow up on one of my colleague's from California's question.

If you have a situation where essentially you have retail wheeling and obviously the first targets are going to be large industrial consumers, and let's say hypothetically—this is an issue that Scott will appreciate—if you have Wisconsin utilities then selling into northern Illinois, not any of Mr. Hastert's customers, of course, or constituents, anyway if you sell into Illinois, it seems to me you have got a problem in northern Illinois because their large industrial customers get siphoned off and then you have got residential customers left to fill all the rates. Do you not see a problem that we are going to have disadvantages for some individual consumers?

Mr. NEITZEL. Well, clearly—I mean, the essence of regulation has been to do things on an average basis, so while it may cost a little bit less to serve me in Madison, Wisconsin, and cost a little bit more to serve in Monticello, we give everybody the same rate.

The trick that we have in Wisconsin, and I think the reason here again why we are looking at this in addition to States that have high cost, is that our concern is that if you move to some regional market price and that reaches some equilibrium, if you were below that equilibrium before, does that mean going to a market based scheme?

While it may be good for customers, industrial customers in Illinois, does that translate into a price increase to customers in Wisconsin?

We have used that advantage in getting customers from Illinois. I think there will be certain customers who will, because of their usage, load factor and just the basic cost, will get a better deal than they have now, because historically there has been some cross subsidy in rates between large users and small users.

The trick is going to be, here again, not disadvantaging what we currently have and also having the courage to let that happen if we choose that to be good public policy. Because right now I can tell you, even at the Wisconsin commission, the real cost of serving a residential customer is a lot more, just the basic cost without the electrons is a lot more than the $3.50 customer charge that we put on people's bills. But that is what we have decided in the name of equity, which you can do in regulation, which may not be available to us under a competitive world.

So it is, how do you pull all these things together and make sure people are better off? Is that responsive?

Mr. KLUG. It is. I think the committee is wrestling with the same issue frankly in telephone deregulation. That is, you allowed more competition in the past because of toll rate averaging, essentially everybody was on a level playing field. And now if you essentially take those competitive barriers—those guarantees down, you now end up in a situation where there is a competitive environment, there have to be winners and losers by definition. So I guess my question is, what happens to the little guys who are the consumers of electric power?

Mr. NEITZEL. I think we are going to have to work very hard and very intelligently to make sure that they are not disadvantaged. One thing that I think is essential is that if we go to a customer choice scenario, that all customers have choice.

My biggest concern is that you have a group of customers who have choice and a group of captive customers, and just the way the world works, I will guarantee that all the costs will be loaded up on those captive customers and somehow it will cost nothing to serve those customers that have choices. So if we are going to go to choice, I think it should be a public policy objective. And I compliment California for this, is that you have to have or assure that if you are going to have choice, if it is good for one set of customers, then you have to figure out a way to give it to the other set of customers, because you can't leave a segment of customers there captive. It just won't work in the long run. They will end up, come hook or crook, they are going to end up taking it, and so you can't leave somebody captive.

That is the other problem we are going to have to struggle with, because you can give choice to a large industrial. It doesn't—the cost of that is small relative to their bill, but the cost of giving choice to a small customer is high relative to their bill, and so we have to deal with the fairness and equity problems as we move towards efficiency.

Mr. KLUG. Go ahead.

Mr. RUSSELL. The potential for harm to come to a segment of ratepayers is real. It won't go away with the evolution that is taking place in the electric industry. However, State commissions have been geared since their inception from the early turn of the century on trying to deal with these problems.

The issue of—you have mentioned the issue of ratepayers, residential ratepayers being subsidized by industrials and commercials. Well, those were policy decisions that commissions had made over the course of time. So those problems have been there since the course of time, since the inception of utility regulation.

My only point here is that those problems exist, they will continue to exist, and the best forum for deciding and weighing the public policy issues that determine who and where costs will be borne is best left to the State commissions who are geared exactly to answer these types of cross questions.

Mr. KLUG. OK. Now let me confuse the mess even further, and now ask Mr. Neitzel this because you have worked both at FERC and now at the State regulator office as well.

In those dual roles then, how are you going to balance out cross-State jurisdictional issues? How do you see the PSC interacting with FERC, and vice versa?

Mr. NEITZEL. I think what I would—we have been in other areas in the wholesale competition area able to use our jurisdiction complementary to FERC's jurisdiction. I think in this debate, I would like to have the State commissions have a chance to work on some bilateral-type agreements or some regional type of regulation, and while that sounds like the panacea, none of us have ever done it.

I mean, I don't know how we would go about picking who represents us at the region and whether or not our States are going to be willing to give up that sovereignty. But I think we have to find a way to make sure that we are allowed, as States—now I am going to sound like a real State's rights person—but we are allowed as States to find our own answers, and to the extent that they don't harm interstate trade and commerce, we should be allowed to pursue those, but to the extent that they harm interstate trade and commerce, then of course the FERC does have to step in. That is what I would like. I don't know if we can figure out how to do it.

Mr. KLUG. Mr. Fessler, I want to follow up with you on this, but don't States have competing interests? If you have got a high-cost producing State and a low-cost producing State, obviously the incentives for a public service commission in a high-cost producing State is to build bricks around it to make sure their producers don't have troubles.

Mr. FESSLER. Well, and if that was a successful strategy I would be here in my mason's apron, but it doesn't work, Congressman.

We are discussing choice as if it is something which, if we became persuaded that it was the wise thing to do, we would allow, and I have a humbler estimate of where we might be. I think the choice is already upon us.

Large industrial customers have choices today that I do not have as a householder in Davis, California. I cannot self-generate. I cannot become a cogenerator easily, and yet the Congress of the United States opened that possibility and encouraged it and now it happens and it goes on. I can leave the State, and lamentably there are those who would help me pack, but if I leave the State, it is the loss of one taxpayer. But if businesses leave my State, it is—as lamentably they have—it is the loss of employers, and then I face people at the supermarket who not only are complaining about jobs, but they are also—or bills. They are also worried about jobs.

So, yes, there are competing interests and to the extent that the market operates to disregard States, that the market operates to take opportunistic efforts to bring about service and to promote efficiencies, then in that sense, I certainly believe that Congress had it right with PURPA. I think you had it right with the Energy Policy Act of 1992. What you have done is set the stage for the emergence of this vibrant wholesale market.

If we can now perfect it and get it right, and it is efficiencies that will have a tendency, because I agree with my colleague that as costs levelize across wholesale markets, then there will be opportunities for sellers that may have been forced to deal in a market in one State where they could find buyers in a neighboring State or four States away willing to pay more for that electricity. But to those buyers, that electricity still represents a very good deal because it is lower than the historic costs that they have been paying for it.

And in my part of the world, those players include—and now Congress has passed NAFTA—they include providers in Alberta and in British Columbia and providers in Sierra Prieto, in Baja

California Norte. So this whole issue, it is here. It isn't whether we will allow choice. It is what we will do now that choice is upon us.

Mr. KLUG. Well, in the interest of time, let me turn it back to the Chair because I know we have some questions from other people. But thank all three of you, Scott in particular.

Mr. SHARP. The gentleman from Illinois is recognized.

Mr. HASTERT. I thank the chairman.

It is interesting to have you gentlemen before us. I guess I go back to the first job I had with public utility regulation, it was rewriting the Public Utility Act in Illinois in 1984. So I have dealt with, gentlemen, the likes of you, even lawyers for a long, long time. And I think, Mr. Russell, we have had discussions before about your view of the States and utility commissions making perfect decisions and holding them forever.

I think it is our job, both in legislatures across this country and certainly the Congress, to try to tweak legislation once in awhile to make the world more perfect if we can do that, and under the leadership of our chairman, I think we have opened up public utilities.

I remember studying the life of Sam Insull, who happened to be a gentleman of some economic push in Illinois at one time, and, you know, looking at what is happening, it is almost a full circle. Insull created electric parks and electric trains and diversities. He created markets for his own utility. And today, all of a sudden, because of the changing in the law, we are almost doing the same thing. We are pushing those people out to get into different businesses so that, again, they in a sense in many ways create markets for their own products.

But one of the things I am really concerned about, and my colleague from Wisconsin brought it up, I represent northern Illinois. I have a community in my district, city in my district, that takes power from Wisconsin, Wisconsin Power, and they have a cadre of old coal-producing plants that keep producing electricity and they can offer electricity a lot cheaper than the 12 nuclear plants that we happen to have in northern Illinois.

So you can cherry-pick and through being able to wheel, you can do that and actually through the provisions that we wrote in 1984, allows you to do that in Illinois as well. But the concern that I have is that as we have more people who are cogenerators out there and in California, you have the windmill plants. In Illinois we have some large corporations that are cogenerators, and you have some people just going out and being cogenerators on their own, PPO's, IPP's, whatever you want to call them, and the law, the requirement that they can buy—sell excess capacity to utilities really pushes the issue that you have a people, a lot of people out there who end up picking that extra cost because there is stranded investment by utility companies in the gas peaker plants that the utility bill, because the law said they had to meet the requirements, that was in their public purpose to do that, all of a sudden that investment is out there and not being used and they don't get a return on it and ratepayers end up making the difference.

How do you see this developing, and what is the check and balance that the States perceive that they can deal with this issue? Yes, anybody.

Mr. FESSLER. Very briefly, because my colleagues are equally if not more capable of addressing this question, first, I absolutely recognize that you have highlighted something which was part of the debate you had last week. When we speak of the consequences of increased competition and look at the circumstances of utilities that made their investments under the aegis of what has been referred to as a regulatory compact, which has clearly changed, the emphasis to date has been talking about stranded assets, generating plants that would be above the clearing price that a market would produce. But you have pointed to something that I think is equally of concern and must be addressed by all of us, and those are what I will call stranded liabilities because Congress and my State legislature and my Commission, make no mistake about it, has ordered utilities in the State of California, and I am sure this has happened in other States, but we took PURPA. It had a Paul of Tarsus conversion in California, and the consequences have been both good and bad. The economic consequences are ones that we must now come to terms with and recognize.

One thing I would ask the Congress to do that we have tried to do in California is as we look at these contracts and we talk about competitive markets, please let us begin to develop some of the psychology of the men and women in the marketplace who know that contracts are the formation of long-term relationships. They are not and cannot be looked upon as a, I happened to have gotcha for a long time in this arrangement.

And so one of the things that will have to happen is some of these contracts are going to have to be modified. As we modify every single commercial relationship in this country on a daily basis between men and women in the marketplace who recognize that no one plays the Sybil for a long-term period of time without getting it wrong, and that we don't have two sticks to rub together in California to waste on anything, and so one of the things we are trying to do is to talk, not only about stranded liabilities, but as we go through this area, to begin to emphasize the notion that long term relationships must be adjusted. No one can ride the horse to the point that it absolutely drops if you are going to need transportation tomorrow.

And so those are very critical questions. I agree with my colleague from Michigan. Those types of issues have been before public utilities commissions from their inception, and while we can point to some success, I also think we must take responsibility for recognizing that we have had some policies which have turned out to have cost consequences and we just cannot now pretend they are somebody else's problem. We must step up to the plate and address those issues in a realistic manner and I think that we shall.

Mr. HASTERT. And the realities is those costs are exacerbated by these types of problems. It makes Wisconsin power look much more attractive in northern Illinois, and again, it continually exacerbates the problem.

Mr. Russell, you are not representing NARUC today, I see, but what is your wisdom here?

Mr. RUSSELL. That is correct. First of all, let me respond that regulations, nor the utilities, nor Wall Street is perfect. We do the best we can. Similarly, the reason why we are here today is to discuss and debate the ambiguities and/or the opportunities that Congress provided us in implementing the Energy Policy Act.

In Michigan, we have tried to deal with this issue as we try to move the question on competition and options to deal with stranded costs, stranded investment with an eye to minimizing the exposure that our utilities would have along that avenue, and we have done that by—in our experimental wheeling program, to put as an order—to order our utilities to provide retail wheeling at the point in time at their next need or the solicitation of need, thus we have eliminated the—or at least reduced or minimized the exposure the utility would have to stranded investment.

To the extent that they need to acquire power means that they are looking either to purchase or to build, not stranding existing assets that are functioning and operating. So to that extent, we have tried to mitigate those types of concerns. And as an effort, again, I would agree with my colleague also, that we do the best we can.

We have done and we have moved issues and moved policies at State levels that have been beneficial to both the interstate commerce issue and to our respective parochialisms in our States, and we have also done some other things that have not been as forward looking or as beneficial to ratepayers, but we are people, we are humans, we do the best we can.

But what you have in front of you are people who, for 8 hours a day, 8 days a week are dealing with this issue, trying with staffs capable of dealing with these issues, of debating where policy should go, and as States, we are your laboratories. We are the ones who do the cutting edge of development of policy, not the Federal Government with respect to these types of issues.

And so what we are asking is give us a shot. Give us a chance. We believe that we do have authority to do these types of things. We have the wherewithal and the inclination to deal with them. Allow us to do it.

Mr. NEITZEL. It is hard to say when you are representing NARUC. I think it gets to the basic crux of this is we talk about competition like somehow it can pass a no losers test. In my mind, any competition, most competitions don't usually end in a tie. There is a winner and a loser, and what regulation has done is made sure that nobody loses very big, but the mirror image of that is that no one wins very big either.

And now we are going to talk about competition because we think there will be more winners, but the fact is that in a competition, somebody wins at the expense of somebody losing, and that the marketplace is a very, very efficient allocator of resources, but it is a ruthless allocator of resources and we have to realize that up front.

I think we just—we really have to get out of the thinking that the—of course, the wires will continue to be a monopoly, but the electrons will not, and in 100 percent market share business, which is what a retail franchise is, we could do a lot of these things, like I mentioned before. Like PURPA.

You can say you must buy this at avoided cost. Now if you can to a market-based system, a utility avoided cost may be here so they may under PURPA rights have to pay this price but the market price may be down here. So you are actually asking them to buy an uneconomic asset. It is more economic than what they have but it is not market clearing economic.

So I think if we are going to go to a competitive market, you can't load up any competitor, any single competitor with costs that aren't associated with another competitor so that everybody has to be treated the same, whether that is in taxation, environment, contribution to social programs, because we will no longer have this hundred percent market share utility to deliver in kind of an off-budget, United Way fashion the social programs that we have used it for in the past. And I think we have to deal with that up front, and if we don't deal with it up front, we are really going to end up hurting the very customers that we ultimately want to help, which is the massive customers, which is the residential customers who are going to be stuck paying these things because they won't have the choice, but they will still be loaded up with paying for all the programs.

Mr. HASTERT. Thank you.

Thank you, Mr. Chairman.

Mr. SHARP. Thank you very much.

Let me indicate for our colleagues and audience that after Mr. Lehman completes his questioning, we will release this panel and we will take a 20-minute lunch break. Some people may want to get out the door first to the carry-out, and then return and try to expeditiously hear, because we want to be sure we hear the important voices on our second and third panels, too, even though most of us would like to continue some of our discussion with these witnesses.

The gentleman from California.

Mr. LEHMAN. Thank you, Mr. Chairman, and I will strive to be brief and first maybe I should just ask if I could submit additional questions in writing beyond what I have today to the panel.

Mr. SHARP. I think that would be certainly—with their cooperation, we would appreciate to the extent they can.

Mr. LEHMAN. Certainly.

Again, Professor Fessler, it has now been 3 months, I guess, since the original proposal was put forward. In your letter to me, you stated that the Commission had yet to make, in your words, an affirmative assertion of its jurisdiction in this matter. To my knowledge, you have not as yet made such an assertion. Could you tell me why?

Mr. FESSLER. Surely. As I had indicated earlier, Congressman, in a colloquy with the chairman, what we have been attempting to do in California is to engage in the broad-ranging discussion which we think is necessary to hold and to hold in public and to hear from stakeholders throughout the industry before we attempt, as the five constitutional officers in the State of California, to come to policy determinations.

If I knew where the public advantage were to be identified within or without the context of the covers of this so-called blue book, then it would not be necessary for us to be engaging in the very lengthy process that we are participating in. But I don't, and I need all of the help that I can get.

We did not undertake this on the concept that we were a sufficient cause for reform. We undertook it on the concept that somebody had to step forward and hold public dialogue over issues which everybody was telling me behind closed doors were pressing issues that were sweeping ahead, had a momentum of their own, and I had no idea where the public advantage was to be found in that context. I know of only one way to do that, and that is for government to take part in the dialogue that you are continuing with me today.

And I would be delighted to respond to the best of my ability to written questions at some juncture, and my State legislature is very interested in this. At some juncture, the Commission is going to have to make a determination as to whether or not it wishes to pursue a proposal that goes beyond attempting to work with everyone else who is a stakeholder beyond California in perfecting wholesale markets and attempt to introduce any form of retail access or any form of allowing the customer greater opportunities to take self help steps to influence the dimension of his or her bill.

Once we have done that, we will hold hearings on whether we believe there is sufficient jurisdiction, and it will not surprise me that the answer to the question will be no, there is not, because California doesn't have the jurisdiction to affect a regional market, and as I have indicated before, not even the Congress of the United States can order and command the outcome of a regional market that includes Canadian and Mexican jurisdictions within it. So it is going to be a process that is going to take time and it will be cooperation rather than command and control from any quarter that I think will bring us the optimal result.

Mr. LEHMAN. In that context, I am curious as to why California would propose to allow others to enter our market without some arrangement for reciprocity.

Mr. FESSLER. We have expressed no view as to whether some form of reciprocity would be absolutely essential, and we have been given conflicting recommendations from Californians and non-Californians alike.

One of the agricultural interests that appeared before us made a following telling statement. “If you mean to me that the thing that could be the worst consequence of the discussions we are having is that sellers from outside the State of California would fall all over themselves to try and sell us lower-cost energy, lead me into temptation.”

Mr. LEHMAN. Commissioner Russell, I understand that the Michigan Attorney General has expressed some, I guess, concerns about the proposal there. Could you tell me about his concerns?

Mr. RUSSELL. We respectfully refer to our Attorney General as the Eternal General. He is, I think, a dean of the attorney generals in the country of 40-some years. His major concern is the cross-subsidization issue. His fear is that what we have talked about here today, is that if you give certain parties choice, others may have to pick up the tab for that choice as leaving resources or leaving assets needing to be paid for.

So basically that is his premise, is that the cross subsidization should not be there because it has the potential of leaving excess cost on those remaining from the system, that other types of planning issues, in the name of all of the panorama of policy issues. The Michigan commission is not willing to take a side on whether it is less filling or tastes great. What we are trying to feel out is whether or not there are existing circumstances under which an efficient and an economic transaction can take place without harming or without cross subsidization to those less able to defend themselves.

Mr. LEHMAN. Maybe I will ask Mr. Neitzel then as well and you can jump in then.

You agree with Professor Fessler then that you can accomplish what you set out to accomplish here in terms of retail wheeling and hold the retail customer harmless in the process? The residential customer.

Mr. NEITZEL. Let me be very careful. OK. I will just talk through it. I think the goal has got to be to hold the residential customer harmless to the greatest extent possible related to their cost of service, which is a little bit different than holding them harmless relative to the current price that they pay, because——

Mr. LEHMAN. Enlighten me as to the difference?

Mr. NEITZEL. The difference is that in most jurisdictions, a small customer does not pay the true cost of service to them. They are subsidized by large users who pay a little bit more per unit that it costs to serve them. So if we are going to go after cross subsidy, that is the way it flows in a lot of jurisdictions. So I think our goal is—I am not sure, and I think, at least this is my perspective and this is in the State that has very low cost, you surely don't want to do something that is going to increase cost, but I think anything you can do to align price with cost is good and more efficient because people will see the real cost of their service, and if we are looking at doing more demand side, then we need to show people the real cost that energy use—you need to give them the real price signal.

So I think we can and have to hold people harmless vis-a-vis their current cost, but I am not sure we can hold everybody harmless vis-a-vis their current price.

Mr. LEHMAN. Anyone else care to comment?

Mr. FESSLER. Again, I simply wish to indicate that it is our objective, and we welcome participation from any quarter in attempting to divine whether or not we can take advantages of the competitiveness in generation to bring to all classes of customers an ability to realize the efficiency gains that are being captured there.

We have had some pretty interesting proposals advanced before us. I am not ready to endorse them, but I do want to draw to your attention, because I know you are a very concerned member, that these factors are taking place.

I would point one thing out to you, Congressman, that you may find interesting. We made a determination shortly after we opened the docket in both proceedings to take the step of putting all of the proceedings on the Internet so that all filings, all pleadings, all submissions, our correspondence is available on Internet. I inquired before leaving California as to the number of access opportunities that we have had this month on the Internet looking for that data. They are in excess of 2300, and it is intriguing to note that 800 of them came from outside of the United States.

So the problems which we are examining and discussing are ubiquitous to people not only within the jurisdictions that are within the domain of the Congress of the United States, but literally around the world, and the issues that you raise and continue to insist upon are part of the focus that I am attempting to keep on these proceedings.

I wish that I could tell you that I had an ideal solution right up my sleeve. I do know that we have a process that is beginning to elicit a great deal of very, very imaginative and affirmative thinking, and as I think you saw last week, also having some realism cast on how we as, quote, lawyers or regulators or public policy types are being brought into communion with the realities of what electricity is like as a physical presence and it does not necessarily match up to the model that is generally discussed around these tables. That, too, I think has been very valuable to understand the industry we are trying to optimize.

Mr. LEHMAN. Thank you very much.

Mr. SHARP. As you can hear, we have a vote and a quorum call on the House Floor. What the Chair wants to do is bring this panel to a close and we will try to begin in 20 minutes, no later than 12:45 p.m.

And, gentlemen, I do want to especially thank you for your time and attention and the enormous efforts you are making in your respective States and we appreciate very much your helping us out here today. Thank you.

[Brief recess.]

Mr. SHARP. The subcommittee will please come to order.

We are very pleased to have with us this afternoon Mr. Steven M. Fetter, the Senior Vice President and Director of Regulatory and Government Affairs with Fish Investors Services, Incorporated; Mr. Paul C. Parshley, a Senior Vice President with Lehman Brothers; and Mr. Mark Seetin, Vice President for Governmental Affairs with the New York Mercantile Exchange.

Gentlemen, I think you are familiar with our processes. We will be happy to make your full testimony a part of our written record, and will be delighted to hear your oral summary at this point.

Mr. Fetter, we will start with you.

STATEMENTS OF STEVEN M. FETTER, SENIOR VICE PRESIDENT AND DIRECTOR OF REGULATORY AND GOVERNMENT AFFAIRS, FITCH INVESTORS SERVICE, INC.; PAUL C. PARSHLEY, VICE PRESIDENT, LEHMAN BROTHERS, INC.; AND MARK W. SEETIN, VICE PRESIDENT OF GOVERNMENT AFFAIRS, NEW YORK MERCANTILE EXCHANGE

Mr. FETTER. Thank you, Mr. Chairman and members of the subcommittee. I appreciate the opportunity to be here today to provide, I guess, highlights of the testimony I filed, and also I think maybe to try to bridge the first panel's discussion into what I filed in that testimony.

I was chairman of the Michigan Public Service Commission at the time we started the retail wheeling case 2 years ago, along with Commissioner Ron Russell, but unlike my former colleagues in the first panel, I had the good sense to leave regulation before it had to be decided. Almost from the day we started that case, what came to my mind were the legal, the jurisdictional, the technical and the operational problems that would crop up in trying to reach that retail wheeling setting. And by the time I left last October, my view was that retail wheeling would probably take 4 to 5 years to come about, primarily because of the care and concern that regulators would bring to the issue to assure that there be a fair balance. And also I guess, referring to Congressman Klug's concept of the city dubbing where if one State does it, all they are able to do is let others—other companies and other States send power in to pick off their customers while not having that reciprocity going back in the other direction.

Not surprisingly, my colleagues who I left in Michigan then went ahead with a rather limited experiment focusing on new load that avoided the stranded investment issue that would have brought the entire electric industry down upon their head.

California, as we all know, is certainly a different situation. I think no one can argue with the policy goals that California is seeking. They want a market-driven, performance-based system. They want protection between classes of customers, and they want protection against stranded investment and to assure the financial integrity of the utilities in that State.

The market reaction to those somewhat conflicting goals and the ability to achieve those goals within the time frame set was viewed as illusory and the equity positions of the three major IOU utilities in California took a beating starting on April 20th.

I think a lot of the problem is how the proposal was portrayed. The Commission put it out there to seek input from all entities, as President Fessler told us this morning, but I think I was in a very opportune place to see how it was portrayed nationally because I was vacationing in Florida on the day that that hit.

And so my only connection to the California order on that day was in The Wall Street Journal and The New York Times, and just to look at the first sentence or two in each of the articles that followed, in The Wall Street Journal, they said, "In what is likely to become a benchmark for electric deregulation nationwide, California proposed to set up head-to-head competition." The New York Times, just as forthright in their certainty, stated, "California took a significant step today toward opening its electric services market to competition in what experts say could become a national model involving billions of dollars in rate payments." They noted the unanimous vote and that it would be starting by August.

Needless to say, when I called into my office that day, they said not only is your 4 to 5-year prediction wrong, but you are about 4½ years off in your prediction. And I think when you look at how it was portrayed and you take the things that move markets, perception, uncertainty and momentum, the movement on the stocks, which had been weak in general because of the weakness of the electric industry since last fall accelerated and the three IOU's have suffered $3 billion in equity damage, and I think the portrayal, rather than leaving the perception that this is what the Commission would like to do or this is what the staff would like the Commission to adopt, it moves it 180 degrees, and that was viewed as what a majority of the commissioners wanted to do by August unless someone could come up with a pretty good reason why they should not do it. And as President Fessler said today, that is not the situation but that is how it was felt around the country and led to the negative actions against the equity positions.

I had the opportunity to testify in California during the first 2 days of testimony, along with 65 other interest groups, and I think I was the only one there who did not bring self-interest along with my testimony. Everyone came to say, it is OK to move to market based, but leave us with the protection we have gotten from the social ratemaking structure.

Especially the environmental groups, which numbered around 15. They came in and said, we need the social backing for the current ratemaking process so if you move to market, just drop on this additional charge so that we can still flow through rates. Because if they were expected to move to market, they would have to defend their policies and concepts on a market-based system or, alternatively, go and seek political support, and what that would mean would be going to the California legislature and asking them to support their programs through taxation, and I think when you put the word taxation in any sentence with the California legislature, you are talking about a very low probability outcome.

The biggest problem the Commission is facing, just as a spectator at those hearings, is that if you move to an across-the-board market system that they want to reach all the way down to the residential level, then you have to get rid of subsidies that have supported some of the residential rates in the past. You likely have to get rid of the traditional obligation to serve so that people leaving the system are then cast free and have to let the market determine any backstop provisions, and even with regard to PURPA, the must purchase requirements cannot survive and should not survive in a market-based system.

What I tried to put forward in California that day was a proposal on behalf of Fitch Investors that was predicated on the concept of respect for the preexisting regulatory compact that came before, and in a way, it also was to respond to some of the members' concerns on the first panel, also probably the only way to try to hold residential customers harmless as we move towards competition.

And that is that at the outset of direct access, it has to be limited to the large user, large industrials and potentially large commercial. In my view, those are the ones most in need and those are the ones who hold out the greatest potential to help the economy of California move forward to the benefit of everyone, including residential customers who would not be included in the direct access tariff.

In order to do this, a much longer transitional period would be required beyond the one set by the California commission that is now set at January 1, 1996, and during that transitional period, an attempt would have to be made to eliminate the potential strand investment problem for the group of customers, the large users, who would be benefiting from direct access, and this could be done through growing capacity need, plant retirements, accelerated depreciation, restructuring, both operationally and financially, including possibly the reallocation of costs based on value of assets utilized by utilities, most especially the transmission grid.

I also think it would be appropriate to put a risk component on the companies themselves through performance-based ratemaking so that part of the stranded investment would have to be eaten up through greater efficiencies on their part or they indeed would suffer the negative impact from that stranded investment.

Thus, when this transitional period would be finished in, then for those large users, they would be facing a true competitive environment where they would seek their supply, pay the competitive rate without regulators adding any add-ons through a competitive transition charge that the CPUC envisions.

What has become clear in the last several months is that no one State can provide market competition.

There are two areas in which I think FERC or you in the Congress can provide guidance. One is, and we have been talking about these issues for years since I last came to testify on PUHCA, and that is some means of deference for State commission decisions when it relates to wholly in State transactions with no interstate commerce impact. And that deference could be based upon a standard that allowed a reasonable range of discretion for State commissioners to determine decisions affecting transmission where it is wholly in State.

Short of that, some delineation between Federal and State jurisdiction would help. As Commissioner Russell mentioned, Michigan is in the process of going through its pricing reopener for the retail wheeling pricing, and out in California when I was there, a representative from the Department of Energy said that they are waiting and as soon as Michigan sets its pricing, then they plan to go in and sue over what Michigan does no matter what Michigan ends up doing. And it struck me that that is not the most welcoming environment in which to have cooperation that many people have talked about.

I think this line of demarcation on jurisdiction could be somewhat similar to the gas industry structure where rather than a city gate barrier between State and Federal, we would be drawing the line based on size of voltage line, possibly somewhere below subtransmission system.

I think alternatively, if there could be Federal guidance on some vehicle for cooperation on a regional basis and problem solving on a regional basis would be helpful. I am not sure RTG's, which are predicated on deciding things based on consensus, are the best means, but perhaps some form of interstate compact where reciprocity could be provided within two or more neighboring States based on supply and demand profiles where the agreements would be mutually beneficial.

I think in any event, it is clear that the threat of retail wheeling has changed the regulatory atmosphere. In the old days, no matter how poorly a utility did, its goal was to get those missteps into rates. Now every utility that comes before us or that we meet with around the country realizes that the marketplace will be much less forgiving than any regulator has been, and they are taking steps to realign.

We referred earlier to the FPL situation in SoCalEd cutting their dividends. Here is a situation where a strong company did it and a weaker company did it, and what is clear is that there are no longer traditional modes of behavior to get the financial house in order and that every utility will look within and make its own estimate as to how fast competition will come and what it will look like so that they can be best situated to survive in the competitive environment.

I think the California case will be watched, the Michigan experiment will be watched, both of them closely, and if either of them figure out a way to achieve competition with protecting the financial capability of utility companies, it will be an impetus for change. However, if they are not able to obtain all those conflicting goals within any reasonable timetable, no doubt it will serve to warn other States away from even thinking about it.

Thank you.

[The prepared statement of Mr. Fetter follows:]

STATEMENT OF STEVEN M. FETTER, SENIOR VICE PRESIDENT, DIRECTOR OF REGULATORY AND GOVERNMENTAL AFFAIRS, FITCH INVESTORS SERVICE, INC.

The electric utility industry has entered a period of unprecedented change grounded in a new competitive environment that is market driven and subject to shifting regulatory policy. Utility executives will be called upon to manage their companies to a degree not required—and not even expected—in the past, and regulators will have to reevaluate the artificial structures they have erected to simulate market competition over the past 100 years.

In this new environment, regulators will be called upon to referee disputes between customer groups, the likes of which can not be easily resolved through creative rate design. They will need to adapt their behavior to changes in market structures and provide leeway for industry participants to experiment to achieve the appropriate relationship between those with competitive choices and those without.

Clearly innovators should not be held to a strict liability standard; commissions, through their policies, should allow room for trial and error. Both energy providers and customers will have to make their own assessment of risks and opportunities, and chart their course based upon their own best estimate as to the timing and scope of competitive reform. Large users, the earliest proponents of competitive choice, are already benefiting from rate design alterations that reflect the economies of scale of their usage, and they soon stand to gain from the ability to procure power directly from suppliers through State-ordered retail wheeling programs.

That all said, the many years of electric industry regulatory inertia will not fall away overnight. Legal, jurisdictional, technical, operational, and sociopolitical issues remain to be reconciled before regulators and legislators will be willing to relinquish the provision of electricity entirely to the competitive arena. Accordingly, the movement away from regulated status will be gradual, with delays from time to time to allow missteps or unforeseen consequences to be ameliorated.

With all the uncertainties resulting from this substantial regulatory reform effort, investors will also have to alter the methods they have used to analyze the electric industry. Traditional measures of financial performance high debt coverage ratios and balanced capital structures while still important, will no longer be sufficient to assess a utility's health in this increasingly competitive environment. Instead, qualitative factors will have to be analyzed along with quantitative data in order to derive a complete picture of how a utility will fare, both during the transition to a new competitive industry structure, as well as after it is substantially in place.

As detailed below, a utility's ultimate success or failure will depend upon the quality of its management team and how capable team members are to respond to the changing environment. Their ability to increase productivity to benefit from incentive-based rate mechanisms will be a plus, especially if they use them as a vehicle to bring their rates to competitive levels. Likewise, efficiency improvements in plant operations and understanding and responding to customer needs within the service area and beyond will be crucial as they position the company to attract future business. Finally, financial integrity, defined in large part by strong cash flow, will assure that a utility will be able to meet its ongoing obligations, including debt service, as well as provide additional operating flexibility to confront new pressures brought on by competition.

It is management's responsibility to develop and implement a comprehensive plan that will allow the utility to become more competitive or maintain its current competitive posture. Whether management chooses to operate its utility in a traditional conservative manner or instead selects an aggressive tack, the key will be the formulation of a well conceived plan to deal with critical competitive issues, including long-term operational and financial planning, customer needs and desires, rate competitiveness, economic development, diversification, and interaction with regulators and other public officials and intervenor groups.

The relationship between a utility and its regulators will continue to be important, even as the traditional "regulatory compact" and "obligation to serve", begin to fade away. As the new utility environment takes hold and Federal and State regulatory policies evolve, all utility managers will have to be able to understand these regulatory issues and incorporate them into their day-to-day decision making. It is clear that over the near term State public utility commissions will continue to maintain significant utility oversight, so managements must balance their movement toward competitive provision of electricity with any restrictions that remain from traditional regulatory doctrine.

Rate competitiveness will be a major determinant of a utility's future success. While it remains uncertain whether industry-wide retail wheeling will come to pass, it is clear that retail competition has arrived and that utilities are obliged to position themselves accordingly. The threat of customer self-generation, municipalization, and retail wheeling (in whatever form it eventually takes) make expense control paramount. To be rate competitive, utilities need to focus on eliminating corporate inefficiencies through downsizing, improving plant performance, refinancing high-cost debt, benchmarking company operations, and taking advantage of cost-sharing opportunities presented by the newly competitive environment.

Rate design continues to gain importance as utilities move to unbundle services to meet individual customers, needs and prepare for competition. The mere threat of retail wheeling already has furnished large industrial customers in several States with the necessary leverage to exact preferential economic development, rates that regulators have been reluctant to provide at the expense of residential customers. Likewise, there has been an industry-wide shift toward eliminating cross subsidies and setting rates based on a truer, cost-of-service basis. Some utilities already have accomplished this, particularly those with large load factor customers.

Through steps like these, the longstanding subsidization of residential rates by larger customers is being reduced or eliminated in anticipation of a system of rates set by market conditions. The ability of management to gain from regulators sufficient flexibility to deal with the often conflicting circumstances of different customer classes will be essential for a utility to be competitive in the new environment.

In the past, most utilities were responsible for providing their customers with complete service—generation, transmission and distribution. With the future obligation to serve in doubt, many utilities will realign their operational structures to meet their customers' needs in the most efficient manner possible.

If a utility is able to navigate successfully through the limitations placed upon its actions by bond indenture provisions, the likely result will be a separation of generation assets from transmission and distribution services. Generation could come from the utility itself, another power provider, or an energy broker, while transmission and distribution services would be supplied by a separate regulated entity responsible for retail operations.

Whether a utility generates its own power or contracts to purchase it from another supplier will be not be significant if the power is produced and provided in a quality manner, is appropriately priced, and is needed. A major uncertainty, however, is how regulators will treat purchased power in the future, whether it will continue to be passed through rates or rather is tied to an incentive-based mechanism. Clearly, some form of incentive rate structure, holding out potential benefits for both ratepayers and shareholders, would be most consistent with a market-based concept.

Nuclear power in recent years has come under increased scrutiny with the extended outages of several plants and early decommissioning of certain units. Management's previous focus on problems with nuclear construction resulting in financial disallowances has now shifted to day-to-day operating and life cycle issues. The ongoing inability by the U.S. Department of Energy and the States to provide either long-term or temporary nuclear waste storage renders the future prospects for nuclear relicensing or new construction an open question.

Other plant-related issues include the extent of regional transmission interconnections, the ability to broker excess power, and the ability to site new transmission lines. Given current environmental considerations, construction of new transmission or baseload plant is likely to be much more difficult than in the past. Utilities already positioned with low-cost production facilities or strategically placed transmission should benefit from more restrictive policies.

Federal Energy Regulatory Commission (FERC) jurisdiction will be very significant in the development of a market-based electric utility industry, especially in the area of regional transmission groups and transmission pricing. The level of State regulatory oversight will be influenced by individual commission policy, interaction with the FERC, and pressures from the marketplace. State and Federal regulators, abilities and willingness to resolve key issues, both individually and cooperatively, will have a substantial say in how far and how fast the electric industry will move away from its traditional monopoly underpinning.

Retail wheeling remains a focal point for many commissions with Michigan and California leading the way. Clearly, the allocation of the enormous costs associated with assets no longer economic in a competitive environment is the most difficult transitional issue facing regulators. To the extent that a formula can be set that respects the assets related to the preexisting "regulatory compact"—rate base assets, deferred and regulatory assets, amortized recoveries, mandated purchases, and the like a transitional period can be initiated that will result in some form of retail wheeling, most likely directed at the outset to the large user.

A further shift away from traditional cost-of-service regulation to performance-based returns, currently in use in many States, is a natural step for the industry. Such incentive-based mechanisms can be utilized both to share burdens of the transition to competition, as well as to provide potential benefits for customers still receiving power under regulated tariffs.

By placing a portion of the above-described "regulatory compact assets" at risk, a commission would be dictating that increased productivity on the part of the electric utility is to be a required element of the move toward competition. Likewise, implementation of an incentive structure for the remaining regulated tariffs could help move them toward market rates by sharing efficiency gains between ratepayers and shareholders.

Finally, fairness concerns relating to reciprocity will likely affect the progress of reform. True competition will only come if electric power can flow across State lines in both directions, and between all types of power providers. Since no one State can order or authorize such a structure, some form of Federal involvement will be necessary to manage interstate transmission issues that rest within the purview of FERC.

Alternatively, some form of regional cooperative mechanism might be structured to alleviate this jurisdictional uncertainty. Early experience with regional transmission groups suggests that an issue as divisive as retail wheeling might not fare well in a body designed to effectuate policy through consensus. It is feasible, however, that two or more neighbor States might find that their electric supply and demand profiles render a cross-border association, perhaps through interstate compact, a mutually beneficial arrangement.

Maintaining good customer relations will be at the core of any successful electric utility. In addition to competitive rates, customers care most about quality of service suited to their particular needs. While the importance of large industrial user satisfaction is well understood, over time other customer classes will be able to access alternative supply from nontraditional power providers. Accordingly, management must be responsive to customer desires at all levels and be proactive in meeting them.

Service area demographics will be important in formulating a strategic plan for meeting customer needs, because competitive success will be driven by a company's proficiency at finding a flourishing market for the sale of electricity. Relevant measures of economic health include employer diversity and employee growth, the impact of governmental policies on business development, cost of living and per capita income, and unemployment.

Utilities with above-average growth also will be well positioned to take advantage of attractively priced energy and capacity that should result from a more competitive wholesale supply structure. Likewise, utilities that lose customers, but have planned for the eventuality, might actually benefit by not having to construct or procure new generating capacity.

While financial performance will remain an important factor in assessing a utility's ability to succeed in a more competitive world, it will have to be considered in light of the factors detailed above. For instance, a utility that possesses high debt coverage as a result of high rates that are noncompetitive within its region will not enjoy that financial status on into the competitive world. Rather, a utility that strikes a balance between quantitative and qualitative factors will be better prepared for the changes coming within the industry.

Significant financial factors include historical and projected debt coverage ratios, free cash flow, dividend policy, capital structure, and size of capital expenditures, including adequate funding for environmental compliance and nuclear decommissioning. Asset quality and the nature and amount of deferred charges as a percentage of assets and equity will be important, particularly for utilities that will not be able to support these items in its rates as they move toward competitive levels.

The more thorough analysis required of investors should not be viewed as signaling that the financial outlook for the entire electric utility industry is negative. In any competitive environment, some companies adapt more quickly and efficiently than others, and these are the ones that benefit from the upheaval that often accompanies regulatory reform. The key to success will be management's ability to analyze, assess, and respond to the changing competitive landscape in a timely manner.

For the near term, the equityholder stands to be at greater risk than the bondholder. Recent dividend cuts, slippage in the overall utility equity market, and incentive-based mechanisms that require company efficiency gains merely to maintain the status quo all stand to have a negative impact on utility shareholders. However, to the extent that these conditions result in management actions to create a competitive entity, all stakeholders will be in a better position to benefit from a strengthened company's ability to thrive in the less regulated, more market driven environment to come.

Finally, the role of the regulator should not be overlooked For a century, regulators have sought to bring stability and fairness to the provision of utility services. Their traditional practice has been to limit the gains of winners and prop up the weaknesses of losers. The result until recently has been the creation of an atmosphere in which innovation and excellence has neither been encouraged nor rewarded, and where deficient performance has hardly been penalized.

This attitude, while serving as an unintended catalyst for regulatory reform, also stands to serve as a damper against precipitous change to the detriment of the electric utility industry as a whole. To the extent that regulators continue to pursue fairness as they loosen their controlling reins, they will require that the "regulatory compact" that came before be respected. The result will be a transitional period structured to provide utilities with close to level playing field status by the time they are ordered to compete against all comers.

Mr. SHARP. Thank you very much, Mr. Fetter.

Mr. Parshley, very pleased to have you with us.

STATEMENT OF PAUL C. PARSHLEY

Mr. PARSHLEY. It is nice to be here. Thank you for the invitation. I appreciate the opportunity to offer an investment community perspective on the electric utility industry as it moves toward the more competitive business environment that the Energy Policy Act envisioned.

I learned this morning that my job is to try to interpret for investors how a plow horse metamorphosizes into a thoroughbred, and sometimes that sounds more interesting to me than just following the utility companies.

Let me throw some more numbers on the pile of numbers that you have collected in the past 1½ weeks or so of hearings. It is true that while the capital markets have been aware for several years that competition and secular change inevitably are coming to the traditional electric utility sector, the market did not begin to reflect the associated risks and uncertainties until last fall. Since then, however, the market's reaction has been strong and it has been negative.

I should also add that I am talking primarily about the equity markets, but I think that my comments do bear some relevance to the fixed-income markets as well.

Since last fall when the market began to react, the common stock performance of the utilities in the congressional districts represented by the members of this subcommittee illustrate this point. By my count, there are 28 such investor-owned electric utilities. Since September 15th of 1993, the common shares of those 28 companies at that time, they had a combined market value of about $120 billion. Today the market value of those same companies is about $85 billion. That represents a $35 billion or 30 percent loss in market value in less than 1 year. That is unprecedented. In contrast, during this same period, the broader market represented by the Standard and Poor's 500 index is down only about 3 percent.

While I would agree with Chairman Fessler that some of this decline can be attributed to rising interest rates and other cyclical forces, there is also a very strong undercurrent of secular change in the electric services industry which suggests that these losses are not likely to be recouped fully.

There is no question that investors, including electric utility investors, assume certain risks of economic loss when they choose to play the market. They knowingly expose themselves to economic cycles and other market risk and must live with the consequences. Similarly, they generally should bear losses from company specific risks, such as bad management decisions, poor plant operations, adverse rate case decisions and the like. But it seems unfair for existing electric utility investors to be forced to shoulder most of the transition costs as a traditional regulatory compact is being unwound.

The obligation to serve has required utilities to raise huge pools of capital over the years. Last week, Elizabeth Moler, the Chair of FERC, pointed out to the subcommittee that generating capacity can be built and operated at costs that are less than many utilities' current embedded costs. I believe that is a very true statement.

The Energy Policy Act of 1992 and subsequent State retail wheeling initiatives, such as the ones in California and Michigan that we heard about this morning, provide a new public policy base which seems to be accelerating this process, and this economic reality is breaking down the traditional regulatory compact as customers gain the ability to choose their suppliers of power.

Estimates of presumably prudent investment incurred by utilities under the obligation to serve, which could be rendered uneconomic by competition, range as high as $200 billion according to the testimony that I think you heard last week. It is not reasonable to assume that electric utility investors could have or should have anticipated this huge financial burden. And the market currently does not reflect that expectation, even after the 30 percent drop that I described. Utility shareholders will have—if the utility shareholders will have to absorb that cost, you should expect another devastating decline.

I think it is important to keep in mind who these investors are for the electric utilities, because they tend to be different from the broader market. Individual or retail investors represent more of the ownership base of electric utilities than they do of the broader market.

Ownership of the S&P 500 comprises about 50 percent retail and about 50 percent institutional. In contrast, about two-thirds of the common shares of the 28 investor-owned companies serving this subcommittee's constituents are owned by individual investors and only about one-third are owned by institutional investors.

A relatively high proportion of these individual investors probably live in the districts that are represented by the members of this subcommittee. Over the years, many of them have been attracted to the high dividend utility stocks as relatively safe sources of income. Traditionally, the industry shareholders' base has included the so-called widows and orphans, that I am sure you have heard about in other hearings, who are living on fixed incomes.

I also think it is important to keep in mind who some of the institutional investors are. Some State employee and teacher pension funds have suffered significant losses in utility holdings attributable to the change overtaking the electric services business.

The recent experience in California is noteworthy. Since the California PUC floated its controversial retail wheeling proposal on April 20th, the price of SCE Corp. stock has plummeted about 21 percent and the company has cut its dividend by 30 percent. During this same period, Pacific Gas & Electric, the largest investor-owned utility in the country, their stock has fallen over 16 percent. During this same 3-month period, the S&P utility index is down just 2 percent. That indicates to me that it is not related to a movement in interest rates, at least these more recent changes in the prices of California utility stocks.

The California public utility—I am sorry—the California Public Employees' Retirement System, the regents of the University of California and the California State teachers' retirement system own a total of about 11.5 million shares of SCE Corp. and about 9.2 million shares of Pacific Gas & Electric.

While they are institutional investors, the fiduciary rules that guide their operations force them to own these stocks without being able to unload large positions like that if they see a storm brewing on the horizon. As a result, in the last 3 months, these combined holdings have lost about $95 million in market value and nearly $5 million in annual income as a result of the SCE dividend cut.

I think it is fair to say that the market nervously is awaiting the answers to several questions regarding the handling of transition costs, particularly of stranded investments. Now we know that there are also stranded liabilities. Unfortunately they don't really offset each other.

The most important question relates to the allocation of stranded costs that Commissioner Massey of FERC has called the elephant in the living room. That is pretty telling description.

While the definitions of stranded cost and uneconomic investment are unclear and therefore the magnitude of the problem is a matter of some conjecture, investors clearly are worried about the potential impact. The matter is further muddied by the jurisdictional uncertainties surrounding the extent to which this is a State or Federal problem that was described in some detail by the panel of public utility commissioners this morning.

Last week, Larry Hobart of the American Public Power Association told this subcommittee he thought the potential problem of stranded investment, and there is a typo in my statement that says standard, which may be the way that Larry would rather have it read, was vastly overblown, and probably amounted to only about $10 to $20 billion, by my calculation, for all investor-owned electric companies. While a precise number is unknowable, I think this estimate grossly underestimates the prudently incurred uneconomic investment that is out there.

Let me give you an example. Recently we gathered some data about the hundred or so operating nuclear plants that are out there that have an ownership stake with investor-owned utilities. At the time those plants were put into service, they had a total book value or investment of about $125 billion. Today those plants on average are about 15 years old and the book value is about $110 billion. In over half the cases, today's book value is higher than it was at the time those plants went into service.

Well, that value, if you are going to recover the full cost of the investment, has to be depreciated. By my calculation, if you assume that all of those plants reach the end of the 40-year lifetime that they are allowed under their NRC licenses, you are still going to be left with an undepreciated amount of about $85 billion. I should add that I don't believe that all those plants are going to live for 40 years of operations, which would only tend to make the problem somewhat worse. So the number associated with stranded investment I think is a big one.

A second question which I think is important that flows from the California PUC debate is how certain important concepts are going to be defined. Chairman Fessler talked about preserving the financial integrity of the utilities, and that is certainly something that offers some degree of comfort to investors in the California companies, but as I described, in the performance of those stocks, it doesn't offer all that they would like to see.

A couple examples of things that could be better defined. Does financial integrity in the minds of the commissioners mean the preservation of triple B or investment grade credit ratings? In which case, that suggests that dividends may continue to be cut and certainly won't grow?

If that is the case and if that is a standard that gets applied more generally in other States, you should expect to see the price of utility stocks continue to decline quite precipitously.

If, on the other hand, in keeping with the regulatory compact under which these facilities were built, there is some effort to allow investors to earn—recover their investment and earn a return on it, then that would assure the markets that there will be a transitional period of some equity where the burden of those costs are not going to be borne entirely by the equity investor.

The best mechanism I can think of to protect utility investors from shouldering an unfair amount of this stranded cost would be some kind of an exit fee that would be tailored to match the actual cost of service that was incurred by those customers who might be dropping off the system.

I think that this will slow down the arrival of competition. But one thing that I don't believe Chairman Fessler mentioned, something that he did say when he was briefing the analysts in New York soon after the proposal was floated, is that the costs of power in California aren't expected to come down very soon anyway under the proposal that the Commission is looking at. So perhaps a little bit more delay, or at least some treatment of its stranded cost, would serve a variety of parties well and it would, I think, help address the issue that some of your witnesses have raised about transferring the burden to other classes of customers as well.

Are utilities taking steps to be competitive?

There is one more point I would like to make. They are cutting costs. They are trying to be more efficient. The 28 companies that serve the constituents of this subcommittee have eliminated 20,000 jobs since the beginning of 1993. Because those same employees very well may be participants in the 401(k) plans, some of the members of this committee may be running into voters this fall who have lost their jobs with their local utilities, seen their 401(k) plan cut by a third, and are wondering when they are going to start seeing the benefits as an electric consumer.

Let me close with the comment that my view of the industry going forward is that there will be winners, there will be losers, and that is how it should be. I think that there are some things that could be done to ease the transition. I think the transition is inevitable.

Finally, let me end with the immortal words of Mo Udall, which is—or to paraphrase them, "These are my views, and if you don't like them, I will try and change them."

[The prepared statement of Mr. Parshley follows:]

STATEMENT OF PAUL PARSHLEY, SENIOR VICE PRESIDENT, LEHMAN BROTHERS, INC.

I appreciate this opportunity to offer an investment community perspective on the electric utility industry as it moves toward the more competitive business environment envisioned by the Energy Policy Act of 1992. As Chairman Sharp noted in his opening statement at last Wednesday's hearing, changes in wholesale electricity markets were underway prior to enactment of this landmark legislation.

In recent years, investors have grown increasingly wary of the electric utility industry's eroding fundamentals. Since 1989, industry earnings have suffered as State regulators across the Nation began lowering authorized rates of return to keep them in line with falling interest rates. New challenges to compete in power generation and to increase transmission access were created by the Energy Policy Act of 1992. This double whammy of increasing competition and constrained ability to earn has hit the utility industry at a time when demand growth for electricity is slack. Some self-inflicted wounds also have contributed to the industry's weakened financial condition. For example, most utility managements until recently continued to raise dividends at a steady pace despite deteriorating prospects for earnings growth. And, some holding companies made poor management decisions in their diversified and unregulated business activities.

In the past year the financial markets have reacted strongly and negatively to the secular changes in the electric utility industry

Yet, while the capital markets have been aware for several years that competition and secular change inevitably are coming to the traditional electric utility sector, the market did not begin to reflect the associated risks and uncertainties until last Fall. Since then, however, the market's reaction has been strong and negative. The common stock performance of the utilities in the congressional districts represented by members of this subcommittee illustrate this point. By my count, 28 investor-owned electric utilities, or "IOU's", provide service to your constituents.

On September 15, 1993 the common shares of those 28 companies had a combined market value of about $120 billion. Today, those shares are worth about $85 billion which means they have lost $35 billion, or nearly 30 percent, in market value in less than 1 year. In contrast, during this same period, the broader market represented by the Standard & Poor's 500 Index is down only about 3 percent. While some of this decline can be attributed to rising interest rates and other cyclical forces, there is a strong undercurrent of secular change in the electric services industry which suggests that these losses are not likely to be recouped fully.

Utility investors could not have anticipated the huge financial burden of breaking the regulatory compact

There is no question that investors, including electric utility investors, assume certain risks of economic loss when they choose to play the market. They knowingly expose themselves to economic cycles and other market risk, and must live with the consequences. Similarly, they generally should bear losses from company-specific risks such as bad management decisions, poor plant operations and adverse rate case decisions. But, it seems unfair for existing electric utility investors to be forced to shoulder most of the transition costs as the traditional regulatory compact is being unwound.

The traditional compact which has shaped the utility industry for most of this century is composed of several key components. First, it grants the utility monopoly franchise rights. Second, it ensures the utility's financial integrity by granting it an opportunity to recover reasonably incurred expenses and earn a fair return on its investment. In return for these privileges, the investor-owned utility has been subject to regulation. Typically, the regulators duty has been to ensure the utility provides safe, reliable and reasonably priced service to all consumers within its monopoly franchise according to terms and conditions which do not unduly discriminate against any consumer.

The obligation to serve has required utilities to raise huge pools of capital over the years. Now, as Elizabeth Moler, Chair of FERC, pointed out to the subcommittee, generating capacity can be built and operated at costs that are less than many utilities' current embedded costs. The Energy Policy Act of 1992 and subsequent State retail wheeling initiatives provide a new public policy base which is accelerating the process. This economic reality is breaking down the traditional regulatory compact as customers gain the ability to choose their suppliers of power. In the words of the California PUC, "the duty to serve must evolve." The market expects these changes to occur. The key question is: who will pay the transition cost?

Estimates of presumably prudent investment incurred by utilities under the obligation to serve, which could be rendered uneconomic by competition range as high as $200 billion for the industry. It is not reasonable to assume that electric utility investors could have, or should have, anticipated this huge financial burden. And, the market currently does not reflect the expectation even after a 30 percent drop—that utility shareholders will have to absorb these costs.

Who are the electric utility investors?

It is important to keep in mind as the rules of the electric services game are being changed that the profile of electric utility shareholders differs from the market in general. Individual or "retail" investors represent more of the ownership base of electric utilities than they do of the broader equity market. Ownership of the S&P 500 comprises about 51 percent retail and 49 percent institutional.

In contrast, about two-thirds of the common shares of the 28 IOU's serving this subcommittee's constituents are owned by individual investors, and only about one-third are owned by institutional investors. A relatively high proportion of these individual investors probably live in your districts. Over the years, many of them have been attracted to high-dividend utility stocks as relatively safe sources of income. Traditionally, the industry shareholder base has included the so-called "widows and orphans" living on fixed incomes. More recently, the purchase of 100 shares of the local utility's stock has been a convenient way for many private investors to make their annual IRA contributions. Utility employees, another subset of the private investor base, tend to own between 5 percent and 15 percent of each company's outstanding common shares through 401-K programs, leveraged ESOP's and other stock purchase plans.

Among institutional investors, some State employee and teacher pension funds have suffered significant losses in utility holdings attributable to the changes overtaking the electric services business. The recent experience in California is noteworthy. Since the California PUC floated its controversial retail wheeling proposal on April 20, 1994, the price of SCEcorp stock has plummeted 21 percent and the company has cut its dividend by 30 percent. During this same period, Pacific Gas & Electric stock has fallen over 16 percent while the S&P Utility Index is down just 2 percent and the S&P 500 is up 3 percent. The California Public Employees Retirement System, the Regents of the University of California and the California State Teachers Retirement System own a total of about 11.5 million shares of SCEcorp and 9.2 million shares of Pacific Gas & Electric. In the last 3 months, these combined holdings have lost about $95 million in market value, and nearly $5 million in annual income from the SCEcorp dividend cut.

The market nervously is awaiting the answers to several questions regarding the handling of transition costs and stranded investment

The most important question relates to the allocation of stranded costs, or what FERC Commissioner Massey has called "the elephant in the living room". While the definitions of "stranded cost" and "uneconomic investment" is unclear, and therefore, the magnitude of the problem is a matter of conjecture, investors clearly are worried about the potential impact. The matter is further muddied by the jurisdictional uncertainty surrounding the extent to which this is a State or Federal problem.

Last week, Larry Hobart of the American Public Power Association told this subcommittee he thought the potential problem of standard investment was "vastly overblown" and probably amounted to only about $10-$20 billion for an investor-owned electric companies. While a precise number is unknowable at this time, I think this estimate grossly underestimates the prudently-incurred uneconomic investment that is out there.

Recently, I gathered data on the current investment in operating nuclear power plants in the United States which helps illustrate this point. Investor-owned utilities have ownership stakes in 100 power reactors. The original investment in these facilities was approximately $125 billion. The current depreciated book value of these plants is about $110 billion, and their average time in service is just under 15 years. If this effective depreciation rate were carried forward, the industry would have an unrecovered investment of about $85 billion when all the reactors have reached the end of their 40-year NRC operating licenses. My strong sense is that a significant number of these plants will be shutdown early because competition renders them uneconomic.

A second question creating concerns among investors is: what does the California PUC mean by its expressed intent to preserve the "financial integrity" of that State's IOU's during the transition to a direct access, or retail wheeling system. The market regards the California rulemaking as an indicator of where other States might go. If financial integrity means that equity investors will be allowed to recover and earn a "return on" historic and prudently incurred uneconomic investment, then the equity market will be reassured. If, on the other hand, financial integrity means only "return of" the investment and maintenance of "BBB" credit ratings, then the equity market may experience another devastating decline. Current pricing of electric utility stocks seems to indicate that investors do not expect much dividend growth, but large write-offs and widespread dividend reductions are not reflected at current levels.

The best mechanism I can think of to protect utility investors from shouldering unfair amounts of stranded costs is to charge exit fees to those retail customers who choose to drop their local utility for a cheaper supplier of electricity. The fees could be designed to cover the costs incurred by the utility to serve the customer under its former obligation to provide that service. This approach has the added advantage of not transferring those costs to the utilities remaining customers.

Utilities are taking steps to become more competitive, and investors are feeling the effects

Most electric utilities are taking aggressive steps to control costs, unprove service and become more competitive. Many have initiated programs to reduce staffing levels including early retirement incentives and lay-offs. Since the beginning of 1993, the 28 IOU's serving this subcommittee's constituents have eliminated about 20,000 jobs, which represents an 8 percent reduction. I would not be surprised to see this number double over the next couple of years.

On May 9, 1994, FPL Group stunned the beleaguered electric utility market with the surprising announcement that it was slashing its common dividend by one-third. The dividend cut was the result of a strategic decision by FPL management, rather than because the company had suddenly encountered a financial crisis. The decision to cut the dividend reflected management's intent to reposition FPL in the increasingly competitive electric power business as a company with more financial flexibility to react to new challenges and opportunities than it had with its precut 90 percent payout ratio.

While we do not expect FPL's dividend policy to become the industry norm, we suspect the dividend cut will tend to accelerate trends already beginning to reshape the electric utility industry and its traditional shareholder base. For example, we believe the FPL dividend cut increases the likelihood that other companies not previously considered to have imminent dividend risk may choose to lower their dividends in the next couple of years. The subsequent decision by SCEcorp to cut its dividend supports this view.

Also, we would not be surprised to see a significant downward trend in dividend growth rates among those companies currently capable of growing their common dividends under previously accepted dividend growth and payout norms. Entergy's announcement on June 15, 1994 that its dividend growth rate would be substantially below market expectations triggered a 6 percent drop in that company's stock price. On an industry-wide basis, we would not be surprised to see a negative dividend growth rate over the next 5 years.

What lies ahead for electric utilities and their investors?

My general view of the utility world going forward is that the healthier companies will become effective competitors and grow stronger, and the "fundamentally challenged" companies will face an increasingly difficult struggle.

Mr. SHARP. Thank you very much.

Mr. Seetin, we are very pleased to hear from you now.

STATEMENT OF MARK W. SEETIN

Mr. SEETIN. Thank you very much, Mr. Chairman.

Mr. Chairman and members of the subcommittee, my name is Mark Seetin. I am Vice President of Government Affairs for the New York Mercantile Exchange. Unfortunately, illness has prevented our President, Pat Thompson, from testifying, and he sends his regrets.

On behalf of the exchange, I wish to thank you for the opportunity to participate in today's hearings. You and the members of the subcommittee are to be commended for your foresight in seeking to better understand the rapid and significant changes occurring in the electric power industry.

As a public regulated market, NYMEX appreciates the opportunity to put our viewpoint and to discuss these changes. Our first-hand experience in the recent evolution of competitive markets in the energy industry will serve as the basis for our observations. As we have said before, the electricity market is not the first to undergo these types of changes and they are wrenching changes and we will bring our views forward as you requested.

Futures markets provide two important economic functions: price transparency, which is the public reporting of the price; that is when you look on the writer's screen or television screen that is reporting the price of oil or other energy products, that is a function of the futures markets. The second function is risk shifting, which is the ability to shift price risk from the individual or the entity into the marketplace.

Earlier discussions before this committee related to the concerns regarding volatility in price shifting. The reason marketplaces such as futures markets exist is because of indirect response to that volatility to serve as a venue for dealing with that.

Highly competitive and transparent marketplaces are the best determinants of price. NYMEX's long and ongoing experience in market development is deregulation and restructuring occurred in other energy markets. Industry and consumers have benefited from that deregulation and competition through increased efficiency and lower prices.

Recently, even U.S. governments and government regulated utilities have begun to utilize any energy risk management tools. A couple of examples of these are the State of Texas which, through legislation in 1991, established a hedging program to protect its 4.7 percent revenue stream share, the tax on oil revenues.

Delaware, the State of Delaware on the consuming side, began a program about 4 years ago which they have expanded designed to hedge their heating oil expenses, first for State purchases. Later, they have expanded it to gasoline purchases for their State fleet. And most recently they have added a hedging function for their LIHEAP program as well, to protect their Low-income Home Energy Assistance Program participants from unexpected price shocks.

The electric power industry itself is undergoing significant and rapid transition. Changes occurring in the transition can translate to a more active and a competitive wholesale market where low cost means to manage risk enables market participants to be more competitive and to improve service.

Futures contracts are a necessary component of an active and a competitive market. As such, the public market shares with producers and consumers a vested interest in that competition and in economic efficiency in the wholesale market.

Over the past several years, NYMEX has met with and sought input from a broad range of entities within the electricity industry, including investor-owned utilities, municipal, non-utility generators, industrial users, academics and others in the development of an electricity futures contract. Those individuals and utilities have led us to conclude that the electricity industry believes that the market changes currently underway are widespread and are irreversible.

We also have come to appreciate the challenges faced by regulators during the transition period, as well as those by policymakers addressing the social and environmental objectives. Our observation from the standpoint of market efficiency is that such goals are best attained through incentive mechanisms, not market distorting command and control methods.

Our specific interest in the electricity market derives, as mentioned earlier, from our efforts to develop an electricity futures contract or contracts that are a necessary component of an active and competitive market. We think that a successful futures contract, we believe that a successful futures contract must provide a low-cost means to manage risk. And consequently, we also have a direct interest in these changes which are leading to demand for our product.

The guiding principle of our efforts and our suggestions are that economic efficiency in the wholesale cash market must be respected. The more efficient the underlying cash market is, the more effective our risk management tools, such as futures contracts, which are overlaying that. We all, as I said before, have a vested interest in competition and economic efficiency.

To summarize, I want to make just a few points in response to some of the questions that came forward in the formation of this hearing. Dramatic changes are going on. Regulators are struggling. You as policymakers are struggling.

One thing that we want to make clear is that unfettered competition will not hinder delivery and grid integrity. And we do have some recent experience in that. The gas industry faced it last January, in fact, approximately January 23rd, 24th when we had severe winter weather conditions. The system under 636 was probably put to its most severe test. It did respond and it responded very well. Again, the competitive nature of the market, we think, served both sides of that issue.

Again, from the side of fostering market efficiency, social, environmental goals should be attained by direct subsidies, and I will give you an example of this. LIHEAP is probably one of the best examples.

As a marketplace, we don't tend to make judgments on whether or not a particular social or environmental goal is put into place, but our observation is that a participant with LIHEAP is that it really is run very well and does not interfere with the marketplace, simply because it operates as a direct subsidy.

The LIHEAP funds are transmitted to the States. The States administer them directly to the end users. That way they are direct, they are efficient and, in fact, do not serve to distort the market.

I acknowledge also, though, from Mr. Fetter that the political side of that is one that is a reality as well. The problems that we face with mandated mechanisms is that once a mandated mechanism is put in place, it is difficult to undo vis-a-vis the last 60 years of arguing over PUHCA.

A significant part of our commentary and our interest is in the discussions on creation of the power pools that are underway. We do have some views on that and some concerns about some of the trends that may be going on.

We are familiar, because of our study and development of the electric power futures contract, with a couple of the pools that are working, including the one in the UK. The UK 4 years ago deregulated from the standpoint of government ownership of its electric power market and created a pool mechanism whereby electricity is metered out in half-hour increments.

The manager of the pool basically polls producers of electricity to determine what their capacity to produce is, polls the users of that electricity to determine what their needs are and basically sets a market clearing price incremental in half-hour intervals.

Some have said that that deregulated market is something the United States ought to adapt. We think that that would be a step backward for the United States from the standpoint that the Western power pool already is operating much more efficiently than that. It is the beginnings of a true competitive market with bilateral negotiations.

Bilateral negotiations and standardization of contract terms are two things that happen in the cash marketplace that allow for the development of the futures market, that allow for efficiency in that marketplace. We feel very strongly that that Western power pool ought to be studied very closely because, in fact, it does have the beginnings. That is in fact where we are concentrating most heavily as we develop our futures contract.

The northeastern pools, which have a much tighter operation as we describe, already have their participants going outside of the pools and doing that bilateral thing outside of that tight pool, so I guess the summary there is that when you take a look at the mechanisms, be careful that those mechanisms provide and foster truly competitive conditions in that cash marketplace.

The changes that are occurring in the electricity marketplace offer significant challenges and substantial opportunities, and you have heard from people concerned about both sides of that as these hearings have gone on.

Our experience in other sectors of the energy industry undergoing restructurings clearly demonstrate that the increasing opportunities for competition benefits both industry and consumers. Important in the restructuring process is that it not inadvertently hamstring the ability of cash market participants to innovate.

An example that may be helpful in this regard relates to the PBR in California where utilities have negotiated with the CPUC a band of benchmark for purchases of fuel in the natural gas area. If they do well and beat that benchmark and actually do a better job of purchasing, they are incentivized by receiving half of the benefits and passing half on to the customers of that utility.

On the other side, if they don't do as well, they exceed that band. In the old days, they were able to pass that directly on to the ratepayers, to the customers. Now they have to have it evenly split between the shareholders and the ratepayers. Therefore, it is an incentivized form that does benefit and does have the great potential to benefit the consumers.

In the Northeast United States, even more history with the heating oil market. As the heating oil market went through deregulation as a result of the deregulation of the oil industry in the 1970's, heating oil dealers, because they were forced to deal with fluctuating prices, came to our marketplace.

In fact, the first successful energy futures contract was in heating oil in 1978. Heating oil dealers first began to use those futures contracts to protect themselves from price volatility. Market changes occurred from the period 1984 to 1989 when price spikes in the industry gave a couple of those dealers the idea that said, Gee, you know, this might be a good marketing tool because we know we can go to the NYMEX and lock in our price. We can turn around and market to our customers fixed-price heating oil contracts, or, using options, a price cap. So they could go to their customer and say, You are guaranteed, let's say, 80 cents a gallon for all of your heating fuel needs for this season, or if you wish, we can guarantee you that the heating oil will not exceed 85 cents per gallon, but if the price goes down, you will pay less.

That was a very, very effective marketing tool and in fact competition spread it like wildfire throughout the Northeast. Yankee Oilman survived their dealers and found that 73 percent of them were utilizing the futures markets to hedge their oil needs, and turning around and two-thirds of them were offering those types of contracts to their customers.

So what we are saying as an observer of this is that there is a very good chance that as this competition goes into the marketplace, that we could see this as well. Price transparency and risk management opportunities provided by futures markets interacting with the cash marketplace can provide the mechanism of tools for an efficient and competitive market so necessary in today's global economy, and I think the California, again, analogy of recognizing their competitors and neighbors in Canada and Mexico are another example of that.

Mr. Chairman, I want to thank I again for the opportunity to testify.

[The prepared statement of Mr. Thompson and responses to subcommittee questions was submitted for the record. Attachments to the prepared statement are retained in subcommittee files.]

STATEMENT OF R. PATRICK THOMPSON, PRESIDENT, NEW YORK MERCANTILE EXCHANGE

Mr. Chairman, and members of the subcommittee, my name is R. Patrick Thompson. I am President of the New York Mercantile Exchange ("NYNEX" or the "Exchange"). On behalf of the Exchange, I wish to thank you for the opportunity to participate in today's hearing. You and the members of the subcommittee are to be commended for your foresight in seeking to better understand the rapid and significant changes occurring in the electric power industry. As a public, regulated market, NYMEX appreciates the opportunity to discuss these changes. Our first-hand experience in the recent evolution of competitive markets in the energy industry will serve as the basis for our observations.

Futures markets provide two important economic functions; price transparency (price discovery) and risk shifting (risk management). Price transparency is the constant reporting to the world of the prices of actual trades being made at the exchange. With tens of thousands of energy contracts traded daily, price information is available on a virtually continuous basis. Risk shifting, in the secure liquid market that NYMEX provides, allows commercial interests to transfer their risks to a marketplace willing to accept those risks in return for the opportunity to profit from price changes. For the commercial participant, the result is a form of insurance against the financial adversity that can result from volatile energy prices.NYMEX guarantees all contracts traded on the exchange. The strength and liquidity of NYMEX is found in the financial integrity of the exchange, its commitment to the provisions of fair and orderly markets, and its clearing system. NYMEX has taken strong steps to ensure integrity, including daily price limits, customer margin requirements, speculative position limits, market surveillance and strict financial requirements for all NYMEX members.

Well functioning, transparent markets are the best determinant of price. Recent experience with energy provides some clear examples of the value of the marketplace. The impact of the oil embargoes of the early and late 1970's and the Gulf War on oil prices provides a telling study in market performance. The oil market of the 1970's was, if not monopolistic, oligopolistic, with OPEC being the major controlling factor. When the oil embargo was imposed, the price of oil tripled between October and December 1973, sending the entire world into "economic shock" for most of the decade. A second shock occurred with the embargo of 1979, and again prices tripled. A remarkable contrast occurs when comparing those events with the invasion of Kuwait and the Gulf War. There was no public oil or energy market during most of the 1970's (The NYMEX heating oil contract was launched in 1978, crude oil in 1983). By the time of the invasion of Kuwait, deregulation in the oil sector had allowed the establishment of a well functioning futures market to provide a public window to oil pricing. While oil prices increased dramatically immediately after the invasion, the market worked—prices soon adjusted to supply/demand factors, falling a record $10.55 per barrel the day after the onset of the war on January 17, 1991. The conclusion that the marketplace was reflecting supply/demand factors was arrived at after the NYMEX markets were examined and investigated by the Commodity Futures Trading Commission (CFTC) (Attachment 1), the Energy Information Agency (EIA) (Attachment 2) and the General Accounting Office (GAO) (Attachment 3). The functioning of the market allowed for the public exchange of price information, and removed the possibility of "backroom" or cartel established oil prices.

The benefits of the public marketplace are not just confined to industry, but extend to consumers as well. Again, the energy industry has an example with the experience in natural gas. Beginning with the passage of the Natural Gas Policy Act in 1978, continuing to the Federal Energy Regulatory Commission (FERC) Order 636, deregulation in the natural gas industry allowed for the functioning of the public marketplace (NYMEX launched its natural gas futures contract in April, 1990). The market received perhaps its greatest test to date last January, with the record cold, and severe winter conditions. Again, it responded to the challenge. In an interview in the Monday, January 24, 1994 Gas Daily, (Attachment 4) Transco Gas Marketing Vice President H. Dean Jones observed:

"Significantly more transparent price information compared to December 1989 (prior to the NYMEX contract) contributed to everyone's improved confidence in the system. The New York Mercantile Exchange, day trading deals on electronic bulletin boards and industry trade publications kept everyone informed about price movements at various points in the market."

As the natural gas market transition continues, industry and consumers will continue to benefit. Recently, precedent setting pilot programs have been approved and implemented in California by several public utilities for their natural gas operations. The utilities negotiated a performance benchmark with the California Public Utilities Commission (CPUC). If the utility does better in managing its gas purchases, the benefits are evenly divided between shareholders and utility customers. If the utility does worse than the benchmark, instead of passing on the increased costs to customers, the increased costs are evenly divided between shareholders and customers. This innovative structure is a "carrot and stick" approach which rewards utility managers for price management, and passes on benefits to consumers as well. Utilities in New York and New Jersey are developing similar programs for their customers. Massachusetts has legislation under consideration which calls for development of an incentive based rate system for utilities (Attachment 5).

86-340 95 - 11

Vast segments of the energy industry and the public are realizing benefits from the competitive marketplace. The heating oil industry has applied the competitive tools of the marketplace in innovative ways as well. As heating oil distributors and dealers learned how to use the energy futures marketplace to "lock in" the price of fuel, some began to use the ability as a marketing tool. Dealers began offering their customers a season long fixed fuel price, or a guaranteed "price cap" which would pass the benefits of lower prices to consumers, yet provide an upper price limit which they would not exceed. A survey of heating oil dealers in the Northeast conducted by the New England Fuel Institute in 1992 reported that 73 percent were involved in hedging heating oil costs in order to provide guaranteed price programs to their customers (Attachment 6).

Since 1991, the State of Delaware, through its Office of Management and Budget has developed and expanded a market based risk management program to protect the State budget from unexpected energy price spikes. Originally limited to State purchases of heating oil, the program has been expanded to include State vehicle fleet gasoline purchases and the Low Income Home Energy Assistance Program (LIHEAP).

The State of Texas has utilized the public marketplace to protect State revenues received from the oil severance tax. In 1991, the Texas legislature passed legislation establishing a pilot program overseen by a board and administered through the treasurer's office. The success of the pilot program in protecting the State from unexpected price shocks led to its being made permanent and expanded to natural gas severance tax revenues in 1993. Alaska, deriving nearly 85 percent of State revenues from a 12.7 percent oil severance tax is exploring the establishment of a program similar to that of Texas.

Restructuring, the breakup of monopolies, and deregulation in the energy sector have resulted in significant benefits to the United States. Those benefits are realized not only by business and industry, but by consumers as well. NYMEX's markets are widely utilized by the domestic and international oil and gas industries, by major banking institutions, and a diverse array of other participants. NYMEX's prices are disseminated to the public in every major industrialized nation on a real time basis. The transparency of NYMEX's prices, and the integrity of its markets, makes NYMEX a benchmark for energy pricing around the country and the world. The electric power sector is currently in the process of undergoing restructuring—the topic of this series of hearings. Our experience in the energy marketplace is the basis upon which we make our observations and suggestions.

NYMEX's specific interest in the electricity market derives from our efforts to develop an electricity futures contract (or contracts) which are a necessary component of an active and competitive market. In pursuit of this goal we have actively met with hundreds of members of the industry and discussed with them: changes in regulation, increasing competition, greater access to transmission, and increasing participation in the wholesale market by all participants including newcomers. All of these translate into a more active competitive wholesale market, where a low cost means to manage economic risk enables a market participant to be more competitive and improve service. A successful futures contract must provide a low cost means to manage risk, and, consequently, we have a direct interest in these changes in the industry which are leading to greater demand for our prospective product.

As such, the changes that are occurring in the wholesale cash market are of keen interest to us, and, based on our earlier described experience with deregulation and increased competition in other critical energy goods industries—crude oil, heating oil, gasoline, and natural gas—as well as meetings with electricity industry members, we offer a seasoned hands-on perspective regarding many of the issues before this committee today. We are accustomed to working closely with dozens and more participants from the underlying industry which our products serve. We have been doing this with electricity just as we have previously and continue to do so with the other energy product futures contracts. Currently, our Electricity Advisory Committee process includes active participation from about four dozen different entities within the electricity industry, including investor owned utilities, municipals, nonutility generators, electricity brokers, electricity marketers, industrial users, academics, consultants, and others.

The one guiding principle of our comments will be economic efficiency in the wholesale cash market because the more efficient the underlying cash market, the more effective are risk management tools such as futures contracts. We all have a vested interest in competition and economic efficiency in the wholesale cash market. Below, we address the questions the committee has submitted for comment.

1. To what degree do you believe the electric utility industry has entered a period of irreversible change, and how widespread do you feel change will be?

Based on the many meetings we have held with industry participants, including the active dialog we maintain with dozens, we believe that the electricity industry thinks that change has occurred, is irreversible, and is widespread. Based on our experience in other industries and knowledge of the electricity industry, we believe that this change is and will continue to be positive for the industry and its customers, including consumers.

In our research, we have already seen a solid beginning of a thriving competitive wholesale cash market for electricity. A greater variety of products is beginning to be offered providing wholesale customers with choice. Where regulation permits, competition governs price. For municipals, non-utility generators, and those regulated utilities pursuing performance based rates, the incentive of profit and risk of loss discipline decisions. Those decisions include generation, transmission, and distribution which result in allocating the highest valued product at a competitive price supplied at the lowest achievable cost. This is the essence of economic efficiency, and its beginnings are alive and well in the electricity market.

2. Do you have any recommendations as to how regulators should handle the transition period to a more competitive market?

We recognize the sensitivity of these issues and the importance of transition in general. To a large degree, our role with respect to transition is that of an observer. Those that are directly impacted have much more at stake here, and we respect that their concerns must be dealt with fairly.

However, in industries where markets determine allocation and competition drives the market, transition seems to be never ending because there is constant incentive to innovate and pressure to respond to competitors' innovations providing end-users variety, quality, and low prices. Once the transition to a competitive market begins, market based risk management tools such as futures contracts begin to have the potential to add value and further lower costs in the underlying industry. We believe we are at the onset of just such a point in electricity, and this is why we have been actively pursuing development of our electricity futures contract. It is no accident that we have concentrated much (but not all) of our attention on the western US electricity market, which, currently, embodies the most sophisticated commercial practices in terms of variety, activity, and number of participants amongst wholesale cash market for electricity we have observed. Included in that analysis is our observation of the remainder of North America as well as the United Kingdom market.

3. How can social and environmental objectives be adapted to a "restructured" industry, in an environment in which services once provided by a single vertically integrated utility are increasingly "unbundled"? To what degree should these objectives be retained, changed, or abandoned?

We respect the role of policymakers to determine those social and environmental goals they deem important to achieve. We urge them however, to pursue those goals through the most efficient means possible, because pursuit through inefficient means simply results in additional waste which benefits no one. Our comments in this area will be limited for there is much expertise upon which to draw which could sustain a very lively, long and detailed discussion in and of itself We offer, though, the observation that supporting social and environmental objectives through command and control is usually more costly and less effective than directly subsidizing them. The transition to a competitive market place represents an opportunity to pursue these types of goals more efficiently and, thus, at a lower cost.

4. How should pricing mechanisms be adapted to further carry out Congress' goal, embodied in EPAct, of a more competitive wholesale electricity market facilitated by open access to transmission services?

This is an area where we would caution policymakers and others to "look before they leap." We are concerned that this traditional warning is not being adequately heeded by many who are advocating new ideas for this industry. Our experience in other energy markets as well as the electricity market has let us to conclude electricity is evolving along a traditional path of market development that begins with bilateral negotiations of every term and condition and, eventually, results in establishment of standard sets of commercial terms and conditions actively used by most market participants in the determination of competitive wholesale cash market prices. This type of standardization has already begun in the electricity market in the US. We have noticed it in all regions of the US. In the western US, the industry commonly cites specific locations for pricing and delivering electricity in wholesale contracts. Among the most active pricing points are the California Oregon Border ("COB") and Nevada Oregon Border ("NOB"), which served several dozen participants actively and others as well. Although COB and NOB are distinct, because they are near each other and because electricity market participants are sophisticated, it is common for parties to deliver electricity to one point and counter parties to take delivery at the other point through exchanges. Consequently, in many respects COB and NOB serve as one pricing and delivery point. This reminds us of the crude oil wholesale cash market's most active pricing point, Cushing, Oklahoma, where storage and pipeline terminals separately operated by Arco and Texaco serve the oil industry as one effective pricing and delivery point through exchanges as well.

There are other pricing points in the western US such as Palo Verde, Mead, Big Eddy, McCullough, in the prospective "marketplace," which is under development in Nevada, and, when completed in its entirety (including proposed transmission lines from the Northwest), will provide a pricing and delivery point that consistently reflects the competitive wholesale market value for electricity for most of the western US, British Columbia and Alberta, and parts of Mexico.

At the same time as standardization has grown for pricing and delivery points, standardization has also grown over the definition of electricity as a commodity. Under the Western Systems Power Pool ("WSPP"), which operates under authorization granted by the Federal Energy Regulatory Commission ("FERC") and which includes participation by about sixty utilities, there are several specified definitions for electricity as a commodity. Among others, these include firm capacity, firm energy, and economy energy. In the WSPP transactions are individually negotiated principal to principal either bilaterally or multilaterally.

The variety of pricing/delivery points and definitions of commodities in the western US signify commercial innovation which, through competition, provides the electricity market with a choice of quality products at competitive prices. This is what economic efficiency is all about. Commercial innovation such as this, derived from the wholesale cash market, has prodded and spurred operators of transmission and distribution to consider corresponding innovations in operations and policies to enable transactions to be completed. Sometimes, these innovations made sense and transmission and distribution benefited by a new idea. We should all be reminded that the birth of the modem transmission market derived from changes in commercial opportunity from the bulk power market, not from a change in technology. In recent years, electricity market participants responding to commercial pressure has been a greater source of innovation in transmission and distribution than scientific advance alone. At the same time as this process of innovation has been taking place operators of transmission and distribution (control area operators and local utilities respectfully) have adopted only those innovations that preserved integrity of the grid and overall performance.

Let us now contrast what has just been described with some other markets where the mechanism for pricing and physical delivery is more regulated by government. In the Northeast US there are three "tight" power pools: NYPOOL in New York State; NEPOOL in New England; PJM in the Mid-Atlantic States. The Western Systems Power Pool previously described is known as a "loose" pool because buyers and sellers negotiate between each other to complete a transaction. In "tight" power pools, by contrast, pricing and delivery are determined by an automatic formula that calculates the cost of producing electricity by each of its members and then directs lower cost producers to sell to higher cost producers at a price that intends to split the difference in those costs. Because of growing competition in the wholesale electricity market in other regions that are nearby or which neighbor these "tight" power pools, members are choosing to buy and sell electricity outside the pool apparatus at competitive market prices. Our understanding is that in many cases most of their transactions are outside the pool apparatus, and, more importantly, for many of these participants, most of their transactions with other participants is outside the formal apparatus of the pool at competitive market prices. Even while this is going on, integrity of the grid is being preserved as is overall performance.

Just a little over 4 years ago the United Kingdom transformed from a completely government operated electricity industry to one where private ownership exists in the wholesale market. The U.K. has established and maintained a pool mechanism which oversees the delivery of all wholesale cash market electricity and pricing. There is no bilateral or multilateral negotiations that take place in the wholesale cash market in the UK. All physical supplies of electricity (wholesale cash market electricity) is sold through the pool, which aggregates producers willingness to supply into schedules for every half hour period of the day, estimates buyers' needs for electricity for each half hour of the day and clears the wholesale market for electricity for each half hour of the day at the price along the supply schedule that corresponds to the estimated buyers' needs. All wholesale electricity clears at that one price for each half hour. All wholesale electricity sales are to the pool. All wholesale electricity purchases are from the pool. Buyers and sellers do not interact with each other in the wholesale market. They, as well as others, have the option to deal with each other in "paper" instruments that need have no connection to the wholesale market whatsoever. The problem is that there is a clear disconnect between these "paper" instruments and the physical wholesale cash market, and there is nothing that can bridge this outside of buyers and sellers transacting directly with each other in the wholesale market. This disconnect is problematic in and of itself, but the UK pool suffers from a number of significant other shortcomings which are worth noting as well. These include criticisms from most major end user groups that the pool is biased towards sellers, and earlier this year price ceilings were introduced because of lack of confidence by the regulator that the pool mechanism was providing a competitive price.

When mandated market mechanisms do not work properly, it requires a substantial amount of effort and time to try and adjust the regulation to correct the shortcoming, and, even then, there is no guarantee the intended correction will work. Regulation cannot hope to replicate the discipline that arises within an unfettered market where innovation and competition are permitted to operate. Because, under the UK pool, all commercial transactions between buyers and sellers are restricted to purely financial obligations, there is no spark forthcoming from commercial innovations which would to lead to corresponding innovations in the operations or policies governing the physical constraints on system, transmission and distribution. Contrast that to the western U.S. market where, as described above, there is a healthy flow of innovation originating with commercial pressures and resulting in transmission or distribution flexibility. In a market free of unnecessary regulation, where competition reigns, if buyers are dissatisfied with what sellers have to offer, a seller responds by changing the product to better meet buyers needs. They need not go through a lengthy and potentially wasteful process to do this.

For the UK, the pool signifies a step forward from 100 percent control. However, by comparison to the western United States, it has a long way to go before becoming a truly competitive efficient wholesale cash market. We urge all policymakers to understand that mandatory pool mechanisms are steps backward for the United States, and are not required to maintain the integrity of the grid and performance overall.

5. What are the most important questions involving State and Federal regulatory authority? Please feel free to offer any opinion you may have on the state of the law, or on the need for changes to, or clarification of, the respective role of State and Federal regulators.

We have previously commented before the FERC on transmission pricing as well as before State regulators. We continue to support changes in regulation that would increase competition, including better access to transmission and greater opportunity for participation in the electricity cash market.

We have noted that as changes in regulation are considered, there is some question as to the regulatory boundary lines between State and Federal authorities. We understand that it is important to provide clarity to those regulated and defining boundary lines may contribute to this. However, we caution all regulators to consider that in clarifying the division of responsibilities they not inadvertently hold up the tides of progress to a more competitive market place. The dynamics that have provided the undercurrent for the move to greater competition do not reside in regulation. They reside in economics, and that undercurrent will continue to get stronger regardless of regulation. The proper changes in regulation enable us to move to competition more efficiently, but move we shall whether with good changes in regulation or not so good changes in regulation. Major industrial customers would probably prefer the option of access to the grid and to choose among a variety of suppliers to provide their electricity. However, if regulation will not permit them this route, they will either choose to self-generate, be designated a municipality for direct access to the wholesale market, or move their investment and jobs to someplace else that provides them more competitive electricity. From the vantage of Federal officials, as along as that investment in those jobs stay in the US, there may not be the same level of concern individual States have. But, it must be remembered that there is no guarantee that investment and jobs will stay in America. It would certainly be unfortunate if the United States lost economic wealth inadvertently while regulators with perceived overlapping responsibility argued over boundary lines. We support clarification, but urge that it not be at the expense of progress.

Mr. Chairman, as you noted in your letter of invitation to testify, "the pace of change towards a 'restructured' electric industry operating in a far more competitive marketplace has been far more rapid than many had anticipated." Changes occurring in the marketplace offer both significant challenges and substantial opportunities. The recent experience of NYMEX in other sectors of the energy industry which have undergone restructuring demonstrates that increasing the opportunities for competition in the marketplace benefits both industry and consumers. Important in the restructuring process is that it allow for market based risk management tools such as futures contracts which have the potential to add value and further lower costs in the underlying industry. The price transparency and risk management opportunities provided by the futures market, interacting with the cash marketplace provide the mechanism and tools for an efficient and competitive market so necessary in today's global economy.

Mr. SHARP. Well, thank you very much.

Let me ask, in terms of the amount of—this is speculation of course—but the amount of stranded investment, Mr. Parshley, when you were giving us some of the figures that—some of which we referred elsewhere, some of which were your own as I recall, are you compiling the entire asset that—in other words, the entire power plant, or are you looking at the increment of that investment that really is above market clearing price, or is that a useful distinction?

Mr. PARSHLEY. I think there is a distinction. The increment above the market clearing prices is the one that you should be focusing on. I think, however, though that the definition of stranded investment comes along in a long line of terms in this industry that are used frequently and frequently misunderstood.

I don't think there is a clear definition of what uneconomic investment or stranded investment is, and that would be something that would be, I think, appreciated. It might be something that might come out of some of the State PUC proceedings where the issue is being addressed.

Mr. SHARP. Because, obviously, while I think most intelligent people realize you have to be careful about what they read into the gross figures that are given, nonetheless those figures have an implication for our thinking about how disruptive or not certain policy changes can be.

But if we are focused on what is above the clearing market, then that is the quantity that either the regulator must seek to pass through and spread out if they are trying to, in this instance, not damage the credit rating or which the investor has to eat, and that might well be a much smaller figure.

And while there may be serious problems with the system it might not be quite as terrorizing to some people when some of those gross figures suggest that half of this industry is at risk. And it seems improbable that the country doesn't need half of the industry, that we have so much surplus capacity that half of it really we are going to just can it.

Mr. PARSHLEY. I think that is the right way to think about it. I think it applies to not just power plants that are out there but to some of the contracts like the SO-4 contracts in California which may be escalating to prices of 12 or 13 cents when the market clearing price is three or three and some fraction, and the stranded part I would imagine would be the differential.

Mr. SHARP. It does strike me that—and you might be interested whether you think the investing community—many folks in that community or in the Wall Street community that advise a lot of investors, whether or not they genuinely fear that State and Federal regulators will be indifferent to stranded investment and will not take it into account in the sense that they will not just dump it on the investor?

Mr. PARSHLEY. Well, I think there is some concern about that. There are degrees of indifference.

One of the issues that I think needs better clarification in California is what is meant by preserving the financial integrity, and I think that is related to the concept of stranded asset.

Do you think that as a commissioner in California that financial integrity means recovering that part of the investment that is uneconomic?

Because when the plant was built presumably there was an element in the return that was allowed to the utility and to the former regulatory regime that was a return for equity. And if that level is being eliminated and simply the cost of construction is being recovered—I am oversimplifying, of course—then I don't believe currently that the equity market is anticipating the full loss of that layer of return.

And if that becomes the way in which this is handled, then I believe the market would further downgrade the price of many utility stocks, and you might see that the credit quality of the companies would erode as well.

Mr. SHARP. I mean, obviously, you are saying whether we can expect this market to go up or down remains heavily conditioned on what answers come from some of the—in particular, California, but some of the other commissions as well as FERC.

Let me ask you if you think——

I am sorry. Mr. Fetter or Mr. Seetin, if you wish to respond.

Mr. FETTER. Let me just follow up on what Mr. Parshley said. I think the view of most financial analysts on Wall Street is that most regulators are trying to do the right thing and be true to the regulatory compact and treat utilities fairly from a financial basis.

And I think if you use Michigan and California as two examples of how they reacted during the pendency of the Michigan case there was a lot of interest on Wall Street but not negative equity activity with regard to the Michigan utilities. Because I think the feeling was that the Michigan regulators would structure something that provided a—if they structured anything at all, it would be fair—a fair experiment that treated the companies in an appropriate way.

And I think the suddenness and the scope and the timing of the California proposal hitting changed the presumption on the part of Wall Street from the view that this is sitting out there. Will the commission adopt it to this is what the commission wants to know. Will they be able to do it? And I think it changed the optimistic view to a negative one and led to a different result in California than with regard to the Michigan utilities.

Mr. PARSHLEY. I think that is fair.

Let me add one thing. The California proposal came connected to a very aggressive schedule for implementation. Under the way it was proposed on April 20th, we right now would be less than a month away from when they originally expected to have the thing up and running. And so that certainly got the investment community's attention, and it certainly did raise concerns about how fast should these guys go and in what direction.

Mr. SHARP. Mr. Seetin.

Mr. SEETIN. I am not an expert in this field, but in an earlier life I was a banker and maybe I will offer an observation in this regard. I was a banker in the Midwest in the mid-1980's, and we had a substantial number of real estate loans out in the particular bank that I was in that were real estate valued at $3,000, $4,000 per acre.

As the economy collapsed, that real estate dropped by 50 percent or 60 percent in value. And, in fact, we were left with a significant stranded investment in that case, and we charged off about $3 or $4 million of stranded investment.

It was a case not unlike this in which the State and Federal Government got involved through changes in the Bankruptcy Code and others that basically decided who was going to eat that lunch. And, in fact, that is the situation that you are looking at here. And, just as an observer, probably the guys with the deepest pockets are going to dig down the deepest as well.

Mr. SHARP. Obviously, Mr. Fetter, in your opening description of how you learned about the California proposal, Mr. Fetter at least has a fair point that there were some folks who quickly interpreted the finality of it. Whether the commission might have done a better job of warning that this is not the final decision, perhaps they deserve some criticism, but at least some folks clearly jumped to the conclusion that this was a very——

Mr. FETTER. I tend to think that on April 20 had they put out the same document and stated that it was a reformulation of prior stated staff views, we would not be in the same situation we are today.

Mr. SHARP. Well, there have been a few voices, of course, I mean both in testimony before us that simply sort of say, look, we are going to the marketplace, and let's get rid of all these niceties about regulatory compact and past commitments. Everybody has made mistakes in the private sector and had to eat them, so let's take it raw. That tends to come from people that don't have to eat much investment, of course, but it is a luxurious position to be in.

But it does strike me that everything that we are hearing from most everybody else suggests that there are no staunch advocates in the regulatory system for just taking it raw, I mean, so there ought to be some sense that this is being taken seriously at the State and Federal level.

Let me ask you on the proposed rule at FERC whether you see much utility or it makes much difference—and they try to draw a distinction between an existing contract and future contracts as how they might be treated giving, obviously, some protection to the existing contract in the regulatory compact. Do you think that has any utility or relevance to the investor folks? Does it mean something?

Mr. FETTER. I think the investment community will assess the future based on the ground rules in place. I think the comments of all of us are based on if we change the ground rules for the last 20 years of activity then it causes chaos in the investment community. Basically, the investment community lives with whatever rules are put out there as long as they know what the rules of the game are when they go into it.

Mr. PARSHLEY. I certainly agree with that statement. I think we are talking about a transitional issue here, and I think that it is going to be—the market will figure out how to value companies going forward under a new set of rules.

One of the things to keep in mind is that the degree of chaos, to use Mr. Fetter's term, that is created may affect the ability to achieve some of the objectives that you had in mind with the Energy Policy Act and some of the States had in mind. It may delay the arrival of more competitive pricing. It may raise the cost of capital.

I still think that building a power plant to the extent that new ones are going to have to be built sometime, someplace, is going to be a capital-intensive undertaking, and so the cost of capital is going to be a relevant question at the time.

Too much chaos is going to eat away at some of the benefits of competition that may be captured for consumers down the road under a more rational system than perhaps the one we have in place right now.

Mr. SHARP. Well, obviously, the State regulators or FERC or ourselves can either add to or perhaps dampen a little bit the chaos, but we are not in control of it at any level.

I might add—I mean, I am always torn in my own mind between the fact that what I might not describe as chaos but intense pressure tends to also create intense hustle and creativity. And I must say both in oil and gas, as we saw the regulatory systems change and people warned us that the world—it would be a nightmare. And, of course, it was for some individuals in some companies a nightmare, but, in fact, what we saw were unbelievably rapid adjustments in management.

Now, there were some managements that could not adjust, and they are not there today. There are different folks running the show because they simply lived in a different world and had difficulty adjusting to that. That is the way it is in Congress, too, believe it or not. People think it is eternal here.

So, I mean, I think that we want to be careful. I mean, at least my own sense is that, while we want to be sensitive to past commitments in various policies and regulatory compacts both for ratepayers and investors, that there is a danger in being so caught up in that argument that then you think you can hold back change or you prevent change, that in fact there are gains to have been made that would be lost out because you are so afraid of changing the system.

It is clearly going to change. It is changing on us all over the place.

Mr. FETTER. In fact, Mr. Chairman, in my testimony, I noted that, given a fair transition period, I believe the electric utility industry could take a lot of the burden of the potential stranded investment and undo it through greater efficiencies.

But they can't do it if you tell them that it happens next week. Give them 2, 3, 4 years with that hanging over their head, and they will get probably most of the way towards undoing that, their burden of it, and they will eat the rest. And I think they are willing to do that, given a fair transmission period.

Mr. SHARP. Like you, I am certainly, without being so intimately involved, struck by the exceptional speed, let's put it that way, with which California appeared that it wanted to both make the decisions and implement them.

One gets the impression that since no final decisions have been made that that has been brought to their attention in a vigorous way, and it may not hold.

Mr. FETTER. And then potentially lock it down to the residential level before the first direct access transition has occurred at the large user level. Why tie your hands on flexibility on the day you issue the final order? It seems like get some experience with those most able to access the market and, if it works, then transition all the way down to the residential user eventually.

Mr. SHARP. Well, the difficulty we are all starting with is most of our ideas are so out of date we cannot figure out what this world will look like, and all it means is we are beginning to function like we have always functioned for 200 years in the country and that is we create it pragmatically as we go and then we intellectually figure out what that model is that we were doing. And we would like intellectually to reverse that process, but it never seems to work where we actually know the model of what we are going to become.

With that useless piece of advice, I don't know if any of you gentlemen have any further comment you wish to make. We certainly appreciate your time and attention and know you bring several different perspectives.

But a critical perspective in this whole debate at the State and Federal level, obviously, is where is the money going to come from and are people going to be willing to put their money into something that was historically very stable. And while at least it might not be the place to make dramatic new bucks, you can at least stably do very well. And, obviously, that is a changing world, that investors are going to adjust just as everybody else.

Mr. PARSHLEY. If I can just make one more comment. It relates to a question that Mr. Lehman asked this morning about the volatility.

And I guess I respectfully disagree with the answer that Mr. Fessler gave. I do think that there is a pretty good chance that the stocks of these companies and the other participants in the electric services business, some of the ones represented on the panel that follows us, will be more volatile. That doesn't mean they can't be good investments, but it does mean they are not going to be the same kind of investments that some of today's investors in electric utilities are used to. Dividends are probably going to come down. Earnings are going to fluctuate. I think there really is under way—there should be a change in the makeup of the traditional investor base.

Mr. SHARP. Well, let me go back to that, because you raised—I believe you were the one, Mr. Parshley, who raised in your testimony the issue of so many institutional investors who are essentially pension funds who are under certain—and I can't remember whether State or Federal obligations as to how they manage their—these assets, and that has led them—and I thought you were implying trapped them into a utility or utility-like investment.

Would you care to comment on both what is required and also whether, as you seem to suggest now, that whole perspective may change so we may see a real shift in who is investing here?

Mr. PARSHLEY. Sure. I am not at all an expert in what some of the fiduciary responsibilities are of a money manager, and there are different ways to handle that job, but my understanding is that the California public employee retirement system, for example, is an indexed fund and that they have to remain invested in various segments of the market in the percentage which that segment composes the market. And so——

Mr. SHARP. In other words, when you say by that you mean that 10 percent or whatever the percentage is must be in utilities.

Mr. PARSHLEY. If utilities represent 5 percent of the broader group, then 5 percent of their holdings probably should be in that group.

Mr. SHARP. Sure.

Mr. PARSHLEY. And so that got them, with some local loyalty, to become large owners of a couple of large California utilities.

As a practical matter, it is not easy to unwind a position of that size without destroying the market price through the knowledge that would get into the market that somebody was selling a big position. So they basically became trapped over the past 3 months in their holdings, and they had to ride the value of those stocks down.

I think that oversimplifies the case, but I don't want to leave you with the impression that institutional investors all are these kinds of stereotypical wheeler-dealer types who can make immediate decisions and act on them in every case.

Mr. SHARP. That is what we may see as a result. You know how the State legislature in California will respond to that kind of an issue over time, and we may see less local loyalty. Even the Japanese are finding that difficult to sustain even though it has been one of their high-mark qualities.

Well, gentlemen, thank you very much. We appreciate your time and attention and your help to us today. Thank you very much.

We will take a 10-minute break and then we will take our third panel. We will be right back.

[Brief recess.]

Mr. SHARP. The subcommittee will please come to order.

We are very pleased to have our third panel of witnesses for the day. We have Mr. Thomas C. Hinrichs, Vice President with Magma Power Company; Mr. John Stauffacher, Manager of Public and Environmental Affairs with Destec Energy, Incorporated; Peter Garam, the Assistant General Counsel with Consolidated Edison Company of New York—I believe here on behalf of EEI; and Mr. Jaime Steve, Washington Representative for the Union of Concerned Scientists; and Mr. Allen Koleff, Division Vice President with Stone Container Corporation.

Gentlemen, I think you are familiar with how we operate. We will be glad to make your written statement documents a part of our printed record, but we would be pleased to hear your oral summary at this point.

Mr. Hinrichs, we will start with you.

STATEMENTS OF THOMAS C. HINRICHS, VICE PRESIDENT, MAGMA POWER CO.; JOHN J. STAUFFACHER, MANAGER, PUBLIC AND ENVIRONMENTAL AFFAIRS, DESTEC ENERGY, INC.; PETER GARAM, ASSISTANT GENERAL COUNSEL, CONSOLIDATED EDISON COMPANY OF NEW YORK, INC., ON BEHALF OF EDISON ELECTRIC INSTITUTE; JAIME C. STEVE, WASHINGTON REPRESENTATIVE, UNION OF CONCERNED SCIENTISTS; AND ALLEN M. KOLEFF, DIVISION VICE PRESIDENT, STONE CONTAINER CORP., ON BEHALF OF AMERICAN FOREST AND PAPER ASSOCIATION

Mr. HINRICHS. Thank you, Mr. Chairman.

My name is Tom Hinrichs, and I am Vice President of the Magma Power Company, a geothermal energy development company based in San Diego. I am also testifying today on behalf of the Geothermal Energy Association, which involves all the companies engaged in the production of geothermal energy within our country.

We greatly appreciate your leadership in the support of the 80 megawatt legislation before the subcommittee today. As this may be the renewable energy industry's last appearance before you as chairman of this subcommittee, I would be remiss if I did not take this opportunity to tell you of the great appreciation and respect that the renewable energy industry has had for you, Mr. Chairman. Along with the Energy and Commerce Committee ranking minority member Moorhead, you have been the most effective advocates for clean energy in the U.S. Congress. We will greatly miss the leadership, intelligence, fairness and courage you have long displayed as a member and as the chairman of this subcommittee. You have been greatly involved in the development of the renewable energy industry, and because of the legislation that has occurred in this Congress we are able to look to a competitive environment in the future.

PURPA created that environment in the——

Mr. SHARP. Thank you very much for your kind comments. If we allow too much of that I will have to give equal time to those who don't share that perspective.

Mr. HINRICHS. You have played a very leading role, and we appreciate it.

PURPA was the basis for the development of the independent power industry, which is what allows movement to a more competitive electric supply industry. It was the forerunner of transition access mandates established in the 1992 Energy Policy Act, and, most importantly, it moved the renewable power industry from an R&D boutique of the utilities to a thriving commercial independent industry which is becoming very competitive.

In sponsoring your legislation to permanently lift the 80 megawatt cap on renewable energy production, you are once again taking up the fight to maintain some common sense in our government's regulatory policies. At a time when we should be doing all that we can do to reduce greenhouse gas emissions and to increase the availability of clean and domestic U.S. energy supplies, arbitrary size limitations on America's cleanest and most secure forms of electric energy make no sense at all.

And at a time when ratepayers are beginning to see greatly reduced costs for renewable energy as a result of larger, more efficient power plants that enjoy economies of scale, this cap would deny America the clean energy it needs at the lowest possible cost.

At a time when U.S. renewable energy companies are boosting U.S. export of clean energy technologies around the world and creating U.S. jobs in the process, this cap could cripple our ability to stay competitive in international markets.

Mr. Chairman, it has been the policy of our Nation since 1978 that renewable energy developments should be encouraged. In enacting PURPA Congress clearly stated that it is in America's best interests to rely more on domestic renewable energy resources. I believe that all Members of Congress on both sides of the aisle, particularly members of this subcommittee, strongly believe that America should include domestic renewable energy as an important component of our Nation's energy portfolio. Diversity of our electric supply has been a guiding principle of both Republican and Democratic administrations over the past two decades.

I have been participating in President Fessler's restructuring hearings in California, and I have not seen any one of the 95 people that have been in that arena that are opposed to maintaining a healthy, renewable presence in any new system.

There can be legitimate debate over how we structure a system of incentives for renewable energy development, but there should be no debate over the fact that arbitrary caps on renewable project capacity make no sense under any scenario. Ratepayers deserve to benefit from the lowest cost renewable power available, and that is only possible when we are allowed to build power plants that are the most cost-effective and make the most economic sense.

My own company, Magma Power, has at present a generating capacity of 240 megawatts of geothermal power, enough to serve the needs of a quarter of a million Americans. We have been looking to add another 160 megawatts of domestic power production over the next several years.

Additionally, we are working in nations such as the Philippines and Indonesia to develop several hundred megawatts of new geothermal power which should help meet the critical need for power in those nations as well as help reduce global emissions of atmospheric pollutants and greenhouse gases.

In my State of California alone, geothermal electric generation totals some 16 billion kilowatt hours annually with an installed capacity of over 2,500 megawatts. Over the last 30 years, about $5 billion has been invested in constructing geothermal electric facilities in California, and the projects have supported $3 billion in payroll and $50 million in property taxes.

As a result of the continuing efforts of leaders such as you, Mr. Chairman, the U.S. renewable energy industry has come of age in this country, and the cost of our power has continued to fall. As a result of our increased efficiencies and economies of scale, we are now able to produce clean power for America at lower and lower costs.

For example, in California we have been working through the State resources acquisition process for the past 4 years. Under this process, qualifying facilities under PURPA, such as renewable energy companies, were allowed to bid for a small portion of the State's projected future power needs. The utilities established the cost of the power if they develop the project themselves, and this avoided cost was used as the cap on the bidding so ratepayers would not pay any more for power from the private bidders than they would if the utility built the projects themselves.

Taking into account the amount of megawatts in each bid, both gas-fired—there were gas-fired bids and renewable bids. The gas-fired and renewable bids averaged one third less than the indicated utility cost. This was a tremendous outcome for the people of our State and for the entire Nation. It underscores the powerful benefit of allowing renewable power generation to grow without arbitrary size caps.

I have included some graphs at the end of my testimony to highlight the results of that.

In short, PURPA has worked to help bring down costs for America's cleanest and most secure forms of energy: our renewable resources. That success story has only been possible through the gains in efficiencies and economies of scale our industry has been allowed to develop. Your efforts to lift the cap in 1990 on renewable power was an important element in our success.

Magma Power successfully bid in the California process with 160 megawatts of geothermal power and at a cost as low as 5 cents per kilowatt hour in 1990. Such a low bid was possible because of the Sharp-Moorhead bill which passed in 1990 which allows us to build power plants above 80 megawatts. If this production cap is reimposed, we and others like us may not be able to produce the low-cost, renewable power for the future. Companies such as Calpine Corporation, UNOCAL and California Energy Company are all operating or supplying steam to power plants that produce over 80 megawatts of clean geothermal power today.

Reimposing this cap would force our industry to lose economies of scale by preventing us from building or buying renewable energy plants above 80 megawatts. Reinstating this cap will destroy the jobs that are being created in this growing sector of U.S. economy.

Returning this cap would increase the production of harmful atmospheric pollutants, including greenhouse gas emissions. Just as our government is spending millions of dollars to fight greenhouse gas emissions through the President's Climate Change Action Plan, we would be forced to holster one of the most effective weapons against these emissions.

Allowing this cap to return would also erode the global leadership position of U.S. renewable energy companies in the export of large-scale renewable energy plants. Companies such as Magma are working hard to keep our competitive edge in developing overseas markets. These exports help us create jobs here at home.

Domestic production caps would seriously diminish U.S. technological expertise and competitiveness in design construction and operation of larger-scale renewable energy plants. Larger plants are desired in most nations in order to reduce power costs as much as possible. How can U.S. companies compete against Japanese and European firms in building these larger plants when we are prevented from constructing them and perfecting our technology here at home?

Mr. Chairman, opponents of your legislation raise issues related to the underlying justification of PURPA as a reason to oppose this sensible measure. I agree that there may be reasonable areas of debate on some of the issues of PURPA, and certainly as restructuring occurs there will be need to revisit PURPA. I am sure that the future Congresses will look into this area of the law more closely.

However, there can be no debate over the fact that arbitrary production caps on renewable power serve no interest today but to deny America the benefits of lower-priced renewable power, a cleaner environment and safer and more secure domestic energy supplies.

Holding America's renewable resources hostage in order to force a future Congress to examine PURPA and the Federal Power Act is not a responsible position for anyone interested in seeing our Nation develop a strong, diverse and environmentally beneficial mix.

Under your leadership, the House unanimously passed this legislation just 4 years ago. As a result of the compromise with the Senate, a 4-year sunset was put in place. Now it is more important than ever to permanently remove this costly and counterproductive shackle on America's renewable energy industry.

We look forward with you, Mr. Moorhead, and other members of the committee on both sides of the aisle to move this legislation forward with the necessity to enact it into law this year. Thank you very much.

Mr. SHARP. Thank you very much, Mr. Hinrichs.

[The charts referred to by Mr. Hinrichs follows:]

Statewide Resource Plan

MW Need v. MW Subject to Competition

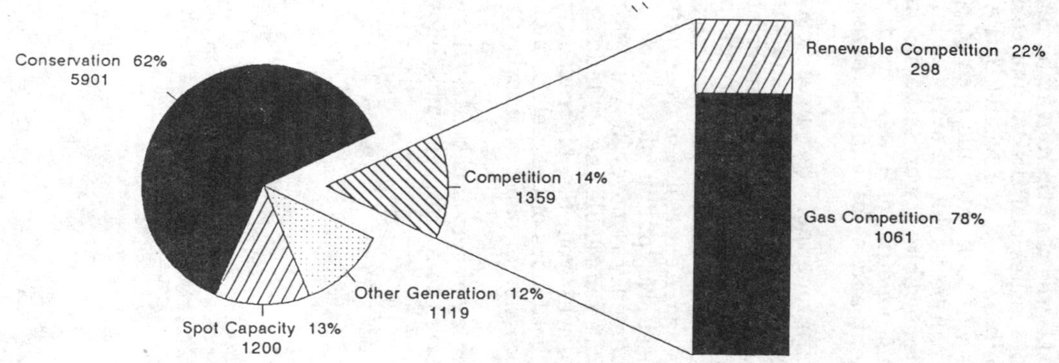

Figures from CPUC D. 92-04-045, as subsequently modified to reduce the MW subject to bid.

PG&E Ratepayer Savings

Utility Cost v. Winning Price

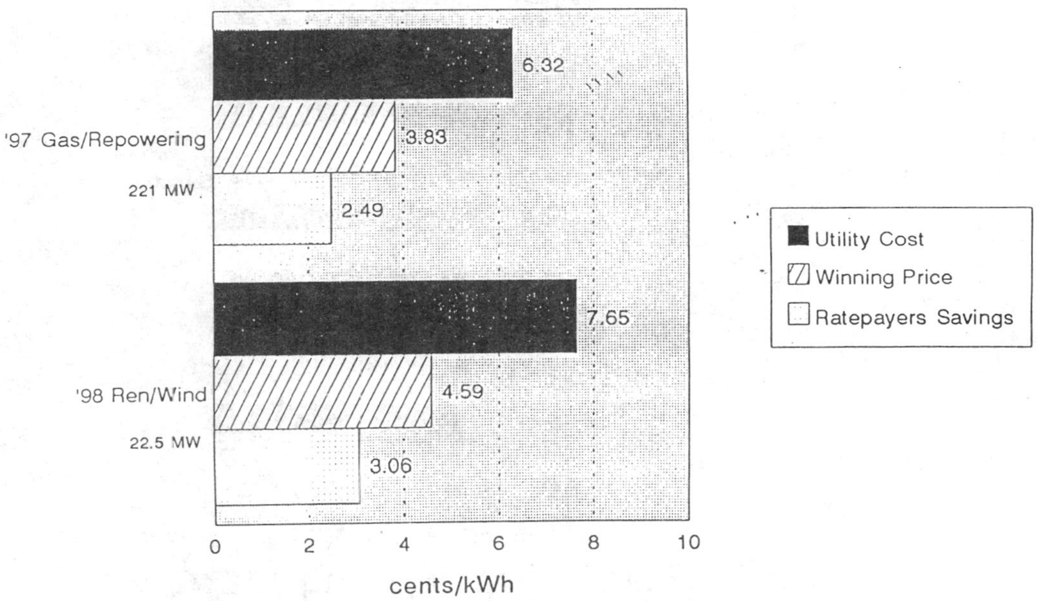

SCE Ratepayer Savings

Utility Cost v. Winning Price

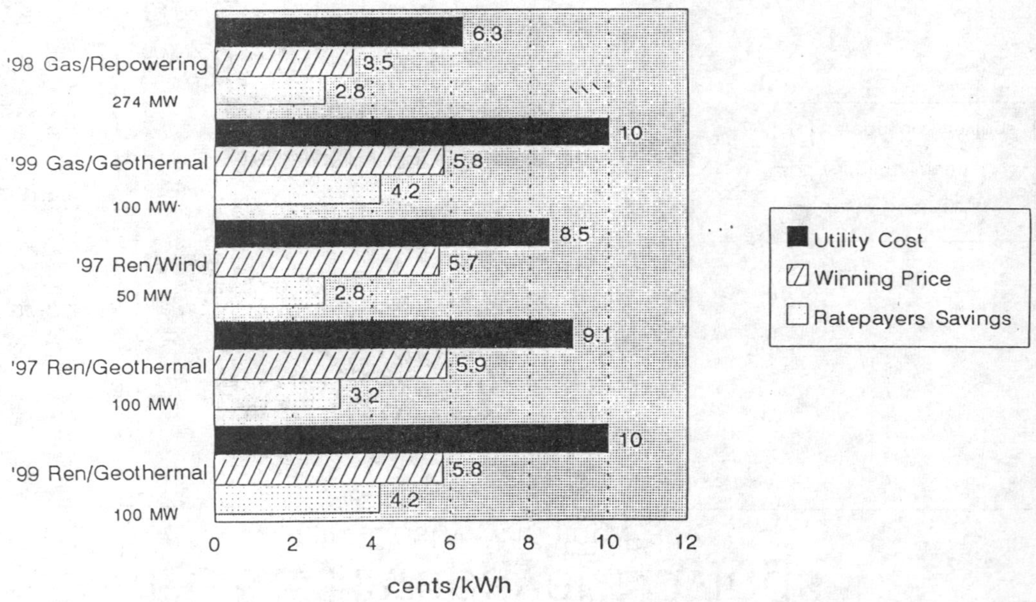

SDG&E Ratepayer Savings

Utility Cost v. Winning Price

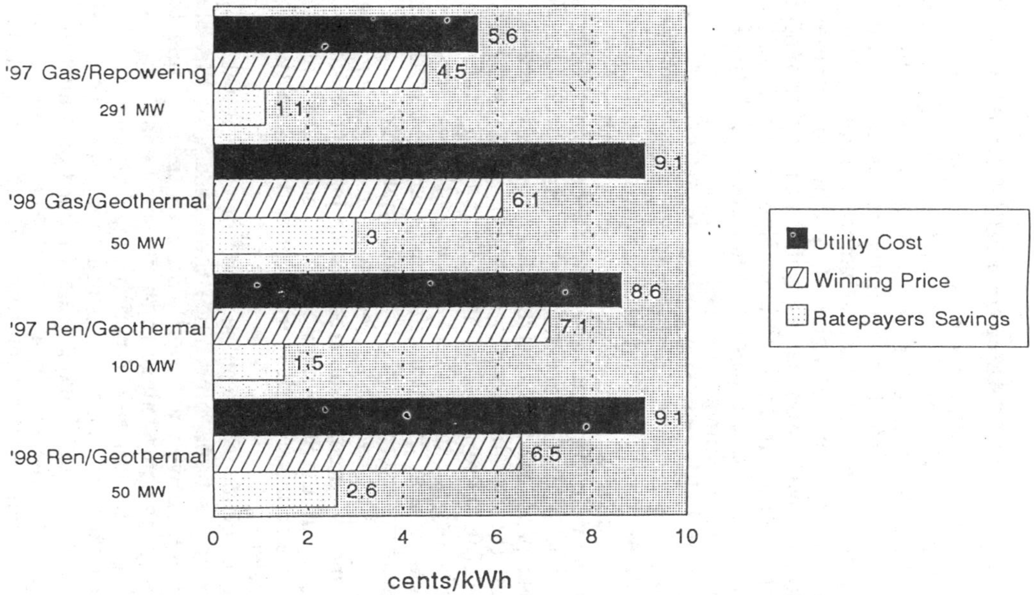

Mr. SHARP. Mr. Stauffacher, we are happy to hear from you now.

STATEMENT OF JOHN J. STAUFFACHER

Mr. STAUFFACHER. Thank you, Mr. Chairman.

Rather than spend too much time on my pre-filed comments, I would like to this afternoon give a response to some of the comments and information you have received not only today but from some of the earlier witnesses before the subcommittee.

Again, my name is John Stauffacher. I represent a major independent power producer as well as am chairman of the task force of the National Independent Energy Producers Task Force on Transmission and a past chairman of LCON, the Industrial Energy Consumers Council.

I think, with that background, it gives me the experience to look at what is going on in the industry right now from a couple perspectives: one from that of a large industrial user which is facing a great deal of competition not only within the United States at this point in time but also worldwide competition as well as from the viewpoint of a very aggressive, large, well-financed independent power producer that is struggling to work its way through the transition that we have right now to a more competitive marketplace.

I think there are some myths out there that we need to address, but before we do that let me echo the comments of Mr. Hinrichs that said that, if nothing else happens out of this hearing, I think a clear message needs to come that PURPA has been a success story. It has been a success story from saving of energy, of minimizing oil imports, of creation of jobs, of creation of tax base, of making industries which buy both steam and electricity from these reliable sources much more competitive than they would have been otherwise.

A lot of focus, of course, appropriately is taking place right now on what is happening since the passage of the Energy Policy Act in 1992. As you are well aware in the key role that you played, there were some things in there, some enablements which certainly will lead to a more competitive marketplace, but I think the importance there is tense. We heard today from the investment community about what is happening during this transition period.

Let me tell you a story about a large industrial-based and now independent power producer and what our transition experience has been.

Destec Energy lost over 40 percent of its stock market value since the 15th of September last year, and I would attribute a significant portion of that to the uncertainty that we are facing right now in the transition that we are going through to get to a more competitive marketplace for electricity.

What has happened is approximately one-third of the assets that we have are committed to QF facilities in Texas which had signed short and intermediate term contracts with the utilities. These contracts are expiring. They were based on PURPA rates that were selling, in the last years of the contract, in the area of 5 or slightly over 5 cents a kilowatt hour.

The market has moved down, and we understand that, and that is the competitive marketplace. But as we go out and try to place that power in the marketplace, we are finding that the 3.5 and 4 cent offers that we are making can't get to the market.

At home in Houston, Texas, I am paying 9.5 cents for my power. The industrials are paying 4 to 5 to 6 cents, depending on tariff and load factor.

The utility, meanwhile, has put in some very expensive nukes. Those nuclear facilities, as we heard earlier on one of the panels, not only were very massive expenditures in the past but aren't shrinking. There is massive additions that are being made to keep them viable assets.

Those viable assets, by the way, are not very economic. They are costing in the Texas market at least 12 to 13 cents a kilowatt hour. And here I am with a stranded investment and not an uneconomic investment, unlike you heard from WebCo last week that certainly in a marketplace one needs to expect if one has an uneconomic asset that the marketplace may well have some repercussions on your ability to earn a return on that investment.

What we are finding is that 3.5 cents, that there are market barriers out there that we cannot get to the market. If I am in New York and I am on Long Island—we have been talking to the Suffolk County Electrical Agency. They are paying 16, 17 cents a kilowatt hour. I would love to get to that market. I would love to get to the market that—my own house where I am paying 9.5.

The point is there are significant barriers out there. What does PURPA have to do with this? What PURPA has to do with this is PURPA is still a very effective lever. It can be used in States and at commissions where only now, 16 years after the passage of PURPA, are some States now looking at how they will add alternatives to the utility-rate-based addition to meet the resource needs of that State.

Certainly, the prices that we are getting today, with over 30 States having competitive solicitation, are not PURPA prices that are based on someone's wild assumption of where oil prices may go in 5 or 10 or 15 years. Certainly, there were some excesses.

I must contradict the comment of Mr. Parshley on the earlier panel who suggested that the Standard Offer Fours in California, which are a good example of what it cost, was based on assumptions that later turned to be erroneous. His point was that these contracts are still escalating. That is not so.

But if you look at now—in fact, within the industry, within the independent power industry they talk about the cliff problem, that the prices are dropping off dramatically in the 10th year. Some of us might say appropriately so. Because now, in year 11 and beyond, the prices will reflect the actual energy inputs that the utilities that are purchasing this power are facing.

The point is that these prices and early excesses of some State commissions, in attempting to put in avoided cost measures, are being worked out in the system. And what we are seeing today is examples like Mr. Hinrichs mentioned in California where, after a 4-year process of determining the BRPU solicitation needs, the bids came in, and these bids were compared to repowering—repowering, not greenfield kind of alternatives—that the utilities offered and came in at discounts between 17 and 40 percent, significant ratepayer savings.

Will that mean savings for all ratepayers? Certainly my contention and the independent power industry's contention is that it will.

We have heard already again frcm the investment community that in the 28 utilities that are represented by the subcommittee's area of interest, that over 20,000 jobs have been eliminated. Certainly, the competitive forces that are out there are proving to the utilities as well as the independents that it is the agile and, as you said earlier, the party with the hustle who is going to get that new job, who is going to get to supply the power to the consumer.

Let me briefly point out some of the ways that we have seen market barriers still existing within the framework that we have in the industry today. In Wisconsin, we found that the utility—because, as in most States, utilities still have a high incentive to build rate base. This is how they grow their earnings. And in most States—or in very few States, at least—are there mechanisms that have been set up which reward the utility for adding lease cost options that are not their own build options. We are starting to move on that in California.

Commissioner Fessler's actions have two tracks to them. One is for performance-based rates that does indeed reward the utility if it adds least-cost alternatives to its mix. The other is direct access.

My concern, my fear is that somehow, if we take these two parallel tracks and we allow them to diverge—we all know what happens to a train if the tracks diverge. We need to keep both the performance-based rates and direct customer access as critical aspects as we move forward towards a more competitive world.

One of the things that strikes me as I have been listening to the testimony that you have heard over the last 2 weeks, that is if indeed what we had was a competitive market, my advice to you would be real simple, and it would be the same advice that my industrial friends give me when I now come in as an independent power producer. Their rejoinder is, "It is the price, stupid." It isn't the price. I wish it was that simple.

And indeed when we get to the day that we can say that then I think we should be back in here, and we need to talk about is there a need for mandatory purchase obligations? Is the lever to open the door to have both utilities and State commissions look at alternative sources of power to be included in the mix? Is the marketplace adequate to make that happen and do we not need the additional help and incentives that PURPA gives us to be able to open that door?

I think, unfortunately, the answer is no right now. Certainly Destec shareholders hope that the answer will work through more rapidly than more slowly because we already have taken our hit on stranded investment, and we need to have the ability to take economic resources that are out there—they have proven reliability and high efficiency and very low impact environmentally—and be able to get them to the market.

We can't do that today. I am optimistic that I will be back here and will be able to say at some later date that indeed the marketplace has been opened and that we can get directly to the customer. Thank you.

Mr. SHARP. Thank you very much.

[The prepared statement and attachments of Mr. Stauffacher follows:]

STATEMENT OF JOHN J. STAUFFACHER, MANAGER, PUBLIC AND ENVIRONMENTAL AFFAIRS, DESTEC ENERGY, INC.

Since the passage of the Energy Policy Act in 1992 (EPAct), a new class of electric generation resources has been added to the mix. The flexibility provided by these entities will increase supply options to wholesale customers. The new authority provided FERC to order transmission services is increasing the options of wholesale purchasers to reach distant sources of supply. While EPAct, a preactive FERC, and efforts by some State Commissions are pushing the electric industry toward a more competitive framework, much still needs to be accomplished before any semblance of a truly competitive electric marketplace evolves. Any effort, at this time, to remove the provisions within PURPA that encourage efficiency and fuel conservation as enforced by the mandatory purchase provisions of the Act, should be rejected.

Despite the creation of EWG's and the new transmission authority granted by EPAct, significant market barriers continue to exist. The mandatory purchase provisions of PURPA continues to provide an effective tool to open electric markets to competitive forces across the country. Numerous cases exist in a variety of States where franchised utilities continue to effectively block new entrants to the market through a variety of actions such as use of discriminatory "competitive" evaluations to favor their own projects, manipulation of avoided costs and capacity need assessment, refusal to cap their own proposals at the bid price, and post contract harassment of QF's by attempts to unilaterally change contract interpretation.

PURPA has met and in many ways exceeded the expectations of its originators. Despite a few early implementation problems, competitive procurement procedures are providing substantial cost savings to ratepayers as well as continued fuel conservation and resultant environmental and efficiency advantages. In California, a State often used to slander the QF industry, cogenerators have created and/or retained a substantial core of jobs and tax base, provided sorely needed cost benefits to hosts as diverse as heavy oil production, hospitals and local school districts, and made significant contributions to improving the air quality of the State.

In short, while progress has been made, much remains to be accomplished to make the electricity market more competitive and responsive to customer needs. The encouragement of cogeneration provided by PURPA remains an important tool to assist in moving towards this goal.

With the passage of PURPA in 1978, Congress provided an avenue for non-utility generators to gain entrance into a market previously controlled solely by utilities. The establishment of qualifying facilities (QF's) that met prescribed efficiency standards, utilized alternative fuels, and minimized imports of foreign oil engendered substantial benefits to all stakeholders. An important fact often missing from the current PURPA debate is that PURPA in fact has and continues to work. Significant amounts of energy have been saved, lowering our dependency on imports. U.S. industry utilizing cogeneration have created new jobs and the United States is now the international leader in the development of renewable resources.

California provides an excellent example of how PURPA works—over 60,000 jobs have been retained or created by use of cogeneration and renewable resources. California QF's represent approximately $13 billion in private investment capital and contribute approximately $142 million a year in local property taxes.

Most of California oil production is dependent upon cogeneration. This oil production decreases the need of the United States to import oil. Other sectors of California's economy in manufacturing, refining, food processors and forest products employ cogeneration which has helped keep these sectors internationally competitive. Even in the service economy cogeneration has played an important role. For example, the La Canada School District has several small cogeneration plants (120 KW capacity). According to their facilities manager, "The cogeneration plant is one of the few areas where we can generate some revenue through our own efforts to keep up the school's quality of education." Similarly the U.S. Navy utilizes cogeneration at its North Island base in San Diego, California which cuts power costs for U.S. taxpayers by 30 percent or $10 million per year. Moreover, it reduces energy requirements by 30 percent which enhances the energy efficiency of the facility.

Independent energy producers have made important contributions to the environment. Biomass energy producers convert, into energy, agricultural and forest product waste, which traditionally were open field burned or placed in landfills. The 800 MW of biomass in California such as the Chowchilla II facility in Fresno County, California convert 7.5 million tons of agricultural residue, urban wood waste and forest management residues into energy every year. This is the equivalent of nearly

20 percent of all waste that goes into California landfills on an annual basis. If agricultural residues alone were open field burned, they would create 131,000 tons of criteria air pollutants per year. Combustion in biomass power plant reduces this by 98.4 percent. This is the equivalent of taking 1.8 million (1994 model) cars off the road.

Similarly, cogeneration using state of the art technology have helped improve air quality. Destec built the Corona Cogeneration facility at the Golden State Cheese plant in Corona, California, which is located in South Coast Air Basin, most heavily polluted air basin in the United States. The Destec facility, which won an award from the South Coast Air Quality Management District, has now set the standard for state of the art control technology. In fact, the air coming out of the facility is cleaner than the air going into it. What this demonstrates is that cogeneration can help create jobs even in areas where due to air quality problems, economic growth is difficult.

In short, PURPA is an example of an energy policy that far surpassed its original goals. Congress should be proud of this achievement and should look for ways to enhance this success.

Several inquires have been heard lately, regarding whether PURPA is still needed. A brief review of the problems that led to the enactment of PURPA in 1978 shows that many such problems are still present even in today's "market"—16 years after PURPA was enacted.

Passed shortly after the fuel crisis of the 1970's, PURPA provides an economically neutral purchase preference for generation in which the maximum energy is extracted from the fuel consumed. Using a given fuel to generate both power and thermal energy at the same time—cogeneration—is still the optimum method to maximize efficiency in power generation.

The ominous but little known fact of the matter is that our country is considerably more dependent today on imported oil than we were when PURPA was passed. Fuel resources have become more scarce since then and despite the gas "bubble" of the 1980's, fuel will become more scarce in the future. Using scarce energy resources to their maximum efficiency should be as much a priority today as it was when PURPA was passed.

Cogeneration is the means by which power can be generated with minimum emissions per unit of useful energy produced. Utilization of cogeneration facilities—especially those already in existence—is the way to generate power and thermal energy with minimum adverse effect on our environment.

Although PURPA provides preferential purchase rights on an economically neutral basis relative to utility generation, the facts have been that cogenerators typically agree to sell their power to utilities at prices well below the utility's cost of construction, providing a significant net savings to ratepayers.

In this regard, it is ironic to note the "complaints" of a few utilities and even some regulators that PURPA has "led to" high-priced contracts. In fact, the state of technology at the time would have caused utility constructed plants that would cost more than cogeneration purchases and usually are much less than the cost the utility would have incurred on its non-cogeneration alternative. In hindsight, cogeneration facilities have in general delivered superior performance and reliability at guaranteed prices—an advantage not found in utility constructed facilities.

Cogenerators provide important competition to utilities and have spearheaded the drive to ever more efficient generating equipment. However, the fact that utilities are not allowed to earn a profit on "investments" in out-sourced cogeneration facilities—while given non-performance based profit on their own facilities—means that most utilities continue to resist adding cogeneration to their resource mix. In some areas, utilities have helped ratepayers to realize savings available through competition (e.g., California). However, the opposite is true in other areas. An alarming situation is developing in these areas, (e.g., Texas), where the amount of cogeneration, both as a percent of total and in actual megawatts is declining. If cogeneration is to continue to make a significant contribution to the supply mix, a continued Federal mandate for cogeneration is required—at least until the market changes sufficiently to ensure that all historic disincentives to purchase from QF's (as well as other non-utility generators) are removed from the marketplace.

The industry still has numerous structural and regulatory impediments to truly competitive procurement of generating resources: (i) the vertical integration of generation and transmission ownership; (ii) rate base rate-of-return regulation which rewards investment rather than efficiency; (iii) a general lack of recognition of cost effective purchase strategies by the franchised supplier, etc. Perverse signals are sent to the service provider by factors such as utilities returns that are not performance-based (i.e., as measured by achievement of lowest-cost power delivery), but rather, based upon investment in generating capacity as noted above—or, in some jurisdictions, based upon investment in DSM programs.

All in all, utilities thus face tremendous incentives to build their own plants—regardless of whether their own plants truly minimize costs of generation for the ratepayers. Ultimately, utility managers answer to their shareholders—not their ratepayers. Unless and until regulation is modified to ensure that all market signals appropriately reflect a level competitive marketplace, utilities must continue to be required to purchase from QF's at appropriately calculated avoided costs. Utilities remain monopsony purchasers of power from QF's and independent power producers in the vast majority of jurisdictions. The ability to leave a region to find another utility purchaser does not change the basic fact that within a given jurisdiction utilities remain the sole potential purchaser of non-utility generating output—and they know that as well as you or I do. The EPAct gives non-utilities the ability to widen their market scope somewhat based upon the economics and availability of wheeling services—but such limited access does not a free market make. The continuing potential for restrictive practices of the vertically integrated monopoly supplier have been dramatically borne out by recent experiences of Destec and other independent power producers in States such as West Virginia, Wisconsin, and Georgia.

In Wisconsin, the State regulations and (supposed) PURPA enablements still do not recognize QF's as parties who can independently bring issues to the Commission in a docketed proceeding. Some progress has been made in recent integrated resource planning proceedings to establish and solidify the presence of QF power, but only the compelling Federal mandate to purchase and the ability to establish a legally enforceable obligation with the utility has predicated this establishment of independent power. The utility still has and continues to use an impressive arsenal of weapons to repel the attempts of competitors to create alternatives to the utility-built option. In a recent process to determine the least cost alternative for needed additional capacity, the utility, while "negotiating" with numerous independent power project developers, developed its own rate-based cogeneration project. To ensure that it would be able to site the project, the utility convinced local residents to move from their homes by "explaining" how the power of eminent domain would entitle the utility to condemn the homes if the residents did not move. Despite utility claims that no QF's could better the costs of the utility self-build option, when the PSC staff evaluated the results of competitive procurement submittals (following three PURPA complaints), the utility proposal placed ninth in overall present value revenue requirements. Thus, only after use of enablements provided by PURPA, was the Wisconsin PSC convinced that only a competitive procurement policy with Commission staff judging the results would lead to a fair result—the first real QF obtained a contract with a Wisconsin utility.

Despite this first brush with success, a similar story has emerged with a second utility in the State—it too decided it desired to meet its capacity needs solely through building of rate-base. During negotiations with QF developers, capacity needs miraculously disappeared—until WPS determined that it wanted to build a rate-based cogeneration facility itself. In a subsequent self-evaluation of generation alternatives, the utility ranked its own projects first, second, third and fourth.

In West Virginia, Destec spent 2 years "negotiating", as well as a long time waiting for a PSC hearing to determine if APS had an avoided cost, etc. In the end, our QF was defeated as a result of a utility and Public Service Commission determination that avoided cost was only in the 1.5 cent/kWh neighborhood—despite substantial evidence that the calculation of avoided cost was flawed on the low-side.

Even in States where some significant QF presence has been established, there exists substantial market barriers and utility resistance to competition, even at the wholesale level. New York, California, Florida, and Michigan are States where Destec has witnessed substantial utility resistance to change.

In Florida, a striking example of utility market power is shown in the case of Nassau Power and Florida Power & Light (Petition for Determination of Need for Electric Power Plant by Nassau Power Corporation (Docket No. 910816-EQ)). In the instant case, Nassau Power (Falcon Seaboard Power) signed an approved standard offer contract and submitted it to FPL for 1996 C.O.D. Following submission of the contract, Nassau delayed approximately 2 years before beginning its plant siting process. Facilities larger than 75 MW steam must be approved pursuant to the Florida Power Plant Siting Act (FPPSA) prior to construction. As a component of the FPPSA, Nassau was required to receive a FPSC "need determination." One would assume that such a determination was merely a formality since Nassau had signed an approved standard offer—approved following an evaluation by the FPSC of statewide needs. However, during the time between submission of the standard offer and the filing of the petition for need, FPL had negotiated and consummated the purchase of Plant Sherer from Georgia Power (600+MW by 1996), as well as signed a negotiated contract for approximately 300 MW from Indiantown (U.S. Generating—a utility affiliate). In addition, the FPSC had changed its Cogeneration Rules—eliminating the "statewide" standard offer and replacing it with utility specific avoided units. Thus, FPL was effectively able to argue during the need hearing that they no longer needed the Nassau capacity, and, although the FPSC had already approved the contract, the PSC ruled that the approval was only for purposes of pricing and denied Nassau Power's need certificate. Nassau Power was unable to begin construction of its project. This example clearly illustrates the inherent bias against non-utility generators.

An additional Florida case of interest is the Cypress Energy case (In Re: Joint Petition to determine need for electric power plant to be located Okeechobee County by Florida Power & Light Company and Cypress Energy Partners, L.P. (Docket No. 920520-EQ)). FPL and Cypress petitioned the FPSC for approval and need determination of an IPP project to be located on Lake Okeechobee in South Florida. During the proceeding, two intervenors supplied substantial evidence that they could supply the capacity and energy at enormous savings compared to the Cypress project. The FPSC ruled in a Catch-22 that all they could do was deny Cypress' contract and "need" since neither of the intervenors (QF's) could be "applicants" under the FPPSA since they did not have utility contracts and therefore could not be awarded utility contracts.

In New York, aggressive anti-cogeneration actions have been of such high profile and vehemence that it has been labelled a "mug-a-NUG" program. The utility monopoly purchasers are attempting to (i) unilaterally reinterpret contracts to require discriminatory tracking accounts for future revenue mismatches despite already having the right to take over such a facility; (ii) obtain new curtailment rights; (iii) manipulate computer runs used to calculate long run avoided costs; and (iv) require regulation of any electricity sales to thermal hosts of the QF.

In Michigan, the ownership interest in the QF matters. A clear display of the market power impact of vertical integration is demonstrated in a recent filing by Consumers Power (In the matter of the applicant of Consumers Power Corporation for approval of a competitive "Rate K " tariff and for certain accounting and ratemaking approvals in connection with such service (Case No. U-10625)). Consumers has requested the ability to make retail sales at market-based rates from 325 MW of nonjurisdictional capacity from a QF in which they have 49 percent ownership interest. Meanwhile, the MPSC retail wheeling "experiment" limits to 60 MW the amount of non-utility-owned retail wheeling.

In the first years of PURPA implementation, some utilities had relatively high avoided costs, leading to some high-priced contracts based on assumptions which later proved to be erroneous. In addition, some States allowed for unlimited capacity to be signed regardless of the proven capacity needs of the purchasing utility. Most, if not all, of these excesses have been worked out of the system and now currently over 30 States require competitive bidding for selection of additional generating capacity.

California is often given as an example where early excesses were prevalent. Often, however, lost sight of is the fact that Standard Offer No. 49 (the highest priced contracts awarded in California), typically had a 10 year initial period where prices were high, but prices are now falling and are below the current generation cost of both SoCal Edison and PG&E. Other contracts were signed under Standard Offer No. 2 provisions, which specified that contract price never exceed the average production cost of the utilities. These SO_2 contracts also have exhibited a downward price trend due to downward trending of the fuel component in the contracts, and the overall system efficiency which independents have brought to the utility mix. See Attachment A—graphs which demonstrate the significant decline in cogeneration payments in California versus the continued, much higher retail prices charged by both SoCal Edison and PG&E.

During the EPAct debate leading to establishment of EWG's, interest groups that were attempting to limit the establishment of EWG's, succeeded in adding section 712 to the EPAct. Section 712 required State commissions to consider the need to adopt four new rulemaking standards regarding the purchase of wholesale power. Generally, State commissions rejected adopting new standards and for the most part determined that at most, the impacts should be considered on a case-by-case basis. A recent study entitled Financial Impacts of Nonutility Power Purchases on Investor-Owned Electric Utilities, (June 1994) by DOE's Energy Information Administration (EIA) concluded:

The results indicate that nonutility power purchases did not raise a utility's cost of equity capital. In fact, there was more evidence to support the notion that utility construction raises the cost of capital more than nonutility power purchases do—there is no conclusive evidence that power purchases from nonutility generators raised the cost of capital to the utilities which purchase the electricity. (See Attachment B).

Competitive procurement procedures (or bidding) have taken center stage for resource procurement in most jurisdictions. Few QF contracts now contain prices based on forecasted fixed fuel factors. However, the concept of avoided costs is still an important benchmark to ensure cogenerators a reasonable opportunity to sell their power at costs approaching, but not exceeding, the utilities' own costs to procure the power. Absent this standard, market access could be denied to cogenerators on a cost pretext that fails to accurately reflect the actual costs production. Indeed, the increasingly competitive marketplace has driven the price that utilities claim they themselves could built the capacity for, so far down that in many cases the cost for new capacity is below the embedded cost of the utility. For example, in a recent first bid program in Pennsylvania, Destec was the successful bidder and was awarded a contract which was priced approximately 30 percent below the price that the utility said it could produce the power for itself. Another example is the recent bidding experience in California where bids were awarded at 17–40 percent less than utilities' estimates of costs for such power.

For utilities having stated that a competitive market place now exists, it is rather puzzling to see the same utilities now coming to both State and Federal regulators and legislators, to question whether the mandatory purchase requirements of PURPA should be maintained. At the same time, and often in the same presentation, these utilities are urging the banning of retail sales or retail wheeling for fear that they may lose market share and not be able to compete in a more open marketplace. At Destec, we have yet to see this open marketplace—"the competitive" monopolist has yet to exist. Many utilities appear to be intent on convincing legislators and regulators that there is a threat of competition, but that does not justify succumbing to threats of, or to pleas of, protectionism due to the threat of sham competition and continue to construct market barriers that preclude real competition from being accomplished. Until real competition is obtained, the protection provided by PURPA needs to be reminded so that the dominant utility generation resource is only one option that is considered when looking at cost effective reliable ways of meeting capacity needs for this country.

Attachment A
Page 2 of 2

The Forces for Change

Southern California Edison

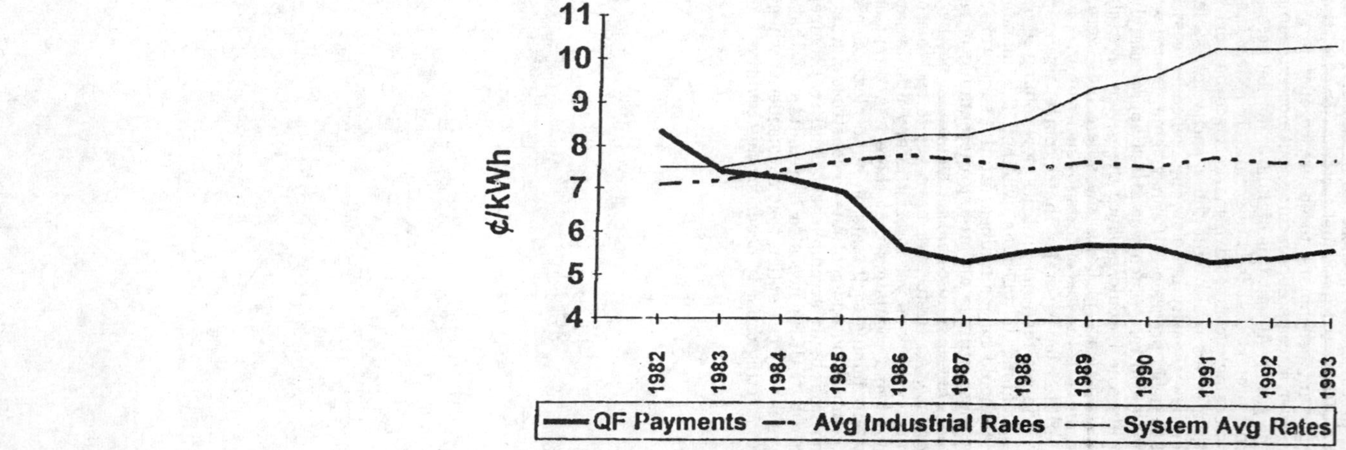

* Assumes $173/kW levelized capacity payment and 80% capacity factor

MRW & Associates, Inc.

Attachment A
Page 1 of 2

The Forces for Change

Pacific Gas and Electric

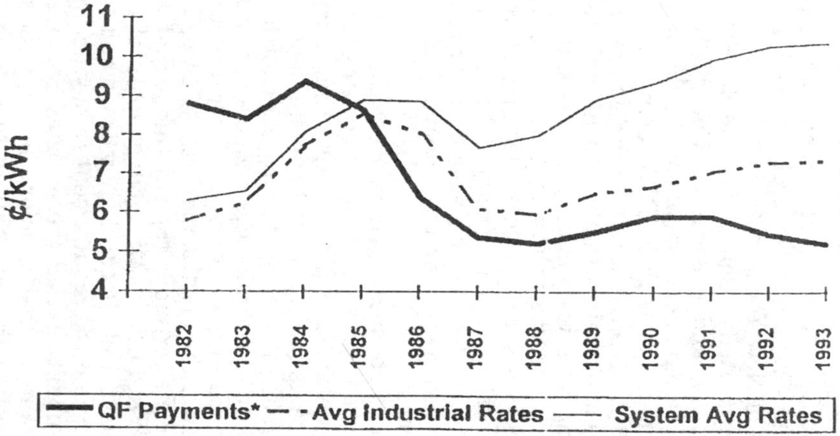

* Assumes $173/kW levelized capacity payment and 80% capacity factor

MRW & Associates, Inc.

Mr. SHARP. Mr. Garam.

STATEMENT OF PETER GARAM

Mr. GARAM. Good afternoon, Mr. Chairman. My name is Peter Garam, Assistant General Counsel of Consolidated Edison Company of New York, Inc.

Very briefly, Con Edison serves 8.7 million customers in New York City and Westchester County. We are proud of providing the most reliable electric service in the world. We have also had a lot of experience with PURPA, being one of the largest purchasers of PURPA power in the country. It is a real honor for me to appear today on behalf of the Edison Electric Institute.

The first two panels, Mr. Chairman, talked about the global issues of electric utility industry restructuring. The subject of this panel is a very innocuous looking or apparently innocuous looking piece of legislation of only 14 lines. But let me suggest that there is a very close tie between the topics that were discussed in the first two panels and the legislation that we are discussing this afternoon.

Mr. Chairman, you started the discussion with your opening statement this morning by raising the issue of conflict between competition and other goals. That was followed by Mr. Moorhead who talked about a level playing field.

Mr. Fessler then discussed distortions in the market, the high rates in California, the tremendous loss of jobs in California, and his hope to lower costs to electric customers and increase the efficiency of the electric market.

Mr. Fetter made the statement that the "must buy" provisions of PURPA must go. Mr. Neitzel spoke of a work horse that is so loaded down that it can barely walk and again of buying uneconomic power and the requirement to buy uneconomic power.

I submit that all these issues are directly relevant to the PURPA amendment that is before us today. So for that reason I very much appreciate the opportunity to express our concerns with the proposed legislation and our more fundamental concerns with PURPA in today's more competitive electric environment.

Let me start with a summary of our position. Then I will elaborate on it. Briefly, EEI strongly opposes the proposed expansion of PURPA. EEI believes this bill is a step in the wrong direction, and we believe that PURPA's mandatory purchase obligation has outlived its usefulness and is inconsistent with the increasingly competitive wholesale electric markets that we are seeing today.

Let me explain these positions, if I may. When PURPA was enacted in 1978, we had a vertically integrated electric utility industry. There was no independent power market. Today, in contrast, there is a highly competitive wholesale electric market, and PURPA is largely responsible for this change.

PURPA has changed the face of the industry and indeed the entire structure of the electric utility industry. Today, there is over 55,000 megawatts of independent power on line, and most of the new generation is coming from independent producers.

Many utilities have stopped building power plants altogether. For example, utilities in New York have not built any new plants since 1988 and have no plans to build any more in the foreseeable future.

The Energy Policy Act of 1992 which originated in this subcommittee, was premised on even greater competition in the electric industry. EPAct created a whole new class of generators in the electric market, exempt wholesale generators, so it brought additional competition to the generation sector. It also increased competition by expanding FERC's authority to order utilities to provide transmission service.

This subcommittee held 2 days of hearings last week on the changes that have taken place in the electric industry since EPAct's enactment. I believe that all the testimony you received can be summarized in just two words: increasing competition.

We believe that PURPA's mandatory purchase obligation and, by extension, the enactment of this bill is fundamentally inconsistent with today's highly competitive electric markets for several reasons.

First, PURPA's mandatory purchase obligation has resulted in billions of dollars of excessive costs and unneeded power. For example, Con Edison's customers are facing an almost $1 billion overpayment on account of a single energy-only contract that we were required to enter into under PURPA.

Second, PURPA's mandatory purchase obligations skews competition by imposing costs and burdens on utilities that are not imposed on their competitors.

You cannot expect utilities to compete effectively if at the same time you force them to buy uneconomic sources of power. This is particularly a severe problem for utilities such as those in New York that operate in areas with weak economies.

We heard this morning about the state of the economy in California. It is unfortunate that I have to say that we face similar problems in New York. New York has lost over 500,000 jobs in the last few years. And Con Edison alone has over one million people on some form of public assistance.

Utilities are working hard at controlling their costs and improving the competitiveness of businesses in their areas and to make energy more affordable for the residential customers. These efforts would be frustrated by the imposition of additional costs and mandates on utilities.

Third, I would suggest that there are more efficient and direct ways to encourage the development of renewable energy resources than through ratepayer subsidies. Under EPAct, any generator can obtain EWG status regardless of size. Therefore, renewable technologies that claim to need to achieve economies of scale can do so without PURPA's mandatory purchase obligation.

Another point is that significant renewable energy R&D efforts are being undertaken by utilities across the country. For example, New York utilities have budgeted $50 million for the 1993-1998 period. In addition, they have committed to acquiring over 300 megawatts of renewable resources in a market test demonstration program.

These activities, along with tax credits and incentive payments authorized by the Energy Policy Act, allow the promotion of renewable energy without imposing direct and uneconomic costs on individual utilities and their customers.

Finally, in a competitive market, suppliers should be selected on the basis of who offers the best deal.

I would like to respond directly to Mr. Hinrichs' assertion a few minutes ago that we need PURPA to get the cost of renewable energy down. On the contrary, we don't need PURPA to get the of cost of renewable energy down. PURPA interferes with that objective and let me offer to you an example from Con Edison's experience.

We had a bidding solicitation a couple of years ago when we were looking for 200 megawatts of power. We chose the winners without regard to their qualifying status under PURPA, and some of the winners turned out to be renewable resources.

So the best way to get—take advantage of the lower prices, if indeed prices are going down, is to let them compete and let the customers get the benefit of that competition rather than distort the market through the artificial means of PURPA.

Today, you will hear testimony as to the alleged benefits of expanding PURPA. We urge you to examine these claims very carefully. We believe you will find that this bill would impose higher cost on utility customers by continuation of mandatory purchase obligation coupled with the system of administratively determined avoided costs.

In sum, industry's opposition to this proposal does not indicate an unwillingness to embrace renewable energy technologies, because we have—our position is based on the fact that competitive forces have been unleashed in the wholesale electric markets. We urge you to let competition work and not turn the clock back by approving an expansion of an outdated model that is inconsistent with today's realities.

I therefore recommend that, rather than creating a new mandatory purchase obligation under PURPA, Congress undertake a fundamental examination of whether we need this provision at all in light of today's highly competitive electric markets.

Thank you very much, Mr. Chairman. And I would be pleased to answer any questions.

[The prepared statement of Mr. Garam follows:]

STATEMENT OF PETER GARAM, ASSISTANT GENERAL COUNSEL, CONSOLIDATED EDISON COMPANY OF NEW YORK, INC., ON BEHALF OF THE EDISON INSTITUTE

Good morning. Mr. Chairman and members of the subcommittee, my name is Peter Garam. I am assistant general counsel for the Consolidated Edison Company of New York, Inc. I am appearing here today on behalf of the Edison Electric Institute (EEI), the national association of the investor-owned electric utility industry. Its members generate approximately 79 percent of all the electric in the country and serve 76 percent of all ultimate customers in the Nation. Investor-owned electric utilities purchased 156,473 GWh, or 51 percent, of all power generated by non-utility producers as of December 31, 1992.[1] The balance was consumed by the non-utility producers themselves.

Con Edison provides the most reliable electric service in the world to the 8.7 million people in it's service territory, which covers New York City and Westchester County. Con Edison has had extensive involvement with the Public Utility Regulatory Policies Act of 1978 (PURPA). Con Edison is one of the largest purchasers

[1] Source: "Capacity and Generation of Non-Utility Sources of Energy," Edison Electric Institute, 1992.

of PURPA power in the country. The company has entered Into contracts for almost 3000 MW with PURPA qualifying facilities (QF's). Unfortunately, as many utilities around the country can attest to, the administratively determined avoided costs which formed the basis for these contracts turned out to be too high, and have already and will for the lives of these contracts cost our consumers dearly in premiums for electric power. Con Edison is also an active participant in the very competitive wholesale energy market; it now purchases about half of its energy requirements on the open market at substantial savings to its customers.

We want to thank the subcommittee for holding this hearing today and appreciate this opportunity to express our concerns regarding the proposed legislation to remove permanently the size cap on QF's under PURPA that use certain renewables as sources of generation. We are most concerned that the proposed legislation's removal of the size cap, in combination with PURPA's mandatory purchase obligation, while appearing to some as a simple extension of current law, is in reality a contradiction of the principles of competition in the electric generation sector that this same subcommittee embraced when it passed the Energy Policy Act of 1992 (EPAct). Our testimony today will focus on this contradiction and on the industry's concerns regarding PURPA in general.

Enactment of EPAct fundamentally changed the nature of competition affecting the electric utility industry. Just last week, the subcommittee conducted 2 days of hearings on the conditions within the electric industry since enactment of EPAct; the two previous panels continued that important discussion. As the hearings on July 13 and 14 pointed out, EPAct encouraged new participants in the wholesale generation market by creating a new class of wholesale power supply entities, known as exempt wholesale generators (EWG's). Under EPAct, EWG's receive exemptions from restrictions imposed by the Public Utility Holding Company Act of 1935 (PUHCA) and may sell their power at market-based rates. Any power producer can request EWG status, including developers of renewable resources.

In order to help assure the efficient development of a more competitive wholesale power market, EPAct also expanded the authority of the Federal Energy Regulatory Commission (FERC) to order transmission access for wholesale power buyers and sellers, Including QF's to ensure that power developed in the competitive generation market could reach the widest market possible, so long as system reliability and the public interest were not harmed. FERC is moving quickly to implement its responsibilities regarding the terms and conditions for transmission access and pricing. As a result of these provisions, QF's have the ability to obtain transmission access to wholesale customers, the same as any other wholesale generator. Their potential market is thus far larger than the utility to which they are interconnected. Moreover, recognizing the competitive nature of today's wholesale generation market, FERC recently announced that it will no longer conduct a generation market power analysis to determine whether or not to approve market-based rates for sales from new generation capacity.

While designed to encourage competition in wholesale generation markets, EPAct was unfortunately silent on the obligations imposed on electric utilities to purchase power from QF's, as required under PURPA. EPAct's legislative changes, and the changes implemented and endorsed by FERC and the State regulatory agencies, call into question the need to retain PURPA's mandatory purchase obligation for any electricity generators, including those using renewable energy sources. In light of these fundamental changes in the industry, we urge the subcommittee to examine fully whether an appropriate role for PURPA exists in the current competitive environment.

In response to the energy crises of the 1970's, the 95th Congress passed PURPA as part of the National Energy Act. Five key goals stand behind Title 11 of PURPA, which established the foundation for the development of a cogeneration and small power production industry: to promote energy independence and diversify the energy supply; to encourage the development of cogeneration and small power production facilities; to improve the overall efficiency of the electric supply; to conserve electric energy; and to provide equitable rates to electric consumers. As encouragement, Congress provided significant benefits to these generators: QF's are exempted from regulation as electric utilities and are provided a guaranteed utility market for their electrical output.

Title II of PURPA also brought about a commingling of Federal and State authority. FERC was to promulgate the regulatory guidelines for purchases from QF's; the development of backup, standby, and maintenance rates, terms, and conditions; and the development of interconnection standards. Under section 210, the rates paid by utilities for purchases of electricity from QF's must be just and reasonable to the utilities' customers and in the public interest, and may not exceed the purchasing utility's "avoided cost." FERC, in implementing these provisions, chose to require that rates for purchases from QF's be based on full avoided cost. Avoided cost is defined by FERC as the incremental cost to the utility of energy or capacity or both, which the utility would otherwise incur by generating power itself or purchasing it from another source. FERC's full avoided cost standard was upheld by the U.S. Supreme Court in 1983 in American Paper Institute v. American Electric Power.

PURPA left it up to the States to implement the provisions of section 210 and FERC regulations dealing with utility purchases from QF's and sales to QF's. State translation of the avoided cost principle into practice has been a major area of controversy over the years. In addressing the avoided cost question, State commissions have faced the issue of how to balance the competing objectives of encouraging QF development while providing rate equity to utility customers. In addition, some States, acting under their own authority, have adopted rates that clearly exceed the Federal ceiling of full avoided cost for QF purchases (discussed below).

PURPA has succeeded in bringing more than 55,000 MW of QF capacity on-line by the end of 1992. In 1992, the year for which we have the latest data, all non-utility power producers accounted for 55,188 MW of installed generating capacity in the United States. Of that figure, 40,674 MW (or 74 percent) came from cogenerators and 10,154 MW (or 18 percent) came from small power producers (SPP's), including renewables. In total, these figures represent a 64 percent increase over 1988 in the installed generating capacity of such generators.[2] In addition, it is estimated that over 50 percent of the capacity additions between now and the end of the decade will be built by non-utility generators, accounting for approximately 33,000 MW.[3] In New York, for example, utility plants have not been built since 1988 and none are planned for the foreseeable future. These figures provide evidence as to the development and growth of non-utility generators, including the renewable energy industry, since the passage of PURPA.

While PURPA in theory requires utilities to pay no more to QF's than they would pay for alternative sources of power, in practice PURPA has forced utilities to pay billions of dollars in excessive costs for power. This not only harms utility customers but also does not fit with the competitive wholesale electric generation market encouraged by EPAct.

In California alone, electricity consumers have had to pay at least $6 billion to QF's above actual avoided costs determined by the market. In New York, the excess payments are projected to exceed $1 billion per year.

Electric utilities and their customers in at least fifteen other States are being forced to purchase QF power at rates that exceed what would be an accurately determined avoided cost. The problem occurs in several guises.

First, some State legislatures have deliberately required utilities to pay QF's more than the utility's avoided cost. Until 1992, for example, New York State required payments to QF's of a minimum of six cents per kilowatt hour, which was far above actual avoided cost. Because these contracts covered long terms, utilities in New York will be paying excessive costs for many years to come. Also, Connecticut still requires utilities to purchase power from municipal solid waste facilities at rates that exceed the actual avoided costs of the utilities.

Second, FERC's regulations require long-term purchases, even of energy only, to be based on estimates of avoided energy costs, rather than actual avoided energy costs. This is a significant and unusual departure from the workings of the highly competitive energy markets where prices can change several times an hour. This FERC requirement can result in huge overpayments because of the volatility and unpredictability of fuel prices and is no longer necessary due to technological advances. For example, under a single contract that Con Edison was required to enter into in 1991, the company and its customers will be subjected to overpayments of almost $1 billion.

Third, some States require utilities and their customers to pay capacity costs even when they have no need for capacity.

Fourth, several States use levelized rates or front-end loaded contracts. In this case a utility is required to sign a long-term contract at fixed rates that can exceed the utility's true avoided costs over the term of the contract because there is a risk that the front-end load would not be paid off.

A fifth way that QF purchase rates can be distorted is when the capacity rate is based on an avoided unit (proxy plant) that does not match the lowest cost alternative available to the utility.

Each of these cases demonstrates that the original benefits PURPA was intended to achieve and share with electric consumers are instead being denied them. Utilities and their customers are being hamstrung with excessive costs for uneconomic

---

[2] Ibid. The remaining 8 percent are classified as other non-utility producers.
[3] Source: UDI State Directory of Electric Utility Plants, 1994.

and, in many cases, unnecessary power. These excessive costs are imposed directly on customers in violation of the avoided cost ceiling established by PURPA. It is not just and reasonable for a State commission to require a utility or its customers to pay more than its actual avoided costs. While many States have moved toward competitive bidding to insure that all generation rates are based on real alternatives available to the utility, not all States have taken that step.

As the Agency with the lead responsibility for implementing section 210 of PURPA, FERC should ensure that excessive and illegal pricing practices under PURPA are eliminated. We encourage FERC to correct the overcalculation of avoided cost.

The investor-owned electric utility industry remains opposed to the Committee Print (dated June 24, 1994) that has been circulated by the subcommittee. The proposed legislation would remove permanently the 80-MW size limitation for QF's that use solar, wind, waste, or geothermal energy. It also, by its nature, would retain and expand the mandatory purchase obligation for power from these facilities. Under current law, the 80 MW size limitation will be reinstated on January 1, 1995.

The industry's concerns with the proposed legislation center around its retention and implicit expansion of PURPA's mandatory purchase obligation. We offer the following points to demonstrate the real problems this will cause:

—The mandatory purchase obligation, coupled with improperly calculated avoided cost, can be harmful to utilities and their customers, resulting in billions of dollars in excessive costs and unneeded power.

—The mandatory purchase obligation is no longer needed, since most States have competitive bidding programs under which all competing power suppliers can be treated equally. Markets can, and should be allowed to, set avoided costs, and State regulators should not be required to guess what prices the wholesale markets would set.

—As the industry changes, it will be faced with the risk of stranded investment, which is only exacerbated by the mandatory purchase obligation.

—The mandatory purchase obligation skews competition by providing QF's guaranteed markets and exposes utilities and their customers to risks and costs that are not borne by the utilities' competitors.

—EPAct created EWG's to encourage power producers of all types, without burdening utilities with a mandatory purchase obligation.

—There are other more efficient and direct methods of encouraging the development of renewable energy resources, rather than relying on the mandatory purchase obligation to provide subsidies at the expense of utility ratepayers.

As stated in our June 28, 1994, letter to the subcommittee, our concerns are based on the lack of a demonstrated need for the legislation in light of the competitive electric generation markets that have emerged even more strongly since EPAct. Recent efforts in some States to promote direct retail access heighten our concerns. In a competitive environment where there is increased retail competition, PURPA's mandatory purchase obligation can result in utilities incurring increased costs or excess generating capacity without choice, which must be borne by a shrinking base of utility customers. This problem was recognized in last week's testimony by the FERC commissioners before this subcommittee. Utilities and their customers will not only be faced with possible stranded costs of existing PURPA contracts, but may bear further uneconomic costs due to new purchase obligations as proposed by the legislation. These excessive, uneconomic costs will exacerbate the demand for retail access, while making utilities less able to compete in electricity markets. The changes in the industry require a thorough examination of the rationale behind PURPA.

The industry's opposition to this proposal does not indicate an unwillingness to embrace renewable energy technologies. The industry is aggressively supporting renewable energy collaboratives with a wide range of stakeholders to develop promising technologies without forcing the costs of uneconomic choices on ratepayers. EEI led the development of the Utility PhotoVoltaic Group (UPVG; with over 80 utility members), underwrote preparation of the Utility Wind Resource Assessment Program (UWRAP), is supporting the Utility Biomass Energy Commercialization Association, and has hosted meetings with the geothermal community to begin the development of a geothermal collaborative. Furthermore, EEI and utility industry employees are active participants in the photovoltaic and wind national collaboratives to help those technologies penetrate the power supply market and resolve technical, environmental, and economic issues associated with them. The substantial response to the UPVG's early opportunity notice of Federal cost sharing for photovoltaic projects indicates a willingness by our industry to advance that technology further. In addition, utilities around the country are devoting significant resources to the development of renewable energy technologies. For example, utilities in New York have budgeted $50 million for renewables R&D for the 1993-1998 time period. In addition, they recently agreed to conduct a market test/demonstration program to acquire 300 MW of renewable resources.

Mr. Chairman, you were present at the Earth Day ceremony during which the electric utility industry's commitment to advancing renewable energy technologies was further demonstrated through the various initiatives announced as part of the administration's Climate Challenge program. Specifically, the EnviroTech Investment Fund is a venture capital fund that is dedicated to support both renewable technologies and electrotechnologies, which may make it possible for some renewables to achieve the economies of scale of mass production that often have been projected. The Earth Comfort initiative promotes the development of geothermal heat pumps to use sustained temperatures from the earth for hearing and cooling as an environmentally acceptable HVAC technology. Our industry also is an active participant in implementing Option 26 of the President's Climate Change Action Plan to commercialize photovoltaics, wind, biomass, and geothermal through the collaborative processes described above.

For a competitive electric industry to work effectively, a single set of "rules of the road" must be applicable to all entities. Without consistent rules and obligations for all electricity providers, competition will be skewed, inefficiency will reign, and a few customers and producers will benefit at the expense of all the rest. As stated in EEI's testimony before the subcommittee on July 13, there must be a consistent set of rules for all competitors.

The competitive EWG market facilitated by EPAct is flourishing. As of the end of June, there have been 148 applications for EWG status, 82 of which are for domestic facilities. It must also be pointed out that there are no size restrictions on any facility that wishes to receive EWG status. This avenue is available to renewable energy facilities, just as it is to any other type of generating facility, including cogenerators.

Last year, for example, FERC granted EWG status to a hydroelectric facility that had been certified under PURPA. In that case, Malacha Hydro Limited Partnership sought and received the flexibility to produce and deliver in excess of 30 MW capacity to its purchaser. Malacha requested a determination that it was an EWG under section 32 of PUHCA, arguing that it could qualify as an EWG under PUHCA without affecting its QF status under PURPA. FERC agreed, stating that there was "nothing in the Energy Policy Act or its legislative history to indicate that a facility that satisfies the requirements for QF status may not also be an eligible facility. . . . Thus, the Commission is of the view that an EWG may own an electric generating facility that qualifies as an eligible facility under PUHCA at the same time that such facility also satisfies the requirements for a QF under PURPA."[4]

This case is cited to point out that any QF entity that wishes to expand its output above the 30 MW (for hydro) and 80 MW size restrictions can do so by becoming an EWG under PUHCA and can sell its output at market-based rates. Therefore, the barrier to large-scale renewable energy project development presumed by the proposed legislation simply does not exist. Thus, there is no real reason to extend the removal of the size limitation beyond the end of 1994 for renewable facilities, unless the sole reason is to continue government-mandated subsidies for such facilities at the expense of certain utilities and ratepayers.

It also should be noted that, under EPAct, Congress provided both tax credits and incentive payments to encourage ownership of certain renewable energy facilities. The establishment of these incentive mechanisms enables Congress to continue to promote such technologies as it chooses, without imposing direct costs on individual utilities and their customers in a manner that results in electric rates far in excess of market value for those utilities who, by mere accident of location, have service territories that are favorable to renewable energy development.

Much of PURPA is simply inconsistent with the current competitive climate within the wholesale electric generation industry. A thorough review of both the implementation of PURPA by FERC and the States, as well as a more fundamental review of the statute itself, is needed. In today's competitive electricity markets, electricity suppliers should be selected on the basis of who offers the best deal, not on the basis of who can get a special preference by legislative mandate.

We encourage FERC to respond to the Orange and Rockland petition for declaratory order and the more recent petition filed by Connecticut Light & Power Company. In addition, we believe the Commission should eliminate the unnecessary and inefficient requirement for buying energy only at long-term estimates of avoided cost. Moreover, FERC should review its regulations and update them in light of the

---

[4] 62 FERC Para.61,184,62,216,62,217 (1993) (citing Richmond Power Enterprise, L.P., 62 FERC Para.61,157, 62,096, 62,098 [1993]).

passage of EPAct. Lack of clear requirements and enforcement of existing and more appropriate QF criteria continues to contribute to the development of uneconomic QF power at the expense of utility ratepayers and investors.

Some State public utility commissions are beginning to recognize that their implementation of PURPA has contributed to high utility costs and thus are taking steps to correct this problem. However, such efforts are only just beginning, and the high costs will continue for many years. Many utilities are expecting to pay large sums to buy out or buy down QF contracts which are priced above avoided cost in order to save their customers hundreds of millions of dollars in uneconomic costs. For example, Con Edison has spent over $150 million to buy out unneeded and uneconomic QF contracts.

One mechanism that would be helpful to State regulatory commissions would be the ability to monitor effectively QF compliance with FERC regulations. EPAct provides State regulatory commissions with access to an EWG's books and records, and we believe that providing similar access to a QF's books and records would permit State regulatory commissions to monitor more effectively the development of all power supply options within their jurisdictions.

This subcommittee has an especially important role to play in deciding the future of PURPA. We urge this subcommittee to make a complete review of PURPA one of its highest priorities. In addition to the electric utility industry's concerns with the mandatory purchase obligation, we also note that the benefits of competition envisioned under EPAct are being denied to utility affiliates because of the ownership restrictions placed on them under PURPA.

EEI and its member companies strongly encourage a reexamination of the policy rationale for PURPA in light of today's competitive power market, with particular scrutiny on the need for the mandatory purchase obligation, the administrative determination of avoided costs, and the limitation on utility ownership of QF's. Further, we believe continuing the mandatory purchase obligation is not necessary to encourage the development of renewable resources. To the extent that Congress believes that incentives still are needed to make renewable resources cost effective and attractive in a competitive environment, we believe that R&D assistance, tax credits, and other programs already in place are more equitable. Finally, we encourage the subcommittee not to expand the problems created by PURPA by permanently removing the size limitations on certain facilities. Thank you for the opportunity to present our views on this important matter. I will be happy to respond to any questions.

Mr. SHARP. Thank you very much, Mr. Garam.

As you heard, we have a notice of a vote underway in the House of Representatives, so what I think we will do, rather than have to break up during Mr. Steve's testimony, is take a break now. Hopefully we can start again in 10 minutes, but certainly no later than 15 minutes, and I appreciate your patience.

[Brief recess.]

Mr. SHARP. The subcommittee will please come to order.

Mr. Steve, we are very pleased to hear from you now

STATEMENT OF JAIME C. STEVE

Mr. STEVE. Good afternoon, Mr. Chairman.

My name is Jaime Steve and I serve as Washington, D.C. representative for the Union of Concerned Scientists here in D.C. Established in 1969, UCS has created a unique alliance between many of the Nation's leading scientists and thousands of committed citizens. One of UCS's main missions is promoting energy technologies that are renewable, safe and cost effective.

My background in PURPA law comes from working as litigation counsel for the New York City Department of Energy. I am also a former staff writer for Congress' own Environmental and Energy Study Conference, where I specialized in energy legislation and seriously enjoyed covering this committee in 1990 and 1991 and early 1992 when you were chairing it.

Mr. SHARP. Not many people express pleasure at that. They just express that it was a lot of work for most of you.

Mr. STEVE. You will seriously be missed.

Thanks for this chance to testify regarding the committee's draft print of legislation to permanently remove the 80 megawatt size cap on power plants receiving favorable treatment under the 1978 PURPA law. For 16 years, PURPA has worked to encourage the production of electricity through renewable energy resources. PURPA has also encouraged greater use of domestically produced natural gas.

My message today is simple. The Union of Concerned Scientists supports permanent removal of the PURPA size cap for qualifying facilities. We support this legislation because its passage will allow continued growth of the market for electrically produced power through solar, wind and geothermal resources. The cost competitiveness of these facilities is greatly enhanced when they are allowed to gain PURPA benefits when sized above the old 80 megawatt cap.

While only a limited number of renewable energy plants are currently sized above the old cap, facilities larger than 80 megawatts are in the planning stages. One particularly promising facility is the 210 megawatt wind generating station under development by the KENETECH Wind Power Company. KENETECH is a San Francisco-based company and they are intending to build this facility in the Boundary Mountains area of Northwest Maine, up near the Canadian border.

A key benefit of continuing the present PURPA policy is that it would help reduce our ever-increasing dangerous dependence on imported oil from unstable regions of the world. In addition, passage of this legislation would help reduce utility companies' reliance on domestically produced oil and coal, two fossil fuels that when burned release significant levels of air pollutants.

For example, in 1988, U.S. coal-fired power plants alone released into the air about 7.4 million tons of sulfur dioxide, 2.6 million tons of nitrogen oxides, 150,000 tons of particulate matter, and 130,000 tons of partially burned fuel, known as volatile organic compounds. These coal-fired plants also emitted about 575 million tons of the greenhouse gas, carbon dioxide. That was—in 1988 that was one-tenth of the Nation's total for that year of carbon dioxide greenhouse emissions. That number has only gone up in recent years. Chipping away at these emission levels would help States meet their obligations under the 1990 Clean Air Act amendments and help every U.S. citizen to breath easier.

UCS's only criticism of the committee print is that it also would permanently remove the size cap on power plants that burn items such as coal waste and discarded rubber tires. These waste-burning plants should not be included in the draft bill because such facilities emit large amounts of CO_2 and air toxics.

Quite plainly, our air is already too dirty, therefore we should not encourage loading up the atmosphere with more pollutants that are responsible for a host of environmental problems ranging from human respiratory ailments to acid precipitation and long-term global climate change.

Lastly, maintaining current PURPA law, and I stress that all this legislation would do is maintain current PURPA law, which has been in effect for 4 years; this is not an expansion of PURPA law as it stands right now. Maintaining current PURPA law beyond the end of this year should not become entangled with efforts to reexamine the entire structure of PURPA law. As we all know and we heard from this morning's testimony, that is an extremely complicated barrier when links—especially when linked to competitive emerging markets.

Limiting today's discussion to the simple matter at hand, making permanent the already agreed to removal of the PURPA size cap is important so as not to get bogged down in emerging and highly complex issues raised by these attempts.

Finally, Mr. Chairman, UCS urges this subcommittee to embrace a cleaner, more sustainable energy future and assist President Clinton in carrying out his Climate Action Plan by adopting your committee print minus the waste-to-energy language. By doing so, this panel will maintain current law while taking a giant step toward leaving our children a cleaner planet and a more sustainable way of life.

Thanks for the chance to testify.

Mr. SHARP. Thank you very much, Mr. Steve.

Mr. Koleff, happy to hear from you.

STATEMENT OF ALLEN M. KOLEFF

Mr. KOLEFF. Thank you, Mr. Chairman.

My name is Allen Koleff. I am Division Vice President for Stone Container Corporation, responsible for environmental process and energy technology for that company. Our company, incidentally, is a major producer of pulp and paper, the largest producer in the world of corrugated boxes.

We operate pulp and paper mills in 12 States, six of which are represented on your subcommittee. All of those paper mills cogenerate electric power and in total we cogenerate about 72 percent of the total electric power used by our pulp and paper mills.

I have been associated with this part of the business for about 13 years now and have thoroughly enjoyed watching the evolution of PURPA and being involved with it. I am not here today on behalf of my company, however. I am here on behalf of the American Forest and Paper Association, the National Trade Association for the Forest, Pulp, Paper, Paper Board and Wood Products Industry. We certainly appreciate this opportunity to present to this subcommittee the views of our industry on PURPA and the Solar, Wind, Waste and Geothermal Power Production Act of 1994.

I am here specifically to talk about PURPA and specifically section 210. Other speakers who have appeared before you have attacked PURPA and suggested that its time has passed.

In my comments today, I hope to show you that PURPA was a good law when it was passed, the reasons for its enactment are equally valid today, and in fact I would go so far as to say that PURPA is probably a good example of a great law.

Before discussing PURPA, however, let me explain why our industry is interested. The forest and paper industry is the third most energy-intensive manufacturing industry and the third largest industrial purchaser of electricity in the United States, buying over 50 billion kilowatt hours annually at a cost of over $3 billion.

The industry is also the leader, a leader in industrial cogeneration, accounting for over 50 billion kilowatt hours annually. Most of this self-generated electricity is consumed at the mills and the industry sells only about 14 percent of its output to the grid. The industry thus has the unique perspective of viewing PURPA both as a consumer and a generator, though we are far larger consumers than we are generators.

Overall, industrial cogeneration is far more efficient than public utility generation because of the sequential use of energy. To produce both electricity and steam sharply reduces fuel usage.

A public utility converts approximately 35 percent of its energy Btu input into available electric power. In a typical pulp and paper plant, over 80 percent conversion of Btu input into useful energy, both electric power and process steam, is achieved.

In paper mills, the savings of fossil fuels, often oil and gas, are compounded by the use of pulp mill residues, such as bark or spent liquors from the pulping process as fuel. Today these biomass residue fuels constitute over 56 percent of the total energy consumed at our mills.

We need to go back and recall, I think, what the primary purpose of Congress was in establishing the programs under PURPA to promote the increased development of cogeneration and small power production. One of the primary purposes envisioned was the advantage of the greater efficiency inherent in cogeneration facilities over conventional electric utility facilities and the substitution of renewable resources for fossil fuels.

Benefits to society include a reduction in domestic fossil fuel consumption and petroleum imports with the concomitant improvement in balance of payments, increased national security and a better environment.

Additional benefits to the electric generating system are shorter lead times in adding generating capacity to the grid and increased reliability of the system in providing a variety of renewable technologies. All of these benefits accrue to the country at no additional cost to electric consumers. I would like to enumerate the benefits of PURPA based on the results now of over a decade of experience.

First, PURPA allows power to be sold to the grid under contract with all of the construction risk, financial risk and operating risk on the cogenerator, not the ratepayer. Under the section 210 program, the cogenerator provides power under a contract, thus fully accepting the risk for its supply. If we don't supply the power, we don't get paid.

The ratepayer avoids the risks of using utility power plant construction and associated continually increasing costs, because any cost overruns are borne by the QF's. While much has been made of the full avoided cost in contracts, most contracts for power sale in PURPA-qualifying facilities are usually negotiated at less than the utility's full avoided cost, so ratepayers see savings from the first kilowatt hour.

Second, PURPA results in a generally more reliable supply of power to the utility grid. This comes in three forms: First, the relatively smaller size and multiple units in the private system immunize it from potential systemwide failure when compared to the loss of one central station facility. In fact, during the San Francisco earthquake a few years ago, it was cogenerated power, including some from paper mills, that was a mainstay of the California electric grid. The forest and paper industry's multifuel capability is another reason for the high availability of our units.

Industrial cogeneration, because of the extremely high reliability required for industrial processes, is often much more reliable than utility central fossil fuel or nuclear stations. By virtue of their continuous operation to supply our process needs, cogeneration facilities in our industry typically produce power 98 percent of the time compared to the approximately 80 percent availability of large utility coal-fired stations and even lower availability of nuclear power stations.

Private generation sources also use a broader spectrum of energy sources. The private system can therefore insulate the utility grid to some extent against increases in the price of fossil fuels, as well as from unexpected fluctuations in the availability of these fuels.

PURPA also provides for increased electrical generating capacity without utility capital expenditures for new plant and equipment, and PURPA reduces the financing risk for utilities, and it also reduces pressure for ratepayer contributions to utility capital programs.

It is interesting to note that a recent Department of Energy study showed that utilities with significant power purchases from non-utility generation have slightly lower cost of capital as compared to utilities with little non-utility power purchase.

PURPA allows growth in generating capacity to be more closely matched with growth in demand for electricity. Utilities that purchase power from QF's, which are generally smaller facilities than central station plants, are able to match more closely the growth of their capacity with growth and demand.

Also, cogeneration capacity can be brought on line with lead times much shorter than those which typically accompany the construction of large-scale central station units, significantly reducing a utility's risk in forecasting and meeting its own load growth.

Cogeneration is more energy efficient generation. Generation of both electricity and useful heat through cogeneration is more efficient than the separate production from fossil fuels, thus, fewer fossil fuels are used. A typical energy savings is in the range of 3,000 to 4,000 Btu's per kilowatt hour of electricity generated in cogenerating facilities.

To the extent that private generation uses biomass, small hydro, coal, gas and oil in the cogeneration mode, there can be reductions in oil used for power generation resulting in a lower demand for imported oil.

Since less fuel is burned to produce the combination of electricity and useful heat, there are significantly reduced air emissions. And PURPA results in generally reduced line losses of electric power in distribution because private generation tends to be disbursed rather than centralized, and industry is often located near load centers.

To the extent that cogeneration displaces central power station condensing plants, there will also be a reduction in cooling water demand.

Taking into consideration all of these advantages a QF brings, the forest and paper industry sees no reason to limit small power producers to an arbitrary 80 megawatt size limit and we support the Solar, Wind, Biomass and Geothermal Power Production Act of 1994.

Finally, we might ask, is it any surprise that utilities who oppose PURPA also generally oppose competition? What is ironic is they are called to gut PURPA in the name of competition by eliminating the mandatory purchase provision.

PURPA was the first breach in the utility generation monopoly. Utility past claims that cogenerated power is impractical and would harm the reliability of the system have been clearly debunked in the last 15 years of reliable QF operations, and if the utilities succeed today in their attempts to discredit PURPA and thereby successfully limit competition in those jurisdictions, they will have been able to turn back the clock 20 years and have the playing field once again to themselves.

In a fully competitive market, PURPA's mandatory purchase obligation causes no harm. Where there is competitive bidding, it guarantees that a QF has the right to meet the lowest bid. The ratepayers benefit from the lowest bid, and the Nation benefits from QF's energy efficiency, environmental and economic attributes mentioned earlier. To me that sounds like a win-win situation.

In conclusion then, it is our industry's view that PURPA was a good idea when passed, a better one as implemented, and with our Nation's continued emphasis on energy efficiency, energy security and environmental improvement, every bit as important today.

Thank you very much for the opportunity to appear here.

[The prepared statement of Mr. Koleff follows:]

STATEMENT OF THE AMERICAN FOREST AND PAPER ASSOCIATION

My name is Allen M. Koleff. I am Division Vice President of Stone Container Corporation, for environment, process and energy technology. I am testifying today on behalf of the American Forest and Paper Association (or AF&PA). AF&PA is the national trade association of the forest, pulp, paper, paperboard, and wood products industry. AF&PA represents approximately 550 companies which grow, harvest and process wood and wood fiber, manufacture, pulp, paper and paperboard products from both virgin and recovered fiber and produce solid wood products.

As a single national trade association, AF&PA represents a vital national industry which accounts for over 7 percent of the total U.S. manufacturing output and 90 percent of domestic recycled paper manufacturing capacity. Employing some 1.4 million people, this industry ranks among the top 10 manufacturing employers in 46 States. The forest and paper products industry generates sales of $200 billion annually, and, as one of the largest exporters to global markets, makes a positive contribution to the U.S.balance of payments.

I appreciate the opportunity to present to the House Subcommittee on Energy and Power the views of the U.S. forest and paper Industry on the Public Utility Regulatory Policies Act (PURPA) and the Solar, Wind, Waste and Geothermal Power Production Act of 1994.

I'm here today to talk about the Public Utility Regulatory Policies Act, or PURPA, particularly section 210. Other speakers who have appeared before you have attacked PURPA and suggested that its time has passed. In my comments today I hope to show you that PURPA was a good law when passed, and the reasons for its enactment are equally valid today. In fact, PURPA is probably a good example of a great law.

All laws have unintended consequences and PURPA's unintended consequence was the creation of the independent power industry, which has stopped the upward spiral of utility rates and brought the first breath of competition to the utility industry. But PURPA was not passed for these reasons. Before discussing PURPA, let me explain why our industry is interested in PURPA.

The forest and paper industry is the third most energy intensive manufacturing industry and the third largest industrial purchaser of electricity in the United States, buying over 50 billion kilowatt hours annually, at a cost of over $3 billion. The industry is also the leader in industrial cogeneration, accounting for over 50 billion kilowatt hours annually of electric generation. Most of this self-generated electricity is consumed at the mill and the Industry sells only about 14 percent of Its output to the grid. The industry thus views PURPA more as an electric consumer than an electric generator, but has the perspective of both.

Cogeneration in the paper industry usually consists of the production of steam at high temperature and pressure, which is first expanded through a steam turbine to generate electric power and then extracted or exhausted from the turbine at lower temperatures and pressures, to be used in the industrial process for such purposes as mechanical drives, pulping, drying, bleaching and space heating.

Overall, industrial cogeneration is far more efficient than public utility generation, because Its sequential use of energy sharply reduces fuel usage. A public utility converts approximately 35 percent of its energy Btu input into available electric power. In a typical pulp and paper plant, over 80 percent conversion of Btu input into useful energy—electric power and process steam—is achieved. In paper mills, the savings of fossil fuels (often oil and gas) are compounded by the use of pulp mill residues, such as bark or spent liquors from the pulping process, as fuel. Today these biomass residue fuels constitute over Am of the total energy consumed at U.S. pulp and paper mills.

The primary purpose of Congress in establishing a program under PURPA to promote increased development of cogeneration and small power production was to take advantage of the greater efficiency inherent in cogeneration facilities over conventional electric utility generating facilities, and to substitute renewable resources for fossil fuels. Benefits to society include a reduction in domestic fossil fuel consumption and petroleum imports, with the concomitant improvement in balance of payments, increased national security, and a better environment. Additional benefits to the electric generating system are shorter lead times in adding generating capacity to the grid and increased reliability of the system provided by a variety of renewable technologies. All of these benefits accrue to the country at no additional cost to electric consumers.

I would like to enumerate the benefits of PURPA based on the results of over a decade of experience.

1. PURPA allows power to be sold to the grid under contract with the construction risk, financial risk and operating risk on the cogenerator, not the ratepayer. Under the section 210 program, the cogenerator provides power under a contract, thus accepting risk. The ratepayer avoids the risks of utility power plant construction and associated continually increasing costs, because any cost overruns are borne by the QF's. While much has been made of the use of "Full Avoided Cost" in contracts, most contracts for power sale from PURPA qualifying facilities are usually negotiated at less than the utilities avoided cost, so ratepayers see savings from the first kilowatt-hour.

2. PURPA results in a generally more reliable supply of power to the utility grid. The increased reliability comes in three forms:

(a) The small size and multiple units in the private system immunizes it from potential system failure when compared to the loss of one large central station. (In fact, during the San Francisco earthquake, it was cogenerated power, including some from paper mills, that was a mainstay of the California electric grid. The forest and paper industry's multi-fuel capability is one reason for the high availability of its units.)

(b) Industrial cogeneration, because of the extremely high reliability required for operation of industrial processes, is often much more reliable than utility central fossil fuel or nuclear stations. By virtue of their continuous operation, cogeneration facilities in the pulp and paper industry typically produce power 98 percent of the time compared to the approximately 80 percent availability of large-utility coal fired stations and even lower availability of nuclear power stations.

(c) Private generation resources use a broader spectrum of energy sources. The private system can insulate the utility grid to some extent against increases in the price of fossil fuels, as well as from unexpected fluctuations in the availability of these fuels.

3. PURPA provides for increased electrical generating capacity without utility capital expenditures for new plant and equipment. If private industry builds electrical capacity, the public utilities are able to scale down or defer the construction of new capacity and in turn reduce capital requirements.

4. PURPA reduces financing demands for utilities. Since utilities will require no new financing to obtain this additional capacity, they will avoid any negative reper-

cussions that might otherwise be experienced in going to the financial markets. In fact, a recent DOE-EIA study shows that utilities with significant power purchases from non-utility generation have slightly lower cost of capital as compared to utilities with little non-utility power purchases.

5. PURPA reduces pressure for ratepayer contributions to utility capital programs. As private development of cogeneration reduces the need for utility financing of major construction programs, it also reduces the need for ratepayer contributions to the financing of utility constructions through devices such as construction work in progress (CWIP).

Thus, the sound traditional utility regulatory practice of the ratepayer paying for facilities that are "used and useful" is buttressed by private development of generation. Use of CWIP is not at all applicable in private power generation development.

6. PURPA allows growth in generating capacity to be more closely matched with growth in demand for electricity. Utilities that purchase power from QF's are able to match more closely the growth of their capacity with growth in demand, because cogenerators usually provide capacity in smaller increments than would be cost effective for utilities to build. Also, cogeneration capacity may be brought on line with lead times much shorter than those which typically accompany the construction of large-scale central generation stations. This significantly reduces a utility's risk in forecasting and meeting load growth.

7. Cogeneration is more energy efficient generation of useful energy by cogeneration. Generation of electricity and useful heat through cogeneration is more energy efficient then separate production from fossil fuels; thus, less fossil fuels are used. The energy savings is typically 3,000–4,000 Btu per KWH.

8. PURPA results in reduced fuel oil demands. To the extent that private generation uses biomass, small hydroelectric, coal, gas, and non-oil based energy sources and oil in the cogeneration mode, there can be reductions in oil used for power generation, resulting in a lower demand for imported oil.

9. PURPA results in reduced emissions to the air. Since less fuel Is burned to produce electricity and useful heat (generally steam) in cogeneration plants than would be burned to separately produce the electricity and heat from fossil fuel, emissions to the air are reduced.

10. PURPA results in generally reduced line losses of electric power in distribution. Because private generation tends to be disbursed rather than centralized, there are opportunities for reduction in line losses in distribution. This is particularly true with industrial generation since industry is often located near load centers.

11. PURPA reduces water use. To the extent that cogeneration displaces central power station condensing plants, there will be a reduction in cooling water demand. The use of cogeneration, in its pure sense, eliminates the need for condensing steam with cooling water as is done in a central power station.

This reduction in water use should be beneficial in States with water shortages.

Taking into consideration all of these advantages a QF brings, the forest and paper industry sees no reason to limit small power producers to an arbitrary 80 MW size limit and we support the "Solar, Wind, Biomass and Geothermal Power Production Act of 1994."

Finally, is it any surprise that utilities who oppose PURPA also oppose competition? What is ironic is their call to gut PURPA in the name of competition, by eliminating the mandatory purchase provision.

We must not lose sight that with all the talk about competition, almost 90 percent of generating units supplying the grid are owned by regulated electric utilities.

PURPA was the first breach in the utility generation monopoly. Utility past claims that cogenerated power is impractical, unreliable, and would harm the reliability of the system have been clearly debunked by the last 15 years of reliable QF operations. If the utilities succeed today in their attempts to discredit PURPA, and then successfully limit competition in most jurisdictions, they will have been able to turn back the clock 20 years, and have the playing field to themselves.

In a fully competitive market, PURPA's mandatory purchase obligation causes no harm. Where there is competitive bidding, it guarantees that a QF has the right to meet the lowest bid. The ratepayers benefit from the lowest bid, and the Nation benefits from the QF's efficiency, environmental and economic attributes mentioned earlier. That sounds like a win-win situation to me.

In conclusion, it is the forest and paper industry's view that PURPA was a good idea when passed, a better one as implemented, and with our Nation's continued emphasis on energy efficiency, energy security, and environmental improvement, every bit as important today.

I appreciate this opportunity to present my industry's views to this subcommittee, and I will gladly answer any questions you may have.

Mr. SHARP. Thank you very much, Mr. Koleff.

Let me raise that question that you have most recently alluded to there with Mr. Garam regarding the question of just how harmful utilities really think PURPA is today where you have States that have gone to competitive bidding. I can appreciate in a more competitive market, and some just don't think we are there yet in terms of a pure competitive market, but I can appreciate their current concern if a State commission or a bureaucracy were determining a price and then—and I know historically California initially set a very high avoided cost price and that unquestionably stimulated some activities. I happen to think that was wonderful for the country, but the consumers of California must bear the marginal increment in that.

But the point is, isn't today with competitive bidding, isn't that—are you likely to be committed to above-market costs there more than you are likely to be in any other decision you make?

Mr. GARAM. Mr. Chairman, I believe the short answer is yes, and let me explain why. First of all, let me try to frame the issue by saying that we do not believe that the issues are, are cogenerators reliable, are they efficient, can they be sized to match the load? We don't dispute any of that.

We have cogeneration plants—in fact, again, we have the distinction of buying power from one of the largest cogeneration plants in the country. It is a 600 megawatt cogeneration plant in Linden, New Jersey across the river from us. That plant, we have great admiration for the operators of the plant.

They constructed that plant very quickly in a very short time. They are operating it with far fewer people than any of our plants are being operated with. It is highly efficient. It uses a most modern technology, and we have fully integrated it into our system.

So we are not debating the merits of cogeneration. What we are talking about is the merits of pricing cogeneration at administratively determined avoided cost.

The theory is great, but our practice, what has happened, Mr. Chairman, is that we have been saddled with billions of dollars of overpayments.

Now, your question, I think very importantly gets to the point that we made in the written statement. As long as the markets can set the price as in a competitive bidding situation, there is no need for regulators to guess what prices the market would set. Therefore, there is no need for the mandated PURPA obligation coupled with administratively determined avoided cost.

Now, we in New York do have bidding and that has eliminated much of a problem because the best price—the facility with the best attributes does win. For example, if you have a need for 200 megawatts, the top 200 megawatts will end up winning.

Now, we do take exception to the notion that if the winning bid, let's say, is 5 cents a kilowatt hour and we need 200 megawatts, and another 10,000 megawatts of QF power, we ought to be obligated to buy. That, I suggest, should not happen. It would be a disaster and a misinterpretation of PURPA, a misapplication of PURPA.

I can speak to the way—the harms that are left after bidding in the New York context. The New York Commission has determined that once our capacity needs are met, let's say 200 megawatts, we are under no more obligation to buy any more capacity. So that problem is solved.

But it did say that we have an obligation to buy energy only at administratively determined avoided cost to satisfy PURPA, and that, in fact, is how we got into a billion dollar overpayment with an energy-only contract that we had to enter into after a competitive solicitation for capacity was completed.

Mr. SHARP. I am not sure I follow what you are saying. Would you run through that again? I am not clear on what——

Mr. GARAM. Certainly. Let me try to be clearer.

The New York Commission has determined that we should—we ought to meet our capacity requirements through competitive bidding, and we have. Once we get a solicitation, we select the winners, that fills our capacity needs and we are relieved of the obligation at that point under the New York Commission rules from purchasing any more capacity, so we will have no more overpriced capacity at that point. And that isn't an answer to your question, but the New York Commission said that we still have a PURPA obligation to buy whatever energy is offered to us, not capacity, but energy, to meet the PURPA, and that energy is priced under the old administratively determined avoided cost scheme, and the overpayments under energy-only contracts can be as large or larger than under a capacity contract, so still we are still saddled with the overpayments under PURPA because that residual obligation to purchase the energy without any capacity value still remains under PURPA.

Mr. SHARP. These are under previous contracts, not new contracts?

Mr. GARAM. We are still under obligation to buy all the energy that——

Mr. SHARP. Anything that is offered from those——

Mr. GARAM. Anything that is offered for energy. Any QF can offer to us as much as energy as he has available and we are still obligated to buy as much energy as is offered to us.

Mr. SHARP. I guess I am asking, are those QF's that have been in existence, or are they new QF's this year and new QF's next year that would fall under that policy? Do you understand what I am——

Mr. GARAM. Any QF's would fall under that policy, Mr. Chairman, as long as they are QF's and we would be obligated, all the energy, whether from existing QF's or future QF's.

Now, it so happens that this unfortunate experience we had with this very large cogenerator, they built the plant, oddly enough, or shockingly enough, they were able to build a plant on the basis of a 5-year, energy-only contract. They had no capacity sales committed and they were able to finance and construct the plant—they are constructing the plant on the basis of a 5-year, energy-only contract that we had to enter into. After all, our capacity requirements were met.

So it is still—I believe it is still a very serious problem to have this mandatory purchase obligation because it still requires us to buy all the energy and all the output of QF plants that are offered to us.

Mr. SHARP. Could the State—under the current law, could the State of New York, the Commission, have decided in other ways so that that would not, in fact, be the case?

Mr. GARAM. They believe—we have been concerned about this for some time, Mr. Chairman, and we have had extensive discussions with the New York Commission, and they believe that their hands are tied and they are required to implement PURPA in a way so that we buy all the energy that is offered to us, and what they have also ruled is that we have to——

Mr. SHARP. No matter what your need is?

Mr. GARAM. Well, energy is—always can displace energy basically.

Mr. SHARP. So it can displace your nuclear power or other power if it is more expensively priced in your system?

Mr. GARAM. Well——

Mr. SHARP. Or what does it displace?

Mr. GARAM. It would displace other energy, and so we have to buy the PURPA energy and we have to shut down or not operate other plants——

Mr. SHARP. At the same level?

Mr. GARAM. [continuing] that may or may not be less or more economical than the energy that we are buying because it is bought at a predetermined arbitrarily determined, regulatory determined administrative cost. So it may fall and very likely will fall out of the normal dispatch, economic dispatch, that plants are operating under normal circumstances.

Mr. SHARP. Obviously, I mean, the State commission could have an effect on that in terms of pricing, however. I mean, if it made a pricing decision that changed what you had to pay for that power, that might affect whether those generators, you know, want to sell or not or expand or whatnot. Am I not correct?

Mr. GARAM. Absolutely, and that has changed over time, Mr. Chairman, but the problem is, they have determined that the QF's are entitled to a minimum of 5 years worth of predetermined prices. So they have to estimate and they do estimate 5 years out and——

Mr. SHARP. Well, let me let some of your colleagues at the table—I don't know how collegial you all are feeling, but let me let some of them address the particular issue that you have identified, and then I would like to turn to a little bit of historical comparison between where we were 10 years ago and where we are now.

Mr. Stauffacher, I see you.

Mr. STAUFFACHER. Let me try a couple more examples, Mr. Chairman, because I think as we well understand that each State approaches this in a slightly different way and there undoubtedly are States that, just as Mr. Garam has pointed out, one, that he feels has disadvantaged the investor-owned utility or the monopoly franchise, I could point out numerous other States, and let me, a couple, that have a very different approach.

The State of Texas, for example, the very facility that I was talking about earlier that I would claim is now my own stranded investment, a contract that I had expired April 30th of this year. I am now selling spot-market energy only. The utility must buy the energy, but the pricing mechanism is one where what we get paid in night hours is incremental avoided coal. It is what the utility would back down to take that power.

I would and have debated with the utility that if it is indeed 100 percent of wooded coal, I would contend it is a mix of a few other things that have a more attractive price from my standpoint, but basically I am getting 10 and 11 mils for that power in the middle of the night. During the day, it is based on a weigh cog times a heat rate. That weigh cog changes every month depending on what the as-purchase experience of that utility is.

Take California; California not only are the standard offer ones which are energy only, is the energy readjusted on a monthly basis with a very complex formula that includes all the various purchase prices of the gas, its transportation costs, average heat rates, the amount of hydro that gets into the system.

If it is a dry year in California, it is a good deal for me as a cogenerator. If it snows or rains a lot in the Sierra Nevadas, I can expect that my rates are going to go down because what they would be avoiding would be some hydro for taking my power in the middle of the night.

You take a State like—well, Texas and California, I think are probably the best two examples that I can give, but certainly there are many States where the price for the energy is based on a trued-up avoided energy price that the utility would burn itself.

Is there a solution to this? Certainly is, and one that I would like to see that I think—it is probably not one that I would bring to the Congress to solve, but it is a State issue, and what I would do and what I would suggest would be the solution would be to allow me to be a member of the club. Let me get into the power pools.

If I am in PJM, I know on an instantaneous—well, actually on an hourly basis, because they redetermine the price every hour, what the energy price is. I will know whether I should gear my unit up or if I should take it down, and we have operated plants that way.

We know that we limp along with a plant until 11 o'clock at night until the on-peak hour is done, just like a utility would, then we take it down and do the maintenance. Because they don't need the power after 11 o'clock and before 7 in the morning.

We will tailor our operations to meet the price signal that we get. If we get a price signal that says, look, the utility really doesn't care, or as was pointed out, if someone has been able to convince the Commission that this is sort of a simple world and you should set one price and don't get too sophisticated and there is no variation in price signal, we will operate that way too, but we won't vary our output to maximize it as driven by a market price.

If I had access to that market, if I could move through the power pool, get on the electronic bulletin board, I probably wouldn't have an awful lot of interest in going to the individual utility and saying, you must take that.

Mr. SHARP. Mr. Steve, did you want to?

Then we will go to Mr. Koleff.

Mr. STEVE. Without jumping into too much detail, I wanted to refer back to what Mr. Garam said about the one particular qualifying facility in New York State. This was one that I worked on.

It was built by the Sythe Energy Company, and that is located in upstate New York in, I believe it is—was it Oneonta? I forget the name of the town. You can tell we are both downstate New Yorkers.

But the two quick points I want to make was, that was a 1,000 megawatt facility which ConEd had to purchase the power for. As the New York City Department of Energy, we kind of somewhat brokered a better deal for ConEd out of that.

They wanted to break the entire contract. We kind of helped them to get something out of it and that was a capacity component as well.

But here for a $1 million overcharge, that was an extremely soft figure that was put out at the time by Richard Flynn, who was chairman of New York Power Authority, really for political reasons. They were trying to kill the plant as well.

The other point is that purchasing that power, ConEd's purchasing of that power has allowed the company to shut down some of its older more polluting facilities located within the Boroughs of Manhattan.

And lastly, earlier Mr. Garam mentioned the economic downturn in New York City as a reason for not requiring ConEd to make these purchases. When I used to argue in front of the Public Utility Commission in rate cases representing the ratepayers of the City of New York, that was an argument that we made all the time, was that there is an economic problem in New York, therefore your rates should not be as high as they are. We did not prevail.

Mr. SHARP. Mr. Koleff.

Mr. KOLEFF. Mr. Chairman, I think it is important to try and separate what may be a unique situation in the State of New York from generally what is happening around the country and what has happened around the country.

To the best of my knowledge, New York is the only State that has fixed the avoided cost at 6 cents administratively. California had its standard offer but it wasn't fixed in the same way, and that is a—that is something that was done by the Commission in New York within their jurisdiction for reasons which some people have suggested may have been the result of the New York utility industry's reluctance to responsibly and cooperatively determine their own avoided costs and was done out of a sense of frustration by the New York Commission. So I don't think we should extend that to a general concern over the whole country.

On a more general basis, the fixed price contracts that were signed during the early years of PURPA before the bidding became a more common thing were done on the basis of the utility's own long-range plan and the best information available at that time on what the avoided costs were, and had the contracts not been signed with cogenerators, because there was a need demonstrated for that power, probably the utilities would have built their own facilities at no guarantee that the cost would have stayed the same. But I think there are some important issues to keep in mind as we analyze this problem.

Mr. SHARP. Mr. Hinrichs.

Mr. HINRICHS. There is a lot of baggage associated with PURPA, and I am sure there are many stories that can be told. I would just like to be sure that the purposes of the 80 megawatt extension don't get lost in that. That is very important, and I think probably even Mr. Garam would agree that that is a minor issue associated with the whole involvement.

Mr. SHARP. Well, I appreciate that. It is ultimately going to bring us back to the only thing that had been legislatively considered, was obviously a select issue to continue the policy of not having that cap apply.

But let me ask you, Mr. Garam, and we didn't mean to structure this so you didn't have other colleagues that were—some of them testified last week who share your perspective that they don't like underlying the fundamental question of the purchase requirement, the basic structure of PURPA today.

I can't resist not aiming this at you personally, but when we were considering PURPA in 1978, the generous comments you had made about your cogenerators and others were not generally heard at that time. The presumption was they would be unreliable, that they—we could not manage them, they would run whenever they wanted to, they wouldn't be interested in us and it was not a manageable proposition and it was crazy.

Now, I am sure everybody in the industry didn't recite that, but that was one of the advocacy positions of some in the IOU system. They, like everybody in this world, has changed their view of things, so I am not trying to hold just them to account, but——

Mr. GARAM. Let me just interrupt, Mr. Chairman.

You might as well make it personal because I was personally involved with some of the debates on PURPA back in 1977, and those were some of the views that I advanced. But let me just say, we have learned with time and experience and we are living with cogenerators in our system now.

Mr. SHARP. Well, I don't like to get too personal because then it can understandably be aimed at me as well and my consistency is not at the—a standard that one might always wish to be, but—and we do all learn, fortunately, or most people do. We have a few around here that that seems to be beyond them, but let me ask you this, though.

Isn't it fair to also say that—just as a matter of the general debate, because I share some of your logical concerns of the models of commanding one thing over here and then competing over there, and the difficulty we have as we are in a transition era, where some things are commanded and some things are competition in many realms, and here we are dealing with the PURPA power.

But it strikes me that I guess I feel strongly that in this debate, PURPA power figures are a little more easy to identify. We have specific contracts. We have specific increments of power. And so one can take those and compare them to the marginal cost of new power today, just as one can the entire system average compare it, but it gets a little more difficult when we start identifying, well, which piece of that is the old coal-fired plant and which piece of that is the nuclear power plant.

I mean, isn't it fair to say that in your system—and again I am not trying to pick on ConEd, but you would know that system best, I assume—that you are contending with a number of investments which may well have been prudent at the time; I am not challenging that question, a number of investments which, in this new climate, become less economical, just as any industry must, but you make those decisions under the regulatory requirements that you had 5 years or 10 years ago, just as you do at PURPA.

My point being that it really is inappropriate, and I am not suggesting you were, but to single out that we have PURPA power above market but we don't seem to have much else above market.

I mean, some people suggest to me that nuclear may well fall big in that category, and I am not trying to dump on nuclear here. It is just that the range of decisions made were very wide, and a number of them are in a more competitive bulk power market, and when interest costs and other costs are down, simply are not nearly as attractive and make it very difficult for some systems to compete.

Mr. GARAM. Well, you are right. We do have, unfortunately, a lot of other power other than PURPA power that is—has a cost above market.

I would like to say a couple of things about that, and I don't mean to pick on PURPA power, but what—we are saying a couple things.

One, it is a problem and it has been a problem and we have suffered from a problem. Please do not extend a problem and take a look at the problem before stepping off the cliff and continuing the problem that we have experienced.

The industry is going in one direction. Let's not continue the same old ways of looking at things at PURPA as we have for the last 16, 18 years.

To address the broader questions of stranded investment that is underlying your question, and this came up in the earlier panel, I would like to say that there are two things that I believe we need to do with stranded investment. As you suggested, those investments were incurred at a time under a different regime.

Now, this is not like an investment risk, a normal investment risk. There is a drastic fundamental change in the industry that is taking place. And as Mr. Parshley suggested, I don't believe that this change—a change of this magnitude was what was anticipated by investors and taken into account by commissioners when they were setting the returns of equity in electric utility rate cases.

So if you go from a regulated to a vastly different structure, we need time to adjust to it. So we need some transition time, and also, I believe, we need to change the rules of the game so we don't continue to operate under the mandates that resulted in the prices of power and the relationship of cost to power that we are seeing today.

So that brings me directly to PURPA. We cannot continue PURPA, and we cannot continue the mandatory purchase obligation and all the prior mandates that the utilities are still under, they still have an obligation to serve.

There is a lot of talk of competition, but nothing has changed as far as utility obligations are concerned. We are still subject to all the social obligations.

In fact, again, I regret to say, the New York Public Service Commission continues to add to those obligations. I mean, they have a program of externalities, renewables.

They are now talking about undergrounding certain transmission lines at costs of half a billion dollars over the next 5 years that have been overhead for longer than I can remember, and the PURPA is just a continuation of this baggage that is going to kill this horse.

So I think if utilities can't compete fairly in a level playing field, I think customers and the public interest will suffer. So——

Mr. SHARP. Well, let me ask. Obviously, the component we are dealing with, I mean, some of us see it as small. You are articulating a much larger picture, and I realize part of that argument is aimed at the entire structure of PURPA.

But let me ask you on the cap itself. I mean, do you actually envision or is it a problem you simply cannot know how many actual facilities over the next 2 or 4 years this might entail? You are not going to change your mind here. I am just trying to get a picture of——

Mr. GARAM. Well, are you saying is this going to be a real big problem?

Mr. SHARP. Yes. It strikes me that it isn't that big of a problem, and that the State of New York, if it wished—and I understand they didn't help you if you can't convince the State of New York, but can take steps so that if it feels like these are uneconomical decisions, that it can, through its pricing powers, make them so they are not so much of a burden.

Mr. GARAM. Well, if you are getting at the question of whether a short extension, maybe 2 years, would be harmful or not, I would argue it would be harmful in that if you are talking about an 80 megawatt facility with a misestimated avoided cost, you are going to have 10 times the problem if you have an 800 megawatt facility that you have to deal with.

I would also suggest that it is not a 2-year problem but a 20-year problem, because in the 2 years, you are going to enter into a 20-year contract. So we are going to be stuck with the consequences of that decision for a long time to come.

And more fundamentally what our testimony says, Mr. Chairman, is that in a competitive environment, these market distortions really have no place, and a short extension, maybe it is a small problem; we think it is a step in the wrong direction. It is a wrong thing to do.

It is inconsistent with the direction the industry is moving in, and it is inconsistent with the direction that you have set the industry on a couple years ago in sponsoring the Energy Policy Act.

So we would ask you to respectfully consider the problems that would be caused by a continuation, even for a short time, of the removal of the cap.

Mr. SHARP. Well, I appreciate that.

Let me just indicate—I mean, the practical problem—we had both 2 years ago, and even more so today, is that obviously many have multiple goals. I mean, just competitive efficiency is not the only goal. It is a very useful tool, and we never had the expectation that everything would become competitive, given at least the—we were hopeful the power market could be—competition could be enhanced.

Many of us were not certain that the distribution system or the wires wouldn't remain a monopoly, though people do come up with interesting new theories all the time as to how to manage the pricing and operation so that you can make these things in a sense open access and whatnot, and so I am quite open to these possibilities.

But the practical problem is how to sustain other societal goals in a fair way, obviously, and many of the States are going to be grappling with this, and PURPA obviously established a societal goal. Ironically, it served two goals that are coming into conflict now.

One was to actually break open the monopoly power of the system, and the intent then was, because it seemed difficult to get these large centralized systems to buy into the smaller power purchases and especially an alternative source of energy, and of course now that the two are coming into conflict, I recognize that.

Let me see—yes.

Mr. GARAM. Well, our testimony points out a number of other ways that renewables can be encouraged, and as far as the time line now for the extension and the expiration at the end of the year, I believe we have the luxury of time, and rather than rushing ahead and saying that let's do the extension now because it is going to expire at the end of the year and we have got to do something and we don't have time to discuss PURPA now, the broader issues, I would suggest just the opposite.

I think, as you know, in many parts of the country, there is a fair amount of excess capacity. So I don't think we are going to suffer by delaying dealing with the cap issue until we can deal with PURPA in a comprehensive way, and if that takes 2 years, well, I don't think a great deal of damage will be done by not removing that cap for the couple of years and letting it lapse until it just can be looked at in a comprehensive way.

Mr. SHARP. Let me—some of the others might want to take a shot and respond.

Mr. HINRICHS. Mr. Chairman, our company has, for 4 years, fought to get additional geothermal plants contracted in the State of California, and on—the April 20th day is in everybody's mind, but it was 2 weeks later, the Commission did vote out the issuance of the contracts, and there will be plants in that over 80 megawatts.

In the event that we delay, I can certainly see the utility industry utilizing the decline of the 80 megawatt exemption at the end of this year as another reason to hold up on those contracts.

It is a continual battle, and I can see our industry in great decline if we are unable to continue right through the end of this year with the knowledge that the 80 megawatt cap is not there.

Mr. SHARP. Are you saying that specifically of this round of contracts you were just talking about——

Mr. HINRICHS. Yes, yes.

Mr. SHARP. [continuing] still could be in jeopardy because it is possible they wouldn't get signed by January?

Mr. HINRICHS. That is right. That is right.

Mr. SHARP. Mr. Stauffacher.

Mr. STAUFFACHER. Let me talk just for a second about the characterization of PURPA as the problem. And I think as we break open the market, what we really have here is a mechanism for us to take a look at highlighting some of the barriers that are still out there and, indeed, at least if history is any guidance, a way to look at how we might solve some of the problems.

For example, if you look at some of the things that PURPA has led to, we have had take or pay—I am sorry, we have had take and pay contracts, which have been performance-based. Typically, these were initially put in place as cogeneration contracts where we, the developer, would take the risk and if we didn't deliver, we did not get paid.

Now we are finding, as California goes forward with pay—with performance-based rates, as other States, New York being one; Rochester Gas and Electric has put in place some performance-based rates, I would contend that some of the examples of the benefit in how that can be done came from the utility commissions around the country looking at how to put some teeth into performance for cogeneration contracts the same way we take the risk on cost caps.

We can't come back into our utility and say, gee, we forgot to put in a SRC, selected catalytic reduction, on our project and we would like to rebid this contract and have you pay us an extra $5 million so we can put in that SCR to make the air cleaner and meet the requirements, our permitting requirements.

I would expect that you might see that indeed that kind of thing may come into competitive solicitation in States. There are a number of States right now that are talking about cost caps so that if the utility is allowed to bid in its own solicitation, that it too would have to live with that bid.

Now, there needs to be a flip side to that. A utility, if it comes in under its bid, needs to be able to keep the rewards of that just as it would be penalized or would be capped, as we are capped, if they come in higher.

Another way that we have learned by the implementation of PURPA is through something that we find frankly rather onerous. It is a trended capacity payment.

The thesis was that cogeneration isn't very reliable, that you better not pay them very much in the early years because they may not be there in the far out years to be there when you need them. Therefore, what we will do is use a net present value calculation, pay them the same net present value, but we will scallop the payments such that the payments in the far out years are much higher than in the early years.

I would contend that that is somewhat what happened for PG&E in Diablo Canyon. They looked at it both from a performance standpoint and from we cannot stand the rate shock and we will use some kind of a trended cost recovery for the revenue requirements.

The point here is that PURPA has been a solution. It has pointed State commissions as to mechanisms as to how costs—ratepayers can be protected, costs could be recovered. And I think as we move through this transition to a more competitive marketplace, and it is—indeed, that is what we are doing right now. We are moving, but we are not there.

As we move through that, there are some things that we can continue to learn by having the lever of PURPA so that in States like Wisconsin, where recently we finally got the first QF contract signed—this was after the utilities said, no, we should rate base. We should build the QF—not a QF—a cogeneration facility and rate base it. The State commission staff evaluated that proposal as ninth on present value versus the other proposals that it had before it.

If we hadn't had PURPA, if we hadn't had the ability to force consideration of alternatives to the rate-based projects of the utility, we would not have gotten to that evaluation. We need at least through this transition to maintain that lever we have.

Mr. SHARP. Yes. Mr. Steve.

Mr. STEVE. If I could just add two very brief points.

Con Ed might not agree with this, but it appears the price that Con Ed is paying is for the power that comes from these facilities and not the facilities themselves.

In New York State—sorry to be so parochial on this again—but in New York State, the avoided cost has already been reduced because the State Public Service Commission is very familiar with some situations where even they agree that perhaps there has been some overpayment, maybe not to the level that has been described earlier.

And, right now, the New York State Public Utility Commission is also currently involved in a case where all interested parties are participating on what should be the future policy of avoided costs in the State of New York.

So that issue is being dealt with up there.

And, second, just for the record, the site plant we were discussing earlier was in the town of Oswego, New York, a lovely town.

Mr. SHARP. Those folks will appreciate that.

Mr. Koleff, I couldn't tell if you were—as you could hear, we have another vote under way in the House of Representatives and this probably would be a good place to bring the conversation to a close. Not that this will stop the conversation, I can assure you.

But I thank you very much for your time and attention, your help for our hearing today. It is very, very useful to all of us. Thank you very much.

[Whereupon, at 3:50 p.m., the subcommittee was adjourned.]

[The following statements were received for the record:]

STATEMENT OF THE AMERICAN WIND ENERGY ASSOCIATION

The American Wind Energy Association has represented all facets of the US. wind energy industry since 1974. Its more than 800 members from 49 States include turbine and turbine component manufacturers, project developers, utilities and academicians.

The wind energy industry in the United States has become a quiet success story for the 1990's. Only a few years ago, three utilities in two States were actively involved in wind energy; today, 27 utilities in 12 States are either negotiating contracts, building wind power plants or purchasing energy from utility-scale wind projects. The most recent wind energy contracts have been awarded at costs as low as five cents per kilowatt-hour or less, depending on the quality of the wind source. This represents a tremendous reduction from costs as high as 30 to 35 cents per kilowatt-hour in the early 1980's.

Although the Europeans and Japanese have been very active in wind energy technology development, the U.S. industry has fought diligently to remain competitive. Companies in Washington, California and Vermont have introduced (or will shortly introduce) new, utility scale turbines, bringing wind energy costs down even further.

In areas with a strong resource, wind energy has become fully competitive with new coal plants and nearly competitive with cogeneration and combined-cycle gas turbines. Wind energy has quietly become the most cost-affective renewable energy technology and, within a decade, is expected to be the most cost-effective source of electricity generational.

Mr. Chairman, on behalf of the American Wind Energy Association, I want to express strong support for your work to remove permanently the arbitrary 80 Megawatt (MW) size cap on renewable power producers under the Public Utility Regulatory Policies Act of 1978 (PURPA). We applaud not only your efforts, but also those of the full Energy and Commerce Committee's ranking minority member, Mr. Moorhead. Together, you have shown that promoting the use of our Nation's renewable energy resources can be done rationally, effectively and in a bipartisan manner.

Periodic broadside attacks on PURPA by a small, yet vocal, minority are as predictable as the swallows returning to Capistrano. Regardless of the context or the specific issue within PURPA that may be up for discussion, there will always be those who oppose the Nation's growing commitment to renewable energy technologies as an unwanted challenge to the status quo.

It is all the more discouraging to see the rearguard of the power industry attempt to derail this much-needed legislation. The U.S. wind energy industry believes that PURPA represents the single-most important legislation enacted by the U.S. Congress to recognize the national energy, economic and environmental benefits that come from increased use of renewable energy technologies. Enactment of PURPA sent a powerful message to the country's utility industry and to the capital markets of the United States. More than any single action at the Federal level, PURPA has been responsible for the creation and development of our country's wind, solar, biomass and geothermal energy resources.

There may be legitimate points of disagreement on how PURPA achieves its goals of encouraging efficiency and renewable power. There can be, however, no legitimate disagreement that the 80 MW size cap is arbitrary and impedes the development and use of renewable power by increasing transaction costs, preventing economies of scale and marginalizing renewable industries.

Section 1201 of the Energy Act makes national energy policy unequivocal regarding renewable energy:

The purposes of this title are to promote: (1) increases in the production and utilization of energy from renewable energy resources; and (2) further advances of renewable energy technologies . . . Permanent removal of the size cap will surely help us to achieve these goals. By contrast, allowing the cap to be reinstated will stymie their attainment.

The cap impedes renewable power development at virtually every stage of its development. Initially, during conceptualization of a project, the cap forces engineers to eschew large projects, even where optimal use of the available renewable resource would suggest a project larger than the cap, for example in the 100 to 200 megawatt range. Alternatively, if not downsized, the project is forced to divide itself into a hydra, with separate owners of each ungainly head. Loan documents, power sales agreements, title and liability insurance packages, and management and operating agreements proliferate, as do project costs.

As this subcommittee is aware, many if not most energy projects today enjoy economical project financing because the revenue strewn from a power sales contract is predictable and typically stable over time, thereby reducing risk. The cap, however, reintroduces pointless risk. In the experience of AWEA members, such as Zond Systems of Tehachapi, California, it tells financial institutions—which instinctively shun the cutting edge technologies that characterize renewable power—that the revenue stream of a power sales agreement under PURPA may be lost if a project ventures too close to the cap. This increases a lender's perception of risk. As institutional lenders are perhaps the most risk averse breed in existence, this risk perception, in turn, increases project costs and forces otherwise unjustifiable downsizing of renewable projects to levels comfortably below the cap.

Project construction similarly suffers as economies of scale are sacrificed to the 80 megawatt cap. If size is absolutely limited, then there are fewer megawatts and proportionately higher costs and lower revenues per megawatt. Alternatively, if the cap forces only multiple, separate ownership of what should be a single project, then the result is typically duplicative capital investments. Each separate project constructs its own, smaller substations and interconnecting transmission line. Each project must be separately metered and, frequently, redundant collector and gathering lines may need to be installed to avoid having the power of the separate projects combined prior to delivery to the substation for interconnection with the electrical grid. To the extent substation or collector line assets are shared by separately-owned projects, determinations of how to share transmission and line losses can become very complicated.

The harmful effects of the cap extend yet further into project operation. Because projects are forced into multiple separate ownerships, separate financial and business books and records must be kept and separate tax returns and financial statements need to be prepared and filed. Separate records must also be kept on energy production, revenues from sales of electric energy, equipment maintenance and service and related costs, and on other matters pertaining to the costs and revenues of producing electric energy.

In short, Mr. Chairman, the cap makes no sense. The cap is an arbitrary rule from the nascent days of renewable power, when virtually all technologies were small. Today the cap needlessly increases the costs of all phases of renewable power project development and operation. What is particularly ironic is that, without this legislation, the cap will be reinstated right at a point in time when renewable technologies, particularly wind-powered ones, are just beginning to be able to compete effectively with conventional fossil fuels on the basis of cost and reliability.

No, this does not make sense. It flies in the face of national energy policy. It flies in the face of economy and efficiency. It also impedes environmental policy. Renewable power is relatively, if not totally, non-polluting. Surely, the hobble of the 80-megawatt cap on renewable power does not serve the Nation's environmental goals at a time when there is no parallel limit on fossil fuel plants, whether they cogenerate or not. Nor does it serve the administration's Rio commitments—embodied in the administration's Global Climate Action Plan—to reduce and stabilize the emission of greenhouse gases, which destroy the world's atmosphere.

Congress has previously recognized that the cap has outlived any useful purpose it may once have served. In 1989, Congress removed the cap, but placed a sunset on this action to ensure that further removal of the cap would precipitate a future debate and legislation on the future of the Nation's electric power industry. That goal was achieved in 1992 when the full Congress enacted Title VII of the Energy Policy Act, which grew in large measure out of the efforts of this subcommittee and particularly your efforts, Mr. Chairman, and the efforts of the ranking member, Mr. Moorhead.

Mr. Chairman, beyond the multitude of sound policy and financial reasons for eliminating permanently the cap, an additional reason is the chilling message that reinstatement of the cap will send to the renewable power industry and its lenders. It will relegate renewable power to the sidelines of national energy policy, disregarding the very impressive performance gains and cost reductions achieved in recent years by each of the four principle renewable energy technologies—wind, solar, geothermal and biomass. In short, reimposition of the cap at this critical point in the maturation of these technologies would significantly impede market growth and thwart attainment of the Nation's energy and environmental goals.

AWEA stands ready to discuss with subcommittee the benefits of PURPA and ways that PURPA can be strengthened to accomplish its original intent.

The U.S. wind energy industry strongly supports legislation to eliminate permanently the size cap on renewable power production under PURPA for the following reasons: (1) The size cap is an arbitrary artifact dating from the infancy of the renewable power industry, which serves no legitimate purpose today; (2) existence of the size cap makes many lenders apprehensive toward renewable projects that even begin to approach the 80 megawatt limit; (3) the size cap discourages economies of scale and increases; transaction and operating expenses; (4) weakening PURPA at a time when our Nation is seeking ways to reduce air pollution and greenhouse gas emissions is counterproductive, at best; and (5) reinstatement of the PURPA size cap runs counter to the national goals stated in section 1201 of the Energy Policy Act of 1992 and the administration's Global Climate Action Plan.

PURPA remains the cornerstone of our Nation's commitment to clean, renewable energy. The Congress could do few things more damaging to the advancement of renewable power in this country than allowing the cap to be reinstated.

STATEMENT OF J.C. ROBERTS, PRESIDENT, NATIONAL RURAL ELECTRIC COOPERATIVE ASSOCIATION

Chairman Sharp and Members of the Subcommittee, I am J.C. Roberts, President of the Board of Directors of the National Rural Electric Cooperative Association (NRECA) and I am pleased to submit this testimony on behalf of NRECA's member systems, 1,000 consumer-owned, not-for-profit rural electric systems, most of them cooperatives, which provide electric service to more than 25 million rural Americans. This testimony is submitted for the record of the subcommittee's hearings on July 13, 14 and 21 on electricity issues of the 1992 Energy Policy Act. We look forward to being included in future hearings.

The Act obviously touches on myriad issues of interest to electric utilities and to rural electric systems in particular. NRECA generally supports the aims and objectives of the Act and has been active in responding to the many proposed rules to implement the Act.

In this testimony, I would like to address three issues of particular concern to rural electric systems nationwide: competition, retail wheeling and size limitation for mandatory purchases of energy from renewable resources.

NRECA's comments involve our concerns over the problems affecting movement toward a more competitive electric power industry. As you know, NRECA has been a strong advocate of an increasingly competitive bulk power marks stressing open transmission access on reasonable rates, terms and conditions. NRECA also takes partial credit for the Regional Transmission Group (RTG) concept, having developed and worked with others toward RTG's which fostered competitive bulk power markets, but prevented their subversion into monopolistic barriers to competition.

First among our concerns is the problem of competing for and maintaining loads when others can use their monopoly powers to appropriate those loads, without any competition or review by relevant State commissions to ensure that the public interest is being protected. Municipal systems in the majority of States have the right to annex and condemn the service territories of cooperative and investor-owned utilities without any consideration other than their own self interest. While the American Public Power Association (APPA) alleges that virtually every State has laws which govern who shall serve whom, and how disputes are to be resolved, a recent survey by the National Association of Regulatory Utility Commissioners proves otherwise. The survey points out that in the majority of States, the public utility commissions have no right to approve or disapprove of municipal condemnation of cooperative service territories (survey attached). We cannot move toward a competitive electric industry when one player has the absolute right to take the customers of the other without competition or consideration of the public interest.

We agree with the statements by APPA when dealing with retail wheeling, that "[t]he worst of all worlds for a public power utility would be to have retail customers picked off by competitors when, at the same time, you were prevented by one mechanism or another from competing for others retail customers." Ironically, this is exactly the dilemma faced in many States by electric utilities serving an area which, when annexed by municipals, is taken in the name of "local control." Where municipal systems are not subject to State commission oversight there is no hearing, no ability to object, and certainly no competition. Conversely, no investor-owned utility or electric cooperative may condemn the service territory of a municipal system; there is no competition. It is hypocritical for municipals to ask for retail load protection when it is convenient, then take over utilities' customers under the guise of "competition" when it suits their purposes.

To "level the playing field;; and protect the public interest, NRECA supports Rep. Jill Long's (D—IN) proposal (H.R. 3790, the Rural Consumer Protection Act of 1994) to allow municipal condemnation only in those cases where an independent third party has determined that such condemnation is in the public interest.

NRECA's second concern involves retail wheeling. As the attached set of comments which NRECA filed before the California Public Utilities Commission points out, we have serious concerns as to whether retail wheeling will ever be truly in the public interest. NRECA questions whether providing some short-term benefits to a few very large industrial customers is worth the cost of disrupting a regulatory regime which has been tasked with protecting the public interest, by balancing the needs of shareholders/consumer-owners with those of ratepayers. Public utility commissions balance these needs in every rate proceeding; the issues of global competition, prudence and lowest rates are issues that are discussed and briefed before these commissions. Retail wheeling is an end run around the regulatory regime which considers the "public interest," as opposed to the "self-interest" of the large industrials. It is most likely that small industrials, commercial and especially residential consumers will be the losers in a race to obtain lower cost alternative sources of power. While deregulation has come to the gas industry, residential consumers still cannot choose their gas supplier.

NRECA believes that retail wheeling will have a significant adverse impact on the cost of capital to all electric utilities. As we have seen in California, the mere proposal has reduced the market value of utility stocks by over $3 billion. Regardless of assurances that remaining customers will not have to pick up stranded costs of idled generation, the increased risk for electric utilities will raise the very cost of capital which these utilities must pass on to all their customers. Additionally, NRECA is not persuaded that retail wheeling can have a positive impact on the Congressional purposes of increasing efficiency and conservation, as pursued through the use of Integrated Resource Planning. At a minimum, demand-side management 9DSM) and other conservation programs will no longer be weighed against the marginal cost of generation experienced by utility generators, but will only be pursued to the extent that the cost of the DSM measures is lower than the average cost of generation.

Finally, NRECA is concerned that retail wheeling will undermine the highly coordinated system of inter-utility operation. Utilities share obligations to provide reserve capacity, operate joint economic dispatch, and buy and sell excess capacity, all on a basis which benefits both utilities and their ratepayers.

If adjacent systems begin raiding their neighboring system's large loads, the logical result would be less coordination and cooperation. No utility will want to help its neighbor lower costs when both utilities are in head-to-competition for loads.

The end result of retail wheeling may well be that efficiencies which currently result from the coordinated operation of adjacent systems would be lost.

NRECA's third concern is the proposal to permanently remove the 80 Megawatt size limitation for mandatory purchases from solar, wind, waste and geothermal power plants, pursuant to the Public Utility Regulatory Policies Act of 1978 (PURPA). NRECA's members, along with other sectors in the electric utility industry, are currently configuring themselves to meet the challenges and opportunities of competitive bulk power markets, as well as increased retail competition. This requires some very hard choices, as evidenced by the constant reporting in the trade press of the wholesale downsizing of utilities, reduction or elimination of dividends, and the sale of power at rates which are below the actual cost of service.

To require ratepayers of cooperative-, municipal- and investor-owned utilities to purchase uneconomic power could have a very negative impact on the very existence of some utilities. Problems which have resulted from State determinations of avoided costs and mandatory power purchases at these rates to date are all too evident. The California Order Instituting Rulemaking and Order Instituting Investigation, issued April 20, 1994, identifies Qualifying Facility (QF) purchases as one of the problems behind California's high utility rates which have contributed to a significant loss of industry. The Order states in footnote 30: "Many argued that . . . the contracts utilities signed with qualifying facilities over the past decade as part of this Commission's competitive procurement program are uneconomic." To force small entities such as co-ops to purchase power without size limits could spell the end of their competitive existence. At least, with some size cap, the impact on smaller systems could be limited.

NRECA is not just saying "No," but has suggested an alternative way to achieve a national policy without substantially impacting the competitive status of electric systems. In the attached June 28th letter to the subcommittee, NRECA suggests that to the extent that Congress seeks to foster renewables as a national policy, it should consider some modification of sections 1212 and 1914 of the Energy Policy Act of 1992. These sections, which provide production incentives and tax credits for electricity from certain renewables, provide an appropriate national incentive for national policy agendas.

I appreciate the opportunity to provide these comments and look forward to working with the subcommittee on these and other issues in the future. I will be happy to respond to any questions.

○

9780160466779
TP
APEX